Lecture Notes in Computer Science

Lecture Notes in Computer Science

Edited by G. Goos and J. Hartmanis

450

T. Asano   T. Ibaraki
H. Imai   T. Nishizeki   (Eds.)

Algorithms

International Symposium SIGAL '90
Tokyo, Japan, August 16–18, 1990
Proceedings

Springer-Verlag
Berlin Heidelberg New York London
Paris Tokyo Hong Kong Barcelona

Editors

Tetsuo Asano
Department of Applied Electronics
Osaka Electro-Communication University
Hatsu-cho, Neyagawa, Osaka 572, Japan

Toshihide Ibaraki
Department of Applied Mathematics and Physics
Kyoto University
Kyoto 606, Japan

Hiroshi Imai
Department of Information Science
Faculty of Science, University of Tokyo
Hongo, Bunkyo-ku, Tokyo 113, Japan

Takao Nishizeki
Department of Information Engineering
Tohoku University
Sendai 980, Japan

CR Subject Classification (1987): F.1–2, G.2–3, H.3, I.3.5

ISBN 3-540-52921-7 Springer-Verlag Berlin Heidelberg New York
ISBN 0-387-52921-7 Springer-Verlag New York Berlin Heidelberg

Printing and binding: Druckhaus Beltz, Hemsbach/Bergstr.
2145/3140-543210 – Printed on acid-free paper

Preface

The papers in this volume were presented at the SIGAL International Symposium on Algorithms. The workshop took place from August 16 to August 18, 1990, at CSK Information Education Center in Tokyo, Japan. It was organized by the Special Interest Group on Algorithms (SIGAL) of the Information Processing Society of Japan (IPSJ) in cooperation with the Technical Group on Theoretical Foundation of Computing of the Institute of Electronics, Information and Communication Engineers.

SIGAL was organized in IPSJ in 1988 to encourage research in the field of discrete algorithms in Japan and the world, and held 6 to 8 research meetings in each year. This symposium was the first international symposium on algorithms organized by SIGAL, and will be held regularly in future.

In response to the program committee's call for papers, 88 papers were submitted. From these submissions, the program committee selected 34 for presentation at the symposium. In addition to these papers, the symposium included 5 invited talks and 10 invited papers.

August 1990

Tetsuo Asano
Toshihide Ibaraki
Hiroshi Imai
Takao Nishizeki

Symposium Chairs:
Akihiro Nozaki (International Christian Univ.)
Takao Nishizeki (Tohoku Univ.)

Program Committee:
Toshihide Ibaraki (Kyoto Univ., Chair)
Tetsuo Asano (Osaka Electro-Comm. Univ., Co-Chair)
Takao Asano (Sophia Univ.)
Yoshihide Igarashi (Gunma Univ.)
Hiroshi Imai (Univ. Tokyo)
Kazuo Iwano (IBM Tokyo Research Lab.)
Tsutomu Matsumoto (Yokohama National Univ.)
Tetsuo Mizoguchi (Mitsubishi Electric Co., Ltd.)
Katsuhiro Nakamura (NEC Co., Ltd.)
Hideo Nakano (Osaka Univ.)
Masafumi Yamashita (Hiroshima Univ.)
Hiroto Yasuura (Kyoto Univ.)

Finance:
Katsuhiro Nakamura (NEC Co., Ltd.)

Local Arrangements Committee:
Yoshihide Igarashi (Gunma Univ., Chair)
Takao Asano (Sophia Univ., Co-Chair)
Kazuo Hirono (CSK Corporation)
Kazuo Iwano (IBM Tokyo Research Lab.)
Mario Nakamori (Tokyo Univ. of Agri. and Tech.)
Takeshi Tokuyama (IBM Tokyo Research Lab.)
Shuji Tsukiyama (Chuo Univ.)
Shuichi Ueno (Tokyo Inst. Tech.)
Osamu Watanabe (Tokyo Inst. Tech.)

Publication, Publicity:
Hiroshi Imai (Univ. Tokyo)

Supported by
CSK Corporation
International Communications Foundation
International Information Science Foundation
Japan Society for the Promotion of Science
Shimazu Science Foundation
Telecommunication Advancement Foundation

Table of Contents

---

† invited papers

Recent Progress in String Algorithms

Zvi Galil

Columbia University and Tel Aviv University

Abstract

The talk will survey recent results in string algorithms including:

1. approximate string matching
2. parallel string matching
3. dynamic programming speed ups
4. exact complexity of string matching
5. tree pattern matching

SELECTION NETWORKS

Nicholas Pippenger*
Computer Science Department, University of British Columbia
Vancouver, BC V6T 1W5, Canada

1. INTRODUCTION

The selection networks of which we speak are comparator networks (see Knuth [K]) that classify a set of n values into two classes, with each of the values in one class being at least as large as all of those in the other. In this paper we shall confine our attention to the simplest case, in which n is even and the two classes each contain $n/2$ values, but similar methods apply to classes of unequal cardinality, as well as to the problem of selecting the value having a prescribed rank.

We shall present an upper bound asymtotic to $2n \log_2 n$ for the number of comparators need to construct such a network. Alekseev [Al] has given a lower bound asymptotic to $(n/2) \log_2 n$.

The classifying problem has traditionally been considered in connection with the problem of sorting n values into order. In 1983, Ajtai, Komlós and Szemerédi [Aj] showed that $O(n \log n)$ comparators are sufficient for sorting, and this bound obviously applies to classifying as well. The constant factor implicit in their original proof is enormous, however, and further efforts to refine their ideas have not brought it below 1000 (see Paterson [Pa]). Our classifiers are based on the same fundamental idea as their sorters; our only contribution is to show that in the context of classifiers, it yields both a much simpler proof and a much smaller constant.

Though we shall confine ourselves to proving the result stated above, two additional points should be mentioned. First, we prove the existence of classifying networks without giving an explicit construction. This situation arises from the use of expanding graphs; by exploiting known explicit constructions for expanding graphs (see Pippenger [Pi], Section 3.2), and by accepting a somewhat larger bound (the best we have been able to obtain is slightly less than $6n \log_2 n$), we could give a completely explicit construction. Second, the networks we describe have depth $\Omega((\log n)^2)$; with more care in the construction and proof, we could establish a bound of $O(\log n)$. Our method does not seem well suited to optimizing the depth, however, and we have not made any attempt to obtain the sharpest possible result in this direction.

* This research was partially supported by an NSERC Operating Grant and an ASI Fellowship Award.

2. EXPANDING GRAPHS

We shall need some results concerning expanding graphs; these will be obtained as special cases of a general result due to Bassalygo [B], to whom we refer for the proof.

A bipartite graph with n "left" vertices and m "right" vertices will be called an (α, β)-*expander* if any k of its left vertices ($k \leq \lfloor \alpha n \rfloor$) are connected to at least $\lfloor \beta k \rfloor$ right vertices (the set A of left vertices is connected to the set B of right vertices if at least one edge from A leads to each right vertex b, $b \in B$; $\lfloor x \rfloor$ is the integer part of x).

LEMMA 2.1: (Bassalygo) For any positive integers q and p, any reals α and β ($0 < \alpha < p/\beta q < 1$), and any sufficiently large n ($n \geq n_0(\alpha, \beta, q, p)$), there exists an (α, β)-expander with qn left vertices and pn right vertices, for which the number of edges does not exceed $spqn$, where s is any integer greater than

$$\frac{H(\alpha) + (p/q)H(\alpha \beta q/p)}{pH(\alpha) - \alpha \beta q H(p/\beta q)},$$

$$H(x) = -x \log x - (1-x) \log(1-x), \quad 0 < x < 1.$$

The proof of Lemma 2.1 considers only graphs in which every left vertex meets sp edges and every right vertex meets sq edges. This observation will be important when we consider the depth, rather than merely the size, of networks.

We shall use this lemma with $p = q = 1$. We shall let α and β depend on a new parameter ϑ by $\alpha = \vartheta$ and $\beta = (1 - \vartheta)/\vartheta$. For every $\vartheta > 0$ there is a value of s that satisfies the hypothesis of Lemma 2.1.

The proof of Lemma 2.1 may be regarded as considering a probability distribution graphs, and when $p = q$ this distribution is invariant under the exchange of left and right vertices. The proof also shows, not merely that there exists a graph with the prescribed expansion properly, but that almost all considered graphs have this property. In particular, a majority of the considered graphs have this property. It follows that there exists a graph such that both it and the graph obtained from it by exchanging left and right vertices have the prescribed expansion property.

Combining these elaborations of Lemma 2.1 we obtain the following corollary.

COROLLARY 2.2: For every $\vartheta > 0$ there exists an s such that for all sufficiently large n (depending on ϑ) there exists a bipartite graph with n left vertices, n right vertices and s edges meeting each vertex such that (1) every set of $k \leq \vartheta n$ left vertices is connected to at least $(1 - \vartheta)k/\vartheta$ right vertices, and (2) every set of $k \leq \vartheta n$ right vertices is connected to at least $(1 - \vartheta)k/\vartheta$ left vertices.

Returning to Lemma 2.1 with $p = q = 1$, if we take $s = 4$ and choose $\beta < 3$, then the hypothesis is satisfied for all sufficiently small $\alpha > 0$ (depending on β). Thus we also obtain the following corollary.

COROLLARY 2.3: For every $\beta < 3$ there exists an $\alpha > 0$ such that for all sufficiently large n (depending on β) there exists a bipartite graph with n left vertices, n right vertices and 4 edges meeting every vertex such that (1) every set of $k \leq \alpha n$ left vertices is connected to at least βk right vertices, and (2) every set of $k \leq \alpha n$ right vertices is connected to at least βk left vertices.

3. CLASSIFIERS

A comparator network with $2m$ inputs and two sets of m outputs, the "left" outputs and the "right" outputs, is an α-*weak approximate classifier with tolerance* ϑ (or simply an α-*weak* ϑ-*classifier*) if, for any assignment of values to the inputs and positive integer $k \leq \alpha m$, (1) at most ϑk of the k smallest values appear at right outputs, and (2) at most ϑk of the largest values appear at left outputs. A 1-weak approximate classifier with tolerance ϑ will be called an *approximate classifier with tolerance* ϑ (or simply a ϑ-*classifier*). An approximate classifier with tolerance 0 will be called a *classifier*.

Classifiers are the goal of our construction. Approximate classifiers and weak approximate classifiers are the ultimate building blocks of our construction. These building blocks are secured by a lemma that is a slight generalization of the most basic lemma of Ajtai, Komlós and Szemerédi [AKS1], from whom the proof is easily adapted (see also Pippenger [Pi], Section 3.2).

Let G be a bipartite graph with n left vertices and n right vertices, in which every vertex meets s edges. The edges of G may be decomposed into s perfect matchings $E_1, \ldots, E_s$ between the left and right vertices. We may regard each perfect matching E_r as a comparator network, by taking a comparator for each edge in E_r, labelling the inputs of the comparator with the vertices met by the edge, labelling the smaller output of the comparator with the left vertex met by the edge, and labelling the larger output of the comparator with the right vertex met by the edge. We may then combine the comparator networks $E_1, \ldots, E_s$ into a single comparator network by identifying the outputs of E_r with the corresponding inputs of E_{r+1} for each $1 \leq r \leq s - 1$. We shall denote the resulting comparator network by G; this notation is ambiguous, since different decompositions of the bipartite graph yield different comparator networks, but this ambiguity will not be important to us.

LEMMA 3.1: (Ajtai, Komlós and Szemerédi) Let G be a bipartite graph with n left vertices and n right vertices and s edges meeting every vertex in which (1) any set of $k \leq \alpha \vartheta n$ left vertices is connected to at least $(1 - \vartheta)k/\vartheta$ right vertices, and (2) any set of $k \leq \alpha \vartheta n$ right vertices is connected to at least $(1 - \vartheta)k/\vartheta$ left vertices. Let the left and right outputs of the comparator network G be those labelled by the left and right vertices, respectively. Then G is an α-weak ϑ-classifier.

4. RECURSIVE CONSTRUCTION

A comparator network with $2m$ inputs, l outputs labelled as "low", l outputs labelled as "high" and and $2m - 2l$ outputs labelled as "middle" is a *strong partial classifier* if, for any assignment of values to the inputs, only values among the $m/2$ smallest appear at low outputs and only values among the $m/2$ largest appear at high outputs. A strong partial classifier is less than a classifier in that there are some outputs, the middle outputs, at which any value may appear; but it is more than a classifier in that fewer values can appear at the low and high outputs. Our goal in this section is to show how strong partial classifiers can be assembled to form a classifier.

It will be convenient to use strong partial classifiers for which the number of inputs is of the form 2^ν or $3 \cdot 2^\nu$, with ν a positive; numbers of this form will be called *magic*. If n is any even positive integer, the largest magic number not exceeding n will be called the *magic part* of n; it is even and at least $2n/3$.

Consider the following recursive construction for a classifier with n inputs. Let $2m$ denote the magic part of n. Feed $2m$ of the inputs into a strong partial classifier with $2m$ inputs. Feed the remaining $n - 2m$ inputs, together with the $2m - 2l$ middle outputs of the strong partial classifier into a classifier with $2 - 2l$ inputs. The left and right outputs of the combined network will be the low and high outputs, respectively, of the strong partial classifier, together with the left and right outputs, respectively, of the constituent classifier.

A value appearing at a low output of the strong partial classifier must be among the $m/2$ smallest of the $2m$ values at its inputs, and thus among the $(m/2)+(n-2m) = n - 3m/2$ smallest of all n values. Since $2m \geq 2n/3$, it must be among the $n/2$ smallest of all n values. Similarly, a value appearing at a high output of the strong partial classifier must be among the $n/2$ largest of all n values. Thus, of the $n/2$ largest and $n/2$ smallest values, equal numbers appear at the inputs of the classifier with $n - 2l$ inputs. It follows that any value appearing at a left output of this classifier must be among the $n/2$ smallest, and any value appearing at a right output must be among the $n/2$ largest. Thus the combined network is indeed a classifier.

LEMMA 4.1:   Let $C > 0$ be a constant. Suppose that, for every $\varepsilon > 0$ and all sufficiently large m (depending on ε) there exists a strong partial classifier with $2m$ inputs, l low outputs and l high outputs, and size at most $(C + \varepsilon)l \log_2 m$. Then for every $\varepsilon > 0$ and all sufficiently large n (depending on ε), there exists a classifier with n inputs and size at most $(C/2 + \varepsilon)n \log_2 n$.

PROOF: Apply the recursive construction until the number of inputs of the strong partial classifier that is needed is too small for the hypothesis to apply. Terminate the recursion with a sorting network using $\binom{m}{2}$ comparators. The size of this final sorting network depends only on ε, and thus is at most $(\varepsilon/2)n \log_2 n$ for all sufficiently large n (depending on ε).

Let $2m_1, \ldots, 2m_s$ denote the numbers of inputs of the strong partial classifiers, and let $l_1, \ldots, l_r$ denote the numbers of low outputs. We have $2m_r \leq n$ for all $1 \leq r \leq s$ and, since each left output of the combined network is a low output of at most one strong partial classifier, $\sum_{1 < r \leq s} l_r \leq n/2$. Thus the total size of all the strong partial classifiers is at most $\sum_{1 < r \leq s}(C+\varepsilon)l_r \log_2 m_r \leq (C/2+\varepsilon/2)n \log_2 n$. Adding the bound $(\varepsilon/2)n \log_2 n$ for the final sorter yields $(C + \varepsilon)n \log_2 n$. $\triangle$

5. CRUDE CLASSIFICATION TREES

This section introduces classification trees, the basic tool we shall use to construct strong partial classifiers. We shall begin with a crude version of the construction, and later refine it to obtain our final bound.

Set $\vartheta = 1/8$. Corollary 2.2 and Lemma 3.1 then yield constants s_0 and n_0 such that for all $n \geq n_0$, there is a ϑ-classifier with $2n$ inputs and depth s_0. (A simple calculation shows that $s_0 = 28$. The determination of n_0 would require scrutiny of the proof of Lemma 2.1, but this proof consists of explicit estimates, so that n_0 is at least effectively calculable. The actual values of these constants will not be important to us.)

Suppose that we wish to construct a strong partial classifier with $2m$ inputs, where $2m$ is a magic number. Feed the $2m$ inputs into a ϑ-classifier with $2m$ inputs. This approximate classifier has m left outputs and m right outputs. Feed each of these sets of outputs into a ϑ-classifier with m inputs. These two approximate classifiers have four sets of outputs. Feed each of these sets into a ϑ-classifier with $m/2$ inputs, and continue in this way until the sets of outputs of the approximate classifiers have cardinality less than $2n_0$. The result is a tree of approximate classifiers that we shall call a *classification tree*. At its root are $2m$ inputs, and at its leaves are sets of outputs each containing fewer than $2n_0$ outputs.

The next step will be to label the outputs as low, high and middle in such a way that the result is a strong partial classifier. When as we do this we assign the same label to all the outputs in a subtree, we may prune away that subtree, and affix the label to the outputs of the approximate classifier feeding the subtree. A large fraction of the tree will be eliminated in this way.

We begin by labelling as middle the right outputs of the left child of the root, and the left outputs of the right child of the root (and pruning away the subtrees below). We shall label as low some of the outputs in the subtree fed by the left outputs of the left child, and as high some of the outputs in the subtree fed by the right outputs of the right child. We shall now describe which outputs are to be labelled as low. The mirror image of this procedure will label an equal number of outputs as high.

Consider the $m/2$ smallest values assigned to the inputs, since it is these that are eligible to appear at an output labelled as low. We shall call these $m/2$ values *good*, and the other $3m/2$ values *bad*.

At most a fraction $\vartheta = 1/8$ of the good values can appear at right outputs of the approximate classifier at the root, and at most a fraction $1/8$ can appear at right outputs of the approximate classifier that is its left child. Thus at least a fraction $1 - 1/8 - 1/8 = 3/4$ of the good values appear at left outputs of the left child. Since the number of good values equals the number of left outputs of the left child, at most a fraction $1 - 3/4 = 1/4$ of the values appearing at these outputs are bad.

We may characterize the set of left outputs of the left child by its cardinality $m/2$ and its "impurity" $1/4$ (the largest possible fraction of its values that could be bad). Suppose now that we have a set of outputs of some approximate classifier with cardinality k and impurity η. Firstly, if $\eta > 1/2$, we shall label these outputs as middle (and prune away the subtree below). Secondly, if $\eta k < 1$, then not a single bad value can appear at one of these outputs; thus we shall label them as low (and prune away the subtree below). Finally, if $\eta \leq 1/2$ and $\eta k \geq 1$, then we shall consider the sets of outputs of the child. The set of left outputs has cardinality $k/2$ and (by Lemma 3.1) impurity $2\vartheta\eta = \eta/4$, and the set of right outputs has cardinality $k/2$ and impurity 2η (the factors of 2 in the impurities arise because we are considering a fraction of half as many things). We may continue in this way along each path in the tree until we assign a label or reach a leaf. If we reach a leaf, we shall label its outputs as middle if $\eta k \geq 1$, and as low if $\eta k < 1$.

The first question we shall ask is: what fraction of the outputs are labelled as middle by being in a set with impurity exceeding $1/2$? To answer this question, we shall consider the following random walk on the integers. Start at the position 2, $Z_0 = 2$. At each step independently move to the psotion one smaller, $Z_{t+1} = Z_t - 1$, or two larger, $Z_{t+1} = Z_t = 2$, with equal probabilities. What is the probability of ever reaching a position smaller than 1? Since the walk is confined to the integers, this is the probability of ever reaching the psotion 0, $Z_t = 0$. The answer to this question is an upper bound to the fraction of the outputs that are labelled as middle by being in a set with impurity exceeding 1/2, as can be seen by considering the correspondence between paths in the tree and walks, where the number of levels from the root corresponds to time in the walk, and the negative of the logarithm (to base 2) of the impurity corresponds to position in the walk.

In the present instance, the probability of ever reaching the position 0 can be determined explicitly and is $(3 - \sqrt{5})/2 = 0.382\dots$. To see this, let $f(x)$ denote the power series in x in which the coefficient of x^t is the number of walks that start at 1 and reach 0 for the first time at time t. Then $f(x)^2$ is the power series for walks that start at 2 and reach 0 for the first time at time t, since each such walk can be uniquely parsed into two subwalks according to the time at which it first reaches 1, and the numbers of possibilities for both subwalks are counted by $f(x)$. Thus the probability we seek is $f(1/2)^2$, so it will suffice to show that $f(1/2) = (-1 + \sqrt{5})/2 = 0.618\dots$. Let $g(x)$ count the number of walks that start at 1 and return to 1 for the first time at time t. Then $g(x) = x f(x)^2$, since such a walk must go to 3 on the first step, then return to 1 for the first time in $t - 1$ more steps. On the other hand, $f(x) = x + x g(x) + x g(x)^2 + \cdots = x/(1 - g(x))$, since the walks counted by $f(x)$ may be classified according to the number of times they visit 1 before reaching 0. Thus $f(x)$ satisfies the equation $x f(x)^3 - f(x) + x = 0$, so that $f(1/2)$ satisfies $f(1/2)^3 - 2f(1/2) + 1 = 0$, which yields the stated result.

Next we shall ask: what fraction of the outputs are labelled as middle by being in a leaf that is not pruned away? Such a leaf has cardinality at most $2n_0$, and thus it must have impurity at least $1/2n_0$ to avoid being labelled as low. Let d denote the number of levels of approximate classifiers in the tree. Rephrased in terms of random walks, our question becomes: what is the probability of being at a position at most $c_0 = \log_2 n_0$ at time $d - 2$? (The first two levels of the tree do not correspond to steps of the random walk.) We shall answer this question with a lemma that goes beyond our present needs, but which will be applied repeatedly later.

We shall consider random walks with discrete time indexed by the natural numbers and discrete positions indexed by the integers. We shall assume that the steps are independent and identically distributed, but we shall allow the steps to have any probability distribution on a finite set of integers. We shall say that such a random walk is *positively biased* if the expectation of a step is positive. (In the present instance, the step is uniformly distributed on the set $\{-1, 2\}$, and the expectation is $(-1 + 2)/2 = 1/2 > 0$.)

LEMMA 5.1: Let Z_t, $t = 0, 1, 2, \dots$, be a positively biased random walk starting at 0, and let c be any position. Then there exist constants A and $b < 1$ such that for all t, the probability that Z_t is at most c does not exceed Ab^t.

PROOF: Let $\Phi(\xi) = \mathrm{Ex}(\exp -(\xi Z_1))$. The power series expansion of $\Phi(\xi)$ is $1 - \xi \mathrm{Ex}(Z_1) + O(\xi^2)$. Since $\mathrm{Ex}(Z_1) > 0$, we can choose $\xi_0 > 0$ sufficiently small so that $\Phi(\xi_0) < 1$. Since the steps are independent and identically distributed, we have $\mathrm{Ex}(\exp -(\xi_0 Z_t)) = \Phi(\xi_0)^t$. If $Z_t \leq c$, then $\exp -(\xi_0 Z_t) \geq \exp -(\xi_0 c)$. Thus, by Markov's inequality, the probability that $Z_t \leq c$ is at most $\Phi(\xi_0)^t / \exp -(\xi_0 c)$, so we may take $A = \exp(\xi_0 c)$ and $b = \Phi(\xi_0)$. $\triangle$

We may now apply Lemma 5.1 with $t = d - 2 \geq \log_2(m/4n_0)$ and conclude that the fraction of outputs that are labelled as middle by being in a leaf that is not pruned away is at most $Ab^d \leq Cm^{-e}$, where C and $e > 0$ are constants. The only feature of this bound that is relevant to our present purposes is that it tends to zero, even when multiplied by $d \leq \log_2 m$.

Finally, we shall ask: what is the size of the strong partial classifier constructed in this way? To answer this question, we shall again transform it into a question about random walks. In a "synchronous" comparator network (in which the two inputs of any comparator are at the same depth), each comparator contributes 2 to the sum of the depths of the outputs. Thus the number of comparators is $n/2$ times the average depth of the outputs. Since the depth of each strong partial classifier is s_0, the number of comparators is $s_0 n/2$ times the average level at which the outputs are labelled. For outputs labelled as low or labelled as middle by being in a leaf that is not pruned away, the level at which they are labelled is at most $d \leq \log_2 m$. (For outputs labelled as low, it is in most cases substantially less than this, but we shall not attempt to exploit this effect, since later optimizations will render it negligible.) For outputs labelled as middle because their impurity exceeds $1/2$, the level at which they are labelled is two more than the number of steps taken by the corresponding walk to reach position 0 for the first time. (Again, the first two levels of the tree do not correspond to steps of the random walk.)

In the present instance, this average number of steps can be calculated explicitly and is $4/\sqrt{5} = 1.788\ldots$. It is obtained from the power series $f(x)^2$ that counts the walks by evaluating $x\, d(f(x)^2)/dx$ at $x = 1/2$ or, equivalently, evaluating $f'(x)f(x)$ at $x = 1/2$. This evaluation is most conveniently accomplished by dividing the equation $xf(x)^3 - f(x) + x = 0$ by x, differentiating with respect to x, multiplying by x^2, solving for $f'(x)$ in terms of $f(x)$ and x, multiplying by $f(x)$ and evaluating the result at $x = 1/2$. Taking account of the equation $f(1/2)^2 = (3 - \sqrt{5})/2$ derived earlier yields the stated result.

We can now sum the contributions to the size of the strong partial classifier. The l outputs labelled as low contribute at most $(s_0 l/2) \log m$, and those labelled as high contribute equally. The outputs labelled as middle by being in a leaf that is not pruned away contribute at most $Fm^{1-e} \log_2 m$ for some constants F and $e > 0$, and the outputs labelled as middle because their impurity exceeds $1/2$ contribute at most Gm for some constant G. Since the $l = \Omega(m)$, we conclude that for every $\varepsilon > 0$ and all sufficiently large m (depending on ε), there exists a strong partial classifier with $2m$ inputs, l outputs labelled as low and an equal number labelled as high, and size at most $(s_0 + \varepsilon)l \log_2 l$. It follows from Lemma 4.1 that for every $\varepsilon > 0$ and all sufficiently large n (depending on ε), there exists a classifier with n inputs and size at most $(s_0/2 + \varepsilon)n \log_2 n$. The remainder of this paper is devoted to refining the construction just given to reduce the constant $s_0/2$ to 2.

6. REFINED CLASSIFICATION TREES

If we ask what properties were essential to the construction in the preceding section, we find three. First, the probability that the random walk ever reaches the position 0 is strictly less that 1. Second, the probability that the walk is near position 0 after t steps decreases exponentially with t. Third, the expected number of steps needed to reach 0 (with no contribution from walks that never reach 0) is finite.

The second property is a consequence of the random walk being positively biased. Thus it is natural to seek ways to reduce the number of comparators while preserving the property that the corresponding random walk is positively biased. When this is done, the explicit calculations by which we established the first and third properties will no longer be feasible, but we will see that these properties are consequences of the second property.

The property that the random walk was positively biased follows from the inequality $\vartheta < 1/4$ (so that the geometric mean of the factors 2 and 2ϑ, by which impurities change from parent to child, is less than 1). We shall arrange for ϑ to vary from level to level in such a way that the average (again in the sense of the geometric mean) of ϑ is strictly less that $1/4$, but by a very small margin. We shall also exploit the fact that for most of the approximate classifiers in the classification tree, the number of bad elements is very small, so that we may substitute weak approximate classifiers (as defined in Section 3).

Let h be a positive integer. Set $\vartheta_h = 2^{1/h}/4$ and $\beta_h = (1-\vartheta_h)/\vartheta_h$. Since $\vartheta_h > 1/4$, we have $\beta_h < 3$. Corollary 2.3 and Lemma 3.1 establishe the existence of $\alpha_h > 0$ such that, for all sufficiently large n (depending on h), there exists a α_h-weak ϑ_h-classifier with $2n$ inputs and depth 4.

Let us now define a *gadget* to be a tree comprising h levels of α_h-weak approximate classifiers, which therefore approximately classifies the values assigned to its inputs into 2^h classes. The α_h-weak approximate classifier at the root will have tolerance $1/8$, and those at the remaining $h-1$ levels will have tolerance ϑ_h.

Suppose that at most a fraction η of the values assigned to the inputs of the gadget are bad, where $\eta \leq \alpha_h$. We may determine the impurities of the 2^h sets of outputs by proceeding through successive levels of the tree as before. The root multiplies the impurity by a factor of 2 or $1/4$, and every other node multiplies it by a factor of 2 or $2^{1/h}/2$. A simple calculation shows that the geometric mean of the impurities of the 2^h sets of outputs is $1/2^{1/2h}$, which is less than 1.

Since all changes to impurities are by factors of $2^{1/h}$ we may consider a random walk on the integers by taking the position to be h times the negative of the logarithm (to base 2) of the impurity, and letting the probability distribution for each step correspond to the changes to impurities for a gadget. The expectation for a step is h times the negative of the logarithm of the geometric mean of the changes, which is $1/2$. Thus the random walk is positively biased.

Since every path through a gadget passes through one weak approximate classifier with tolerance $1/8$ and through $h-1$ with tolerance $\vartheta_h > 1/4$, we can construct a gadget with depth at most $4(h-1) + s_0$.

The gadget we have constructed is composed from α_k-weak approximate classifiers, and thus is only useful when the number of bad values is small. We shall can the gadgets described above *cheap* gadgets. We shall also define *dear* gadgets, which have the same tree structure, but with approximate classifiers (rather than weak approximate classifiers) at all nodes, and with all classifiers having tolerance $1/8$. We shall use the same step probability distribution for dear gadgets that we used for cheap gadgets (though of course dear gadgets do much more). The depth of a dear gadget is at most hs_0.

We shall now construct a refined classification tree using gadgets rather than approximate classifiers as nodes below the root and its children (so that the tree is 2^h-ary rather than binary below the first two levels). We assign impurities to each set of outputs of each gadget as before. We shall use a cheap gadget when the impurity of the set of inputs is at most α_h, and a dear gadget when the impurity exceeds α_h.

We shall label outputs as low, high or middle, and prune away subtrees as before. It remains to estimate the number of outputs labelled as low, high or middle, and to estimate the size of the resulting strong partial classifier.

The fraction of the outputs that are labelled as middle because their impurity exceeds $1/2$ can be bounded as before by the probability that the corresponding random walk reaches a non-positive position. (We must consider non-positive positions, rather than just the position 0, because a single step may now decrease the position by more than 1.) For this purpose we shall use the following lemma.

LEMMA 6.1: Consider a positively biased random walk starting at a positive position. Then the probability that the walk ever reaches a non-positive position is strictly less than 1.

PROOF: By Lemma 5.1, the probability that the walk is at a non-positive position at time t is at most Ab^t for some constants A and $b < 1$. The series $\sum_{t \geq 1} Ab^t$ converges, so we may choose T sufficiently large that $\sum_{t > T} Ab^t \leq 1/2$. Let $-\Delta$ denote the most negative step that is taken with positive probability, and let $P > 0$ denote the probability that a step is positive. Then $Z_{\Delta T} \geq \Delta T$ with probability at least $P^{\Delta T}$. If this event occurs, the walk cannot reach a non-positive position in fewer than T additional steps, and the probability of it reaching a non-positive position in T or more additional steps is no larger than the probability of reaching a position at most ΔT in T or more additional steps, and this is at most $1/2$. Thus the probability of ever reaching a non-positive position is at most $(1 - P^{\Delta T}) + P^{\Delta T}/2 < 1$. $\triangle$

Using Lemma 5.1 as before, we can again show that the fraction of the outputs that are labelled as middle by being in a leaf that is not pruned away can is at most Cm^{-e}, where C and $e > 0$ are constants.

To estimate the size of the strong partial classifier we have constructed, we again begin by considering the average level at which outputs are labelled as middle because their impurity exceeds $1/2$. This is bounded by the expectation for the corresponding random walk of the number of steps needed to reach a non-positive position (with no contribution from walks that never reach a non-positive position). We shall use the following lemma.

LEMMA 6.2: Consider a positively biased random walk starting from a positive position. Let U denote the number of steps needed to reach a non-negative position, if the walk ever reaches a non-negative position, or 0 if the walk never reaches a non-negative position. Then the expectation of U is finite.

PROOF: Applying Lemma 5.1 with c the negative of the starting position, we have that the probability of being at a non-positive position at time t is at most Ab^t for some constants A and $b < 1$. The convergent series $\sum_{t \geq 1} Ab^t t$ bounds the sum over all visits to non-negative positions of the times at which the visits occur. This in turn bounds the sum over first visits, which is exactly U. $\triangle$

For refined classification trees, we shall need an additional estimate concerning the total size of dear gadgets. Since a dear gadget is used precisely when the impurity exceeds a threshold α_h, then total size of dear gadgets can be bounded in terms of the average time spent by the random walk at positions less than a corresponding constant $c_h = -h \log_2 \alpha_h$. By Lemma 5.1, this average time is bounded by a convergent series $\sum_{t \geq 1} Ab^t$, for some constants A and $b < 1$, and thus is finite. It follows that the total size of dear gadgets is at most Hm, for some constant H.

Summing the contributions to the size as before, we find that for every h and all sufficiently large m (depending on h), there exists a strong partial classifier with $2m$ inputs, l outputs labelled as low and an equal number labelled as high, and size at most $(4(h-1) + s_0 + 1)l \log_{2^h} l = (4 + (s_0 - 3)/h)l \log_2 l$. Letting h tend to infinity, we see that for every $\varepsilon > 0$ and all sufficiently large m (depending on ε), there exists a strong partial classifier as above with size at most $(4 + \varepsilon)l \log_2 l$. It follows from Lemma 4.1 that for every $\varepsilon > 0$ and all sufficiently large n (depending on ε), there exists a classifier with n inputs and size at most $(2 + \varepsilon)n \log_2 n$. Thus we have achieved the goal of this paper.

7. CONCLUSION

We have established the existence of classifiers with many fewer comparators than those previously known. For most constructions that rely on expanding graphs (such as those given by Bassalygo [B]), the constant factor depends on the current state of technology of expanding graphs, and can be expected to improve with further advances in the state of this art. The result of this paper, however, will not be improved in this way: the constant in the leading term of the size depends only on the degree needed for expanding graphs to expand very small sets, and this aspect of expanding graphs is understood completely (graphs of degree s can expand small sets by any factor up to $s - 1$, but not by more).

It would be of interest to see if a similar situation can be brought about for other applications of expanding graphs, perhaps even for the most celebrated application of all, the sorting networks of Ajtai, Komlós and Szemerédi.

8. REFERENCES

[Aj] M. Ajtai, J. Komlós and E. Szemerédi, "Sorting in $c \log n$ Parallel Steps", *Combinatorica*, 3 (1983) 1–19.

[Al] V. E. Alekseev, "Sorting Algorithms with Minimum Memory", *Kibernetica*, 5 (1969) 99–103.

[B] L. A. Bassalygo, "Asymptotically Optimal Switching Circuits", *Problems of Info. Transm.*, 17 (1981) 206–211.

[K] D. E. Knuth, *The Art of Computer Programming*, v. 3, *Sorting and Searching*, Addison-Wesley, Reading, MA, 1973.

[Pa] M. S. Paterson, "Improved Sorting Networks with $O(\log n)$ Depth", *Algorithmica*, to appear.

[Pi] N. Pippenger, "Communication Networks", J. van Leeuwen (editor), *Handbook of Theoretical Computer Science*, North-Holland, Amsterdam, 1990.

Computing Edge-Connectivity in Multiple and Capacitated Graphs

Hiroshi NAGAMOCHI   Toshihide IBARAKI

Dept. of Applied Mathematics & Physics, Kyoto University

Kyoto 606 Japan

1. Introduction

In this paper, a graph G=(V,E) stands for an undirected multiple graph that satisfies $|V| \geq 2$. Note that it may have multiple edges but has no self-loop, unless otherwise specified. Let λ (G) denote the edge-connectivity of G. We consider the problem of computing λ (G) for a given G. Most of the known algorithms [2,4,7,10] are based on the fact that λ (G) can be computed by solving $O(|V|)$ max-flow problems.   The best known bounds are $O(\lambda (G)|V|^2)$ (D.W.Matula [7]) if G is simple and $O(|E|^{3/2}|V|)$ (S.Even and R.E.Tarjan [2]) if G is multiple, respectively. In this paper, we present an $O(|E|+\min\{\lambda (G)|V|^2,p|V|+|V|^2\log|V|\})$ time algorithm for computing λ (G), where $p(\leq |E|)$ is the number of pairs of nodes between which G has an edge. Furthermore, this method is extended to a capacitated undirected network N=(G,c), where G=(V,E) is a multiple undirected graph and c is an $|E|$-dimensional vector of positive real capacities $c(e), e \varepsilon E$. The minimum cut capacity can be computed in $O(|E|+p|V|+|V|^2\log|V|)$ time.

2. Some Definitions

Let G=(V,E) be an undirected multiple graph. The set of edges whose end nodes are u and v is denoted by E_{uv}. Throughout this paper, unless confusion arises, an edge $e \varepsilon E_{uv}$ is denoted by e=(u,v). A graph G'=(V',E') is called a subgraph of G=(V,E) if $V' \subseteq V$ and $E' \subseteq E$. It is a spanning subgraph if V'=V. For a given graph G (possibly not connected), a spanning subgraph G' is maximal if the subgraph of G' induced by each connected component of G is connected. Denote the graph obtained by removing from a connected graph G=(V,E) a subset $F \subseteq E$ by G-F. Such F is called a cut if G-F is disconnected. A cut F is a minimum cut if $|F|$ is minimum among all cuts of G.

Denote the graph obtained from G by contracting nodes x and y in G by G/{x,y}, in which all self-loops (if edges e=(x,y) exist) resulting from contraction are deleted.   For a subset $X \subseteq V$, define

$E(X) \equiv \{e \varepsilon E |$ one end node of e is in X, while the other end node is in V-X$\}$

(=E(V-X)).

Any $E(X) \neq \phi$ is a cut. In particular, $E(\{x\})$ for $x \varepsilon V$ is denoted by $E(x)$. The minimum degree of G, i.e., $\min_{x \varepsilon V} |E(x)|$, is denoted by $\delta(G)$. G is called k-edge-connected if G-F is connected for any $F \subseteq E$ with $|F| \leq k-1$. In other words, G is k-edge-connected if and only if $|E(X)| \geq k$ for all $X \subseteq V$ with $X \neq \phi, V$. The edge-connectivity $\lambda(G)$ of a graph G is defined to be k if G is k-edge-connected but not (k+1)-edge-connected. In other words, a cut F is minimum if and only if $|F| = \lambda(G)$ holds. The local edge-connectivity $\lambda(x,y;G)$ for $x,y \varepsilon V$ with $x \neq y$ is defined to be the least number $|F|$ such that x and y are disconnected in G-F, where $F \subseteq E$. Clearly, $\lambda(G) = \min\{\lambda(x,y;G) | x,y \varepsilon V, x \neq y\}$ and $\lambda(G) \leq \delta(G) \leq 2|E|/|V|$ hold. For example, a graph G shown in Figure 1 has $\delta(G)=7$ and $\lambda(G)=6$. See [1] for other basic terminologies.

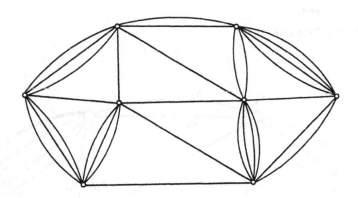

Fig.1 A multiple graphs G with $\delta(G)=7$ and $\lambda(G)=6$

3. Finding k-Edge-Connected Spanning Subgraph

We briefly describe some properties and an algorithm related to k-edge-connected spanning subgraphs, which were derived in [8].

Lemma 3.1 [8] For G=(V,E), let $H_i=(V,E_i)$ be a maximal spanning forest in $G-E_1 \cup E_2 \cup \ldots \cup E_{i-1}$, for $i=1,2,\ldots,|E|$, where possibly $E_i = \phi$ for some i. Then each spanning subgraph $G_i=(V,E_1 \cup E_2 \cup \ldots \cup E_i)$ satisfies that

$$\lambda(x,y;G_i) \geq \min\{\lambda(x,y;G),i\} \quad \text{for all } x,y \varepsilon V \text{ with } x \neq y. \quad \square \quad (3.1)$$

A partition E_i $(i=1,2,\ldots,|E|)$ of E as stated in this lemma can be obtained by the next algorithm FOREST [8].

Procedure FOREST; {input: G=(V,E), output: $E_1, E_2, \ldots, E_{|E|}$}

begin

1   Label all nodes $v \varepsilon V$ and all edges $e \varepsilon E$ "unscanned";

2   $r(v) := 0$ for all $v \varepsilon V$;

3   $E_1 := E_2 := \ldots := E_{|E|} := \phi$;

4   **while** there exist "unscanned" nodes **do**

      begin

5   Choose an "unscanned" node $x \varepsilon V$ with the largest r;

6   **for** each node y adjacent to x by an "unscanned" edge **do**

        {y is "unscanned" and every edge in E_{xy} is "unscanned"}

7   **for** each "unscanned" edge $e \varepsilon E_{xy}$ **do**

            begin

8              $E_{r(y)+1} := E_{r(y)+1} \cup \{e\}$; $r(y) := r(y)+1$; Mark e "scanned"

            end;

9   Mark x "scanned"

      end;

end.

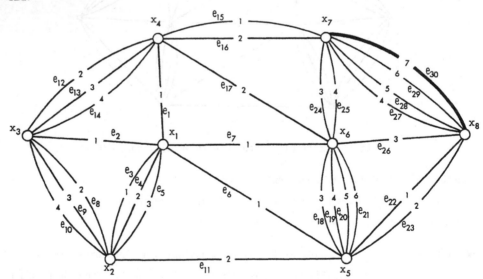

G^1: $\delta(G^1) = 7$

$\circ\!\!-\!\!-i\!\!-\!\!\circ$: edges in E_i

$\circ\!\!\blacksquare\!\!7\!\!\blacksquare\!\!\circ$: edge to be contracted

Fig.2 Partition E_i $(i=1,2,\ldots,7)$ of G of Fig.1 obtained by FOREST

Lemma 3.2 [8:Th.2.1] Given a graph G=(V,E), a partition E_i (i=1,2,...,|E|) of E stated in Lemma 3.1 can be found by FOREST in $O(|V|+|E|)$ time, where each E_i satisfies $|E_i| \leq |V|-1$. □

As an example, the partition E_i, i=1,2,... obtained by applying FOREST to a graph G of Figure 1 is illustrated in Figure 2. In this figure, node x_i (edge e_j) represents that it is the i-th node (j-th edge) scanned by FOREST.

4. Determining Edge-Connectivity

To find edge-connectivity λ (G) of G with $|V| \geq 3$, the following properties are useful.

Lemma 4.1 [9] For G=(V,E) with $|V| \geq 3$, let x,y be two nodes in V.
(1) If G/{x,y}-F is disconnected, where F is a subset of edges in G/{x,y}, then G-F is disconnected.
(2) λ (G) $\leq \lambda$ (G/{x,y}).
(3) If λ (x,y;G) $\geq$ k, then λ (G) $\geq$ k if and only if λ (G/{x,y}) $\geq$ k. □

To compute λ (G) of G=(V,E), G is reduced in the following to a graph containing only two nodes by applying $|V|-2$ edge contractions. Let $G^1=(V^1,E^1)$ denote G, E_i^1(i=1,2,...,$|E^1|$) be the partition of E^1 obtained by FOREST. By Lemma 3.3(2), there is an edge $e^1=(u^1,v^1) \varepsilon E^1_{\delta(G^1)}$. Based on this {$u^1,v^1$}, $G^2=(V^2,E^2)$ is defined to be the graph $G^1/\{u^1,v^1\}$. In analogy with G^1, we define partition E_i^2(i=1,2,...,$|E^2|$) of E^2 and edge $e^2=(u^2,v^2) \varepsilon E^2_{\delta(G^2)}$. Repeating this, G^j, u^j, v^j (j=1,2,...,$|V|-1$) are obtained.

Lemma 4.2 [9:Lemma 4.2] Let G^j,j=1,2,...,$|V|-1$, be defined as above. Then λ (G)=k holds, where

$$k=\min\{ \delta (G^1), \delta (G^2),..., \delta (G^{|V|-1})\}. \qquad (4.1) \quad □$$

The resulting procedure to compute λ (G) and a minimum cut F is described as follows.

Procedure TESTEC; {input: G=(V,E), output: λ (G) and a minimum cut F $\subseteq$ E of G}
begin
  G':=G; k:=+∞;
  while $|V'| \geq 3$ in G'=(V',E') **do**
    begin

Compute partition $E_1, E_2, \ldots, E_{|E'|}$ of E' by applying FOREST to G';

Choose a node $w \varepsilon V'$ with $|E'(w)| = \delta (G')$;

If $\delta (G') < k$ then let $F := E'(w)$;

$k := \min\{k, \delta (G')\}$;

Let $G' := G'/\{u,v\}$ with an edge $(u,v) \varepsilon E_{\delta (G')}$

   end;

$\{|V'| = 2$ now holds, and hence $\lambda (G') = \delta (G') = |E'|\}$

If $\delta (G') < k$ then let $F := E'$;

$k := \min\{k, \delta (G')\}$;

Conclude that $\lambda (G) = k$ and F is a minimum cut of G

end.

The application of TESTEC to G of Figure 1 is illustrated in Figures 2 and $3(a) \sim (f)$. As a result, $\lambda (G) = 6$ and the minimum cut $F = \{e_6, e_7, e_{11}, e_{15}, e_{16}, e_{17}\}$ are obtained.

Theorem 4.1 [9:Theorem 4.1] Let $G = (V,E)$ be a multiple graph. $\lambda (G)$ and a minimum cut $F \subseteq E$ of G can be found by TESTEC in $O(|E||V|)$ time. $\square$

Corollary 4.1 [9:Corollary 4.1] Let $G = (V,E)$ be a multiple graph.

(1) For a positive integer k, whether $\lambda (G) \geq k$ or not (if not, $\lambda (G)$ and a minimum cut F of G) can be computed in $O(|E| + k|V|^2)$ time.

(2) $\lambda (G)$ and a minimum cut $F \subseteq E$ of G can be found in $O(|E| + \lambda (G)|V|^2)$ time. $\square$

5. Capacitated Networks

In this section, we consider the problem of finding the minimum capacity cut in a capacitated undirected network $N = (G,c)$ with $G = (V,E)$, where G is a multiple undirected graph (without self-loops), and c is an $|E|$-dimensional vector of positive real numbers $c(e)$, $e \varepsilon E$. By letting $c(x,y;N)$ denote the value of a minimum capacity cut separating x and y, respectively. Clearly, the value of a minimum capacity cut $c(N)$ is defined by

$$c(N) \equiv \min\{\sum_{e \varepsilon F} c(e) \,|\, G-F \text{ is disconnected}\} = \min\{c(x,y;N) \,|\, x,y \varepsilon V, \ x \neq y\}$$

R.E. Gomory and T.C. Hu [6] showed that all values $c(x,y;N)$ $(x,y \varepsilon V)$ can be computed by solving $|V|-1$ max-flow problems. Based on this, $c(N)$ can be obtained in $O(|V|T(|V|,|E|))$ time, where $T(|V|,|E|)$ is the time required to find the max-flow in a network with $|V|$ nodes and $|E|$ edges. The best time bound for $T(|V|,|E|)$ known to

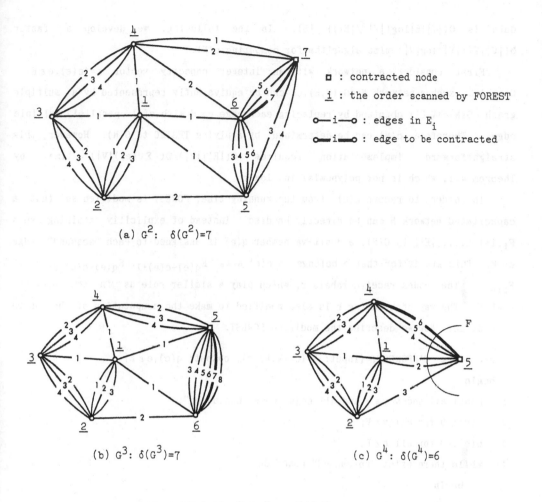

□ : contracted node

i : the order scanned by FOREST

o—*i*—o : edges in E_i

o—*i*—o : edge to be contracted

(a) G^2: $\delta(G^2)=7$

(b) G^3: $\delta(G^3)=7$

(c) G^4: $\delta(G^4)=6$

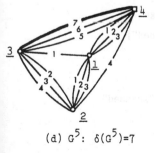

(d) G^5: $\delta(G^5)=7$

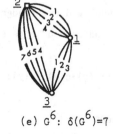

(e) G^6: $\delta(G^6)=7$

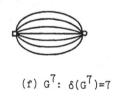

(f) G^7: $\delta(G^7)=7$

Fig.3 Application of TESTEC to G of Fig.1

date is $O(|V||E|\log(|V|^2/|E|))$ [5]. In the following, we develop a faster $O(|V||E|+|V|^2\log|V|)$ time algorithm for computing $c(N)$.

First consider a network with an integer capacity vector $c=(c(e),e\,\varepsilon\,E)$. Clearly, such a network $N=(G=(V,E),c)$ can be equivalently represented by a multiple graph $G(N)=(V,E')$ obtained by replacing each edge $e\,\varepsilon\,E$ with $c(e)$ parallel multiple edges. Therefore, $c(N)$ can be determined by applying TESTEC to $G(N)$. However, this straightforward implementation requires $O(|E'||V|)=O((\Sigma\,c(e))|V|)$ time by Theorem 4.1, which is not polynomial in $\log c$.

In order to remove $c(e)$ from the running time, FOREST is modified so that a capacitated network N can be directly handled. Instead of explicitly retaining sets $E_i, i=1,2,\ldots,|E'|$ in $G(N)$, a positive number $q(e)$ is assigned to each "scanned" edge $e\,\varepsilon\,E$. This stands for that e belongs to $c(e)$ sets $E_{q(e)-c(e)+1}$, $E_{q(e)-c(e)+2}$, $\ldots$, $E_{q(e)}$. The nodes receive labels r, which play a similar role as in the previous FOREST. The way of updating r is also modified to make the computation of the above q possible. Now we describe the modified FOREST.

Procedure CAPFOREST; {input: $N=(G=(V,E),c)$, output: $q(e),e\,\varepsilon\,E$}
   begin
1   Label all nodes $v\,\varepsilon\,V$ and all edges $e\,\varepsilon\,E$ "unscanned";
2   $r(v):=0$ for all $v\,\varepsilon\,V$;
3   $q(e):=0$ for all $e\,\varepsilon\,E$;
4   **while** there exist "unscanned" nodes **do**
      begin
5   Choose an "unscanned" node $x\,\varepsilon\,V$ with the largest r;
6   **for** each node y adjacent to x by an "unscanned" edge **do**
        {y is "unscanned" and every edge in E_{xy} is "unscanned"}
7   **for** each "unscanned" edge $e\,\varepsilon\,E_{xy}$ do
         begin
8           $q(e):=r(y)+c(e)$; $r(y):=r(y)+c(e)$; Mark e "scanned"
         end;
9   Mark x "scanned"
      end;
  end.

The correctness of CAPFOREST is contained in [9]. It is not difficult to see that CAPFOREST can be executed in $O(|E|+|V|\log|V|)$ time by maintaining the set of unscanned nodes X as a Fibonacci heap [3], where each item $x \varepsilon X$ has value $r(x)$.

Now we remove the integrality condition on $c(e)$, and consider that c is a positive real vector. The previous Lemma 3.3(2),(3) can however be generalized under this new setting. For a network $N=(G=(V,E),c)$, let

$$\delta(N) \equiv \min\{\Sigma_{e \varepsilon E(v)} c(e) | v \varepsilon V\}.$$

Lemma 5.1 [9:Lemma 5.1] For a network $N=(G,c)$ with a multiple graph $G=(V,E)$ and positive real valued capacities $c(e), e \varepsilon E$, let $q(e), e \varepsilon E$ be obtained by CAPFOREST.

(1) There exists an edge $e \varepsilon E$ with $q(e) \geq \delta(N)$.   (5.1)

(2) $f(u,v;N) \geq q(e)$ for any $e=(u,v) \varepsilon E$.   ☐   (5.2)

Lemma 5.2 [9] Let $N=(G=(V,E),c)$ be a network such that $|V| \geq 3$ and c is a real valued capacity vector. For two nodes $x,y \varepsilon V$, denote $N'=(G/\{x,y\},c)$. Then the following properties holds.

(1) $c(N) \leq c(N')$.

(2) If $c(x,y;N) \geq k$, then $c(N) \geq k$ if and only if $c(N') \geq k$.   ☐

Lemma 5.3 [9] For a network $N=(G=(V,E),c)$ such that $|V| \geq 3$ and c is a real valued capacity vector, let $N^1=N$ and N^j $(j=2,\ldots,|V|-1)$ be the network obtained by contracting an edge $e^{j-1}=(u^{j-1},v^{j-1})$ that satisfies $f(u^{j-1},v^{j-1};N^{j-1}) \geq \delta(N^{j-1})$ in N^{j-1}. Then $c(N)=\min\{\delta(N^1), \delta(N^2),\ldots,\delta(N^{|V|-1})\}$.   ☐

In analogy with the conversion of FOREST into CAPFOREST, we can get an algorithm (say MINCUT) by modifying TESTEC so as to computes $c(N)$ for a real valued capacity vector c, and the following properties hold.

Theorem 5.1 [9] Let $N=(G,c)$ be a network such that c is a real valued capacity vector. A minimum cut F of N and its value $c(N)$ can computed by MINCUT in $O(|E|+p|V|+|V|^2\log|V|)$ time, where p is the number of pairs of nodes between which G has an edge. If G is a simple graph, the time bound becomes $O(|E||V|+|V|^2\log|V|)$. ☐

Corollary 5.1 [9] The $\lambda(G)$ of a multiple graph $G=(V,E)$ can be determined in $O(|E|+\min\{\lambda(G)|V|^2,p|V|+|V|^2\log|V|\})$ time, where p is the number of pairs of nodes between which G has an edge.   ☐

6. Conclusion

In this paper, we proposed an $O(|E|+\min\{\lambda(G)|V|^2,p|V|+|V|^2\log|V|\})$ time algorithm for determining connectivity λ (G) of a multiple graph G=(V,E), where p is the number of pairs of nodes having an edge between them. This algorithm differs from the previous one [2], and requires no max-flow algorithm. Based on this method, we then showed that the minimum cut in a capacitated network can be found in $O(|E|+p|V|+|V|^2\log|V|)$ time.

References

[1] C.Berge, Graphs and Hypergraphs, North-Holland, 1973.

[2] S.Even and R.E.Tarjan, Network flow and testing graph connectivity, SIAM J.Computing 4 (1975), pp.507-518.

[3] M.L.Fredman and R.E.Tarjan, Fibonacci heaps and their uses in improved optimization algorithms, 25th Symp. Found. Compt. Sci. (1984), pp.338-346.

[4] Z.Galil, Finding the vertex connectivity of graphs, SIAM J.Computing 9 (1980), pp.197-199.

[5] A.V.Goldberg and R.E.Tarjan, A new approach to the maximum flow problem, Proc. 18th ACM Symp. on the Theory of Comput., (1986), pp.136-146.

[6] R.E.Gomory and T.C.Hu, Multi-terminal network flows, J. SIAM 9(4) (1961), pp.551-570.

[7] D.W.Matula, Determining edge connectivity in O(nm), 28th IEEE Symp. Found. Compt. Sci. (1987), pp.249-251.

[8] H.Nagamochi and T.Ibaraki, Linear time algorithms for finding k-edge-connected and k-node-connected spanning subgraphs, Technical Report #89006, Dept. of Applied Mathematics and Physics, Faculty of Engineering, Kyoto University (1989).

[9] H.Nagamochi and T.Ibaraki, Computing edge-connectivity in multiple and capacitated graphs, Technical Report #89009, Dept. of Applied Mathematics and Physics, Faculty of Engineering, Kyoto University (1989).

[10] C.P.Schnorr, Bottlenecks and edge connectivity in unsymmetrical networks, SIAM J.Computing 8 (1979), pp.265-274.

Efficient Sequential and Parallel Algorithms for Planar Minimum Cost Flow

Hiroshi Imai

Department of Information Science, Faculty of Science
University of Tokyo, Hongo, Bunkyo-ku, Tokyo 113, Japan

Kazuo Iwano

IBM Research, Tokyo Research Laboratory
5-11 Sanban-cho, Chiyoda-ku, Tokyo 102, Japan

Abstract

This paper presents efficient sequential and parallel algorithms for the minimum cost flow problem on planar networks. Our algorithms are based on the interior point method for linear programming, and make full use of the planarity of networks in solving a system of linear equations in sequential and parallel ways. For the planar minimum cost flow problem with n vertices and integer costs and capacities on edges whose absolute values are bounded by γ, we give a sequential algorithm with $O(n^{1.594}\sqrt{\log n}\log(n\gamma))$ time and $O(n\log n)$ space and a parallel algorithm with $O(\sqrt{n}\log^3 n\log(n\gamma))$ parallel time using $O(n^{1.094})$ processors. These algorithms are currently best for $\gamma = \text{poly}(n)$. These results can be generalized to the minimum cost flow problem on $s(n)$-separable networks such as three-dimensional grid networks.

1. Introduction

In these years, much research has been done on the minimum cost flow problem on networks. The minimum cost flow problem is the most general problem in network flow theory that admits strongly polynomial algorithms, and should be investigated further from various viewpoints. Recently, the interior point method for linear programming has been applied to the minimum cost flow problem, which is a very special case of linear programming, by Vaidya [13] from the viewpoint of sequential algorithms and by Goldberg, Plotkin, Shmoys and Tardos [6] from the viewpoint of parallel algorithms. In this paper, we consider the minimum cost flow problem on networks having good separators, where the most typical network with good separators is a planar network. We present a unified way of applying the interior point algorithm to the planar minimum cost flow problem to get efficient sequential and parallel algorithms.

We summarize the complexity of existing algorithms before describing our results. Consider a network with m edges, n vertices where costs and capacities of edges are integers whose absolute values are at most γ_{cost} and γ_{cap}, respectively. Let $\gamma = \max\{\gamma_{cost}, \gamma_{cap}\}$. As sequential algorithms for the minimum cost flow problem on the network, there is now an $O(m(m + n\log n)\log n)$-time strongly polynomial algorithm by Orlin [10]. There are also two efficient algorithms, which are not strongly polynomial but are polynomial with respect to the input size: one is an $O(\min\{n^3, mn\log n\}\log(n\gamma_{\text{cost}}))$-time algorithm by Goldberg and Tarjan [7] and the other is an $O(n^2\sqrt{m}\log(n\gamma))$-time algorithm by Vaidya [13]. The algorithm in [7] can be parallelized, which yields an $O(n^2\log n\log(n\gamma_{\text{cost}}))$-time parallel algorithm using n processors. For the zero-one minimum cost flow problem, an $O(\sqrt{m}\ \text{polylog}\ n)$-time parallel algorithm using $O(n^3)$ processors is given by Goldberg, Plotkin, Shmoys and Tardos [6] (the number of processors may be reduced to $O(n^{2.376})$ if the fast matrix multiplication in Coppersmith and Winograd [2] is used).

In this paper, we consider the minimum cost flow problem on networks having good separators, that is, $s(n)$-*separable* networks. Roughly, an $s(n)$-separable graph of n vertices can be divided into two subgraphs by removing $s(n)$ vertices such that there is no edge between the two subgraphs and the number of vertices in each subgraph is at most αn for a fixed constant α. A planar grid graph is easily seen to be $\sqrt{n}$-separable [5]. General planar graphs are $O(\sqrt{n})$-separable; this is well known as the planar separator theorem by Lipton and Tarjan [9]. For a system of linear equations $Ax = b$ with $n \times n$ symmetric matrix A, if the nonzero pattern of A corresponds to that of the adjacency matrix of an $s(n)$-separable graph, the system $Ax = b$ can be solved more efficiently than general linear systems. This is originally shown for planar grid graphs by George [5], and the technique is called *nested dissection*. The technique is generalized to $s(n)$-separable graphs by Lipton, Rose and Tarjan [8]. For A corresponding to a planar graph, the system can be solved in $O(n^{1.188})$ time and $O(n\log n)$ space.

For the planar minimum cost flow problem, we give a sequential algorithm with $O(n^{1.594}\sqrt{\log n}\log(n\gamma))$ time and $O(n\log n)$ space and a parallel algorithm with $O(\sqrt{n}\log^3 n\log(n\gamma))$ parallel time using $O(n^{1.094})$ processors. When γ is polynomially bounded in n, the sequential time complexity becomes $O(n^{1.6})$ which is best among existing algorithms by a factor $O(n^{0.4}\log^2 n)$ (note that, since $m = O(n)$ in planar networks, the best algorithm that is mentioned above is Orlin's, and has the complexity of $O(n^2\log^{1.5} n)$ when the $O(n\sqrt{\log n})$-time algorithm (Frederickson [3]) for the shortest path problem on the planar graph is used). Also, our parallel algorithm is best possible with respect to the sequential algorithm, that is, the parallel time complexity multiplied by the number of processors is within a polylog factor of the sequential time complexity. These results can be generalized to the minimum cost flow problem on $s(n)$-separable networks such as three-dimensional grid networks.

A main task of the interior point method is to solve a linear system at each iteration. We will show that linear systems appeared in Vaidya's interior point method [13] can be solved efficiently when the method is applied to the minimum cost flow problem on $s(n)$-separable network. However, it does not give the best result to directly apply the nested dissection technique to the linear system. In the interior point method by Vaidya [13], besides solving the linear system, we have to perform rank-one updates, etc., and so we have to analyze the complexity in detail. Furthermore, when parallelizing Vaidya's interior point algorithm, we have to tune up the algorithm by slowing down some part of it and scheduling tasks nicely. It should be mentioned that the parallel interior point algorithm in [6] only parallelizes a simple interior method and uses the large number of processors, so that the parallel time complexity times the number of processors does not match the sequential time complexity within a polylog factor. We will parallelize Vaidya's interior algorithm, which is the best known sequential algorithm and has not yet been parallelized, and attain the minimum number of processors with respect to the parallel time complexity within a polylog factor.

2. Interior Point Algorithm

In this section we briefly describe Vaidya's interior point algorithm for the minimum cost flow problem on a connected network. For simplicity, we will consider the minimum cost circulation problem, which finds a minimum-cost flow satisfying the flow conservation constraint at each vertex and the capacity constraint at each edge. The problem can be written as follows:

$$\min \ \tilde{w}^{\mathrm{T}}\tilde{z}$$
$$\text{s.t.} \ \ \tilde{G}\tilde{z} = 0$$
$$0 \le \tilde{z} \le \tilde{c}$$

Here, $\tilde{G}$ is the vertex-edge incidence matrix of the network, $\tilde{z}$ is the vector of flow variables, $\tilde{w}$ is the vector of edge costs and $\tilde{c}$ is the vector of edge capacities. $\tilde{G}$ is an $n \times m$ matrix, and its rank is $n-1$. In order to make it easier to get a good initial feasible interior point and also to simplify the following discussions, we add an artificial variable to each row of the equality constraint with an appropriate sign, and consider the following transformed problem:

$$\min \ \tilde{w}^{\mathrm{T}}\tilde{z} + \lambda e^{\mathrm{T}}\overline{z}$$
$$\text{s.t.} \ \ \tilde{G}\tilde{z} + \overline{G}\overline{z} = 0$$
$$0 \le \tilde{z} \le \tilde{c}$$
$$0 \le \overline{z} \le \overline{c}$$

where $\overline{G}$ is a diagonal matrix whose diagonal elements are ± 1 and $e = (1,\ldots,1)^{\mathrm{T}}$. By setting a scalar λ sufficiently large, this transformed problem becomes equivalent to the original circulation problem. This transformed problem can be simply written as

$$\min \ w^{\mathrm{T}}z$$
$$\text{s.t.} \ \ Gz = 0$$
$$0 \le z \le c$$

where $G = [\tilde{G}, \overline{G}]$, $w = \begin{pmatrix} \tilde{w} \\ \lambda e \end{pmatrix}$, $z = \begin{pmatrix} \tilde{z} \\ \overline{z} \end{pmatrix}$ and $c = \begin{pmatrix} \tilde{c} \\ \overline{c} \end{pmatrix}$. G is an $n \times M$ matrix and its rank is n, where $M = m + n$.

Let $\Phi(z, \beta)$ be the potential function defined by

$$\Phi(z,\beta) = \sum_{i=1}^{M}(\ln(c_i - z_i) + \ln z_i) + M\ln(-w^{\mathrm{T}}z - \beta).$$

At the beginning of the kth iteration, the algorithm maintains a parameter β^{k-1} and a feasible interior point z^{k-1} with $-w^{\mathrm{T}}z^{k-1} > \beta^{k-1}$ and an $M \times M$ diagonal matrix D

such that the ith diagonal entry D_{ii} satisfies the conditions

$$D_{ii} = D_{ii}^+ + D_{ii}^-$$

$$\frac{1}{1.1(z_i)^2} \le D_{ii}^+ \le \frac{1.1}{(z_i)^2}$$

$$\frac{1}{1.1(c_i - z_i)^2} \le D_{ii}^- \le \frac{1.1}{(c_i - z_i)^2}$$

for $i = 1, \ldots, M$. Let r be a positive parameter which will be determined later. Then, the kth iteration performs the following computations.

1. $\beta^k := \beta^{k-1} + \dfrac{1}{30\sqrt{m}}(-w^{\mathrm{T}} z^{k-1} - \beta^{k-1})$.

2. Let μ^k be the gradient of $\Phi(z, \beta^k)$ at z^{k-1}:

$$\mu^k := \sum_{i=1}^{M} \left(\frac{-1}{c_i - z_i^{k-1}} + \frac{1}{z_i^{k-1}} \right) - \frac{w_i}{-w^{\mathrm{T}} z^{k-1} - \beta}$$

$$Q := D + \frac{M}{(-w^{\mathrm{T}} z^{k-1} - \beta^k)^2} w w^{\mathrm{T}}$$

$$\nu^k := (I - Q^{-1} G^{\mathrm{T}} (G Q^{-1} G^{\mathrm{T}})^{-1} G) Q^{-1} \mu^k$$

3. Compute a scalar $t^k > 0$ such that $0.018 \le (t^k)^2 (\nu^k)^{\mathrm{T}} \nu^k \le 0.0196$

4. If $\Phi(z^{k-1} + t^k \nu^k, \beta^k) > \Phi(z^{k-1}, \beta^k)$ then $z^k := z^{k-1} + t^k \nu^k$ else $z^k := z^{k-1}$;

5. For each i, $1 \le i \le M$,

   if $(D_{ii}^+ < \dfrac{1}{1.1(z_i)^2})$ or $(D_{ii}^+ > \dfrac{1.1}{(z_i)^2})$ then $D_{ii}^+ := \dfrac{1}{(z_i)^2}$; $\quad D_{ii} := D_{ii}^+ + D_{ii}^-$

   if $(D_{ii}^- < \dfrac{1}{1.1(c_i - z_i)^2})$ or $(D_{ii}^- > \dfrac{1.1}{(c_i - z_i)^2})$ then

   $$D_{ii}^- := \frac{1}{(c_i - z_i)^2}; \quad D_{ii} := D_{ii}^+ + D_{ii}^-$$

6. If the iteration number k is a multiple of r then

   for each i, $1 \le i \le M$, $D_{ii}^+ := \dfrac{1}{(z_i)^2}$; $\quad D_{ii}^- := \dfrac{1}{(c_i - z_i)^2}$; $\quad D_{ii} := D_{ii}^+ + D_{ii}^-$

The algorithm halts when $-w^{\mathrm{T}} z^k - \beta^k \le 2^{-\alpha' L}$ for a suitably large constant α' where $L = \log(n\gamma)$. About the number of iterations required by this algorithm, the following lemmas hold.

Iteration Lemma. (Vaidya [13]) Starting from an appropriate initial feasible point, which can be obtained trivially, the algorithm halts in $O(\sqrt{M} L)$ iterations. $\square$

Update Lemma. (Vaidya [13]) Between successive resettings of D in Step 6, the total number of modifications to diagonal elements of D in Step 5 is $O(r^2)$, if $r \le \sqrt{M}$. $\square$

3. Sequential Interior Algorithm for Planar Minimum Cost Flow

We will evaluate the time complexity of the interior point algorithm [13] described above when this algorithm is applied to the planar minimum cost flow problem. Since the network is planar, we have $m, M = O(n)$. We will estimate the time complexity per iteration. The computation of ν^k, which involves solving linear systems, is a dominant task. This task will be accomplished in the following two tasks. One is to recompute a factorization of an $n \times n$ matrix at the kth iteration when k is a multiple of r in step 6, and the other is to perform the rank one update operations when D_{ii}'s are modified in step 5. About computing ν^k, the following will be used (Vaidya [13]):

$$Q^{-1} = D^{-1} + u_1 v_1^{\mathrm{T}}$$
$$(GQ^{-1}G^{\mathrm{T}})^{-1} = (GD^{-1}G^{\mathrm{T}})^{-1} + u_2 v_2^{\mathrm{T}}$$

where

$$u_1 = \frac{-M}{(-w^{\mathrm{T}} z^{k-1} - \beta^k)^2 + M w^{\mathrm{T}} D^{-1} w} D^{-1} w, \qquad v_1^{\mathrm{T}} = w^{\mathrm{T}} D^{-1}$$

$$u_2 = \frac{-1}{1 + v_1^{\mathrm{T}} G^{\mathrm{T}} (GD^{-1}G^{\mathrm{T}})^{-1} G u_1} (GD^{-1}G^{\mathrm{T}})^{-1} G u_1, \qquad v_2^{\mathrm{T}} = v_1^{\mathrm{T}} G^{\mathrm{T}} (GD^{-1}G^{\mathrm{T}})^{-1}$$

Unlike Vaidya's algorithm, we will not compute the values of all elements of the inverse $(GD^{-1}G^{\mathrm{T}})^{-1}$ explicitly. This is because $(GD^{-1}G^{\mathrm{T}})^{-1}$ is generally dense even if $GD^{-1}G^{\mathrm{T}}$ itself is very sparse. Therefore, we should be careful in evaluating the time complexity of, say, calculating $(GD^{-1}G^{\mathrm{T}})^{-1}\eta$ for a vector η with a constant number of nonzeros (this can be done in $O(n)$ time if $(GD^{-1}G^{\mathrm{T}})^{-1}$ is explicitly at hand, while this does not necessarily hold when we only have a factorization, say the Cholesky factorization, of $GD^{-1}G^{\mathrm{T}}$).

Suppose that there is an expression for $(GD^{-1}G^{\mathrm{T}})^{-1}$ by which we can compute $(GD^{-1}G^{\mathrm{T}})^{-1}\eta$ for any vector η in $T(n,r)$ time and $S(n,r)$ space. We assume $T(n,r)$, $S(n,r) = \Omega(n)$. The computations required for maintaining the expression are recomputations of some factorization from scratch and incremental updates. The computations during an iteration, except those for maintaining the expression, are computing the gradient μ^k, the direction ν^k, the scalar t^k and the updating of D. For these computations, we have the following.

Lemma 3.1. The computations during an iteration, except those for maintaining the expression, can be executed in $O(T(n,r))$ time and $O(S(n,r))$ space. $\square$

Thus, the remaining problem is to evaluate the complexity for maintaining the expression for $(GD^{-1}G^{\mathrm{T}})^{-1}$. First, we have the following.

Lemma 3.2. Given D^{-1}, a special factorization for $GD^{-1}G^{\mathrm{T}}$ can be computed in $O(n^{1.188})$ time and $O(n \log n)$ space such that, using this factorization, for any vector η, $(GD^{-1}G^{\mathrm{T}})^{-1}\eta$ can be computed in $O(n \log n)$ time. $\square$

The proof proceeds as follows. First, the pattern of nonzeros of $GD^{-1}G^{\mathrm{T}}$ corresponds to that of the adjacency matrix of the planar network. Then, we consider the $O(\sqrt{n})$-factorization of the matrix introduced by Pan and Reif [12]. This factorization is originally considered for parallel computation, but can be employed in a sequential algorithm to incorporate the fast matrix multiplication technique in the nested dissection computation. The details will be described in the full version of this paper.

Now, we evaluate the average number of operations per iteration for incrementally updating $(GD^{-1}G^{\mathrm{T}})^{-1}$ according to changes in diagonal elements D_{ii} in D during the period of r iterations. Let B denote the matrix $GD^{-1}G^{\mathrm{T}}$ at the beginning of the period under consideration. After q updates to D during the period, $GD^{-1}G^{T}$ may be written as

$$GD^{-1}G^{T} = B + UV^{T}$$

where U, V are $n \times q$ matrices. By the Update Lemma, $q = O(r^2)$. Since any column of G has at most two nonzeros and each column of U and V is a column of G multiplied by some constant, any column of U and V also contains at most two nonzeros.

Using the Sherman-Morrison-Woodbury formula, $(GD^{-1}G^{T})^{-1}$ may be written as

$$(GD^{-1}G^{T})^{-1} = (B + UV^{T})^{-1} = B^{-1} - B^{-1}U(I + V^{T}B^{-1}U)^{-1}V^{T}B^{-1}.$$

Since the factorization for B^{-1} is computed at the beginning and is available throughout the period, we have only to maintain U, V and $(I + V^{T}B^{-1}U)^{-1}$.

Lemma 3.3. The average number of operations per iteration for incrementally updating U, V and $(I + V^{T}B^{-1}U)^{-1}$ is $O(rn \log n + r^{5})$. □

We here omit the proof due to the space limitations. We only note that, as mentioned above, $B^{-1}\eta$ can be computed in $O(n \log n)$ time in general from this special factorization, and not in $O(n)$ time from the inverse itself.

Lemma 3.4. $T(n,r) = O(n \log n + r^{4})$ and $S(n,r) = O(n \log n)$. □

Thus, each iteration of the algorithm takes

$$O(\frac{n^{1.188}}{r} + rn \log n + r^{5})$$

on the average, and setting $r = n^{0.094}/\sqrt{\log n}$ we have the following theorem.

Theorem 3.1. The planar minimum cost flow problem can be solved in $O(n^{1.594} \sqrt{\log n} \log(n\gamma))$ time and $O(n \log n)$ space. □

This is the currently best known time bound.

4. Parallel Interior Algorithm for Planar Minimum Cost Flow

In this section, we will consider the zero-one minimum cost flow problem on the planar network. A parallel algorithm for the zero-one minimum cost flow problem is developed by Goldberg, Plotkin, Shmoys and Tardos [6]. Their algorithm is based on some interior point method, not Vaidya's algorithm, and has the finish-up stage of the algorithm which converts a near-optimal solution into an optimal basic (zero-one) solution. In this section, we will parallelize Vaidya's algorithm. Although this new parallelization does not reduce the time complexity of [6], it reduces the number of processors greatly from $O(n^3)$ (or $O(n^{2.376})$) to $O(n^{1.094})$. Moreover, it attains an optimal reduction in the number of processors for this parallel time complexity within a polylog factor with respect to the sequential time complexity given in the previous section.

We begin with mentioning the result about parallel computation of the linear system.

Lemma 4.1. (Pan, Reif [12], Pan [11], Coppersmith, Winograd [2]) A linear system $A\hat{x} = b$ for an $n \times n$ matrix A can be solved in $O(\log^2 n)$ time using $O(n^{2.376})$ processors. Furthermore, if the nonzero pattern of A corresponds to that of a planar network, and good separators of a planar network are given in advance, the linear system can be factorized in $O(\log^3 n)$ time using $O(n^{1.188})$ processors. Given this factorization, the linear system can be solved in $O(\log^2 n)$ time using $O(n)$ processors. $\square$

In the latter half of this lemma, we assume that separators are precomputed. If the planar network is a grid network, $\sqrt{n}$-separators can be easily found. For general planar networks, we have the following.

Lemma 4.2. (Gazit, Miller [4]) An $O(\sqrt{n})$-separator of the planar graph can be found by a randomized parallel algorithm with $O(\log^2 n)$ time and $O(n^{1+\epsilon})$ processors.$\square$

We now consider how to parallelize the interior point algorithm to achieve the currently best parallel time complexity with using a much smaller number of processors.

The computations during an iteration, except those for maintaining expressions of $(GQ^{-1}G^{\mathrm{T}})^{-1}$ and Q^{-1}, are computing the gradient μ^k, the direction ν^k, the scalar t^k and the updating of D. In particular, we compute ν^k in the following way: Given a vector v, we assign a processor to each nonzero element. Let $v(i)$ be the i-th nonzero element in a vector v and $p(i)$ the processor assigned to $v(i)$. Then we maintain pointers from each processor $p(i)$ to $p(i+1)$. We can now compute a multiplication $v^{\mathrm{T}}w$ of two $k \times 1$ vectors v and w in $O(\log k)$ parallel time using as many processors as the number of nonzeros in v. This is because each processor $p(i)$ computes $v(i)w(i)$ and these products can be summed up in $O(\log n)$ steps. Therefore, using Vaidya's techniques for maintaining expressions of $(GQ^{-1}G^{\mathrm{T}})^{-1}$ and Q^{-1}, we can compute ν^k in $O(\log n)$ time using as many processors as the number of nonzero elements in matrices appeared during the computation, which is $O(n \log n + r^4)$.

Lemma 4.3. The computations during an iteration, except those for maintaining expressions of $(GQ^{-1}G^{\mathrm{T}})^{-1}$, (hence, the computation of computing the gradient μ^k, the direction ν^k, the scalar t^k and the updating of D) can be performed in $O(\log n)$ time using $O(n \log n + r^4)$ processors. $\square$

We next consider how to maintain expressions of $(GQ^{-1}G^{\mathrm{T}})^{-1}$. Lemma 4.1 can be applied to the two tasks: one is to compute $(GD^{-1}G^{\mathrm{T}})^{-1}$ at the kth iteration when k is a multiple of r, and the other is to execute incremental updates at each iteration. However, the size of matrices in the former task is much larger than the size of matrices in the latter, and the latter task is executed by r times more than the former. This suggests slowing down the factorization of $GD^{-1}G^{\mathrm{T}}$ by a factor of r, which will reduce the total number of processors required by the algorithm. In fact, we can slow down that process, based on the result of Brent [1].

Lemma 4.4. When the nonzero pattern of A corresponds to that of the planar network and a parameter r is given, a linear system can be solved in $O(r \log^3 n)$ time using $O(n^{1.188}/r)$ processors. $\square$

We now consider how to parallelize incremental updates at each iteration. For changes in diagonal elements of D, $(GD^{-1}G^{\mathrm{T}})^{-1}$ can be computed by

$$(GD^{-1}G^{\mathrm{T}})^{-1} = (B + UV^{\mathrm{T}})^{-1} = B^{-1} - B^{-1}U(I + V^{\mathrm{T}}B^{-1}U)^{-1}V^{\mathrm{T}}B^{-1}.$$

In the sequential case, incremental updates are executed one by one, that is, a column is added to each of U and V and accordingly $(I + V^{\mathrm{T}}B^{-1}U)^{-1}$ is updated at a time.

In the parallel case, we perform incremental updates simultaneously at each iteration. However, this may require a many number of processors to execute the updates at the iteration in a polylog time. Therefore, we slow down this stage if the number of updates at the iteration is comparatively large.

The problem is now how to execute all the incremental updates efficiently at an iteration in parallel. At the beginning of the period of r iterations, the factorization of B is computed, as in Lemma 4.4, and this factorization is available during the period. We will maintain U, V, $V^{\mathrm{T}}B^{-1}$, $B^{-1}U$ and $V^{\mathrm{T}}B^{-1}U$, where, for the former two, only the nonzero elements are maintained, and, for the latter three, the values of all elements are maintained. At the beginning of this period, U and V are empty. Let q be the number of columns of U and V at the beginning of the kth iteration. By the Update Lemma, $q = O(r^2)$. $V^{\mathrm{T}}B^{-1}$ is a $q \times n$ matrix, $B^{-1}U$ is an $n \times q$ matrix, and $V^{\mathrm{T}}B^{-1}U$ is a $q \times q$ matrix.

Suppose that there are q_k updates at the kth iteration, that is, $U_k V_k^{\mathrm{T}}$ is given as an update of rank q_k to $B + UV^{\mathrm{T}}$, where U_k and V_k are $n \times q_k$ matrices. With the matrices mentioned above at hand, we can compute $((B + UV^{\mathrm{T}}) + U_k V_k^{\mathrm{T}})^{-1}$ as follows. Defining $n \times (q + q_k)$ matrices $\tilde{U}$ and $\tilde{V}$ by $\tilde{U} = [U, U_k]$ and $\tilde{V} = [V, V_k]$, we have

$$(I + \tilde{V}^{\mathrm{T}}B^{-1}\tilde{U})^{-1} = \begin{pmatrix} I + V^{\mathrm{T}}B^{-1}U & V^{\mathrm{T}}B^{-1}U_k \\ V_k^{\mathrm{T}}B^{-1}U & I + V_k^{\mathrm{T}}B^{-1}U_k \end{pmatrix}^{-1}$$

Lemma 4.5. Using the factorization and the matrices at hand, $(I + \tilde{V}^{\mathrm{T}}B^{-1}\tilde{U})^{-1}$ can be computed in $O(\log^2 n + \log^2 \tilde{q})$ time using $O(nq_k + \tilde{q}^{2.4})$ processors where $\tilde{q} = q + q_k$.□

The outline of a proof is as follows. First, the values of all elements of $B^{-1}U_k$ and $V_k B^{-1}$ can be computed in $O(\log^2 n)$ time using $O(nq_k)$ processors by utilizing the factorization of B. Adding these two matrices to $B^{-1}U$ and $V^{\mathrm{T}}B^{-1}$, the matrices $B^{-1}\tilde{U}$ and $\tilde{V}^{\mathrm{T}}B^{-1}$ are obtained. $V_k^{\mathrm{T}}B^{-1}U_k$ can be obtained in $O(\log q_k)$ time using $O((q_k)^2)$ processors. Using $V^{\mathrm{T}}B^{-1}$ and $B^{-1}U$, $V^{\mathrm{T}}B^{-1}U_k$ and $V_k^{\mathrm{T}}B^{-1}U$ can be computed in $O(\log q)$ time using $O(qq_k)$ processors, because each column of U_k and V_k contains at most two nonzeros. The matrix $I + \tilde{V}^{\mathrm{T}}B^{-1}\tilde{U}$ can thus be computed in $O(\log^2 n + \log \tilde{q})$ time using $O(nq_k + \tilde{q}^2)$ processors. Since the matrix is a $\tilde{q} \times \tilde{q}$ matrix, its inverse can be computed explicitly in $O(\log^2 \tilde{q})$ time using $O(\tilde{q}^{2.4})$ processors (Lemma 4.1).

Given $(I + \tilde{V}^{\mathrm{T}}B^{-1}\tilde{U})^{-1}$, $(GD^{-1}G^{\mathrm{T}})^{-1}\eta$ can be computed in $O(\log^2 n + \log \tilde{q})$ time using $O(n + \tilde{q}^2)$ processors. During the period of r iterations, $\tilde{q} = O(r^2)$ by the Update Lemma. We thus have the following.

Lemma 4.6. For q_k incremental updates at the kth iteration, the factorization for $(GD^{-1}G^{\mathrm{T}})^{-1}$ can be maintained in $O(\log^2(nr))$ time with $O(nq_k + r^{4.8})$ processors. Given the factorization, $(GD^{-1}G^{\mathrm{T}})^{-1}\eta$ for a vector η can be computed in $O(\log^2 n + \log r)$ time using $O(n + r^4)$ processors. □

The summation of q_k over the period of r iterations is $O(r^2)$. However, this does not imply $q_k = O(r)$. At some iteration, q_k may be $\Theta(r^2)$. Then, the number of processors for maintaining the factorization in Lemma 4.6 becomes $O(nr^2)$, which should be avoided. Hence, we slow down the factorization stage, and have the following lemma, where we assume $q_k \geq 1$.

Lemma 4.7. At the kth iteration, the factorization for $(GD^{-1}G^T)^{-1}$ can be maintained in $O(\lceil q_k/r \rceil \log^2(nr))$ time using $O(nr + r^{4.8})$ processors. $\square$

Since the summation of $\lceil q_k/r \rceil$ over the period of r iterations is $O(r)$, we can modify the first half of Lemma 4.6 to the following:

Lemma 4.8. The factorization for $(GD^{-1}G^T)^{-1}$ can be maintained in $O(\log^2(nr))$ time with $O(nr + r^{4.8})$ processors per iteration on the average. $\square$

Combining Lemmas 4.3, 4.4, 4.6 and 4.8, we see that each iteration of this parallel algorithm can be executed on the average in $O(\log^3 n + \log^2 r)$ time using

$$O(\frac{n^{1.188}}{r} + nr + r^{4.8})$$

processors. Putting $r = O(n^{0.094})$ and applying the result for the finish-up stage in [6], which requires $O(\log n \log(n\gamma))$ time and $O(n)$ processors, we obtain the following theorem.

Theorem 4.1. The planar zero-one minimum cost flow problem can be solved in $O(\sqrt{n} \log^3 n \log(n\gamma))$ time using $O(n^{1.094})$ processors. $\square$

This implies that, using n processors, the planar zero-one minimum cos flow problem can be solved in $O(n^{0.595} \log \gamma)$ time.

Acknowledgment

The work of the first author was partially done while he was at Department of Computer Science and Communication Engineering, Kyushu University, and it was supported by the Grant-in-Aid of the Ministry of Education, Science and Culture of Japan.

References

[1] R. P. Brent: The Parallel Evaluation of General Arithmetic Expressions. *Journal of the Association for Computing Machinery*, Vol.21, No.2 (1974), pp.201–206.

[2] D. Coppersmith and S. Winograd: Matrix Multiplication via Arithmetic Progressions. *Proceedings of the 19th Annual ACM Symposium on Theory of Computing*, 1987, pp.1–6.

[3] G. N. Frederickson: Fast Algorithms for Shortest Paths in Planar Graphs, with Applications. *SIAM Journal on Computing*, Vol.16, No.6 (1987), pp.1004–1022.

[4] H. Gazit and G. L. Miller: A Parallel Algorithm for Finding a Separator in Planar Graphs. *Proceedings of the 28th Annual IEEE Symposium on Foundations of Computer Science*, Los Angeles, 1987, pp.238–248.

[5] A. George: Nested Dissection of a Regular Finite Element Mesh. *SIAM Journal on Numerical Analysis*, Vol.10, No.2 (1973), pp.345–363.

[6] A. V. Goldberg, S. A. Plotkin, D. B. Shmoys and É. Tardos: Interior-Point Methods in Parallel Computation. *Proceedings of the 30th Annual IEEE Symposium on Foundations of Computer Science*, 1989, pp.350–355.

[7] A. V. Goldberg and R. E. Tarjan: Solving Minimum Cost Flow Problems by Successive Approximation. *Proceedings of the 19th Annual ACM Symposium on Theory of Computing*, 1987, pp.7–18.

[8] R. J. Lipton, D. J. Rose, and R. E. Tarjan: Generalized Nested Dissection. *SIAM Journal on Numerical Analysis*, Vol.16, No.2 (1979), pp.346–358.

[9] R. J. Lipton and R. E. Tarjan: A Separator Theorem for Planar Graphs. *SIAM Journal on Applied Mathematics*, Vol.36, No.2 (1979), pp.177–189.

[10] J. B. Orlin: A Faster Strongly Polynomial Minimum Cost Flow Algorithm. *Proceedings of the 20th Annual ACM Symposium on Theory of Computing*, 1988, pp.377–387.

[11] V. Pan: Fast and Efficient Parallel Algorithms for Exact Inversion of Integer Matrices. *Proceedings of the 5th Conference on Foundations of Software Technology and Theoretical Computer Science*, New Delhi, India, Lecture Notes in Computer Science 206 (Springer, Berlin, 1985), pp.504–521.

[12] V. Pan and J. Reif: Efficient Parallel Solution of Linear Systems. *Proceedings of the 17th Annual ACM Symposium on Theory of Computing*, Providence, 1985, pp.143–152.

[13] P. Vaidya: Speeding-Up Linear Programming Using Fast Matrix Multiplication. *Proceeding of the 30th Annual IEEE Symposium on Foundations of Computer Science*, 1989, pp.332–337.

Structural Analyses

on

the Complexity of Inverting Functions

(Extended Abstract)

Osamu WATANABE

Department of Computer Science

Tokyo Institute of Technology

Meguro-ku, Tokyo 152, JAPAN

watanabecs.titech.ac.jp

Seinosuke TODA

Department of Computer Science

University of Electro-Communications

Chofu-shi, Tokyo 182, JAPAN

todacso.cs.uec.ac.jp

ABSTRACT

In this paper we investigate the complexity of inverting polynomial-time computable functions by methods developed in structural complexity theory. We first analyze upper bounds of the complexity of inverse functions by using complexity classes of functions. We prove the following: (1) NP/bit (the class of functions whose *each* bit is NP computable) is an upper bound for inverting honest and *one-to-one* functions, and (2) relative to almost all oracle, the class $\mathrm{PF}_{tt}^{\mathrm{NP}}$ (the class of functions that are polynomial time computable by asking non-adaptive queries to an NP oracle) is an upper bound for inverting honest functions. Next we investigate relative complexity of inverse functions by using polynomial-time reducibility of functions. We prove that an honest function is NP/bit invertible if the class of its inverse functions possesses the least element under polynomial-time non-adaptive one-query reducibility.

1. Introduction and Summary of Main Results

The subject of this paper is to investigate complexity of inverting polynomial-time computable honest[1] functions from Σ^* to Σ^*. We measure "complexity" by methods developed in structural complexity theory.

The complexity of computing an inverse has been studied by several researchers [Al86, GS84, Va76, Wa88a]. In particular, a *polynomial-time one-way function* — a polynomial-time computable honest (one-to-one) function whose inverse is not polynomial-time computable — have received considerable attention in relation to "cryptography"; see, e.g., [Wa89]. However, it seems very hard to prove any nontrivial upper or lower time bounds

---

[1] A function f is called *polynomially honest* (or, *honest* in short) if there exists a polynomial p such that for every x in the domain of f, $|x| \le p(|f(x)|)$ and $|f(x)| \le p(|x|)$. When the polynomial-time computability of inverting functions is concerned, this honest property is usually assumed in order to avoid the trivial case.

for the problem of inverting polynomial-time computable functions in general; we know very little on these lines [Al86]. On the other hand, there are some ways to show difficulty/easiness other than actual resource complexity measures. In structural complexity theory, we have developed several quantitative scales to measure (in)tractability of problems. For example, while we have been unable to prove that SAT, the set of satisfiable Boolean formulas, is not polynomial-time recognizable, we can prove that SAT is the hardest set in NP, where the completeness notion is used to state "hardest". Here by using such structural complexity measures, we investigate complexity of inverting polynomial-time computable functions.

Throughout this abstract, we assume that every function is a honest function from Σ^* to Σ^*. An *inverse function* (or, *inverse* in short) of a function s is a function f such that (i) the domain of f is the range of s, and (ii) for every y in the range of s, $s(f(y)) = y$.

For example, consider the following function:

for every x and y in Σ^*,

$$\text{sat}(\langle x, y \rangle) = \begin{cases} x, & \text{if } y \text{ is a satisfying assignment} \\ & \text{of a Boolean formula encoded by } x, \\ \bot, & \text{otherwise,} \end{cases}$$

where symbol $\bot$ denotes "undefined".

Clearly this function is polynomial-time computable and honest. On the other hand, inverting sat is essentially equivalent to obtaining a satisfying assignment for a given Boolean formula; thus, sat is not invertible unless P = NP. In this paper, we discuss how difficult it is to invert such functions as sat.

Here we should note that a given function may have more than one inverses. For example, the function sat has infinite inverses because it is not one-to-one infinitely often. In such a case, the complexity of the "easiest" inverse is usually regarded as the complexity of inverting s. (Note that the notion of "easiest" may vary depending on the context.) For example, it is natural to say that sat is *polynomial-time invertible* if *some* inverse of sat (i.e., the easiest inverse of sat) is polynomial-time computable. For any function s, let $\mathcal{F}_{s^{-1}}$ denote the class of inverses of s. When s is not one-to-one thus having more than one inverses, we investigate the complexity of the "easiest" inverse in $\mathcal{F}_{s^{-1}}$.

Our first approach is to use complexity classes of functions. We use them to state an upper bound of the complexity of inverting functions. For any complexity classes $\mathcal{CF}$ and any polynomial-time computable function s, if $\mathcal{CF} \cap \mathcal{F}_{s^{-1}} \neq \emptyset$, then we consider that $\mathcal{CF}$ is an upper bound for inverting s (i.e., s is $\mathcal{CF}$ invertible).

Complexity classes used in this paper are from two hierarchies: one studied in [Be87, Be88, Kr86] and one introduced in this paper. Let PF be the class of polynomial-time computable functions. We conjecture that PF cannot be an upper bound for inverting polynomial-time computable functions in general; for example, we conjecture that PF $\cap$ $\mathcal{F}_{\text{sat}^{-1}} = \emptyset$ since otherwise we have P = NP! On the other hand, we know PF^{NP}, the class of functions computable in polynomial-time using some NP set as an oracle, is an upper bound for inverting polynomial-time computable functions; that is, for every polynomial-time computable function s, $PF^{NP} \cap \mathcal{F}_{s-1} \neq \emptyset$. There seems to exist a big gap between PF and PF^{NP}. We consider two hierarchies between PF and PF^{NP} and use classes in them to show tighter upper bounds.

Krentel [Kr86] and Beigel [Be87] considered a hierarchy of function classes between PF and PF^{NP}. They defined interesting function classes by restricting the number of queries and/or the way of asking queries. For example, $PF^{NP}_{k\text{-}tt}$ is the class of functions that are computable in polynomial-time by asking at most k non-adaptive queries (i.e., independent queries) to some NP set. Here we use the following hierarchy:

$$PF \subseteq PF^{NP}_{1\text{-}tt} \subseteq \cdots \subseteq PF^{NP}_{k\text{-}tt} \subseteq PF^{NP}_{k+1\text{-}tt} \subseteq \cdots \subseteq PF^{NP}_{btt} \subseteq PF^{NP}_{tt} \subseteq PF^{NP}.$$

We call this hierarchy the *query hierarchy* of function classes. It is shown [Kr86, Be87] that the above inclusions (except $PF^{NP}_{tt} \subseteq PF^{NP}$) are strict unless P = NP.

In this paper, we propose an alternative way to measure the complexity of functions and introduce another hierarchy. For a given function f, we measure the complexity of f by that of computing each bit of f's values. For any complexity class C of *languages*, we define a class C/bit, which, intuitively, is the class of functions f such that the problem of computing each bit of f belongs to C. For example, NP/bit is the class of functions whose *each* bit is NP computable. Here by using the *query hierarchy* of language classes [Ka88, Wag87] we introduce the following hierarchy:

$$P/bit \subseteq NP/bit \subseteq P^{NP}_{1\text{-}tt}/bit \subseteq \cdots \subseteq P^{NP}_{k\text{-}tt}/bit \subseteq P^{NP}_{k+1\text{-}tt}/bit \subseteq \cdots \subseteq P^{NP}_{btt}/bit \subseteq P^{NP}_{tt}/bit.$$

For example, $P^{NP}_{k\text{-}tt}/bit$ is the class of functions whose each bit is polynomial-time computable by asking at most k non-adaptive queries to some NP set. It follows from the results in [Ka88, Wag87] that the above inclusions are strict unless the polynomial-time hierarchy collapses.

These two hierarchies are incomparable hierarchies under the class PF^{NP}_{tt}. Clearly, $PF^{NP}_{k\text{-}tt} \subseteq P^{NP}_{k\text{-}tt}/bit$ because if f is polynomial-time computable by asking k non-adaptive queries to some $A \in NP$ (i.e., $f \in PF^{NP}_{k\text{-}tt}$), then each bit of f is polynomial-time computable by asking k non-adaptive queries to A. On the other hand, it is unlikely to have $P^{NP}_{k\text{-}tt}/bit \subseteq PF^{NP}_{k+1\text{-}tt}$ or $PF^{NP}_{k+1\text{-}tt} \subseteq P^{NP}_{k\text{-}tt}/bit$. The following figure summarizes those relations. (Those relations are immediate from the results in [Be87, Be88, BH88, He87, Ka88, Kr86, Wag87].)

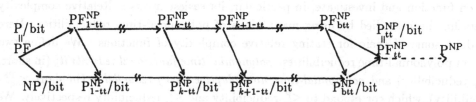

Notation: $C\text{-}\text{-}\text{-}\!\!\rightarrow\!\mathcal{D}$ denotes $C \subseteq \mathcal{D}$, and

$C\!\longrightarrow\!\mathcal{D}$ denotes that $C \subseteq \mathcal{D}$ and that $C \subsetneqq \mathcal{D}$ unless PH collapses.

Now by using these classes, we state the complexity of inverse functions. First we consider one-to-one functions. It is known [Be88, To88] that for any polynomial-time computable one-to-one function s, its (unique) inverse is in PF_{tt}^{NP}; that is, PF_{tt}^{NP} is an upper bound for the complexity of inverting s. While this upper bound seems tight in the PF_{xx}^{NP} hierarchy, we can improve it in the $\mathrm{P}_{xx}^{NP}/\mathrm{bit}$ hierarchy.

Theorem 1. For every polynomial-time computable one-to-one function, its (unique) inverse is in NP/bit.

That is, we have a tighter upper bound, NP/bit, for the complexity of inverting polynomial-time computable one-to-one functions.

Next we investigate polynomial-time computable functions in general. Let us consider the function sat because it is a typical example of polynomial-time computable functions. Following an argument in [Va76] we can prove that $\mathcal{F}_{sat^{-1}} \cap \mathrm{PF}^{NP} \neq \emptyset$. For instance, function max-sat^{-1} defined below is an inverse of sat that belongs to PF^{NP}.

For every x in Σ^*,
$$\text{max-sat}^{-1}(x) = \begin{cases} \langle x, y \rangle, & y \text{ is the largest}^* \text{ satisfying assignment for } x \text{ if it exists,} \\ \perp, & \text{otherwise} \end{cases}$$
($*$ We assume some appropriate ordering on the set of assignments.)

It seems unlikely that max-sat^{-1} belongs to a class lower than PF^{NP} (since max-sat^{-1} is in a sense "hardest" in PF^{NP}). On the other hand, we conjecture that there is an inverse of sat that is easier than max-sat^{-1} and that belongs to at least the class PF_{tt}^{NP}. While we have been unable to prove this conjecture, we show that it holds relative to almost every oracle.

Theorem 2. Let s be any relatively definable function that is polynomial-time computable. Then for almost all oracle A, we have $\mathrm{PF}_{tt}^{NP^A} \cap \mathcal{F}_{(s^A)^{-1}} \neq \emptyset$.

This theorem supports our conjecture that sat is PF_{tt}^{NP} invertible. The question of whether we can tighten this upper bound (relative to almost every oracle) is an interesting open problem.

Our next approach is to develop some methods to discuss relative complexity of functions. By using these methods, we discuss relative complexity between inverses of a given function and investigate, in particular, its easiest inverse. Relative complexity of sets has been studied by using many kinds of polynomial-time reducibilities. Here we adapt some of them for stating relative complexity of functions. We define two kinds of polynomial-time reducibilities: *polynomial-time functional reducibility* (in short, $\leq_{func}^{PF}$-reducibility) and *polynomial-time non-adaptive query reducibility* (in short, $\leq_{q}^{PF}$-reducibility), which correspond to $\leq_m^P$-reducibility and $\leq_{tt}^P$-reducibility respectively. We state relative complexity of functions in terms of these two reducibilities.

Take the function sat as an example again. (Recall that $\mathcal{F}_{sat^{-1}}$ is the class of inverses of sat). By using polynomial-time reducibility, say $\leq_{func}^{PF}$-reducibility, we can define the "easiest" functions in $\mathcal{F}_{sat^{-1}}$ as least functions in $\mathcal{F}_{sat^{-1}}$ under $\leq_{func}^{PF}$-reducibility; that is, a function in $\mathcal{F}_{sat^{-1}}$ that is $\leq_{func}^{PF}$-reducible to *every* function in sat is regarded as one of the easiest functions in $\mathcal{F}_{sat^{-1}}$. Notice that such an easiest function may not exist in $\mathcal{F}_{sat^{-1}}$; its existence is an interesting open problem. We have similar open problems for the other reducibilities between functions.

In order to study the existence or the complexity of such an easiest function, we investigate the class of functions (not necessarily in $\mathcal{F}_{sat^{-1}}$) that are reducible (e.g., $\leq_{func}^{PF}$-reducible) to every function in $\mathcal{F}_{sat^{-1}}$. From our observation, we obtain the following relation.

Theorem 3. There exists an inverse of sat that belongs to NP/bit if and only if $\mathcal{F}_{sat^{-1}}$ has a least element under $\leq_{func}^{PF}$-reducibility if and only if $\mathcal{F}_{sat^{-1}}$ has a least element under $\leq_{1\text{-}q}^{PF}$-reducibility (where $\leq_{1\text{-}q}^{PF}$-reducibility is an abbreviation of polynomial-time non-adaptive one-query reducibility).

Remark. In general we can prove the following: for any polynomial-time computable function s, if $\mathcal{F}_{s^{-1}}$ has a least element under $\leq_{1\text{-}q}^{PF}$-reducibility, then s has an inverse in NP/bit.

Using machinery established in this paper, we can analyze the relative complexity of max-sat^{-1} in $\mathcal{F}_{sat^{-1}}$.

Theorem 4.
(1) If max-sat^{-1} is $\leq_{func}^{PF}$-reducible to *every* function in $\mathcal{F}_{sat^{-1}}$, then NP = co-NP.
(2) If max-sat^{-1} is $\leq_{1\text{-}q}^{PF}$-reducible to *every* function in $\mathcal{F}_{sat^{-1}}$, then $P_{1\text{-}tt}^{NP} = P^{NP}$.

That is, max-sat^{-1} cannot be an easiest inverse of sat. Theorem 2 provides an evidence that there is a function in $\mathcal{F}_{sat^{-1}}$ that is easier than max-sat^{-1}. We have another evidence for it from Theorem 4.

2. Definitions

Here we state the precise definitions for the notions used in this abstract. We follow standard definitions and notation in computational complexity theory (see, e.g., [BDG88]).

We use the alphabet $\Sigma = \{0, 1\}$. By a *language* we mean a subset of Σ^*. We denote by $|x|$ the length of a string x. We consider a standard one-to-one pairing function from $\Sigma^* \times \Sigma^*$ to Σ^* that is polynomial-time computable and invertible. For inputs x and y, we denote the output of the pairing function by $\langle x, y \rangle$; this notation is extended to denote every n tuple. A function from Σ^* to Σ^* may not be total. A special symbol $\perp \notin \Sigma$ is used to denote "undefined"; for example, we write $f(x) = \perp$ if f on x is undefined. Note that if s is one-to-one, then s has a unique inverse; we write s^{-1} to denote the unique inverse.

Computation Model

Our computation model is the standard multi-tape Turing machine. A machine is either acceptor or transducer. We consider only deterministic machines for transducers. On the other hand, an acceptor is either deterministic or nondeterministic.

For a transducer M, M *yields output* y *on input* x (write $M(x) = y$) if y is written on M's output tape when M on x enters an accepting state. By "$M(x) = \bot$" we mean that M on x does not enter an accepting state (and thus yielding no output).

Definition 1. For a transducer M, M *computes a function* f if for every $x \in \Sigma^*$, $M(x) = f(x)$.

We also consider query machines, i.e., machines that ask queries to an oracle set. The above defined notions are extended naturally to query machines. We write, for example, M^A to denote "M relative to A". We can define interesting complexity classes by restricting the number of queries and/or the way of asking queries. For any function t on nonnegative integers, we say that M is a $t(n)$-*query machine* if for every oracle A and every input x, M^A on x asks at most $t(|x|)$ queries (on *each* computation path). Intuitively, a *non-adaptive query machine* is a machine that asks queries in a non-adaptive way; more precisely, M is *non-adaptive* if for every oracle A and every input x, M on x (or, M on *each* computation path for x) first produces all the queries $q_1, ..., q_m$, then consults A on $q_i \in A$, and finally obtains the result.

Complexity Class

For language classes, the classes P and NP are well-known. Besides them we consider some more complexity classes of languages that are defined by polynomial-time deterministic query acceptors and NP oracle sets. For any set A, let P^A (resp., P^A_{tt}) denote the class of languages recognized by polynomial-time deterministic query (resp., non-adaptive query) acceptors relative to A. For any function t on nonnegative integers, let $\mathrm{P}^A_{t(n)\text{-}tt}$ denote the class of languages recognized by polynomial-time deterministic non-adaptive $t(n)$-query acceptors relative to A. Let P^A_{btt} denote the class $\bigcup_{k\geq 0} \mathrm{P}^A_{k\text{-}tt}$. Now we define the following complexity classes: $\mathrm{P}^{NP} = \bigcup_{A \in NP} \mathrm{P}^A$, $\mathrm{P}^{NP}_{tt} = \bigcup_{A \in NP} \mathrm{P}^A_{tt}$, $\mathrm{P}^{NP}_{t(n)\text{-}tt} = \bigcup_{A \in NP} \mathrm{P}^A_{t(n)\text{-}tt}$, and $\mathrm{P}^{NP}_{btt} = \bigcup_{A \in NP} \mathrm{P}^A_{btt}$.

From the definition, we see that the classes defined above consist a hierarchy: $\mathrm{P} \subseteq \mathrm{NP} \subseteq \mathrm{P}^{NP}_{1\text{-}tt} \subseteq \cdots \subseteq \mathrm{P}^{NP}_{btt} \subseteq \cdots \subseteq \mathrm{P}^{NP}_{tt} \subseteq \mathrm{P}^{NP}$. We call this hierarchy the *query hierarchy* of language classes.

Next define complexity classes of functions. A class PF is the class of functions computed by polynomial-time deterministic transducers. Classes such as PF^{NP}, PF^{NP}_{tt}, $\mathrm{PF}^{NP}_{t(n)\text{-}tt}$, and PF^{NP}_{btt} are defined in a similar way as language classes. For example, PF^{NP} is the class of functions computed by polynomial-time query transducers relative to some NP set. Here again we have a hierarchy: $\mathrm{PF} \subseteq \mathrm{PF}^{NP}_{1\text{-}tt} \subseteq \cdots \subseteq \mathrm{PF}^{NP}_{btt} \subseteq \cdots \subseteq \mathrm{PF}^{NP}_{tt} \subseteq \mathrm{PF}^{NP}$. We call this hierarchy the *query hierarchy* of function classes.

In this paper we propose a new way to measure the complexity of a given function. For any function f, we measure the complexity of f in terms of the difficulty of computing each bit of f. More precisely, we state the complexity of functions by using the following new complexity classes.

Definition 2. For any class $\mathcal{C}$ of languages, $\mathcal{C}/\text{bit}$ is the class of honest functions f such that $Bit\text{-}f$ belongs to $\mathcal{C}$. Where the set $Bit\text{-}f$ is defined by

$$Bit\text{-}f = \{ \langle x, i, b \rangle \ : \ \text{the } i\text{th bit of } f(x) \text{ is } b \} \cup \{ \langle x, l \rangle \ : \ |f(x)| = l \}.$$

Polynomial-Time Reducibility

For any functions f and g, we say that f is *polynomial-time functional reducible* (in short, $\leq^{\mathrm{PF}}_{\mathrm{func}}$-reducible) to g if there exist polynomial-time computable functions h_1 and h_2 such that for every $x \in \Sigma^*$,

$$f(x) = h_1(x, \langle g(x_1), ..., g(x_m) \rangle),$$

where $h_2(x) = \langle x_1, ..., x_m \rangle$ (m is *the number of queries* produced by $h_2(x)$).

A pair (h_1, h_2) is called a $\leq^{\mathrm{PF}}_{\mathrm{func}}$-*reduction*. We write $f \leq^{\mathrm{PF}}_{\mathrm{func}} g$ if f is $\leq^{\mathrm{PF}}_{\mathrm{func}}$-reducible to g. Notice that this type of reducibility is called *metric reducibility* in [Kr86].

We assume "strictness" for functions. That is, we assume that every n-ary function f, $f(a_1, ..., a_n)$ is $\perp$ if some a_i is $\perp$. On the other hand, we can consider a pseudo function G such that $G(a_1, ..., a_n)$ may not be $\perp$ even though some a_i is $\perp$. Such a pseudo function is called "nonstrict". In other words, a *nonstrict function* is a pseudo function that can take the special symbol $\perp$ as an input. The polynomial-time computability for those nonstrict functions is defined naturally, where the length of $\perp$ is measured as 1.

In the definition of $\leq^{\mathrm{PF}}_{\mathrm{func}}$-reducibility, strict functions are used. Here we define "non-adaptive query reducibility" by using *nonstrict* functions. For any functions f and g, we say that f is *polynomial-time non-adaptive query reducible* (in short, $\leq^{\mathrm{PF}}_{\mathrm{q}}$-reducible) to g if there exist a nonstrict function H and a function h that are polynomial-time computable such that for every $x \in \Sigma^*$, $f(x) = H(x, \langle g(x_1), ..., g(x_m) \rangle)$, where $h(x) = \langle x_1, ..., x_m \rangle$. Clearly, $\leq^{\mathrm{PF}}_{\mathrm{q}}$-reducibility is a generalization of $\leq^{\mathrm{PF}}_{\mathrm{func}}$-reducibility; i.e., $f \leq^{\mathrm{PF}}_{\mathrm{func}} g$ implies $f \leq^{\mathrm{PF}}_{\mathrm{q}} f$.

For any integer $k \geq 0$, we say that f is $\leq^{\mathrm{PF}}_{k\text{-}q}$-reducible to g (write $f \leq^{\mathrm{PF}}_{k\text{-}q} g$) if $f \leq^{\mathrm{PF}}_{\mathrm{q}} g$ via a reduction (H, h) such that the number of queries produced by $h(x)$ is bounded by k for every $x \in \Sigma^*$.

Acknowledgments

The authors would like to thank Professor R. Beigel and Mr. K. Sakurai for their comments on the earlier version of this paper.

References

[Al86]   E. Allender, Isomorphisms and 1-L reductions, in "Proc. 1st Structure in Complexity Theory Conference", Lecture Notes in Computer Science 223, Springer-Verlag, Berlin (1986), 12-22; the final version appeared in J. Comput. Syst. Sci. 36 (1988), 336-350.

[BDG88]   J. Balcázar, J. Díaz, and J. Gabarró, Structural Complexity I, EATCS Mono-
          graphs on Theoretical Computer Science, Springer-Verlag, Berlin (1988).

[Be87]   R. Beigel, Bounded queries to SAT and Boolean hierarchy, Technical Report
          87-07, Dept. Computer Science, The Johns Hopkins University (1987).

[Be88]   R. Beigel, NP-hard sets are P-superterse unless R = NP, Technical Report
          88-04, Dept. Computer Science, The Johns Hopkins University (1988).

[BH88]   S. Buss and L. Hay, On truth-table reducibility to SAT and the difference
          hierarchy over NP, in "Proc. 3rd Structure in Complexity Theory Conference",
          IEEE (1988), 224-233.

[GS84]   J. Grollmann and A. Selman, Complexity measures for public-key cryptosys-
          tems, in "Proc. 25th IEEE Sympos. Foundation of Comput. Sci.", IEEE
          (1984), 495-503; the revised version appeared in SIAM J. Comput. 17 (1988),
          309-335.

[He87]   L. Hemachandra, The strong exponential hierarchy collapses, in "Proc. 19th
          Ann. ACM Sympos. on Theory of Computing", ACM (1987), 110-122.

[Ka88]   J. Kadin, The polynomial time hierarchy collapses if the Boolean hierarchy
          collapses, in "Proc. 3rd Structure in Complexity Theory Conference", IEEE
          (1988), 278-292.

[Kr86]   M. Krentel, The complexity of optimization problems, in "Proc. 18th Ann.
          ACM Sympos. on Theory of Computing", ACM (1986), 69-76; the final version
          appeared in J. Comput. Syst. Sci. 36 (1988), 490-509.

[To90]   S. Toda, On polynomial time truth-table reducibilities of intractable sets to
          p-selective sets, Theoret. Comput. Sci. (1990), to appear.

[Va76]   L. Valiant, Relative complexity of checking and evaluating, Inform. Process.
          Lett. 5 (1976), 20-23.

[Wag87]   K. Wagner, Number-of-query hierarchies, Technical Report TR-158, Institute
          of Mathematics, University of Augsburg (1987).

[Wa88a]   O. Watanabe, On hardness of one-way functions, Inform. Process. Lett. 27
          (1988), 151-157.

[Wa88b]   O. Watanabe, On 1-tt-sparseness of nondeterministic complexity classes, in
          "Proc. 15th International Colloquium on Automata, Languages and Program-
          ming", Lecture Notes in Computer Science 317, (1988), 697-709.

[Wa89]   O. Watanabe, On one-way functions, in "Combinatorics, Computing and Com-
          plexity", D. Du and G. Hu eds., Kluwer Academic Pub. (1989), 98-131.

Oracles versus Proof Techniques that Do Not Relativize

Eric Allender[1]

Department of Computer Science
Rutgers University
New Brunswick, NJ 08903

ABSTRACT *Oracle constructions have long been used to provide evidence that certain questions in complexity theory cannot be resolved using the usual techniques of simulation and diagonalization. However, the existence of nonrelativizing proof techniques seems to call this practice into question. This paper reviews the status of nonrelativizing proof techniques, and argues that many oracle constructions still yield valuable information about problems in complexity theory.*

1   Introduction

One of the most exciting theorems of this past winter was proved by Adi Shamir in [Sh-89], where it is shown that PSPACE is the class of sets having interactive proof systems. This result is significant for many reasons, but this paper will focus on one aspect of this work: it does not relativize. The results of [Sh-89], along with the related work of [LFKN-89, BFL-90], are the first truly compelling examples of theorems about complexity classes that are known to be true in the unrelativized case, but are false relative to some oracles.

Of course, other examples of nonrelativizing proof techniques have been known for quite some time. These earlier results are surveyed in Section 2.

The survey will be followed in Section 3 by a discussion about what is novel in the proof techniques of [Sh-89, LFKN-89, BFL-90], and how they differ from the nonrelativization results surveyed in Section 2.

In Section 4, we discuss what significance, if any, should be attached to oracle results, in light of the fact that nonrelativizing proof techniques now seem to be available. In particular, arguments will be presented explaining why some oracle results (including some of the author's own results) should still be of interest.

Finally, we conclude with a brief discussion and summary in Section 5.

---

[1] Supported in part by National Science Foundation Grants CCR-8810467 and CCR-9000045 and by a grant from the International Information Science Foundation.

2 A Long History of Nonrelativizing Proofs

Oracle constructions have never been very useful as a tool for talking about possible relationships among space complexity classes, for the simple reason that there is no obvious "right" way to talk about a space-bounded oracle Turing machine. If the oracle tape is subjected to a sublinear space bound, then there are sets that are not even reducible to themselves (because the input can't be written on the oracle tape); this is clearly an anomalous situation. On the other hand, if the oracle tape is not subjected to any space bound, then extremely long queries can be posed to the oracle, which can also lead to anomalies because it represents a way in which the space bound can be circumvented. For instance, there are oracles relative to which NLOG is not contained in P, using this notion of relativization.

These anomalies were first described in [LL-76, L-78], and it was suggested that the proof technique used to prove NLOG ⊆ P fails to relativize because the proof does not proceed via a "step-by-step" simulation. However, an alternative view is that NLOG ⊆ P fails to relativize simply because the wrong notion of relativization is used. Thus a new notion of relativization for space-bounded Turing machines (sometimes called RST relativization) was proposed in [RST-84]; RST relativization avoids at least the most obvious anomalies associated with the other methods of relativization studied in [LL-76, L-78, Sa-83].

Space-bounded AuxPDAs provide an alternative characterization of deterministic time classes, and these characterizations also fail to hold relative to oracles [An-80]. Again, the problem may be blamed on ambiguities about what is the "correct" way to equip a space-bounded machine with access to an oracle.

Time-bounded alternating Turing machines and depth-bounded circuit complexity classes provide useful characterizations of space-bounded computation, and it is proved in [Or-83, Wi-87] that all of the natural methods of relativizing these modes of computation result in oracles relative to which classes that are equal in the unrelativized case are unequal relative to the oracles. It may be argued that these characterizations thus do not relativize; but many researchers are of the opinion that the real issue centers on the question of what is the "correct" way to provide machines such as alternating Turing machines with access to the oracle. Alternative oracle access mechanisms have been presented for space-bounded Turing machines [Wi-88] and alternating Turing machines [Bu-88] that have the virtue of allowing most of the standard characterizations to hold relative to all oracles. A more general mechanism along the same lines was presented in [Bu-87].

The virtues and drawbacks of the various oracle access mechanisms have been debated in several different settings in a number of papers. In addition to the work mentioned above, the interested reader will find these issues discussed in [Si-77, Ha-88, Ha-88a, KL-87, AW-90a]. Let it suffice to say that it is far from clear that the results discussed so far in this section really represent "non-relativizing" proof techniques. It is simply not clear what "relativization" should really mean in this setting.

For time-bounded computation by deterministic and nondeterministic Turing machines, the picture is much less ambiguous. Yet even in this setting there are theorems that have been proved true in the unrelativized case, yet are false relative to some oracles. The simplest example is the linear speed-up theorem; $DTIME(2n) = DTIME(4n)$, but an oracle Turing machine running for $4n$ steps can ask queries that a machine running in time $2n$ cannot ask, and this intuition can be turned into a proof that the linear speed-up theorem is false relative to some oracle [RS-81, Mo-81]. A less trivial example is provided by the proof that $DTIME(n\log^* n) \subseteq \Sigma_2 Time(n)$ [PPST-83]. It is shown in [Ga-87] that there are oracles relative to which this inclusion fails to hold.

However these two examples fail to be very compelling, since both the linear speed-up theorem and the result of [PPST-83] are in some sense automata-theoretic results about the Turing machine model. Certainly the linear speed-up theorem is false in many reasonable models of computation, and the result of [PPST-83] is known only for models of computation that are very similar to the Turing machine model. The questions of complexity theory that are of the most interest are those questions that are independent of the details of the particular models of computation that have been chosen for study; the complexity classes that are of interest are the complexity classes that are invariant under minor changes in the definition. The complexity classes $DTIME(4n)$ and $DTIME(n\log^* n)$ are very sensitive to small changes to the underlying model of computation, and thus the nonrelativizing proofs dealing with these complexity classes seem to be of little help in formulating attacks on the more important questions.

Hartmanis wrote an interesting article [Ha-85] in which a number of nonrelativizing proofs are presented. Some of these results have the flavor of the results cited at the start of this section, in which space-bounded machines are given access to an oracle. However, other results in [Ha-85] involve a novel "double-relativization" trick. For example, take an oracle A such that $P^A = NP^A$. Then there is an oracle B such that $P^{A,B} \neq NP^{A,B}$; that is, the "real-world" result that $P^A = NP^A$ does not relativize. Although this is thought-provoking, it is hard to see how theorems of this sort can lead to the solution of any of the long-standing problems of complexity theory.

Other nonrelativizing proofs are discussed in [DGHM-89] and [IR-89,I-88]. However, the results mentioned in [IR-89,I-88] involve the concept of zero-knowledge, which conceivably should be defined differently in relativized models of computation, and the results of [DGHM-89] involve nonrecursive sets, and it is not clear how to apply those techniques to prove nonrelativizing results about familiar complexity classes.

Even though the proof techniques discussed in this section are all "nonrelativizing" in one way or another, the techniques all are "traditional" in some vague sense. In contrast, there is something fundamentally new about the proofs presented in [Sh-89, LFKN-89, BFL-90]. The next section will focus on one aspect of these proofs that is particularly novel.

3   New Nonrelativizing Proof Techniques

This section begins with a review of the assumptions underlying the use of relativizations as a tool for indicating that certain complexity-theoretic problems are "difficult" in some sense. Then we will show how the proofs of [Sh-89, LFKN-89, BFL-90] violate these assumptions.

Many of the basic results of recursive function theory and complexity theory are based on the simple intuition that it is impossible to say anything about the behavior of a machine other than by simulating the machine. For example, one cannot determine if a machine will halt except by running it until it halts; one cannot determine if a $t(n)$ time-bounded machine will accept or reject its input except by running it for $t(n)$ steps, etc. Of course, the same intuition is also correct when one is trying to predict the behavior of a nondeterministic Turing machine (NTM), using a deterministic Turing machine: it is impossible to determine anything about the acceptance behavior of the NTM without carrying out some sort of simulation. The question of course is: what kinds of simulation are possible?

The P=NP question is really the question of whether or not it is possible, given a configuration of an NTM, to determine anything about the number of accepting paths in the tree, without doing a brute-force search of all the paths in the tree. The conventional intuition is that it is *not* possible to do significantly better than this, and relativization is one way of providing a setting in which this intuition is correct.

A configuration of a NTM is a small object that completely describes a much larger object: the computation tree rooted at that configuration. The results of [BGS-75] were the first to capitalize on the fact that relative to some oracles, the computation tree *cannot* be described compactly, and thus no deterministic strategy for inferring information about the tree can succeed, unless it involves evaluating essentially every branch in the tree. That is, the computation tree is no longer described by its root, as in the unrelativized case; instead it can only be described by the values at each of its leaves.

The proof of [FS-88] is a conventional oracle construction in this regard. In [FS-88], it is shown that, relative to some oracle, coNP is not in IP. (That is, there is a set in coNP for which membership cannot be proved via an interactive proof system.) The coNP language considered is $L(A) = \{1^n :$ every string of length n is in $A\}$, where A is the oracle. The intuition used in [FS-88] is quite simple: if a probabilistic machine communicating with a prover *cannot* accept 1^n with acceptably high probability relative to *any* oracle, then the machine is clearly not usable in an interactive protocol for $L(A)$. Otherwise, if a machine *can* accept 1^n with high enough probability relative to some prover and some oracle, then there has to be some string of length n that is queried on a minority of the paths. Removing that string from the oracle cannot change the probability of acceptance very much; thus in this case it cannot reject 1^n with high enough probability, and thus in this case also it does not define an interactive protocol for $L(A)$.

This relativization argument of [FS-88] is entirely conventional; there is nothing controversial about the method in which access to the oracle is provided. It is difficult to see how any of the objections raised to the "anomalous" relativizations surveyed in Section 2 could be raised here.

And yet, in the unrelativized case, coNP *is* contained in IP. The original proof of this fact, in [LFKN-89], in some sense provides a method of inferring information about the structure of a computation tree *without* doing a brute-force search through the tree. Instead, the configuration describing the tree is manipulated as an object with arithmetic properties that can be efficiently verified in a particular sense.[2]

Let us provide a few more details. (The brief sketch provided here is based on one of the proofs in [BF-90].) The standard reduction given by Cook's theorem [Co-71] shows that any computation tree on inputs of length n can be described with a polynomial-

---

[2] Just as the oracle of [FS-88] shows that the results of [Sh-89, LFKN-89] do not relativize, there is an oracle construction in [FRS-88] that shows that the characterization of NEXP in terms of two-prover interactive proof systems [BFL-90] does not relativize.

sized instance of 3SAT. Replace each of the disjunctions in this 3SAT instance by an equivalent disjunction of 7 conjunctions on the same variables. Now treat this resulting formula as an arithmetic polynomial in m variables, where AND and OR are treated as multiplication and addition, respectively, and $\bar{x}$ is replaced by $1 - x$. As is observed in [BF-90], when this expression is summed over all of the 2^m possible assignments of the variables to $\{0, 1\}$, the resulting number is simply the number of accepting paths in the original computation tree.

An interactive proof protocol is presented in [LFKN-89] that allows a prover to convince a verifier of the value of such an expression. In an oversimplified form, the protocol is as follows. Let the original polynomial in m variables be $p(x_1, \ldots, x_m)$. In the first round, the prover sends the verifier a number v_1 and claims that this number is the value of

$$\sum_{x_1=0}^{1} x_1 \ldots \sum_{x_m=0}^{1} x_m \, p(x_1, \ldots, x_m).$$

The prover also sends the verifier the coefficients of a polynomial q', claiming that this represents the polynomial

$$q(x_1) = \sum_{x_2=0}^{1} x_2 \ldots \sum_{x_m=0}^{1} x_m \, p(x_1, \ldots, x_m).$$

Up to this point, the proof is not so different from several other "relativizing" proofs; all of these polynomials have a straightfoward interpretation in terms of more standard machine-based proofs. However what happens next is a significant break with the past.

The verifier can easily check to see if $q'(0) + q'(1) = v_1$, but the verifier must also have some way to check that the prover is not lying about the claim that $q' = q$. The significant insight of [LFKN-89] is that if the prover is lying, then for a randomly chosen integer r in a given range, $q'(r) \neq q(r)$. Thus the verifier picks r, and asks the prover to prove that

$$q'(r) = \sum_{x_2=0}^{1} x_2 \ldots \sum_{x_m=0}^{1} x_m \, p(r, x_2 \ldots, x_m).$$

This polynomial has one less variable. After m rounds, the verifier will, with high probability, be able to see if the prover has been lying or not. (The reader is referred to [LFKN-89] to see the complete proof.)

The reason that this represents a break with more traditional methods is that the polynomial $p(r, x_2 \ldots, x_m)$ now bears no obvious relation to the original Turing machine whose tree we were trying to evaluate. In fact, the relativization result of [FS-88] implies that, in some sense, there *is* no obvious way to relate $p(r, x_2 \ldots, x_m)$ to this machine; if there were, it would allow us an avenue for making this proof hold relative to all oracles, and this is impossible by [FS-88].

The polynomial p is an arithmetic object that encodes, in a succinct way, all of the information needed to completely evaluate the computation tree of our original NTM. The arithmetic properties of this polynomial can be exploited in the framework of interactive proof systems. This contrasts with the relativization result of [FS-88], which seems to indicate that any evaluation of the computation tree that considers entire paths (e.g. to look at a sequence of queries) cannot be accomplished in polynomial time by interactive proof protocols.

It may be argued that this nonrelativizing proof simply reinforces the intuition of [LL-76] that proofs that proceed via a step-by-step simulation relativize, while proofs that do not proceed in this way do not relativize. However, all earlier examples of such proofs failed to be convincing examples of nonrelativization, because of ambiguity about what is the "correct" way to relativize those classes. (Indeed, as the results of [Wi-88, Bu-88, Bu-87] indicate, it *is* possible to define notions of relativization that defeat those early attempts at nonrelativizing proofs.) The oracle construction of [FS-88] does not seem to suffer from this ambiguity.

The [LFKN-89] technique of evaluating computation trees does not seem to translate to any mode of computation other than interactive proofs. It will be interesting to see if other techniques for deriving information about computation trees shed any light on the DTIME versus NTIME question.

4 The Role of Oracle Results

What place do oracle constructions have in a world with nonrelativizing proof techniques? "None" is perhaps the first answer that springs to mind, but there are in fact a number of situations in which oracle constructions provide a great deal of useful information. In this section we shall consider a few of these situations.

4.1 Properties of Complexity Classes

Many complexity classes are *defined* in terms of oracle computations; thus it is sometimes important to know what properties hold relative to all oracles, and which properties do not.

For example, in [AG-90], it is shown that there are sets in P^{PP} that are immune to uniform AC^0. (An infinite set is *immune* to a class C if it has no infinite subset in C.) The proof proceeds by first showing that there is a set $L \in PP$ and a constant k such that $AC^0 \subseteq DTIME(n^k)^L$; then we note that there is a set in $DTIME(n^{k+1})^L$ that is immune to $DTIME(n^k)^L$, and hence to AC^0. Thus in this instance, it was important that the immunity properties of DTIME hold relative to all oracles.

On the other hand, what if the class under consideration were NP^{PP} instead of P^{PP}? Would similar observations apply? Note that a great many complexity classes of interest are *defined* in terms of relativized computations. (The classes of the polynomial hierarchy are the most obvious examples.) When investigating the structure of these classes, it is thus useful to know, for example, if $NTIME(n^{k+1})^L$ will always contain a set immune to $NTIME(n^k)^L$.

Questions such as these were investigated in [ABHH-90]. There it was shown that, unless $T(n)$ is almost exponentially larger than $t(n)$, then relative to some oracles L, $NTIME(T(n))^L$ does *not* contain any sets immune to $NTIME(t(n))^L$. Other immunity properties of NTIME classes were also investigated in [ABHH-90] in connection with the almost-everywhere hierarchy for nondeterministic time. As illustrated above, such results are often useful when studying complexity classes that are defined in terms of relativized computation.

4.2 Oracles as Circuit Lower Bounds

Some of the most impressive lower bound results in complexity theory are lower bounds for constant depth circuits. This line of research was begun by [FSS-84], and at least some of the motivation for these results came from the connection between constant depth circuits and the polynomial time hierarchy; it was shown in [FSS-84, Si-83] that circuit lower bounds for constant depth circuits translate directly into oracle constructions for the polynomial hierarchy, and vice-versa. This connection was generalized by Torán [Tor-89].

Thus in some sense there is little difference between a circuit lower bound and an oracle construction. For instance, Torán has studied the counting hierarchy (consisting of classes such as PP, PP^{PP}, etc.) and has constructed oracles separating some sublevels of this hierarchy [Tor-89]. He then defines a number of subclasses of NC^1 and shows that, via his oracle constructions, the corresponding subclasses of NC^1 are also distinct.

If an oracle could be found such that all of the levels PP, PP^{PP}, ...of the counting hierarchy are distinct, then this would answer some important open questions about threshold circuits. This connection between circuits and the counting hierarchy is discussed further in the survey [AW-90].

4.3 Oracles in Cryptography

The security of a cryptographic protocol is usually predicated on some complexity-theoretic assumption, such as "one-way functions exist" or "trap-door functions exist". Since it is always desirable to have as weak an assumption as possible, an ongoing theme in cryptographic research is to show new implications and relationships among cryptographic assumptions.

Oracle constructions have been used as a tool to investigate when certain assumptions are *not* sufficient for certain tasks. One of the first examples of this sort of work is [IR-89]. Since many cryptographic constructions use a one-way function as a sort of "black box," an oracle-based approach has obvious applications here.

4.4 Oracles as Indicators

Oracles can also be used as a tool to find questions where progress is still likely to be made. For example, the relationship between PP and the polynomial hierarchy has been a wide-open question for years, and one reason for this is that there are so few oracle results concerning possible relationships between these classes. In the wake of Toda's breakthrough [To-89] in showing that the polynomial hierarchy is contained in P^{PP}, there is even more interest in knowing, for example, whether P^{NP} is contained in PP. Certainly, construction of an oracle relative to which P^{NP} is not contained in PP would be viewed as a very interesting result.

Another example is provided by the Berman-Hartmanis conjecture concerning whether or not all NP-complete sets are p-isomorphic. In spite of all of the work that has been done on this question it still not known if it is "possible" for the conjecture to be true, in the sense that no oracle is known relative to which all NP-complete sets are p-isomorphic. (An oracle relative to which all complete sets for Σ_2^p are p-isomorphic is presented in [HS-89].) In the absence of such an oracle, there is good reason to devote some energy to trying to prove the existence, in the unrelativized case, of non-p-isomorphic NP-complete sets. On the other hand, if someone does construct an oracle relative to which the Berman-Hartmanis conjecture holds, then it would have a substantial impact upon all subsequent approaches to this problem; such an oracle construction would imply that an entire repertory of proof techniques would have to be abandoned in order for further progress to be made.

4.5 Oracles as Guides to Nonrelativizing Techniques

As was mentioned above, there are still no nonrelativizing proof techniques that are suitable for attacking the NTIME versus DTIME question (given that the techniques of [PPST-83] do not seem to be able to provide very significant separations). On the other hand, significant lower bounds have been proved with respect to constant-depth circuits [Ya-85, Hå-86]. A recent oracle construction of [AG-90] seems to indicate that further progress with constant-depth circuits will provide nonrelativizing proof techniques that will be useful in settling the relationship between DTIME and NTIME classes.

More specifically, recall that it was mentioned above that there are sets in P^{PP} that are immune to uniform AC^0. Since P^{PP} is such a "large" complexity class, and AC^0 does not even contain simple sets such as Parity, this may seem like a very weak result. However, if it can be shown that there are sets in NP that are immune to AC^0, then there are statements concerning NE, E, and $\bigcup_k \Sigma_k\text{-time}(n)$ that are true in the unrelativized world, but are false relative to some oracle.

At the very least, the oracle result of [AG-90] points to certain limitations of relativizing proof techniques when it comes to proving the existence of sets that are immune to AC^0. However, given that the program of proving lower bounds for AC^0 circuits has been so successful, this oracle construction could provide an indication of problems to work on, in the hope of developing an approach for finding out more about the NTIME vs DTIME question.

5 Conclusions

The development of nonrelativizing proof techniques will certainly have a profound impact upon research in complexity theory. Although certain types of nonrelativizing proof techniques have been known for quite some time, various considerations caused the significance of these earlier nonrelativizations to be called into question. Those considerations do not come into play with regard to the proofs of [Sh-89, LFKN-89, BFL-90].

In spite of the existence of nonrelativizing proof techniques, there are still a number of situations in which oracle constructions can serve as a useful tool in proving interesting results.

Acknowledgments: I thank Ron Book, Jack Lutz, Pekka Orponen, and Seinosuke Toda for some interesting discussions that took place at the Workshop in Structural Complexity Theory at the University of California, Santa Barbara, in March, 1990. Thanks are especially due to Ron Book for organizing the Workshop. I also thank Osamu Watanabe for arranging support through the International Information Science Foundation, and for his many other efforts.

References

[An-80]   D. Angluin, *On relativizing auxiliary pushdown machines*, Math. Systems Theory 13, 283–299.

[ABHH-90] E. Allender, R. Beigel, U. Hertrampf, and S. Homer, *A note on the almost-everywhere hierarchy for nondeterministic time*, Proc. 7th Symposium on Theoretical Aspects of Computer Science, Lecture Notes in Computer Science 415, pp. 1–11.

[AG-90]   E. Allender and V. Gore, *On strong separations from AC^0*, submitted.

[AW-90]   E. Allender and K. W. Wagner, *Counting hierarchies: polynomial time and constant depth circuits*, guest authors of The Structural Complexity Column (ed. Juris Hartmanis), EATCS Bulletin 40, pp. 182–194.

[AW-90a]   E. Allender and C. Wilson, *Width-bounded reducibility and binary search over complexity classes*, Proc. 5th Annual IEEE Structure in Complexity Theory Conference, 1990.

[BF-90]   L. Babai and L. Fortnow, *A characterization of P by straight line programs of polynomials, with applications to interactive proofs and Toda's theorem*, Technical Report 90-02, University of Chicago.

[BFL-90]   L. Babai, L. Fortnow, and C. Lund, *Non-deterministic exponential time has two-prover interactive protocols*, manuscript.

[BGS-75]   T. Baker, J. Gill, and R. Solovay, *Relativizations of the P=?NP question*, SIAM J. Comput. 4, 431–442.

[Bu-87]   J. Buss, *A theory of oracle machines*, Proc. 2nd IEEE Structure in Complexity Theory Conference, pp. 175–181.

[Bu-88] J. Buss, *Relativized alternation and space-bounded computation*, J. Comput. and System Sci. 36, 351–378.

[Co-71] S. Cook, *The complexity of theorem proving procedures*, Proc. 3rd Annual ACM Symposium on Theory of Computing, pp. 151–158.

[DGHM-89] R. Downey, W. Gasarch, S. Homer, and M. Moses, *On honest polynomial reductions, relativizations, and P=NP*, Proc. 4th IEEE Structure in Complexity Theory Conference, pp. 196–207.

[FS-88] L. Fortnow and M. Sipser, *Are there interactive proofs for co-NP languages?* Information Processing Letters 28, 249–251.

[FSS-84] M. Furst, J. Saxe, M. Sipser, *Parity, circuits, and the polynomial-time hierarchy*, Mathematical Systems Theory 17, 13–27.

[FRS-88] L. Fortnow, J. Rompel, and M. Sipser, *On the power of multi-prover interactive protocols*, Proc. 3rd IEEE Structure in Complexity Theory Conference, pp. 156–161.

[Ga-87] W. Gasarch, *Oracles for deterministic versus alternating classes*, SIAM J. Comput. 16, 613–627.

[Hå-86] J. Håstad, *Almost optimal lower bounds for small depth circuits*, Proc. 18th ACM Symposium on Theory of Computing, pp. 6–20.

[Ha-85] J. Hartmanis, *Solvable problems with conflicting relativizations*, EATCS Bulletin 27, pp. 40–49.

[Ha-88] J. Hartmanis, *New developments in structural complexity theory*, Proc. 15th International Colloquium on Automata, Languages, and Programming, Lecture Notes in Computer Science 317, pp. 271–286.

[Ha-88a] J. Hartmanis, *Some observations about relativizations of space bounded computations*, the Structural Complexity Column, EATCS Bulletin 35, pp. 82–92.

[HS-89] S. Homer and A. Selman, *Oracles for structural properties: the isomorphism problem and public-key cryptography*, Proc. 4th IEEE Structure in Complexity Theory Conference, pp. 3–14.

[I-88] R. Impagliazzo, *Proofs that relativize, and proofs that do not*, manuscript.

[IR-89]   R. Impagliazzo and S. Rudich, *Limits on the provable consequences of one-way permutations*, Proc. 21st Annual ACM Symposium on Theory of Computing, pp. 44–61.

[KL-87]   B. Kirsig and K.-J. Lange, *Separation with the Ruzzo, Simon, and Tompa relativization implies DSPACE[log n] ≠ NSPACE[log n]*, Information Processing Letters 25, 13–15.

[LFKN-89] C. Lund, L. Fortnow, H. Karloff, and N. Nisan, *The polynomial-time hierarchy has interactive proofs*, manuscript.

[LL-76]   R. Ladner and N. Lynch, *Relativization of questions about log space computability*, Math. Systems Theory 10, 19–32.

[L-78]   N. Lynch, *Log space machines with multiple oracle tapes*, Theoretical Computer Science 6, 25–39.

[Mo-81]   S. Moran, *Some results on relativized deterministic and nondeterministic time hierarchies*, Journal Comput. System Sci. 22, 1–8.

[Or-83]   P. Orponen, *Complexity classes of alternating machines with oracles*, Proc. 10th International Colloquium on Automata, Languages, and Programming, Lecture Notes in Computer Science 154, pp. 573–584.

[PPST-83] W. Paul, N. Pippenger, E. Szemerédi, and W. Trotter, *On determinism versus non-determinism and related problems*, Proc. 24th Annual IEEE Symposium on Foundations of Computer Science, pp. 429–438.

[RS-81]   C. Rackoff and J. Seiferas, *Limitations on separating nondeterministic complexity classes*, SIAM J. Comput. 10, 742–745.

[RST-84]   Walter Ruzzo, Janos Simon, and M. Tompa, *Space-bounded hierarchies and probabilistic computation*, J. Comput. and System Sci. 28, 216–230.

[Sa-83]   W. Savitch, *A note on relativized log space*, Math. Systems Theory 16, 229–235.

[Sh-89]   A. Shamir, *IP = PSPACE*, manuscript.

[Si-77]   I. Simon, *On some subrecursive reducibilities*, Ph.D. Dissertation, Stanford University.

[Si-83]   M. Sipser, *Borel sets and circuit complexity*, Proc. 15th ACM Symposium on Theory of Computing, pp. 61–69.

[To-89]   S. Toda, *On the computational power of PP and $\oplus P$*, Proc. 30th IEEE Symposium on Foundations of Computer Science, pp. 514–519.

[Tor-89]   J. Torán, *A combinatorial technique for separating counting complexity classes*, Proc. 16th ICALP, Lecture Notes in Computer Science 372, pp. 732–744.

[Wi-87]   C. B. Wilson, *Relativized NC*, Math. Systems Theory 20, 13–29.

[Wi-88]   C. B. Wilson, *A measure of relativized space which is faithful with respect to depth*, J. Comput. and System Sciences 36, 303–312.

[Ya-85]   A. C. Yao, *Separating the polynomial-time hierarchy by oracles*, Proc. 26th IEEE Symposium on Foundations of Computer Science, pp. 1–10.

20—Relative Neighborhood Graphs Are Hamiltonian

Maw Shang Chang

Institute of Computer Science and Information Engineering

National Chung Cheng University, Chiayi, Taiwan, R. O. C.

C. Y. Tang

Institute of Computer Science

National Tsing Hua University, Hsinchu, Taiwan, R. O. C.

and

R. C. T. Lee

National Tsing Hua University, Hsinchu

and Academic Sinica, Taipei, Taiwan, Republic of China

Section 1: Introduction

In this paper, we shall define the k—relative neighborhood graph first and then prove some properties of this graph. For two points p and q, LUN_{pq} denotes the set of points in the plane enclosed in the region formed by the two circular arcs where one is centered at p and the other is centered at q. Both arcs are with radii d_{pq} which is the Euclidean distance between points p and q. In other words,

$$LUN_{pq} = \{ u \mid u \in R^2, d_{pu} < d_{pq} \text{ and } d_{qu} < d_{pq} \}.$$

LUN_{pq} is also called the lune of p and q. Figure 1—1 shows the lune of p and q. Let V be a set of points in the plane. $LUN_{pq}(V)$ denotes the set of points of V contained in LUN_{pq}. Toussaint [Toussaint 1980] defined a relative neighborhood graph of V (denoted

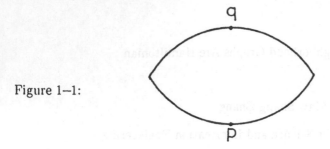

Figure 1–1:

by RNG(V) or simply RNG when V is understood) to be an undirected graph with vertices V such that for each pair $p,q \in V$, (p,q) is an edge of RNG(V) if and only if $LUN_{pq}(V) = \phi$. Figure 1–2 shows a set of points V and its RNG(V). If (p,q) is an edge

Figure 1–2:

of RNG(V), we say that p and q are relative neighbors. Intuitively, two points are relative neighbors if they are at least as close to each other as they are to any other points. The relative neighborhood graph has several applications in pattern recognition and have been studied by Toussaint [Toussaint 1980].

In this paper, we shall generalize the idea of RNG to define the k–relative neighborhood graph of V, denoted by kRNG(V) or simply kRNG, to be the undirected graph with vertices V such that for each pair $p,q \in V$, (p,q) is an edge of kRNG(V) if and only if $|LUN_{pq}(V)| < k$, for some fixed positive number k. Obviously, RNG(V) is 1RNG(V) and kRNG(V) is a subgraph of (k+1)RNG(V). When $k > |V| - 2$, the kRNG(V) is the complete graph induced on V. If (p,q) is an edge of kRNG, we say that p and q are k–relative neighbors. Figure 1–3 shows 2RNG(V) of the set of points V in Figure 1–2. In Section 2, we prove that the number of edges of kRNG is less than O(kn).

Figure 1–3:

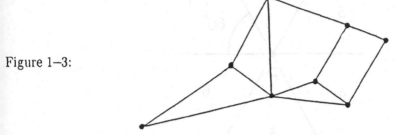

Let $E = \{ e_{pq} \mid p \in V \text{ and } q \in V \}$. Then $G = (V, E)$ is a complete graph. For any subset F of E, define the maximum distance of F as

$$\max_{e_{pq} \in F} d_{pq}.$$

A Euclidean bottleneck Hamiltonian cycle is a Hamiltonian cycle in graph G whose maximum distance is the minimum among all Hamiltonian cycles in graph G. In Section 3, we shall prove that there exists a Euclidean bottleneck Hamiltonian cycle which is a subset of the edge set of 20RNG. The proof is Constructive. It shows that, given a Euclidean bottleneck Hamiltonian cycle which is not a subgraph of 20RNG, we can transform it into another Euclidean bottleneck Hamiltonian cycle which is a subgraph of 20RNG. Hence, 20RNGs are Hamiltonian. This result is theoretically interesting. The following question arises immediately: Given a general graph, can it be embedded in the plane such that a 20RNG is its subgraph? If this question can be answered in polynomial time, then we have a practical sufficient condition for a graph with O(n) edges to be Hamiltonian. Concluding remarks are given in Section 4.

Section 2: On k–Relative Neighborhood Graphs

In this section, we shall show that the number of edges of kGNG, a supergraph of kRNG, is less than O(kn). Let p be a point of V. Divide the plane into regions relative to p as shown in Figure 2–1. The regions are formed by three lines passing through p

Figure 2–1:

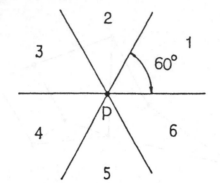

and having angles of $0^0, 60^0$, and 120^0, respectively, with x–axis. The regions are numbered counterclockwise as shown in Figure 2–1, Use $R_i(p)$ to denote the set of points in the ith region of p (including its boundary), for $1 \leq i \leq 6$. For each $p \in V$, let $N_i(p)$ be those points of V which are in the ith region of p (excluding p itself). A point u in $N_i(p)$ is said to be a nearest neighbor to p in the ith region if and only if d_{pu} is the minimum of distances between v and points in $N_i(p)$. Note that such a nearest neighbor does not exist if $N_i(p) = \phi$, and may not be unique when it exists. For example, in Figure 2–2, $N_1(p) = \{ a, b, c, d, e, f, g, h, x, y, z \}$ and the nearest neighbor to p in the first region is point g.

Figure 2–2:

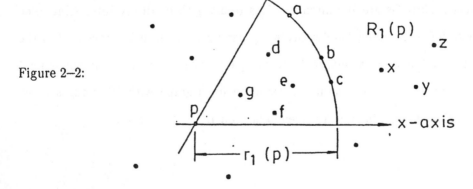

Katajainen and Nevalainen [Katajainen and Nevalainen 1986] defined the geographic neighborhood graph of a set of points as follows: The geographic neighborhood graph of V in the ith region is the undirected graph $GNG_i(V)$ with vertices V and edge set E_i for $1 \leq i \leq 6$ such that for each pair p, q $\in$ V, (p,q) $\in E_i$ if and only if either p is a nearest neighbor of q in the ith region or q is a nearest neighbor of p in the ith region. The union of the above graphs is called the geographic neighborhood graph, denoted by GNG(V)(or simply GNG). That is,

$$GNG(V) = (V, E_G) \text{ where } E_G = \bigcup_{i=1}^{6} E_i.$$

Katajainen and Nevalainen[Katajainen and Nevalainen 1986] proved that the number of edges of GNG(V) is less than $18n - 36$ by showing that $GNG_i(V)$ is a planar graph and the number of edges of a planar graph is less than $3n - 6$.

Now, we extend the concept of geographic neighborhood graph. Let p and q be points of V, and q $\in N_i(p)$. We define $N_i^q(p)$ as the subset of $N_i(p)$ which are closer to p than q. And q is said to be a k–geographic neighbor of p in the ith region if $|N_i^q(p)| < k$. The set of all k–geographic neighbors in the ith region of point p is denoted by $kGN_i(p)$. In Figure 2–2, $N_1^a(p) = \{ d, e, f, g \}$ and $5GN_1(p) = \{ a, b, c, d, e, f, g \}$. The k geographic neighborhood graph of V in the ith region, denoted by $kGNG_i(V)$ (or simply $kGNG_i$), is defined as the undirected graph with vertices V and edge set E_i^k for $1 \leq i \leq 6$ such that for each pair of p, q $\in$ V, (p,q) $\in E_i^k$ if and only if either p is a k–geographic neighbor of q in the ith region or q is a k–geographic neighbor of p in the ith region. Figure 2–3 shows a set of points and its $2GNG_1$ graph. The union of the above graphs is called the k–geographic neighborhood graph of V, denoted by kGNG(V) (or simply kGNG). That is,

$$kGNG(V) = (V, E_G^k) \text{ where } E_G^k = \bigcup_{i=1}^{6} E_i^k.$$

Let the maximum distance between p and points in $kGN_i(p)$ be denoted by

Figure 2–3:

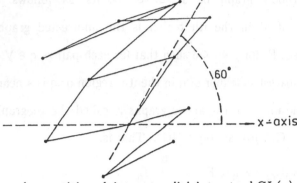

$kr_i(p)$. $kGN_i(p)$ can be partitioned into two disjoint sets, $kGI_i(p)$ and $kGX_i(p)$, such that the distances between p and every point of $kGX_i(p)$ are equal to $kr_i(p)$ and the distances between p and every point of $kGI_i(p)$ are less than $kr_i(p)$. $kGX_i(p)$ is on the arc centered at p with radius $kr_i(p)$. It is obvious that $|kGI_i(p)| < k$ and $|kGX_i(p)| \geq 1$. Sort $kGX_i(p)$ along the circular–arc in clockwise direction. If $|kGX_i(p)| > 2k$, then it can be divided into three parts: $kGX'_i(p)$ contains the first k points, $kGX''_i(p)$ contains the last k points and $kG\hat{X}_i(p)$ contains all other points of $kGX_i(p)$ not in $kGX'_i(p)$ and $kGX''_i(p)$. If $|kGX_i(p)| \leq 2k$, then $kG\hat{X}_i(p) = \phi$.

Lemma 2.1 For $1 \leq i \leq 6$ and for any point p and q, $kG\hat{X}_i(p) \cap kG\hat{X}_i(q) = \phi$.

Proof: We will consider region 1 only. All the other regions can be proved in the same way. If $kG\hat{X}_1(p) = \phi$, then this lemma is true automatically. Without losing generality, assume that $q \notin N_1(p)$. Let L_1, L_2 and L_3 be three lines passing through point p. L_1 is the line parallel to the x–axis. The angle between L_1 and L_2 is $60°$. L_3 is the bisector of the angle between L_1 and L_2. There are three cases:

(1) q is on line L_3: If a point z of $kG\hat{X}_1(p)$ is a k–geographic neighbor of q, then either $kGX'_1(p) \subset kGN_1(q)$ or $kGX''_1(p) \subset kGN_1(q)$ depends on that z is located below or above L_3. Thus either points of $kGX'_1(p)$ or $kGX''_1(p)$ will all belong to $N_1(q)$ and are closer to q than point z. In other words, $|kGI_1(q)| \geq k$ which is a contradiction. Thus, $kG\hat{X}_i(p) \cap kG\hat{X}_i(q) = \phi$.

(2) q is on the half plane below line L_3: In this case, if a point z of $kG\hat{X}_1(p)$ is a k–geographic neighbor of q, then $kGX''_1(p) \subset kGI_1(q)$ because points of $kGX''_1(p)$ will be contained in $N_1(q)$ and are closer to q than z. In other words, $|kGI_1(q)| \geq k$ which is

a contradiction. Thus, $k\hat{GX}_i(p) \cap k\hat{GX}_i(q) = \phi$.

(3) q is on the half plane above line L_3: This case can be proved in the same way as case (2).   Q.E.D.

Theorem 2.1 The number of edges of $kGNG(V)$ is less than $18kn$.

proof: If we can show that $|E_i^k|$ is less than $3kn$ for $1 \le i \le 6$, then $|E_G^k|$ is less than

$18kn$ because $|E_G^k| = \sum_{i=1}^{6} |E_i^k|$. Without loss of generality, we will show that

$|E_1^k| < 3kn$ only. By definition,

$$|kGN_1(p)| = |kGI_1(p)| + |kGX_1(p)|$$
$$\le (k-1) + 2k + |k\hat{GX}_1(p)|$$
$$= 3k - 1 + |k\hat{GX}_1(p)|$$

and

$$|E_1^k| = \sum_{p\in V} |kGN_1(p)| \le (3k-1)|V| + \sum_{p\in V} |k\hat{GX}_1(p)|.$$

From Lemma 2.1, it is easy to see that $\sum_{p\in V} |k\hat{GX}_1(p)| < |V|$.

Therefore, $|E_1^k| < (3k-1)|V| + |V| = 3kn$.

Thus, $|E_G^k| < 18kn$.   Q.E.D.

The following theorem shows that kGNG is a supergrpah of kRNG.

Theorem 2.2 The set of edges of kRNG is a subset of the set of edges of kGNG.

proof: We will prove this theorem by showing that every edge of kRNG is an edge of kGNG too. Suppose that (p,q) is an edge of kRNG, then $|LUN_{pq}(V)| < k$. Assume that $q \in R_i(p)$. Without loss of generality, then we have $N_i^q(p) \subseteq LUN_{pq}(V)$. Therefore $|N_i^q(p)| < k$ and q is a k–geographic neighbor of p. In other words, (p,q) is an edge of kGNG and the proof is concluded.   Q.E.D.

Corollary 2.2 The number of edges of kRNG is less than $18kn$.

Section 3 A Property of the Euclidean Bottleneck Hamiltonian Cycle

In this section, we will show that there exists a Euclidean bottleneck Hamiltonian cycle which is a subgraph of 20RNG. Let V denotes the set of points given in the Euclidean bottleneck traveling salesperson problem and $LUN_{pq}(V)$ denote those points of V contained in the lune of points p and q. The major idea of the proof is to show that a procedure exists which successfully transforms a Euclidean bottleneck Hamiltonian cycle H which is not a subgraph of 20RNG into another one which is a subgraph of 20RNG. We shall informally describe the transformation procedure first. The transformation operations depends on the relationships among edges of H. There are several cases. We shall briefly explain how these cases are classified and then show how to do transformation operations according to the cases.

Let E_r^{20} be the set of edges of 20RNG and $F = H - E_r^{20}$. Let $e_{pq} \in F$ where d_{pq} is equal to the maximum distance of F. Let $LUN_{pq}(V) = \{ u_1, u_2, ..., u_z \}$. Thus, $z = |LUN_{pq}(V)|$. Since e_{pq} is not an edge of 20RNG, $z \geq 20$. There are two major cases:

Case 1 : There exists an edge $e_{u_i u_j} \in H$ where u_i and $u_j \in LUN_{pq}(V)$ and $i \neq j$.

In this case, H can be divided into four parts: (a) e_{pq}. (b) path P_{qu_i} from q to u_i. (c) $e_{u_i u_j}$. (d) path $P_{u_j p}$ from u_j to p. The following transformation operation transforms H into another Euclidean bottleneck Hamiltonian cycle:

$$H = H - \{ e_{pq}, e_{u_i u_j} \} + \{ e_{pu_i}, e_{qu_j} \}.$$

Case 2 : For all $e_{u_i u_j}$ where $u_i, u_j \in LUN_{pq}(V)$ and $i \neq j$, $e_{u_i u_j} \notin LUN_{pq}(V)$.

Let s_k and t_k be the two neighbors of u_k in the Euclidean bottleneck Hamiltonian cycle H for all $u_k \in LUN_{pq}(V)$. Since Case 1 is false, $s_1, s_2, ..., s_z$ and $t_1, t_2, ..., t_z$ are all outside of LUN_{pq}. In this case, H can be divided into the following parts as shown in Figure 3–1:

$$e_{pq}, P_{qs_1}, e_{s_1 u_1}, e_{u_1 t_1}, P_{t_1 s_2}, e_{s_2 u_2}, e_{u_2 t_2}, ..., e_{s_z u_z}, e_{u_z t_z}, P_{t_z p}$$

Figure 3–1:

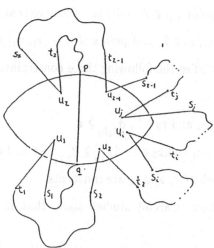

where P_{uv} are defined as the path in H which starts from point u to point v. Note that

any one of $P_{t_k s_{k+1}}$, P_{qs_1} and $P_{t_z p}$ may be degenerated into a point only. In such a case,

then $t_k \equiv s_{k+1}$, $s_1 \equiv q$ and $t_z \equiv p$ respectively. If either Path $P_{t_z p}$ or path P_{qs_1} is empty (

This is not a necessary condition), then, without losing generality, assume that P_{qs_1} is

empty. That is, $q \equiv s_1$. There are two subcases:

Case 2.1 There exists a point t_i where $1 \leq i \leq z - 1$ such that $d_{qt_i} < d_{pq}$:

The transformation operation is as follows:

$$H = H - \{ e_{pq}, e_{u_i t_i} \} + \{ e_{pu_i}, e_{qt_i} \}.$$

Case 2.2 There exist two points t_i and t_j where $1 \leq i < j \leq z - 1$ such that $d_{t_i t_j} < d_{pq}$ or

$d_{t_i t_j} < d_{t_i u_i}$ or $d_{t_i t_j} < d_{t_j u_j}$:

The transformation operation is as follows:

$$H = H - \{ e_{pq}, e_{t_j u_j}, e_{t_i u_i} \} + \{ e_{qu_j}, e_{pu_i}, e_{s_j s_i} \}.$$

Obviously, after every transformation operation as described above, H is still a

Euclidean bottleneck Hamiltonian cycle.

The transformation procedure repeatedly apply the above rules to transform H

until $F = \phi$, that is, H becomes a subset of 20RNG. To show that this transformation

procedure will success eventually, we still need to prove the following two statements:

(i) If $F \neq \phi$, then at least one transformation operation can be applied.

(ii) F will becomes empty in finite number of steps.

We need following lemmas to show that statement (i) is correct:

Lemma 3.1 Let Z be a set of points, $p \notin Z$ and r is a positive real number. Define $\theta(Z,p)$ be the angle of $\angle z_1 p z_2$ where $z_1, z_2 \in Z$, and points of $Z - \{z_1, z_2\}$ are contained in $\angle z_1 p z_2$ (including its boundary). Z satisfies the following three constraints:

(i) For all $z \in Z$, $d_{pz} \geq r$.

(ii) For any two points z_1 and $z_2 \in Z$, $d_{z_1 z_2} \geq r$.

(iii) For any two points z_1 and $z_2 \in Z$, $d_{z_1 z_2} \geq d_{pz_1} - r$ and $d_{z_1 z_2} \geq d_{pz_2} - r$.

If $\theta(Z,p) < c_1$, then $|Z| \leq c_2$ where c_1 and c_2 are constants.

Proof: We omit the proof here. Careful studies show that if $c_1 = \cos^{-1}(7/8)$, then $c_2 = 2$.

Lemma 3.2 For a point v outside of LUN_{pq}, let n_v denote its nearest neighbor on the boundary of LUN_{pq}. If S is a set of points outside of LUN_{pq} and satisfies all the following three properties, then $|S| < 20$.

(i) For every point $v \in S$, $d_{qv} \geq d_{pq}$.

(ii) For every two points u and $v \in S$, $d_{uv} \geq d_{pq}$.

(iii) For every two points u and $v \in S$, $d_{uv} \geq d_{un_u}$ and $d_{uv} \geq d_{vn_v}$.

Proof: We shall show that $|S|$ is less than c, where c is a constant, rather than 20 to simplify the proof. First, divide $R^2 - LUN_{pq}$ into four regions, as shown in Figure 3–2. Region II contains the intersection of $R^2 - LUN_{pq}$ and the interior of $\angle apb$. Region IV contains the intersection of $R^2 - LUN_{pq}$ and the interior of $\angle aqb$. Note that the angles of both $\angle apb$ and $\angle aqb$ are $2\pi/3$. Both regions II and IV include their boundaries. Points of $R^2 - LUN_{pq}$ not contained in regions II or IV are naturally divideed into two connected regions, I and III. We will show that there are at most one points of S in each of regions I, III, at most c_3 points in each of regions II, IV where c_3 is a constant too. We only prove the cases of regions I, II and the other cases can be proved in the same way.

Figure 3–2:

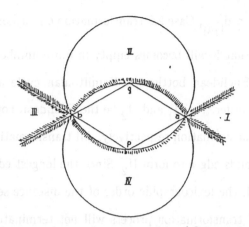

Suppose that there are more than one point of S in region I. Let u and v be any two points of S in region I. Clearly $n_u = n_v = a$ and $\varphi < \pi/3$. Thus $\overline{uv}$ is not the longest edge of $\triangle$ uav. Therefore, either $d_{uv} < d_{ua} = d_{un_u}$ or $d_{uv} < d_{va} = d_{vn_v}$. It contradicts with property (iii) of S. Hence, there is at most one point of S contained in region I.

Next we show that there are at most c_3 points of S in region II where c_3 is a constant. Let $r = d_{pq}$. Since $\angle$ apb $= 2\pi/3$, $c_3 = c_2(\lceil (2\pi/3)/c_1 \rceil)$.

Now we can conclude the proof. There are at most one point in each of the region I and III. There are at most c_3 points in each of region II and IV. Therefore $|S| < c$ where $c = 2c_3 + 3$. The interested reader can extend the proof easily to convince himself that 20 is a tighter bound. Careful studies show that there are at most 7 and 10 points in region II and IV respectively. Thus $c = 20$.   Q.E.D.

If conditions of Case 1 is satisfied, then we use the transformation operation of Case 1. Otherwise, let $S = \{ t_i \mid 1 \le i \le z - 1 \}$, then all points of S are outside of LUN_{pq}. Because $e_{pq} \notin E_r^{20}$, $z \ge 20$. Therefore, $|S| \ge 20$. From Lemma 3.1, we know that at least one of the following three cases is true:

(a) There exists a point $t \in S$ such that $d_{qt} < d_{pq}$.

(b) There exist two points t_i and t_j such that $d_{t_i t_j} < d_{pq}$.

(c) There exists two points t_i and t_j such that $d_{t_i t_j} < d_{t_i n_{t_i}}$ or $d_{t_i t_j} < d_{t_j n_{t_j}}$.

In (a), Case 2.1 can be applied. In (b), Case 2.2 can be applied. In (c), since

$d_{t_i n_{t_i}} < d_{t_i u_i}$ and $d_{t_j n_{t_j}} < d_{t_j u_j}$, Case 2.2 transformation operation can be applied too.

Finally, we show that F will becomes empty in finite number of steps. Define the distance sequence of a Euclidean bottleneck Hamiltonian cycle as its edge distances sorted in non-increasing order. Let H_1 and H_2 be the Euclidean bottleneck Hamiltonian cycles before and after transformation respectively. All transformation operations deletes edges from H_1 and then adds edges to form H_2. Since the longest edge added is less than the longest edges deleted, the lexicographic order of the distance sequence H_1 is greater than that of H_2. Now if transformation process will not terminate in finite number of steps, then there exists an optimal solution H_z which will appear for transformation more than twice due to the fact that the number of distinct H is finite. This implies that the lexicographic order of the distance sequence of H_z is greater than itself, which is a contradiction. Therefore F will becomes empty in finite number of steps.

Now, we can conclude with the following lemma and theorem:

Lemma 3.2 There exists a Euclidean bottleneck Hamiltonian cycle which is a subset of 20RNG.

Theorem 3.1 20-relative neighborhood graphs are Hamiltonian.

Corollary 3.1 20-geographic neighborhood graphs are Hamiltonian.

Section 4 Concluding Remarks

We have shown that there exists a Euclidean bottleneck Hamiltonian cycle which is a subgraph of 20RNG. Hence 20RNGs are Hamiltonian. In other words, we have found a sufficient condition for a graph with $O(n)$ edges to be Hamiltonian. That is, if a graph can be embedded in the plane such that 20RNG is its subgraph, then it is Hamiltonian. This is a theoretically interesting result. We leave the following two problems as open problems to the interested readers:

(1) To recognize graphs which can be embedded in the plane such that 20RNG is its subgraph.

(2) Given a set of points in the plane, find a Hamiltonian cycle which is a subgraph of 20RNG in polynomial time.

Note that if the transformation procedure described in section 3 can be executed in polynomial time, then it can be used to solve problem (2) as follows: First, construct a 20RNG. Next, randomly select a Hamiltonian cycle which may not be a Hamiltonian cycle of 20RNG. Then, transform this Hamiltonian cycle into another which is a Hamiltonian cycle of 20RNG. Thus it is interesting to prove that the time complexity of the transformation procedure is polynomial.

(3) Can the constant 20 be reduced?

Reference :

[Garey and Johnson 1979]

Garey, M. R. and Johnson, D. S., Computers and Intractability: A Guide to the Theory of NP—Completeness, Freeman, San Francisco, CA, 1979.

[Katajainen and Nevalainen 1986]

Katajainen, J. and Nevalainen, O., "Computing Relative Neighborhood Graphs in the Plane," Pattern Recogn. 19 (1986), 221–228.

[Toussaint 1980]

Toussaint, G. T., "The Relative Neighborhood Graph of a Finite Planar Set," Pattern Recong. 12 (1980), 261–268.

The K–Gabriel Graphs and Their Applications

Tung–Hsin Su [*] and Ruei–Chuan Chang [**]

[*]
Tung–Hsin Su is with the Institute of Computer Science and Information Engineering, National Chiao Tung University, Hsinchu, Taiwan, Republic of China.

[**]
Ruei–Chuan Chang is with the Institute of Computer and Information Science, National Chiao Tung University, Hsinchu, Taiwan and the Institute of Information Science, Academia Sinica, Taipei, Taiwan, Republic of China.

Abstract

In this paper, we define and investigate the properties of k–Gabriel graphs and also propose an algorithm to construct the k–Gabriel graph of a points set in $O(k^2 n \log n)$ time. The k–Gabriel graphs are also used to improve the running time of solving the Euclidean bottleneck biconnected edge subgraph problem from $O(n^2)$ to $\theta(n \log n)$, and that of solving the Euclidean bottleneck matching problem from $O(n^2)$ to $O(n^{1.5} \log^{0.5} n)$.

Section 1. Introduction

Let $S = \{p_1, p_2, ..., p_n\}$ be a set of n points in the plane. The Gabriel graph of S(denoted $GG(S)$ or GG when S is understood) is the undirected graph $G(S, E)$ where an edge $\overline{p_i p_j} \in E$ if and only if there does not exist another point $p_k \in S$ such that $d^2(p_i, p_j) > d^2(p_i, p_k) + d^2(p_j, p_k)$ [8, 12]. Here, $d(p_i, p_j)$ denotes Euclidean distance between p_i and p_j. The k–Gabriel graph of S is another undirected graph $G'(S, E')$ where an edge $\overline{p_i p_j} \in E'$ if and only if there are not more than k–1 points of S such that $d^2(p_i, p_j) > d^2(p_i, p_k) + d^2(p_j, p_k)$, for p_k in this $(k$–1)–points subset. Let $circle(p_i, p_j)$ be the interior of the circle centered at the midpoint of $\overline{p_i p_j}$ and with diameter equal to $d(p_i, p_j)$. The geometric interpretation of an edge of k–Gabriel graph is that an edge $\overline{p_i p_j}$ belongs to E' if and only if there are at most k–1 points of S lying inside $circle(p_i, p_j)$, for $p_i, p_j \in S$. Obviously, k–GG is an generalization of GG.

Note that in the original definition of GG, if $\overline{p_i p_j} \in GG$ then not only the *interior* but also the *boundary* of the circle which centered at the midpoint of $\overline{p_i p_j}$ with diameter $d(p_i, p_j)$, are not allowed to contain any other points of S. That is there doesn't exist another point p_k such that $d^2(p_i, p_j)$ " $\geq$ " $d^2(p_i, p_k) + d^2(p_j, p_k)$. But for the needs of this paper we change its definition by *ignoring* the boundary. Nevertheless, we still call it Gabriel graph for convenience.

In this paper, we study the properties of k–Gabriel graphs and propose an algorithm to construct the k–Gabriel graph of S in $O(k^2 n \log n)$ time. The algorithm proceeds in two phases. First, a supergraph of k–Gabriel graph is constructed and then eliminate the superfluous edges from this supergraph. The k–Gabriel graphs are also used to improve the running time of solving the Euclidean bottleneck biconnected edge subgraph problem from $O(n^2)$ to $\theta(n \log n)$, and that of solving the Euclidean bottleneck matching problem from $O(n^2)$ to $O(n^{1.5} \log^{0.5} n)$.

Section 2. Algorithm for K–Gabriel Graphs

In this section, we propose an algorithm for finding the k–Gabriel graph of a points set S with n points in the plane. We shall first find a supergraph of k–Gabriel graph, named k–Delaunay graph. The k–Delaunay graph of S can be extracted from the kth–order Voronoi diagram of S [4, 5, 10, 14]. Then an elimination method is proposed which can be used to prune the superfluous edges from k–Delaunay graph to obtain k–Gabriel graph.

Given a set $S = \{p_1, p_2, ..., p_n\}$ of n points in the plane. The k–Delaunay graph of S(denoted k–$DG(S)$ or k–DG when S is understood) is defined as follows: if the circumcircle of the triangle determined by p_i, p_j and p_k contains at most k–1 points of S in its interior then all the three edges of this triangle belong to the k–Delaunay graph of S, where p_i, p_j, $p_k \in S$. Note that the k–Delaunay graph is an extension of the Delaunay triangulation but it may not be a planar straight–line graph like the Delaunay triangulation [14]. In what follows we will show that the k–Delaunay graph can be extracted from the kth–order Voronoi diagram.

We first define another graph k–DG^* which is similar to k–DG. The k–DG^* is defined as follows: if the circumcircle of the triangle determined by p_i, p_j and p_k contains exactly k–2 or k–1 points in its interior then all the edges of this triangle belong to k–DG^*. Obviously the set of edges of a k–DG is equal to the union of all the edges in l–DG^*'s, for $1 \leq l \leq k$.

In the following, we will discuss the relationship between $k\text{--}DG^*$ and kth--order Voronoi diagram. The kth--order Voronoi diagram which partitions the plane by the perpendicular bisectors of the pairs of points into finite regions, some of which may be unbounded. And each region associates with a subset of k given points. If a test point p lies in a region R then the k nearest neighbors of p is just the points in the associated subset of R. Fig. 2--1 shows a second order Voronoi diagram of eight points in the plane.

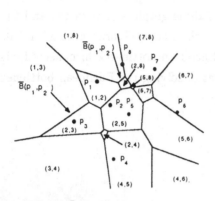

Fig. 2-1 $\overline{B}(p_1, p_2)$ appears twice in the second order Voronoi diagram with eight points.

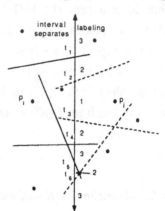

Fig. 2-2 The endpoints of $\overline{B}(p_i, p_j)$'s $(t_1 \sim t_6)$ divide the perpendicular bisector of p_i and p_j into finite intervals.
—— the perpendicular bisectors of p_i and other points.
- - - - the perpendicular bisectors of p_j and other points.

Many properties of the kth--order Voronoi diagram have been discussed by Lee [10]. Let $V_k(S)$ denote the kth--order Voronoi diagram of the given points set S. Also let $\overline{B}(p_i, p_j)$ be an Voronoi edge of $V_k(S)$, where p_i, $p_j \in S$ and $\overline{B}(p_i, p_j)$ is a portion of the perpendicular bisector of p_i and p_j. Note that for a pair of points p_i and p_j, $\overline{B}(p_i, p_j)$ may appear more than once. For example, $\overline{B}(p_1, p_2)$ appears twice in Fig. 2--1. A point in $V_k(S)$ is called a Voronoi point if it is the intersection of two(or more) Voronoi edges. Further, relative to $V_k(S)$, a Voronoi point is called old if it appears in $V_{k-1}(S)$ and $V_k(S)$ simultaneously. Otherwise it is called a new one if it only appears in $V_k(S)$ but not in $V_{k-1}(S)$. The following two lemmas of $V_k(S)$ are from Lee [10].

Lemma 2.1 Let $\overline{B}(p_i, p_j)$, $\overline{B}(p_j, p_k)$ and $\overline{B}(p_k, p_i)$ $be three edges of$ $V_k(S)$ $incident upon a$ $Voronoi point$ q. Let C_q $denote the circle centered at$ q $and passing through points$ p_i, p_j and p_k. $Then$ q $is an old or new Voronoi point of$ $V_k(S)$ $if and only if$ C_q $respectively$ $contains$ $k\text{--}2$ or $k\text{--}1$ $points of$ S $in its interior($ $Assume that not more than three given$ $points are cocircular.$).

Lemma 2.2 $Given a$ $k\text{--}order$ $Voronoi diagram$ $V_k(S)$, q $is a point of an edge$ $\overline{B}(p_i, p_j)$ of $V_k(S)$ $if and only if the circle centered at$ q $with radius$ $d(q, p_i) = d(q, p_j)$ $contains$ $k\text{--}1$

points in its interior.

Lee [10] also proposed an incremental algorithm to construct the $V_k(S)$ which constructs $V_1(S)$, then $V_2(S)$, $V_3(S)$, and so on. Thus with $(k-1)$ iterations of this procedure one can obtain the kth–order Voronoi diagram $V_k(S)$ from $V_1(S)$. In final, Lee showed that this incremental algorithm can construct the $V_k(S)$ in $O(k^2 n \log n)$ and $V_k(S)$ contains at most $O(k(n-k))$ edges.

By using the above properties, we will show that k–Delaunay graph of S can be constructed in $O(k^2 n \log n)$ time and this graph contains at most $O(k(n-k))$ edges.

Lemma 2.3 *If the cardinality of S is greater than or equal to $2k$ then $k-DG = k-DG^*$.*

Proof: Because $k-DG = \bigcup\limits_{l=1}^{k} l-DG^*$, if we can show that $l-DG^*$ is a supergraph of $(l-1)-DG^*$, for $2 \leq l \leq k$, then we have $k-DG^* = \bigcup\limits_{l=1}^{k} l-DG^*$ and $k-DG = k-DG^*$. From Lemma 2.1, the centers of the circumcircles in $l-DG^*$, where each circle of these circumcircles contains $l-2$ or $l-1$ points in its interior, are one–to–one corresponding to the Voronoi points in $V_l(S)$. This implies that $B(p_i,p_j)$ appears in $V_l(S)$ if and only if $\overline{p_i p_j}$ is in $l-DG^*$. To show that $l-DG^*$ is a supergraph of $(l-1)-DG^*$, we only have to prove that if $B(p_i,p_j)$ appears in $V_{l-1}(S)$(it implies $\overline{p_i p_j} \in (l-1)-DG^*$) then it will also appear in $V_l(S)$(this implies $\overline{p_i p_j} \in l-DG^*$).

For a pair of points p_i and p_j, if we record all the $B(p_i,p_j)$'s in $V_l(S)$'s, $1 \leq l \leq n-1$, then the endpoints of $B(p_i,p_j)$'s divide the perpendicular bisector of p_i and p_j into finite intervals. An interval is labeled m, if the circle centered at a point q of this interval with radius equal to $d(q,p_i)=d(q,p_j)$ contains exactly m points in its interior(see Fig. 2–2). Assume that there are not more than three points are cocircular then the difference of the labels between two consecutive intervals is not more than one. If $|S| \geq 2k$ then the line $\overleftrightarrow{p_i p_j}$ divides the plane into two half–planes and there is one half–plane containing at least $(2k-2)/2 = k-1$ points. Furthermore, the interval with maximum label in the half–plane containing at least $k-1$ points is labeled at least $k-1$, because we can always find a circle which covers all the points in the same half plane as the center. The center of this circle is on the perpendicular bisector of p_i and p_j, and far away from p_i(or p_j).

If an edge $\overline{p_i p_j} \in (l-1)-DG^*$ then $B(p_i,p_j) \in V_{l-1}(S)$ and, by Lemma 2.2, the circle centered at a point q of this $B(p_i,p_j)$ with radius $d(q,p_i) = d(q,p_j)$ will contain $l-2$ points. That is the interval between the endpoints of $B(p_i,p_j)$ on the perpendicular bisector of p_i, p_j will be labeled $l-2$. Since the difference of labels between two consecutive intervals is not more than one, there must exist an interval labeled $l-1$ between the above interval which is labeled $l-2$ and the interval with maximum label(.

$\geq k-1$). This implies $\mathcal{B}(p_i,p_j) \in V_1(S)$ and $\overline{p_ip_j} \in l\text{-}DG^*$. ◆

Theorem 2.4 *The k-Delaunay graph of a points set S with n points can be constructed in $O(k^2 n\log n)$ time and the k-Delaunay graph contains at most $O(k(n-k))$ edges.*

Proof: By Lemma 2.3, we can use the incremental algorithm by Lee [10] to construct the $V_k(S)$ in $O(k^2 n\log n)$ time and then extract $k\text{-}DG(S)$ from $V_k(S)$. Because, for a pair of points p_i and p_j, $\mathcal{B}(p_i,p_j)$ may appear more than once(see Fig. 2–1), this will duplicate the edge $\overline{p_ip_j}$ in $k\text{-}DG(S)$. We can eliminate the duplicated edges in $k\text{-}DG(S)$ by using the *union–find* data structure [16] to maintain the extracted edges. Before inserting a new edge, we first apply *find* operation to check whether it is already in the edges set of $k\text{-}DG(S)$, if no then we insert the edge else we ignore it. For all the $O(k(n-k))$ edges in $V_k(S)$, it takes $O(k(n-k)\cdot a(k(n-k),k(n-k)))$ time to eliminate the duplicated edges, which will not influence the running time of constructing the k-Delaunay graph of S.

By Lemma 2.3, $k\text{-}DG = k\text{-}DG^*$ and the number of edges of $k\text{-}DG^*$ is not more than that of $V_k(S)$, so the k-Delaunay graph contains at most $O(k(n-k))$ edges. ◆

To show that k-Delaunay graph is a supergraph of k-Gabriel graph, we first define another graph $k\text{-}GG^*$ which is similar to $k\text{-}GG$. The $k\text{-}GG^*$ is defined as follows: an edge $\overline{p_ip_j} \in k\text{-}GG^*$ if and only if $circle(p_i,p_j)$ contains exactly $k-1$ points, where p_i, p_j are given points. Obviously the set of edges of $k\text{-}GG$ equal to the union of all the edges in $l\text{-}GG^*$'s, for $1\leq l\leq k$.

Lemma 2.5 *The k-Delaunay graph is a supergraph of the k-Gabriel graph.*

Proof: Since $k\text{-}DG = \bigcup\limits_{l=1}^{k} l\text{-}DG^*$ and $k\text{-}GG = \bigcup\limits_{l=1}^{k} l\text{-}GG^*$, if we can show that $l\text{-}DG^*$ is a supergraph of $l\text{-}GG^*$, for $1\leq l\leq k$, then naturally k-Delaunay graph is a supergraph of k-Gabriel graph. By Lemma 2.2, we can easily see that if $\overline{p_ip_j} \in l\text{-}GG^*$ then $\mathcal{B}(p_i,p_j) \in V_1(S)$. From the proof of Lemma 2.3, we know that if $\mathcal{B}(p_i,p_j) \in V_1(S)$ then $\overline{p_ip_j} \in l\text{-}DG^*$. These two facts imply that if $\overline{p_ip_j} \in l\text{-}GG^*$ then $\overline{p_ip_j} \in l\text{-}DG^*$ and therefore $l\text{-}DG^*$ is a supergraph of $l\text{-}GG^*$. ◆

In succession, we show that k-Gabriel graph can be extracted from k-Delaunay graph in $O(k^2(n-k))$ time.

Lemma 2.6 *The k-Gabriel graph can be extracted from the k-Delaunay graph in $O(k^2(n-k))$ time.*

Proof: For a pair of points p_i and p_j, if the edge $\overline{p_ip_j}$ has an intersection point with a $\mathcal{B}(p_i,p_j)$ in $V_1(S)$($\mathcal{B}(p_i,p_j)$ may appear more than once), for $1\leq l\leq k$, then, by Lemma 2.2, the intersection point is the center of $circle(p_i,p_j)$ and $circle(p_i,p_j)$ contains $l-1$ points. Therefore, if the edge $\overline{p_ip_j}$ has an intersection point with the union of all the $\mathcal{B}(p_i,p_j)$'s in $V_1(S)'s$, for $1 \leq l \leq k$, then the $circle(p_i, p_j)$ contains at most $k-1$ points. So to check whether the edge $\overline{p_i p_j}($ ϵ $k\text{--}DG)$ belongs to the k--Gabriel graph or not, we only have to check whether $\overline{p_i p_j}$ has an intersection point with the union of $B(p_i, p_j)'s$ in $V_1(S)'s$ or not, for $1 \leq l \leq k$. Since there are at most $O(l(n-l))$ Voronoi edges $B(p_i, p_j)'s$ in a $V_1(S)($ for all possible pairs of points p_i and $p_j)$, the total number of Voronoi edges $B(p_i, p_j)'s$ in the $V_1(S)'s$, for $1 \leq l \leq k$, is at most $\sum_{l=1}^{k} O(l(n-l)) = O(k^2(n-k))$. To check whether a $B(p_i, p_j)$ and $\overline{p_i p_j}$ have an intersection point or not can be done in $O(1)$ time. Hence, the k--Gabriel graph can be extracted from the k--Delaunay graph in $O(k^2(n-k))$ time. $\blacklozenge$

Summarized the above results we can give the following theorem.

Theorem 2.7 *The k--Gabriel graph of a points set with n points in the plane can be constructed in $O(k^2 n \log n)$ time and this graph contains at most $O(k(n-k))$ edges.*

Section 3. Applications of K--Gabriel Graphs

In this section, we shall use k--Gabriel graphs to improve the upper bounds of two Euclidean bottleneck problems. The first bottleneck problem, Euclidean bottleneck biconnected edge subgraph problem, is defined as follows. Given a set of points S in Euclidean plane, a graph is constructed by connecting some pairs of points of S into edges such that the resulting subgraph is biconnected and the longest edge of this subgraph is minimized. One application of this subgraph is to find an approximation solution of the Euclidean bottleneck traveling salesman problem [13]. The Euclidean bottleneck problem is a variant of the bottleneck biconnected edge subgraph problem which is defined on the graph. Chang *et al.* [3] proposed an algorithm solve this problem in $O(m \log m)$ time, where m is the number of edges of the graph. To solve the Euclidean case, a straightforward method was to compute the complete graph of the given points as an input of the improved algorithm. That would directly derive an $O(n^2 \log n)$ algorithm for the Euclidean bottleneck biconnected edge subgraph problem, where n is the number of given points.

The second bottleneck problem, Euclidean bottleneck matching problem, is defined as follows. Given a set of points S in Euclidean plane, match each point with another point such that the longest matching edge resulting from this matching is minimized. In general this bottleneck matching problem is also defined on the graph. Gabow and Tarjan [7] proposed an $O((n \log n)^{1/2} m)$ algorithm for the general graph, where n and m are the number of vertices and edges respectively. Again, a straightforward method to solve the Euclidean case is to compute the complete graph first and then apply the algorithm of Gabow and Tarjan [7] to the complete graph. Since a complete graph has $n(n-1)/2$ edges, the time complexity of this method is $O(n^{2.5}log^{0.5}n)$.

Recently Chang et al. [2, 3] showed that these two Euclidean bottleneck problems can be solved by using k–relative neighborhood graphs(k–RNG). Given a set of points $S = \{p_1, p_2, ..., p_n\}$, the relative neighborhood graph of S(denoted $RNG(S)$) is the undirected graph $G(S, E)$ where $\overline{p_ip_j} \in E$ if and only if there does not exist another point p_k of S such that $d(p_i,p_k) < d(p_i,p_j)$ and $d(p_j,p_k) < d(p_i,p_j)$ [15, 17]. Let $lune(p_i,p_j)$ be the interior of the intersection of two circles that center at p_i and p_j respectively and with radii equal to $d(p_ip_j)$. A geometric interpretation of an edge of the relative neighborhood graph is that an edge $\overline{p_ip_j} \in E$ if and only if the $lune(p_i,p_j)$ does not contain any other points of S. The k–relative neighborhood graph is an extension of relative neighborhood graph which allows at most k–1 points lying inside $lune(p_i,p_j)$, if $\overline{p_ip_j} \in k$–RNG. Chang et al. [2] proposed a two–phases $O(kn^2)$ algorithm to construct a k–relative neighborhood graph, and showed that this graph contains at most $O(kn)$ edges.

Furthermore, they also showed that there exists a Euclidean bottleneck biconnected edge subgraph in 2–RNG, and a Euclidean bottleneck matching is embedded in 17–RNG. Hence, to solve the first bottleneck problem, instead of constructing a complete graph we can first spend $O(n^2)$ time to construct the 2–RNG with $O(n)$ edges, and then use the improved algorithm of Chang et al. [3] to find a bottleneck biconnected edge subgraph from this 2–RNG. This approach derives an $O(n^2)$ algorithm to solve the first bottleneck problem. To solve the second bottleneck problem, by the same token, we first spend $O(n^2)$ to construct the 17–RNG with $O(n)$ edges then apply the algorithm of Gabow and Tarjan [7] to this 17–RNG. This would improve the running time from $O(n^{2.5}log^{0.5}n)$ to $O(n^2)$.

From the definitions of k–relative neighborhood graph ($lune(p_i,p_j)$) and k–Gabriel graph($circle(p_i,p_j)$), it is obvious that k–relative neighborhood graph is a subgraph of k–Gabriel graph for the same points set. This can easily imply that the first bottleneck problem is embedded in 2–GG and the second one is embedded in 17–GG. Therefore, we can apply the results of Theorem 2.7 to improve the running times of the algorithm by Chang et al. [3] in solving the first bottleneck problem, and the algorithm by Gabow and Tarjan [7] in solving the second bottleneck problem. By Theorem 2.7, a k–Gabriel graph contains at most $O(n)$ edges and can be constructed in $O(nlogn)$ time, if

k is fixed. Hence, we have the following theorems immediately

Theorem 3.1 *The Euclidean bottleneck biconnected edge subgraph problem can be solved in $O(nlogn)$ time.*

Theorem 3.2 *The Euclidean bottleneck matching problem can be solved in $O(n^{1.5}log^{0.5}n)$ time.*

Moreover, we shall prove that our method for the Euclidean bottleneck biconnected edge subgraph problem is optimal. Let's consider an input points set S on the X–axis, the interval between two consecutive points is called a gap. The *maximum gap problem* is to find the length of a pair of consecutive points in S that are farthest apart. The lower bound of this problem is $\Omega(nlogn)$ [11]. We further define another problem, *maximum two consecutive gaps problem*, which is to find the length of a triple of consecutive points in S whose leftmost point and rightmost point are farthest apart. We will prove that the maximum gap problem can be reduced to the maximum two consecutive gaps problem, and the latter problem can be further reduced to the Euclidean bottleneck biconnected edge subgraph problem. If they do then the lower bound of the Euclidean bottleneck biconnected edge subgraph problem is $\Omega(nlogn)$.

Lemma 3.3 *The maximum gap problem can be reduced to the maximum two consecutive gaps problem in linear time.*

Proof: The proof is not hard and is omitted for the space. ◆

Lemma 3.4 *The maximum two consecutive gaps problem can be reduced to the Euclidean bottleneck biconnected edge subgraph problem in linear time.*

Proof: Without loss of generality, assume that a points set S with n points on the X–axis is an input instance of the maximum two consecutive gaps problem, and a triple of consecutive points p_i, p_j and p_k determine the maximum two consecutive gaps of S. We claim that the length of the bottleneck edge $BE($ the longest edge) of a Euclidean bottleneck biconnected edge subgraph of S is equal to the length of the edge $\overline{p_ip_k}$. To complete the proof, we show that there doesn't exist a biconnected subgraph of S in which the length of the longest edge(bottleneck edge) is less than that of $\overline{p_ip_k}$. And we also show that if the length of BE is greater than that of $\overline{p_ip_k}$ then there exists another biconnected subgraph of S in which the length of the longest edge(bottleneck edge) is equal to that of $\overline{p_ip_k}$.

To show that there is no biconnected subgraph of S with bottleneck edge less than $\overline{p_ip_k}$, we divide the points set S into two subsets by the point p_j, the points subset in the left of p_j is called the $L(p_j)$ and in the right is called $R(p_j)$. Assume this biconnected subgraph of S is existent, then there must exist two vertex disjoint paths between $L(p_j)$ and $R(p_j)$ [1], and furthermore the longest edge among the edges of these two paths is less than $\overline{p_ip_k}$. Since p_j is the only point between $L(p_j)$ and $R(p_j)$, there is at least one path(In fact, it is an edge e^*.) that does not pass through p_j and connects one point in $L(p_j)$ and the other in $R(p_j)$. However, the length of this edge e^* must be greater than that of $\overline{p_ip_k}$, since $\overline{p_ip_k}$ is the shortest edge among the edges whose one endpoint in $L(p_j)$ and the other in $R(p_j)$(see Fig. 3–1). This is a contradiction with the assumption.

Assume that the bottleneck edge BE is greater than $\overline{p_ip_k}$. Let's consider the graph in Fig. 3–1 which is the 2–Gabriel graph of S. In Fig. 3–1, each point only connects with the consecutive and the next consecutive points of it. It is not hard to see that the longest edge in this graph is the edge $\overline{p_ip_k}$ whose two endpoints determine a maximum two consecutive gaps of S. Since the 2–Gabriel graph contains a Euclidean bottleneck biconnected edge subgraph of S and the result of above paragraph, so there exists a biconnected subgraph of S with bottleneck edge equal to $\overline{p_ip_k}$. ◆

Theorem 3.5 *The Euclidean Bottleneck biconnected edge subgraph problem can be solved in* $\theta(nlogn)$ *time.*

Proof: It follows directly form Theorem 3.1 and Lemmas 3.3 and 3.4. ◆

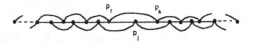

Fig. 3-1 The 2-GG of the points on the X-axis.
The three consecutive points p_i, p_j and p_k
determine the maximum two consecutive gaps.

Section 4. Concluding Remarks

We have studied the properties of k–Gabriel graphs and proposed an $O(k^2nlogn)$ algorithm to construct the k–Gabriel graph for a points set with n points in Euclidean plane. We have also showed that k–Gabriel graphs can be used to improve the running times of the algorithms for the Euclidean bottleneck biconnected edge subgraph problem and the Euclidean bottleneck matching problem.

References

[1]   Aho. A. V., Hopcroft, J. E. and Ullman, J. D.: *The Design and Analysis of Computer Algorithms*, Addison–Wesley, Reading, MA, 1974.

[2]   Chang, M. S., Tang, C. Y. and Lee, R. C. T.: *Solving the Euclidean bottleneck matching problem by k–relative neighborhood graph*, to appear in Algorithmica.

[3]   Chang, M. S., Tang, C. Y. and Lee, R. C. T.: *Solving Euclidean bottleneck biconnected edge subgraph problem by 2–relative neighborhood graphs*, The First Workshop on Proximity Graphs, New Mexico State University, 1989.

[4]   Chazelle, B. and Edelsbrunner, H.: *An improved algorithm for constructing kth–order Voronoi diagram*, IEEE Transactions on Computers, Vol. C–36, No. 11, 1987, pp. 1349–1354.

[5]   Edelsbrunner, H.: *Edge–skeleton in arrangements with applications*, Algorithmica, Vol. 1, No. 1, 1986, pp. 93–109.

[6]   Gabow, H. N.: *A scaling algorithm for weighted matching on general graphs*, Proceedings of the 26th Annual IEEE Symposium on Foundations of Computer Science, 1985, pp. 90–100.

[7]   Gabow, H. N. and Tarjan, R. E.: *Algorithms for two bottleneck optimization problems*, Journal of Algorithms, Vol. 9, No. 3, 1988, pp. 411–417.

[8]   Gabriel, K. R. and Sokal, R. R.: *A new statistical approach to geographical variation analysis*, Systematic Zoology, Vol. 18, No. 2, 1969, pp. 259–287.

[9]   Lawer, E. L.: *Combinatorial Optimization: Networks and Matoids*, Holt, Rinehart and Winston, New York, 1976.

[10]   Lee, D. T.: *On k–nearest neighbor Voronoi diagrams in the plane*, IEEE Transactions on Computers, Vol. C–31, No. 6, 1982, pp. 478–487.

[11]   Manber, U. and Tompa, M.: *Probabilistic, nondeterministic and alternating decision trees*, Technical Report, No. 82–03–01, University Washington, 1982.

[12]   Matula, D. W. and Sokal, R. R.: *Properties of Gabriel graphs relevant to geographic variation research and the clustering of points in the plane*, Geographical Analysis, Vol. 12, No. 3, 1980, pp. 205–222.

[13]   Parker, R. G. and Rardin, R. L.: *Guaranteed performance heuristics for the bottleneck traveling salesperson problem*, Operations Research Letters, Vol. 2, No. 6, 1984, pp. 269–272.

[14]   Shamos, M. I. and Preparata, F. P.: *Computational Geometry–an Introduction*, Springer–Verlag, 1985.

[15]   Supowit, K. J.: *The relative neighborhood graph with an application to minimum spanning trees*, Journal of Association for Computing Machinery, Vol. 30, No. 3, 1983, pp. 428–448.

[16]   Tarjan, R. E.: *Efficiency of a good but not linear set union algorithm*, Journal of Association for Computing Machinery, Vol. 22, No. 2, 1975, pp. 215–225.

[17]   Toussaint, G. T.: *The relative neighbourhood graph of a finite planar set*, Pattern Recognition, Vol. 12, No. 4, 1980, pp. 261–268.

Parallel Algorithms for Generating
Subsets and Set Partitions

Borivoje Djokić

Pediatrics Information Systems
University of Miami, P.O. Box 016820
Miami, Florida 33101, USA

Masahiro Miyakawa

Electrotechnical Laboratory
1-1-4 Umezono, Tsukuba
Ibaraki 305, Japan

Satoshi Sekiguchi

Electrotechnical Laboratory
1-1-4 Umezono, Tsukuba
Ibaraki 305, Japan

Ichiro Semba

Ibaraki University
2-1-1 Bunkyou, Mito-shi
Ibaraki 310, Japan

Ivan Stojmenović

University of Ottawa
Dept. of Comp. Sci., 34 G. Glinski
Ottawa K1N 6N5, Canada

Abstract

We present adaptive and cost-optimal parallel algorithms for generating 1) all subsets of the set $\{1, ..., n\}$, 2) all limited size subsets (each subset has at most m elements for a given m), and 3) all partitions of the set. The algorithms are based on a simple model of parallel computation which assumes the existence of k individual processors operating synchronously without need to communicate among themselves. Parallel ranking and unranking procedures for each case are also presented. Applications of the parallel subset generation algorithm to subset-sum, knapsack and base-enumeration problems are subsequently presented.

1. Introduction

Due to the numerous applications in science and engineering, algorithms for the generation of combinatorial objects have been intensively studied. Many such algorithms rely on some characteristic of enumeration technique. For example, lexicographic enumeration is often emphasized for its speed, simplicity [3] and the possibility to cut or recover some objects easily. Lexicographic enumerations of permutations [8], combinations (i.e. subsets of a fixed size) [23, 24], limited size subsets (each subset has at most $m \leq n$ elements) [18, 25, 27] and set partitions [9, 18, 26] can be found in the literature.

Since the number of different combinatorial objects is typically exponential to the number of elements, there has been increasing focus on parallel generation methods [1, 4, 6, 11].

A parallel algorithm is said to be *adaptive* if it is capable of modifying its behavior according to the actual number of processing elements available on the particular machine being used to execute the algorithm. The algorithm is said to be *cost-optimal* if the product of the number of processors and the runtime matches a lower bound on the number of operations required to solve the problem sequentially. Adaptive and cost-optimal (with a minor restriction) algorithms for generating permutations and combinations are presented in [1]. In this case, a simple model of parallel computation is used. The model assumes the existence of k synchronous processors which do not need to communicate among themselves. The idea is simple: if N is the total number of objects being generated, and the sequential generation algorithm is available, with a procedure $unrank(t, X)$ to discover the tth generated object X, then the job is divided equally into k processors. The distribution is such that processor i generates objects ranked from $(i-1)\lceil N/k \rceil + 1$ to $i\lceil N/k \rceil$, except for the last processor (not necessarily the kth), which should finish enumeration at the object ranked N. Note that this parallelization scheme minimizes the runtime (which is limited by the maximal load on a processor) and there could be at most $\lceil k/2 \rceil - 1$ (for $1 \leq k \leq N$) idle processors that have no object to generate (the minimal value is attained when $k = N - 1$). It is assumed that the sequential algorithm is capable of generating $t+1, t+2, \ldots$ objects for a given tth object X. For some recursive sequential algorithms, this property is difficult to achieve.

We modify this approach to present adaptive and cost-optimal parallel algorithms for the generation of all subsets of the set $\{1, \ldots, n\}$, all limited size subsets, and all partitions of the set. They are produced all in lexicographic order. The central issue is the provision of an unranking procedure for each case.

2. Ranking and unranking subsets in lexicographic order

We consider the problem of generating all subsets of the set $\{1, \ldots, n\}$. Assume that each subset is represented as a sequence $a_1 a_2 \ldots a_r$, where $1 \leq a_1 < \ldots < a_r \leq n$. We call this notation a-notation. We may also represent the same subset by an n-bit string $b_1 b_2 \ldots b_n$ where $b_i = 1$ if and only if $i \in \{a_1, \ldots, a_r\}$ (otherwise $b_i = 0$). We call this notation b-notation.

Recall the definition of lexicographic order of subsets. For two subsets $a = a_1 \ldots a_p$ and $c = c_1 \ldots c_q$, $a < c$ is satisfied if and only if there exists i ($1 \leq i \leq q$) such that $a_j = c_j$ for $1 \leq j < i$ and either $a_i < c_i$ or $p = i - 1$.

If we let $n=5$ then the lexicographic enumeration of all subsets proceeds as follows:

1, 12, 123, 1234, 12345, 1235, 124, 1245, 125, 13, 134, 1345, 135, 14, 145, 15, 2, 23, 234, 2345, 235, 24, 245, 25, 3, 34, 345, 35, 4, 45, 5.

Ranking implies the assignment of ordinal numbers to the subsets from 1 to $2^n - 1$ in lexicographic order. Conversely, an unranking procedure will find the subset with a given rank. In order to divide the job of generating subsets onto multiple processors, an unranking procedure is necessary. Such a procedure is given below and will return the tth ($t > 0$) subset in lexicographic order in both b- and a-notations, where r denotes the number of elements in the subset returned. For example, the 9-th ($t=9$) subset of $n = 5$ elements in lexicographic order is encoded as $b=11001$ or as $a=125$. In view of the large number of potentially idle processors (especially where k is large), performing a range-check based on t is inevitable. We set $r = 0$ when t is out of range so that we can know that current unranking will not generate an object. This is checked each time directly after an unranking procedure is called and if $r = 0$ is detected then the processor stops enumeration. However, in all subsequent enumeration programs we omit this range-check (**if** $r = 0$ **then goto** exit;) to avoid non-essential repetitions (however, in the real programs we need it).

```
program unranksub(t, n, a, b, r);
    begin
        for i:=1 to n do b_i := 0;
        r := 0; k := 1;
        repeat
            if t ≤ 2^{n-k} then begin b_k := 1; r := r + 1; a_r := k end;
            t := t - (1 - b_k) * 2^{n-k} - b_k;
            k := k + 1
```

```
        until k > n or t = 0;
    end;
```

For completeness, we also provide the ranking algorithms. The algorithm ranka(a,r,n,$rank$) uses array a as an input, while the procedure rankb($b, n, rank$) returns the $rank$ of a subset encoded as a bit-string in array b.

```
program ranka(a, r, n, rank);
    begin
        rank := r; a_0 := 0;
        for i := 0 to r - 1 do
            for j := a_i + 1 to a_{i+1} - 1 do rank := rank + 2^{n-j}
    end;
```

```
program rankb(b, n, rank);
        begin
            k := n;
            while b_k = 0 do k := k - 1;
            rank := 0;
            for i := 1 to k do rank := rank + (1 - b_i) * 2^{n-i} + b_i
        end;
```

3. Parallel generation of all subsets

Using the previous unranking procedure, a k processor parallel algorithm is now given to generate all subsets, where k is any integer less than 2^n. It is based on a sequential subset generating procedure [27]. Since there are $2^n - 1$ subsets to be generated, the job is divided equally onto the k processors so that each processor generates $g = \lceil (2^n - 1)/k \rceil$ subsets. Thus processor i ($1 \le i \le k$) will generate subsets which are ranked from $(i - 1) * g + 1$ to $i * g$. The generating algorithm is:

```
program subsetpar(n, k);
    for i := 1 to k do in parallel
        begin
            g := \lceil (2^n - 1)/k \rceil;
            t := (i - 1) * g + 1;
            unranksub(t, n, a, b, r);
            l := 0;
            repeat
                print out a_1...a_r;
                l := l + 1;
                if a_r < n then extend else reduce
            until l = g or a_1 = n;
        end;
```

extend $\equiv$ **begin** $a_{r+1} := a_r + 1; r := r + 1$ **end**
reduce $\equiv$ **begin** $r := r - 1; a_r := a_r + 1$ **end**

A second subset generator algorithm can be described using straightforward ranking and unranking procedures based on the b-notation of subsets. The subsets are generated in order from 1 to $2^n - 1$, and the subset ranked t corresponds to the binary equivalent $b_1...b_n$ of t. This parallel algorithm is based on the sequential subset generating algorithm proposed in [28]. Although it has an advantage of being very simple, there are two disadvantages:

1. The run time per subset is $O(\log n)$, whereas for lexicographic order it is constant [25].

2. The order of generated subsets cannot be easily adapted for applications in solving the subset sum, minimal covering and base enumeration problems [27]. Efficient algorithms for these applications involve a cut technique to skip the generation of unnecessary subsets.

There is a third subset generating algorithm which uses 2^s processors ($s \leq n$). In this approach all 2^s subsets of the set $\{1,...,s\}$ are distributed to 2^s processors, one subset per processor. One processor may be used to generate and distribute these subsets, or an unranking procedure may be employed, so that each processor finds a corresponding subset internally. Each processor can then run a sequential subset generating algorithm to produce subsets of the set $\{s+1,...,n\}$ and append the corresponding subset from $\{1,...,s\}$ to each subset. This algorithm is simple and for small s it does not require an unranking procedure. However, for $s > n/2$ the generation of all subsets of $\{1,...,s\}$ becomes a bottleneck. The primary disadvantage of this algorithm is that it is not adaptive.

We now present a combination of the second and the third algorithms, which needs neither an unranking procedure (other than the simple conversion of an integer to the equivalent base 2 binary string) nor the distribution of the initial subsets from a master processor.

```
program subset(n, s);
    for i := 0 to 2^s - 1 do in parallel
        begin
            find binary equivalent of i
            and print out corresponding subset a_1...a_p;
            a_{p+1} := s + 1; r := p + 1;
            repeat
                print out a_1...a_r;
                if a_r < n then extend else reduce
            until a_{p+1} = n
        end;
```

An adaptive variant of this algorithm can be found by a simple extension that divides the job of processor i ($0 \leq i \leq 2^s - 1$) into 2^{s_i} sub-processes and allocates 2^{s_i} processors to accomplish the task. This extension has a special importance in applications involving the cut technique, since more processors can be assigned to parts where the cut technique is expected to be less effective.

In Section 6 we describe parallel programs for the subset-sum, knapsack and base-enumeration (minimal-cover) problems where efficient generation of subsets plays a central role.

4. Generating all limited size subsets

In some problems it is sufficient to generate all subsets which do not exceed a specified size. For example, all subsets of the 5-element set $\{1, 2, 3, 4, 5\}$ having at most three elements are:

1, 12, 123, 124, 125, 13, 134, 135, 14, 145, 15, 2, 23, 234, 235, 24, 245, 25, 3, 34, 345, 35, 4, 45, 5.

The number of such subsets is $L_m^n := C_1^n + C_2^n + ... + C_m^n$, where $C_m^n = n!/(m!(n-m)!)$ and $u! = 1 \cdot 2 \cdot ... \cdot (u-1) \cdot u$. Note that $L_m^n = L_n^n = 2^n - 1$ for $m \geq n$.

Sequential backtracking algorithms exist for generating all subsets of a set $\{1, 2, ..., n\}$ having at most $m(< n)$ elements in lexicographic order [18, 25, 27]. The running time of the algorithms per subset is shown to be constant [25]. This problem also appears as a subproblem of an approximation algorithm to solve the knapsack problem [2]. Therefore, the subset generation algorithm given below also serves as a parallel approximation algorithm for the knapsack problem.

A pair of ranking and unranking procedures are presented prior to the parallel enumeration procedure. A procedure to rank all subsets of a fixed size of a set is given by Knott [14]. Let rankc($a_1...a_s$) denote the rank of a subset $a_1...a_s$ in the set of all subsets having size s. Then the rank of a subset $a_1...a_r$ is given simply by

$$\text{ranks}(a_1...a_r) \equiv \text{rankc}(a_1...a_r) + \text{rankc}(a_1...a_{r-1}) + ... + \text{rankc}(a_1 a_2) + \text{rankc}(a_1).$$

The following unranking algorithm returns the subset $a_1 ... a_r$ (stored in an array a) for a given rank t ($t > 0$).

```
program unranklim(t, n, m, a, r);
  begin
    r := 0; i := 1;
    repeat
        s := t - 1 - L_{m-r-1}^{n-i};
        if s > 0
            then t := s
            else begin r := r + 1; a_r := i; t := t - 1 end;
        i := i + 1
    until t = 0 or i = n + 1;
  end;
```

Now a parallel subset generation algorithm can be given using this unranking procedure and the sequential algorithm [27] as follows:

```
program subsetlim(m, n, k);
    for i := 1 to k do in parallel
        begin
            g := ⌈L_m^n/k⌉;
            t := (i - 1) * g + 1;
            unranklim(t, n, m, a, r);
            l := 0;
            repeat
                l := l + 1;
                print out a_1...a_r;
                if a_r < n then extendlim else reduce
            until l = g or a_1 = n;
        end;
```

extendlim $\equiv$ **begin if** $r < m$ **then** extend **else** $a_r := a_r + 1$ **end**
extend $\equiv$ as before.

5. Generating set partitions

Let $Z = \{1, ..., n\}$. A partition of the set Z consists of classes $\pi_1, ..., \pi_k$ such that $\pi_i \cap \pi_j = \phi$ if $i \neq j$, $\pi_1 \cup ... \cup \pi_k = Z$, and $\pi_i \neq \phi$ for $1 \leq i \leq k$.

Sequential algorithms for set partitioning are known [7, 9, 18, 26, 28]. The algorithms [7, 9, 18, 26] generate partitions represented by codewords in lexicographic order. A codeword $c_1 c_2 ... c_n$ represents a partition of the set Z if and only if $c_1 = 1$ and $1 \leq c_r \leq \max(c_1, ..., c_{r-1}) + 1$ for $2 \leq r \leq n$, where $c_i = j$ if i is in π_j.

In order to give an unranking algorithm which finds the codeword of the partition having rank t (> 0), we need to pre-compute part of an integer matrix $D_n(r, d)$ for $2 \leq r \leq n + 1$, $1 \leq d \leq r - 1$ (lower triangle). The element $D_n(r, d)$ denotes the number of codewords (partitions) $c_1 ... c_n$ for which $c_1, ..., c_{r-1}$ are fixed and $\max(c_1, ..., c_{r-1}) = d$. These elements are computable using the following recurrence relations: From the definition, $D_n(2, 1) = B_n$, where B_n is the total number of partitions of the set of n elements (called the Bell number); $D_n(n, d) = d + 1$ for $1 \leq d \leq n - 1$ and $D_n(n+1, d) = 1$ for $1 \leq d \leq n$, and $D_n(r, d) = d * D_n(r + 1, d) + D_n(r + 1, d + 1)$ for $3 \leq r \leq n - 1$ and $1 \leq d \leq r - 1$. Using the last recurrence relation the following procedure determines the code c_r for $r = 1, ..., n$ in this order.

```
program unrankpart(t, n, c, r);
   begin
      if not 0 < t ≤ Dₙ(2, 1) then begin r = 0; goto exit end;
      c₁ := 1; d := 1;
      for r := 2 to n do
         begin
            m := 0;
            repeat m := m + 1 until t ≤ m * Dₙ(r + 1, d);
            if m > d + 1 then m = d + 1;
            cᵣ := m;
            t := t − (m − 1) * Dₙ(r + 1, d);
            if m > d then d = m;
         end;
      exit:
   end;
```

For completeness, we also present a simple ranking algorithm:

```
program rankp(t, n, c);
   begin
      t := 1; d := 1;
      for r := 2 to n do
         begin
            t := t + (cᵣ − 1) * Dₙ(r + 1, d);
            if cᵣ > d then d := cᵣ
         end;
   end;
```

A parallel algorithm for generating set partitions based on the sequential iterative algorithm [7] and the previous unranking algorithm follows:

```
program setpartpar(n, k);
    for i := 1 to k do in parallel
        begin
            g := ⌈Bₙ/k⌉;
            t := (i − 1) * g + 1;
            unrankpart(t, n, c, r);
            l := 0; r := n; j := 0; max := 1;
            for s := 2 to n − 1 do
                if cₛ > max then max := cₛ
                            else begin j := j + 1; bⱼ := s end;
            repeat
                while r < n − 1 do begin r := r + 1; cᵣ := 1; j := j + 1; bⱼ := r end;
                repeat
                    l := l + 1; print out c₁...cₙ; cₙ := cₙ + 1;
                until cₙ > n − j or l = g;
                r := bⱼ; cᵣ := cᵣ + 1; cₙ := 1;
                if cᵣ > r − j then j := j − 1;
            until l = g or r = 1;
        end;
```

6. Applications of the subset generating algorithms

6.1. Knapsack and subset sum problems

An input for the subset sum problem consists of $n+1$ positive integers $c_1, ..., c_n$ and C. The problem is to find a subset T of $\{1, ..., n\}$ to maximize $\Sigma_{i \in T} c_i$ subject to the requirement that $\Sigma_{i \in T} c_i \leq C$. This problem is shown to be NP-complete (cf. [2]). A more general formulation is known as the knapsack problem. The input consists of $2n + 1$ positive integers: C and two sequences $c_1, ..., c_n$ and $p_1, ..., p_n$. The problem is to maximize $\Sigma_{i \in T} p_i$ subject to the constraint $\Sigma_{i \in T} c_i \leq C$ where T, as before, is a subset of the indices.

A short survey of parallel knapsack algorithms is given in [22] where additional references to work in this area can be found. We present a parallel solution to the knapsack problem based on a sequential algorithm given in [27] (a similar sequential backtracking algorithm for the subset sum problem is given in [5, p. 445]).

```
program knapsack(n, d, a, p, k);
    begin
    for i := 1 to k do in parallel
        begin
            g := ⌈(2ⁿ − 1)/k⌉;
            j = (i − 1) * g + 1; l := 1; maxsum := 0;
            unranksub(j, n, a, b, r);
            repeat
                S := c(a₁) + ... + c(aᵣ);
                if S ≤ C
                    then begin
                        P := p(a₁) + ... + p(aᵣ);
```

```
                    if P > maxsum then maxsum := P;
                    if a_r < n then extend else reduce;
                    l := l + 1;
                 end;
              else begin l := l + 2^{n-a_r}; cut; end;
         until l > g or a_1 = n;
         if a_1 = n then check last case;
     end;
  find and print out the maximum out of maxsum;
  end;
```

cut ≡ **if** $a_r < n$ **then** $a_r := a_r + 1$ **else** reduce;
extend ≡ as before;
reduce ≡ as before.

The check for the *last case* in the program means examining all conditions for the last subset $\{n\}$. In order to find a maximum of the best result obtained on each processor, processors will need to communicate using shared memory or communication links to find this value (cf. [2]).

6. 2. Base enumeration problems

The application of the lexicographic subset generating algorithm to base enumeration for the set of k-valued logical functions is described in [27]. The base enumeration problem is equivalent to the minimum covering problem (cf. [17]). Here, we provide a description of the problem from a combinatorial standpoint and present a parallel base enumeration algorithm.

An input to the base enumeration problem is an $n \times d$ Boolean matrix $A = [a_{ij}]$, $a_{ij} = 0$ or 1, called a *coefficient matrix* (n and d: positive integers). A Boolean vector $(x_1, ..., x_n)$ is called *complete* in A if $(x_1, ..., x_n)A \geq (1, ..., 1)$ holds, where $(x_1, ..., x_n) \geq (y_1, ..., y_n)$ iff $x_i \geq y_i$ for all $1 \leq i \leq n$ ($x_i, y_i \in \{0, 1\}$). We call a vector $(x_1, ..., x_n)$ *non-redundant* in A if $(x_1, ..., x_n)A > (y_1, ..., y_n)A$ is valid for each vector $(y_1, ..., y_n)$ strictly smaller than $(x_1, ..., x_n)$ (i.e. there is an i such that $y_i < x_i$). A vector is called a *base* in A if it is complete and non-redundant in A. Non-redundant non-complete vectors we simply call *addable*. The rank of a base or an addable $(x_1, ..., x_n)$ is the sum $x_1 + ... + x_n$. The base enumeration problem is to count all bases for each rank. Due to the isomorphism between a vector $(x_1, ..., x_n)$ and a subset of rows of the matrix A, it is easy to see that base enumeration is directly related to subset generation problem. If a subset represents a base or is not addable then no extension of the subset can yield a base. Therefore a cut of enumeration can be applied conveniently to avoid generating such irrelevant subsets. For real cases, where $d \ll n$ the rank of a base (the size of corresponding subset) is at most d. This allows one to use the *unranklim* instead of the *unranksub* procedure, which prevents large exponents of 2 as the increment of l in the algorithm.

The base enumeration algorithm follows (we conveniently represent the i-th row of matrix A as $A(i)$):

```
program baseen(n, d, A, k);
  begin
       for i := 1 to k do in parallel
           begin
               g := ⌈L_d^n / k⌉;
               t := (i - 1) * g + 1; l := 1;
               unranklim(t, n, d, a, r);
```

```
            repeat
                if A(a₁), ..., A(aᵣ) is addable
                    then    begin
                                if aᵣ < n then extend else reduce end;
                                l := l + 1
                            end;
                    else    begin
                                if A(a₁), ..., A(aᵣ) is a base then count it;
                                cut;
                                l := l + L^{n-aᵣ}_{d-r}
                            end
                until l > g or a₁ = n;
                if a₁ = n then check last case;
            end;
        print out the sum of encountered bases of all processors;
    end;
```

The procedures for testing a base and/or a addable vector are described in [27]. The check for the last case is the same as previously described. Note that at the end of the procedure, the sum of bases encountered by each processor should be taken.

In the above algorithm the efficiency of cut depends on the actual data. An additional technique is given in Conclusions to make the load uniform for each processor.

7. Conclusions

Some NP-hardness originates from a necessity to check all subsets of a finite set. Optimal parallel computation can practically resolve this difficulty. Indeed, there are many other applications of subset generation procedures which could also be studied. For instance, tautology checking can be done by a straightforward subset generating algorithm to examine all 2^n values for n-tuples of logical variables. The other application is the enumeration of Boolean clique functions [20]. The minimization of Boolean functions has significant practical importance.

It may be of interest to design techniques and architectures for parallel branch and bound algorithms that will enable processors which have finished their jobs to "help" others still searching or counting. In the knapsack, base enumeration and minimal base problems it is possible since our ranking and unranking procedures provide a way to divide the remaining subsets to be generated by a processor into two equal groups. Each half could then be assigned to the processor which has finished its present task. Using this technique one can eventually achieve optimal speedup on algorithms which utilize cut techniques.

We note that there are other sequential subset generating procedures which can be used for adaptive parallel algorithms. For example, [10] is a survey of shift register cycle algorithms for subset generation. These algorithms have applications in the search for a key sequence where only a fixed number of the most recently entered digits are essential for to break the code. In order to parallelize any of the algorithms described in [10] an effective unranking procedure must be found. This is presently an open problem.

Acknowledgments. We would like to thank Dr Nobuyuki Otsu for providing helpful commentary. Also we appreciate Dr Al Davis for reading the manuscript and improving the English. This work was initiated while the second author visited The University of Miami in June, 1988. We are thankful to the Foundation for C & C Promotion for the partial financial support.

References

[1] S.G. Akl, Adaptive and optimal parallel algorithms for enumerating permutations and combinations. *The Computer Journal* **30**, 5 (1987), 433-436.

[2] S. Baase, *Computer Algorithms, Introduction to design and analysis*, 2nd ed. Addison-Wesley, Reading, Mass. (1988).

[3] M. Carkeet, P. Eades, Performance of subset generating algorithms. *Annals of Discrete Mathematics* **26** (1985), 49-58.

[4] B. Chan and S.G. Akl, Generating combinations in parallel. *BIT* **26**, 1 (1986), 2-6.

[5] K.M. Chandy, J. Misra, *Parallel Program Design, A foundation*, Addison-Wesley, Reading, Mass. (1988).

[6] G.H. Chen and M.-S. Chern, Parallel generation of permutations and combinations. *BIT* **26** (1986), 277-283.

[7] B. Djokić, M. Miyakawa, I. Semba, S. Sekiguchi and I. Stojmenović, A fast iterative algorithm for generating set partitions. *The Computer Journal* **32**, 3 (1989), 281-282.

[8] M.C. Er, Lexicographic enumeration, ranking and unranking of permutations of r out of n objects. *International Journal of Computer Mathematics* **23** (1987), 1-7.

[9] M.C. Er, Fast algorithm for generating set partitions. *The Computer Journal* **31**, 3 (1988), 283-284.

[10] H. Fredricksen, A survey of full length nonlinear shift register cycle algorithms. *SIAM Review* **24**, 2 (1982), 195-221.

[11] P. Gupta and G.P. Bhattacharjee, Parallel generation of permutations. *The Computer Journal* **26**, 2 (1983), 97-105.

[12] E. Horowitz and S. Sahni, Computing partitions with applications to the knapsack problem. *Journal of the ACM* **21**, 2 (1974), 277-292.

[13] E.D. Karnin, A parallel algorithm for the knapsack problem. *IEEE Transactions on Computers* **C-33**, 5 (1984), 404-408.

[14] G.D. Knott, A numbering system for combinations. *Communications of the ACM* **17**, 1 (1974), 45-46.

[15] G.D. Knott, A numbering system for permutations of combinations. *Communications of the ACM* **19**, 6 (1976), 355-356.

[16] G. Li and B.W. Wah, Coping with anomalies in parallel branch-and-bound algorithms. *IEEE Transactions on Computers* **C-35**, 6 (1986), 568-573.

[17] M. Miyakawa, I. Stojmenović, D. Lau and I. G. Rosenberg, Classifications and base enumerations in many-valued logics – a survey –, *Proc. 17th International Symposium on Multiple-Valued Logic*, Boston (May 1987), 152-160.

[18] A. Nijenhuis and H.S. Wilf, *Combinatorial Algorithms*, Academic Press, New York (1978).

[19] J.G. Peters, L. Rudolph, Parallel approximation schemes for subset sum and knapsack problems. *Acta Informatica* **24** (1987), 417-432.

[20] Pogosyan G., Miyakawa M. and Nozaki A. On the number of Boolean clique functions, submitted for publication (1988).

[21] Z.G. Qiang, An $O(\log n)$ parallel algorithm for the subset sum problem. *ACM SIGACT News* **18**, 2 (Fall 86-Winter 87), 57-63.

[22] C.C. Ribeiro, Parallel computer models and combinatorial algorithms. *Annals of Discrete Mathematics* **31** (1987), 325-364.

[23] E.M. Reingold, J. Nievergelt and N. Deo, *Combinatorial Algorithms, Theory and Practice*, Prentice Hall, Englewood Cliff (1977).

[24] I. Semba, A note on enumerating combinations in lexicographical order. *Journal of Information Processing* **4**, 1 (1981), 35-37.

[25] I. Semba, An efficient algorithm for generating all k-subsets ($1 \leq k \leq m \leq n$) of the set $\{1, 2, ..., n\}$ in lexicographic order. *Journal of Algorithms* **5** (1984), 281-283.

[26] I. Semba, An efficient algorithm for generating all partitions of the set $\{1, ..., n\}$, *Journal of Information Processing* **7** (1984), 41-42.

[27] I. Stojmenović and M. Miyakawa, Applications of subset generating algorithm to base enumeration, knapsack and minimal covering problems. *The Computer Journal* **31**, 1 (1988), 65-70.

[28] M.B. Wells, *Elements of Combinatorial Computing*, Pergamon Press, Oxford (1971).

[29] B.W. Wah, G. Li and C.F. Yu, Multiprocessing of combinatorial search problems. *IEEE Computer* (June 1985), 93-108.

Parallel Algorithms for Linked List and Beyond

Yijie Han

Department of Computer Science
University of Kentucky
Lexington, KY 40506, USA

Abstract

Linked list problems are fundamental problems in the design of parallel algorithms. We present design techniques for linked list algorithms and applications of these algorithms.

1. Introduction

Linked list is a basic data structure in the design of computer algorithms. The conceptual simplicity of a linked list often leads algorithm designers to believe that it is a trivial structure to be treated. A text in computer algorithms will not treat problems such as finding a maximal independent set on a linked list as a formal topic for the design of a sequential algorithm because the problem is trivial in the sequential case. However, much research interest arouse in the past few years on the design of parallel algorithms for linked list and many efficient parallel algorithms emerged by using efficient linked list algorithms as a critical subroutine. In this paper we summarize these efforts by outlining the main difficulties once presented in the design of efficient parallel algorithms for linked list, by providing the main thread leading to overcoming these difficulties and by demonstrating the power of linked list algorithms through the applications of these algorithms.

The first step in the design of a parallel algorithm is to choose a model. A PRAM (Parallel Random Access Machine)[BH] is usually preferred because the model allows almost uninhibited communication between processors and memory cells. This enables an algorithm designer to focus his attention on the intrinsic parallelism of the problem instead of the architecture of the model. A PRAM has a set of processors and a set of memory cells. Both processors and memory cells are linearly indexed. In a step, a processor in a PRAM can read or write any memory cell. Conflicts arising from simultaneous read or write to the same memory cell by several processors are resolved by imposing additional rules to the PRAM[Sn]. In an EREW (Exclusive Read Exclusive Write) PRAM, simultaneous access to the same memory cell is prohibited. In a CREW (Concurrent Read Exclusive Write) PRAM simultaneous read to the same memory cell is allowed and all processors reading the same cell simultaneously will obtain the value from the cell. Simultaneous write to the same cell is prohibited on the CREW PRAM. In a CRCW (Concurrent Read Concurrent Write) PRAM both simultaneous read and simultaneous write are allowed. The resulting value in a memory cell after simultaneous write to the cell can be determined according to certain stipulations. In a COMMON CRCW PRAM, simultaneous write is allowed only in the situation that all processors write the same value to the same cell and the resulting value of the cell is this common value. In an ARBITRARY CRCW PRAM, an arbitrary processor among the processors attempting to simultaneous write the same cell succeeds in writing the cell. There are other conflict resolving rules for the CRCW PRAM. Chlebus *et al.* [CDHR] is a paper studying these variations of CRCW PRAMs.

CRCW PRAM represents the strongest version of the PRAM model while the EREW PRAM is the weakest version among three. In certain situations the PRAM models can be further weakened by limiting the number of shared memory cells it can use. A PRAM with p processors allowed only p shared memory cells is equivalent to the local memory PRAM[AM1] or the DCM (Direct Connection Machine) model[KMR].

Two fundamental problems related to linked list are the maximal matching problem and the prefix problem.

A matching set is a set of edges (pointers) with no two pointers incident with the same node. A matching is maximal if it is not a subset of any other matching set. For the input of a linked list the maximal matching problem asks the output of a maximal matching set of the list. For a linked list the maximal matching problem is equivalent to the maximal independent set problem (where a maximal independent set of nodes is to be computed) in the sense that a PRAM algorithm for computing a maximal matching for a linked list can be used for computing a maximal independent set with the same time complexity and vice versa.

For the data items a_0, a_1, ..., a_{n-1} on the input linked list and an associative operation $\bigcirc$, the linked list prefix problem asks the output of prefixes b_0, b_1, ..., b_{n-1}, where $b_j = \bigcirc_{i=0}^{j} a_i$.

The input linked list is assumed to be stored in two arrays $X[0..n-1]$ and $NEXT[0..n-1]$. The data items associated with the nodes of the linked list are stored in array X and pointers of the linked list is stored in array $NEXT$.

Both problems are trivial in the sequential case. They are not trivial in the parallel case where many parallel algorithm design techniques have been invented and tested on the two problems.

2. Linked List Algorithms

2.1. Pointer Jumping

A technique due to Wyllie[W] for handling linked list is the technique of pointer jumping (also called recursive doubling). The application of this technique in solving the prefix problem can be described by the following procedure.

PREFIX$(X[0..n-1], NEXT[0..n-1], HEAD)$
   forall $i : 0 \leq i < n$ **do**
      begin
         $tmp := HEAD$;
         if $NEXT[i] \neq nil$ **then** $NEXT[NEXT[i]] := i$;
         else $HEAD := i$;
         $NEXT[tmp] := nil$;
         for $k := 1$ **to** $\lceil \log n \rceil$ **do**
            if $NEXT[i] \neq nil$ **then**
               begin
                  $X[i] := X[i] \bigcirc X[NEXT[i]]$;
                  $NEXT[i] := NEXT[NEXT[i]]$;
               end
   end.

The effect of pointer jumping is shown in Fig. 1.

This technique is powerful in that it solves the prefix problem in $O(\log n)$ time when n processors are available. However it loses efficiency compared to a sequential algorithm with time complexity $O(n)$ because it uses a total of $O(n \log n)$ operations or instructions.

Substantial research efforts have been put on cutting the factor of $\log n$ in order to obtain a parallel algorithm using optimal number $O(n)$ operations (algorithms using optimal number of operations are called optimal algorithms). All these efforts follow the same approach, namely contracting the linked list by computing a large matching set for the list and pairing off nodes in the list[MR].

2.2. Computing a Matching Set Using Randomization

Miller and Reif used randomization to obtain a large matching set. This is done by using a random number generator at each node of the list. Each random number generator independently generates a 0 or a 1 with equal probability. If the head of a pointer gets a 1 and the tail of the pointer gets a 0, the head and the tail of the pointer can be paired. For other possible random values $((0, 0), (0, 1), (1, 1)$ at the head and the tail) the head and the tail will not be paired. It can be shown[MR] that at least $(1-\epsilon)n/8$ nodes will be in the matching set with probability of failure less than $e^{-3\epsilon^2 n/2^7}$. Due to the power of randomization such a matching set can be obtained in constant time.

2.3. The Matching Partition Function

A function for partition the pointers of a linked list into $O(\log n)$ matching sets can be derived from the following intuitive observation[H1].

For a linked list of n elements stored in an array $X[0..n-1]$, $NEXT[0..n-1]$ is the array of pointers with $NEXT[i]$ pointing to the next element to $X[i]$, as shown in Fig. 2. For a node v in the linked list, we also denote the node following v in the list by $suc(v)$ and the node preceding v by $pre(v)$. We use $< a, b >$ to denote a pointer valued b in location $NEXT[a]$. b is the head of the pointer and a is the tail. A pointer $< a, b >$ is a forward pointer if $b > a$, otherwise the pointer is a backward pointer. By drawing a line c bisecting the array containing the linked list as shown in Fig. 3, we observe that forward pointers crossing line c have disjoint heads and tails. This is because no two pointers can have the same head or the same tail and the head of one pointer can not be the tail of the other pointer because both pointers are forward pointers crossing line c. We associate with bisecting line c two matching sets of pointers, one consisting of forward pointers crossing c, the other of backward pointers crossing c. The linked list array is divided into two sub-arrays by line c. We can draw bisecting lines c_1, c_2 for the two sub-arrays. Forward pointers crossing either c_1 or c_2 but not c have disjoint heads and tails. Continuing in this fashion it is not difficult to see that pointers of the linked list can first be partitioned into two sets, a set of forward pointers and a set of backward pointers, and then each set can further be partitioned into $\lceil \log n \rceil$ matching sets, with pointers in one set having disjoint heads and tails. A close examination of the pointers crossing bisecting lines reveals that the function $g(< a, b >) = \max\{i \mid$ the i-th bit of $a \, XOR \, b$ is 1$\}$, where XOR is the bit-wise exclusive-or operation and bits are counted from the least significant bit starting with 0, characterizes the set of pointers crossing bisecting lines (both forward and backward pointers). Function g can be modified to distinguish between forward and backward pointers. We define function $f(< a, b >) = 2k + a\#k$, where $k = \max\{i \mid$ the i-th bit of $a \, XOR \, b$ is 1$\}$ and $a\#k$ is the k-th bit of a. Note that $a\#k$ denotes whether $< a, b >$ is a forward pointer or a backward pointer.

Theorem 1[H1]: Function f is a matching partition function which partitions the pointers of a linked list into $2\lceil \log n \rceil$ matching sets. $\square$

It is possible to have a matching partition function which partitions the pointers of a linked list into less than $2\lceil \log n \rceil$ matching sets. Let $g(n) = \binom{n}{\lfloor n/2 \rfloor}$. We have

Theorem 2[H5]: There is a matching partition function f which partitions $g(n)$ pointers of a linked list into n matching sets.

Proof: We show, by induction, the existence of a f which partitions $\binom{n}{i}$ pointers of a linked list into n matching sets such that possible outgoing pointers of a node are in no more than i matching sets.

To establish the base of the induction we show the existence of a f which partitions $\binom{n}{1}$ pointers of a linked list into n matching sets such that possible outgoing pointers of a node are in one matching set. We achieve this by putting the outgoing pointer of node v in matching set v, $0 \leq v \leq n-1$. We have used n matching sets and no matter which node is the head of the pointer with its tail at node v, the pointer is always in one set, i.e., set v.

To obtain f for a list with $\binom{n}{i}$ nodes, we first bisect the list into two parts as shown in Fig. 4, the first part L_1 contains $\binom{n-1}{i}$ nodes and the second part L_2 contains $\binom{n-1}{i-1}$ nodes. We obtain two matching partition functions f_1 and f_2 for L_1 and L_2, respectively, by induction hypothesis. We combine f_1 and f_2. The remaining pointers need to be put into matching sets are those pointers with their heads in L_1 and tails in L_2 (or heads in L_2 and tails in L_1). By induction hypothesis, outgoing pointers in L_1 can be assigned to at most i matching sets and outgoing pointers in L_2 can be assigned to at most $i-1$ matching sets. Therefore for a pointer p with head h in L_2 and tail t in L_1, there is a matching set which can be assigned to possible outgoing pointers at h but is never assigned to possible outgoing pointers at node t, we assign this matching set to pointer p. For any pointer with head in L_1 and tail in L_2 we put it into a new matching set which is a set not used in L_1 and L_2. By now we have obtained f which partitions $\binom{n}{i}$ pointers of a linked list into n matching sets such that possible outgoing pointers of a node are in no more than i matching sets.

Setting $i = \lfloor n/2 \rfloor$ proves the theorem. $\square$

By Stirling's approximation we see the existence of a matching partition function f which partitions n pointers of a linked list into $(1 + o(1)) \log n$ matching sets.

Theorem 2 is the best we could possibly do, as shown in the following theorem.

Theorem 3 [H5]: For $g(n)$ pointers in a linked list n is the minimum on the number of matching sets obtainable by any matching partition function $f(x, y)$.

Proof: Given a matching partition function f, let $S(a) = \{f(a, b) \mid 0 \leq b < g(n)\}$ and $P = \{S(a) \mid 0 \leq a < g(n)\}$. For two different nodes a and b, $S(a) \not\subseteq S(b)$, for otherwise there exists a node c such that pointers $< a, b >$ and $< b, c >$ will be put into the same matching set. Therefore P is a Sperner's family[Sp] and the correctness of the theorem follows since a Sperner's family of $g(n)$ sets has at least n elements. $\square$

The matching partition function can be generalized as follows. Define $\log^{(1)} n = \log n$, $\log^{(k)} n = \log(\log^{(k-1)} n)$, $G(n) = \min\{k \mid \log^{(k)} n < 1\}$. Also define $f^{(2)}(a_1, a_2) = f(a_1, a_2)$, $f^{(k)}(a_1, a_2, ..., a_k) = f(f^{(k-1)}(a_1, a_2, ..., a_{k-1}), f^{(k-1)}(a_2, a_3, ..., a_k))$. $f^{(k)}$ represents repeated applications of f to the linked list.

Theorem 4 [H5]: There is a matching partition function $f^{(k)}$ which partitions n pointers of a linked list into $\log^{(k-1)} n(1 + o(1))$ matching sets.

Proof: By Theorem 2 and definition of $f^{(k)}$. $\square$

On the other hand, we have:

Theorem 5[H5]: A lower bound on the number of matching sets any $f^{(k)}$ can obtain is $\log^{(k-1)} n$. $\square$

Lower bounds similar to that shown in Theorems 3 and 5 were obtained by Linial[L]. His bounds are off by a multiplicative factor from optimal. Upper and lower bounds obtained in [H5] are exact for $f^{(2)}$ and tight up to minor terms for $f^{(k)}$ for $k > 2$, as shown in Theorems 2 to 5.

Since $f^{(G(n))}$ gives constant number of matching sets, if node a knows up to $G(n)$ nodes following it in the linked list it can label itself with a matching set number chosen from a constant number of matching sets. This yields a parallel algorithm for maximal matching with time complexity $O(\frac{n \log G(n)}{p} + \log G(n))$, in this paper p is the number of processors used in an algorithm. This algorithm is due independently to Han[H4] and Beame. Han's version provides a scheme for constructing a lookup table for $f^{(G(n))}$ [H4].

Theorem 6: A maximal matching for a linked list can be computed in $O(\frac{n \log G(n)}{p} + \log G(n))$ time. $\square$

2.4. Processor Scheduling

A processor scheduling scheme was provided by Han[H3] which yields processor efficient algorithm for finding a maximal matching set for a linked list. .

To obtain an optimal algorithm with time complexity $O(x)$ using $y = n/x$ processors, we may view the linked list as being stored in a two dimensional array with x rows and y columns. We can thus assign one processor to each column.

It is easy to see[H3] that if all pointers are inter-row pointers, $i.e.$, pointers with their heads and tails on different rows of the two dimensional array, then the pointers can be partitioned into three matching sets by letting the processors walking down the rows of the array. The difficult situation we have to handle is the situation where there are many intra-row pointers, $i.e.$, pointers with their heads and tails on the same row of the two dimensional array.

Assume that all pointers are intra-row pointers and that the pointers are already being partitioned into x matching sets. Further assume that all pointers in a column are in the same matching set. The processor scheduling works as follows. If pointers in column c are in matching set m, the processor assigned to column c will idle for m steps before it walks down column c. This processor scheduling guarantees that no processors will contend on any pointer.

The last paragraph presents the intuition behind Han's algorithm[H3]. To remove the assumptions we observe that inter-row pointers can be handled almost trivially[H3], that the partition of pointers into x matching sets, when $x = \log^{(i)} n$ for any constant i, can be done in $O(in/p)$ time using the matching partition function $f^{(i+1)}$ described in the previous section, and that the pointers can be sorted to ensure that all pointers in the same column belong to the same matching set.

Theorem 7[H3]: A maximal matching for a linked list can be computed in $O(\frac{n \log i}{p} + \log^{(i)} n + \log i)$ time, where i is an adjustable parameter. $\square$

For the linked list prefix problem Wyllie's pointer jumping technique[W] does not yield an optimal algorithm. When the technique of contraction[MR] is applied one can first find a matching set of size n/c for a constant c. For each pointer in the matching set the head and the tail of the pointer can be combined by the $\bigcirc$ operation. This "pair-off" operation contracts the linked list to a list with only $(1 - 1/2c)n$ nodes. $O(\log n)$ stages of this contraction process reduce the input linked list to a single node. After the linked list is contracted to a single node this contraction process can be reversed to expand the single node to a linked list. This contraction and expansion processes can be used to evaluate linked list prefix[MR]. A detailed explanation of these processes can be found in [H1][WH].

A key problem in achieving time complexity $O(\log n)$ using optimal $n/\log n$ processors for computing linked list prefix is how to balance load among processors. Because in the i-th stage we have a list of n/c_1^i nodes for a constant $c_1 > 1$, the number of operations needed for all stages form a geometric series. Therefore the sum of the number of operations is $O(n)$, assuming the load (the number of nodes assigned to each processor) at each stage can be balanced among $n/\log n$ processors.

$O(\log^{(2)} n)$ balance operations are needed for that many stages of contraction of the linked list in order to reduce the size of the input list from n to $n/\log n$, at that point Wyllie's pointer jumping algorithm[W] can be invoked to finish the rest of the contraction process. A deterministic global balance step requires $O(\log n)$ time on the EREW model, thus any algorithm using more than constant number of global balance steps will fail to achieve time $O(\log n)$. This barrier of global balancing can be overcome by the following techniques.

2.5. Randomized Load Balancing

Suppose there are k 1's in an array of n cells. These k 1's can be packed into $2a$ cells by randomly assign k 1's to $2k$ cells. Reif[Reif], Miller and Reif[MR] showed that elaborations of this idea can be used to balance load among processors. Such a load balance step can be done in sublogarithmic time if k processors are available. The following theorem is from [MR], proving it requires probabilistic analysis of random strings.

Theorem 8[MR]: There exist a PRAM algorithm using $O(\log^{(2)} n)$ time and $O(n/\log^{(2)} n)$ processors which for at least $1 - 1/n$ of strings with b zeros discards at least $b/2$ zeros. □

Load balancing algorithm obtained from Theorem 8 is powerful enough to yield an optimal algorithm for linked list prefix with time complexity $O(\log n)$ [MR].

2.6. Global Load Balancing

Although global load balancing can not succeed on the EREW model as we explained above, it works on the CRCW model for achieving optimal linked list prefix algorithm with time complexity $O(\log n)$. The idea originates from Reif's sublogarithmic partial sum algorithm [Reif] with time complexity $O(\log n/\log^{(3)} n)$. Han showed[H1][H2] that Reif's algorithm can be used for globally balancing the load with the same time complexity and that such a global balancing algorithm is sufficient for obtaining a linked list prefix algorithm with time complexity $O(\log n)$ using $n/\log n$ processors.

Parberry[P] and Cole and Vishkin[CV] showed $O(\log n/\log^{(2)} n)$ parallel summation algorithms. Their algorithms will enable faster load balancing.

The above two load balancing techniques are not powerful enough to yield a deterministic EREW linked list prefix algorithm with time complexity $O(\log n)$ using $O(n/\log n)$ processors.

2.7. Dynamic Load Balancing

We can formulate the load balancing problem as follows to understand the key to the solution of the problem. The following formulation is an attempt to offer an intuitive explanation of Anderson and Miller's technique[AM1][AM2].

Initially each processor has a queue of $\log n$ data item[AM1][AM2]. In one step each processor works on the data item at the top of the queue. The data items with processors working on are called the active items. We will use an oracle which provides a pairing of the active items in constant time. Such an oracle is a resemblance and conceptual simplification of the matching algorithms we have discussed above. When two data items a, b are to be paired off as indicated by the oracle, the new data item $c = a \bigcirc b$ can be stored at either the cell where a was stored or the cell where b was stored.

Anderson and Miller's technique points out that c should be stored in the queue with less items, *i.e.*, c should be stored in the cell occupied by a if the queue where a was stored is shorter than the queue where b was stored, otherwise c should be stored at the cell occupied by b. By assigning weight 2^i to item at the i-th position from the end of the queue, one can conclude that in one step the total weight of the data items is reduced by 3/4. Thus after $O(\log n)$ steps the total weight will be reduced to a constant which implies that the linked list has been contracted into a single node.

The above scheme needs to be modified in order to replace the assumed oracle by deterministic matching partition algorithms to obtain an $O(\log n)$ time and $O(n/\log n)$ processor EREW PRAM algorithm for linked list prefix[AM1][AM2].

We note that the crucial information used in Anderson and Miller's technique is the information provided by the pointers which dictates as to where the nodes resulting from pairing-off should be stored. The load balancing could be difficult if the information provided by the pointers is not available. For example, we may assume that the oracle, when invoked, will randomly eliminate half of the active nodes. Such an oracle does not provide the information implied by the pointers. Consequently, Anderson and Miller's technique can not be used for balancing load for such an oracle to achieve time $O(\log n)$ using $O(n/\log n)$ processors. In fact any scheme which balances for the modified oracle must be stronger than Anderson and Miller's technique[AM1][AM2]

From the view of parallel algorithms, computing a maximal independent set of nodes for a linked list is equivalent to computing a maximal matching set of pointers for the list. Treat the problem as computing a maximal matching set has several advantages: it has the intuition of bisecting pointers for the matching partition function $f^{(2)}$ and it has the advantage of discriminating between inter-row and intra-row pointers, and in the case of Anderson and Miller's technique of dynamic load balancing pointers provide vital information as to where load should be moved. These advantages are important in helping us understand the insights behind these algorithms.

2.8. Further Development

Recent results reveals further development of linked list prefix algorithms on weak models. Kruskal *et al.* [KMR], Anderson and Miller[AM1], Han[H4] designed parallel algorithms for linked list which uses $o(n)$ shared memory cells. In particular, it is known[H4] that there exists a deterministic PRAM algorithm for computing linked list prefix with time complexity $O(n/p + \log n)$ using p shared memory cells. Ryu and JáJá [RJ]] developed an optimal linked list prefix algorithm on the hypercube parallel computer.

3. Applications to Tree Problems

There are great many algorithms designed by applying linked list algorithms. In this section we apply linked list algorithms to some tree problems.

3.1. Maximal Independent Set

We first consider the problem of computing a maximal independent set on a rooted tree. Computing a maximal independent set for a rooted tree is equivalent to computing a maximal matching set for the corresponding hypertree $(V, \mathcal{E})$, where V is the set of nodes and $\mathcal{E}$ is a set of pairs (v, W) called pointers, $v \in V$, $W \subseteq V$. Equivalent in the sense that they are reducible to each other on a PRAM in constant time. The problem of computing a maximal matching set for a rooted tree is also intimately related to the problem of computing a maximal independent set.

Computing a maximal independent set for a rooted tree has been investigated by Goldberg *et al.*[GPS] and by Jung and Mehlhorn[JM]. Algorithms given in [GPS] are not optimal. Their technique[GPS] enumerates the independent sets, thus at one step only processors assigned to one independent set are working while other processors are idling. Jung and Mehlhorn gave an optimal algorithm using up to $O(n/\log n)$ processors. We show how to achieve the curve $O(\frac{n\log i}{p} + \log^{(i)} n + \log i)$ for computing a maximal independent set for a rooted tree.

Each node v in the tree can be labeled with number $l(v) = f^{(i)}(v, p(v), p(p(v)), ...)$, where $p(v)$ is the parent of v. This is essentially the same as we did for a linked list. These labels are called l labels. Each node can also be labeled with number $r(v) = row(v)$, where $row(v)$ is the row where node v is stored when the tree is viewed as being stored in a two dimensional array. After the nodes in the tree are labeled turn points[H1][H2][H4] can be identified. For two pointers $< v_1, v_2 >$, $< v_2, v_3 >$, v_2 is called a turn point for v_1 if either of the following is true.

(1). $l(v_1) > l(v_2)$ and $l(v_2) < l(v_3)$.

(2). $r(v_1) > r(v_2)$ and $r(v_2) < r(v_3)$.

(3). $r(v_1) \neq r(v_2)$ and $r(v_2) = r(v_3)$.

In order to obtain a maximal independent set we execute the following steps for an input rooted tree.

(1). Label nodes v with $l(v) = f^{(i)}(v, p(v), p(p(v)), ...)$. They are called the l labels. l is bounded by $m = \log^{(i-1)} n(1 + o(1))$.

(2). Delete v if $p(v)$ is a turn point for v by its l label.

(3). Relabel v. If $l(p(v)) > l(v)$ then $l(v) = 2m - l(v)$. After relabeling the l labels for the nodes in a tree are strictly increasing from root to leaves.

(4). View the tree as being stored in a two dimensional array with x rows and $y = n/x$ columns, where x is bounded by $2m$. Sort the nodes by their l labels such that all nodes in one column are labeled by the same number. Use extra y columns if needed.

(5). Label nodes v with $r(v) = row(v)$.

(6). Delete v if $p(v)$ is a turn point for v by its r label.

(7). Delete root of the input tree.

What we intend to do is to create a forest by deleting a set of vertices such that the resulting forest has the following property. A root-to-leaf path $root = a_0, a_1, ..., a_t = leaf$ in a tree is stored in the form that the subpath $a_0, ..., a_i$ is stored ascending the rows, i.e., $row(a_k) < row(a_{k+1})$, subpath $a_i, ..., a_j$ is descending the rows, i.e., $row(a_k) > row(a_{k+1})$, and subpath $a_j, ..., a_t$ is stored in the same row. We let processors first walk up the rows, then walk down the rows. We then use processor scheduling (Section 2.4) to handle the subpath stored in the same row. During this process we may have to added some of the deleted leaves back to the maximal independent set.

Theorem 9: A maximal independent set of a rooted tree can be computed in $O(\frac{n\log i}{p} +$

$\log^{(i)} n + \log i$) time on a PRAM. □.

Pick i to be an arbitrarily large constant, our algorithm is optimal using up to $O(n/\log^{(i)} n)$ processors.

3.2. Optimal Communication Bandwidth

We now outline how to solve certain tree problems on a PRAM in time $O(n/p+\log n)$ using p shared memory cells. The number of shared memory cells used in a PRAM algorithm is regarded as the communication bandwidth of the algorithm. We show the design of parallel algorithms for some tree problems using optimal communication bandwidth.

We assume that the input tree is stored in a $n/p \times p$ array. Column k of the array is in the local memory of processor k.

It is known[KB][KRS][TV] that many tree problems can be reduced to the linked list prefix problem. We review these reduction techniques[KB][KRS][TV] and show how the reductions are carried on a PRAM with p shared memory cells without losing efficiency (after ignoring multiplicative constant). For these problems the input trees are assumed to be double linked between parents and children.

(1). *Number of descendants in a tree.*

Problem: Label the nodes of a tree with the number of their descendants.

Solution: Convert the tree to a linked list(Fig. 5), label each link in the list with 1. Solve the linked list prefix sum problem. Let $< v, w >$ be the link connecting node v to node w, $n(< v, w >)$ be the result of the prefix sum on link $< v, w >$, and $v_l(v_r)$ be the leftmost(rightmost) son of v, then $\frac{n(<v_r,v>)-n(<v,v_l>)+1}{2}$ is the number of descendants of node v. Linked list prefix sum can be computed in $O(n/p + \log n)$ time [H4]. For the internal nodes, v_l and v_r can be computed in $O(n/p+\log n)$ time by using a straightforward array prefix algorithm.

Time Complexity: $O(n/p + \log n)$ on PRAM using p shared memory cells.

(2). *Preorder traversal labeling of a tree.*

Problem: Label the nodes of a tree with the number of preorder traversal labeling.

Solution: For each leaf v, compute $right(v)$. $right(v)$ is defined as follows. If v is the rightmost leave of the tree, then $right(v) = nil$. If v has an immediate right sibling w then $right(v) = w$, else $right(v) = right(parent(v))$. The definition of $right(v)$ is shown in Fig. 6.

$right(v)$ can be computed using linked list prefix algorithm[H4]. Therefore, we obtain $right(v)$ for all leaves in time $O(n/p + \log n)$.

Now we construct $next(v)$. For each internal node v, $next(v)$ is the the leftmost child of v. For a leaf v, $next(v)$ is $right(v)$. These pointers of $next(v)$ form a linked list. Label every node with 1 and then do a linked list prefix sum[H4]. The resulting numbering is preorder traversal numbering.

Time Complexity: $O(n/p + \log n)$ on PRAM using p shared cells.

Inorder traversal labeling and postorder traversal labeling and many other tree problems can be solved using the same approach.

Certain tree problems are solved more naturally by the technique of tree contraction[MR]. In the rest of this section we show how to adapt tree contraction techniques[MR] to PRAM using only p shared memory cells.

A typical problem in this category is the parallel tree expression evaluation problem[MR]. An expression tree is a tree with leaves labeled with operands and each internal node labeled with constant number of operators and operands. Most expressions involve only unary and binary operators. Corresponding expression trees are binary trees with chains in the tree representing a chain of unary operators.

When a binary operator is associative, we could extend it to be an n-ary operator. Nodes in an expression trees correspond to such operators may have several children.

Parallel tree contraction technique also applies to tree problems with input trees having no links from parents to children. Such input trees are difficult to be processed directly by a linked list prefix algorithm due to the difficulty of constructing lists. To design deterministic algorithms for certain tree problems with such input trees we can apply tree contraction techniques[MR].

Our tree contraction algorithm has two stages. The first stage contracts the input tree from size n to size p. The second stage contracts the size p tree to its root. Notice that in the second stage the number of tree nodes is no more than the number of processors.

We can use a deterministic PRAM tree algorithm for the second stage of our tree contraction algorithm. The PRAM tree contraction algorithm of Miller and Reif[MR] can be used here which takes $O(\log p) \le O(\log n)$ time since the tree has p nodes and the machine has p processors. In the following we only consider the first stage of our tree contraction algorithm.

In order to contract trees properly a node in a tree needs the information when it will become a chain node. Such information is readily obtained if the node has only constant number of children, for its children can inform the node in constant time. When there is a node the number of its children is not a constant, the following strategies can be used.

(a). *The input tree is double linked.*

After each step of RAKE and COMPRESS[MR], perform a packing operation on the number of children for each parent node. The result of this operation tells, for each parent node, how many children have not been contracted to a single node yet. Thus it takes $O(\frac{n}{p} + \frac{\log n}{\log^{(2)} n})$ time to inform the parents. Since the RAKE and COMPRESS operation will reduce the number of nodes in a tree by a constant fraction, $O(\log(n/p))$ executions of RAKE and COMPRESS suffice for the first stage of our algorithm. The total time needed for informing the parents in the first stage of our algorithm is $O(\frac{n}{p} + \frac{\log n}{\log^{(2)} n} \log(n/p)) \le O(n/p + \log n)$ because the number of nodes in the tree being contracted form a geometric series.

(b). *The input tree has no links from parents to children.*

In this case we route the children to their parents and by using concurrent write the parents are informed whether there are children left. Here the routing operation is not a permutation, therefore our permutation algorithm[H4] can not be directly used. Instead we first sort the nodes by the first $\lceil \log(n/p) \rceil$ bits of the addresses in the parent pointer. In doing this we have routed the children into blocks. Here, a block is a continuous address locations containing children whose parents were in the same row before sorting. A block could occupy several rows in the memory and one row could contain several blocks as shown in Fig. 7. If we impose the condition that one row contain one block only, then the number of rows needed is bounded by $\sum_i^{n/p} \lceil \frac{B_i}{p} \rceil \le 2n/p$, where B_i is the number of elements in block i. After sorting we then use concurrent write to inform the parents. The concurrent write is performed one row at a time. Since elements in a row are in the same block, each child sends information via concurrent write to its parent.

Because sorting takes only $O(\frac{n}{p} + \frac{\log n}{\log^{(2)} n})$ time. We obtain the same time complexity as that in (a).

The first stage of our tree contraction algorithm has two steps. In the first step the input tree of size n is contracted to a tree of size $n/\log^{(2)} n$. In the second step the tree of size $n/\log^{(2)} n$ is contracted to a tree of size p. We present the details of these two steps below.

Step 1:

Step 1 is a loop with $\log^{(3)} n/\log c$ iterations. Each iteration reduces the tree from size s to s/c, where $c > 1$ is a constant.

We use schemes outlined above to inform each tree node whether it has children left. A node with no child is a leaf. These leaves then delete themselves, therefore realizing the RAKE operation. These operations take $O(\frac{n}{p} + \frac{\log n}{\log^{(2)} n})$ time for a tree of size n.

A node with one child left is a chain node. After identifying chains in the tree, we compute $f^{(4)}$ for chain nodes and label each chain node v with $l(v) = f^{(4)}(v, ...)$. The computation yields a partition of $c_1 \log^{(3)} n$ matching sets (labeled from 0 to $c_1 \log^{(3)} n - 1$), where c_1 is a constant. We then pack these chain nodes and sort them by 3-tuples $< l(v), l(child(v)), l(parent(v)) >$, where $l(child(v)) = -1$ if v has no child as a chain node and $l(parent(v)) = c_1 \log^{(3)} n$ if v has no parent as a chain node (Fig. 8). After sorting we then let processors walk up these chains. To walk up from nodes with $l(v) = i$ to their parent nodes with $l(v) = j$, $j > i$, we pack source nodes with 3-tuples $< i, s, j >$, $s = -1, 0, 1, ..., i - 1$ together and destination nodes with 3-tuple $< j, i, t >$, $t = j + 1, j+2, ...$ together. A permutation then routes source nodes to their parents (destination nodes). Let n_{ij} be the number of nodes with 3-tuple $< i, s, j >$ or $< j, i, t >$. The time needed for walking up from $l(v) = i$ to $l(v) = j$ takes $O(\frac{n_{ij}}{p} + \frac{\log n}{\log^{(2)} n})$ time. Sum for all possible values of ij the time becomes $O(\sum_{ij} \frac{n_{ij}}{p} + \frac{\log n(\log^{(3)} n)^2}{\log^{(2)} n})$. Because a node with 3-tuple $< i, j, k >$ is counted at most twice, once in n_{ik}, once in n_{ji}, we know $\sum_{ij} n_{ij} \leq 2n$. So the timing for one iteration of the loop of step 1 is $O(\frac{n}{p} + \frac{\log n(\log^{(3)} n)^2}{\log^{(2)} n})$, if the size of the tree is n.

Since each iteration of the loop of step 1 reduces the size of the tree to a constant fraction, therefore the time complexity of step 1 is $O(\sum_{i=1}^{O(\log^{(3)} n)} (\frac{n}{c^i p} + \frac{\log n(\log^{(3)} n)^2}{\log^{(2)} n})) \leq O(n/p + \log n)$.

Step 2:

Step 2 is a loop with $\frac{\log(n/p)}{\log c}$ iterations. Each iteration reduces the tree from size s to s/c, where $c > 1$ is a constant.

We use the same scheme to inform each tree node whether it has children left. Then RAKE operation is performed in the same way as that in step 1.

Maximal matchings are found for the chain nodes. This requires $O(\frac{sG(s)}{p} + \frac{\log s}{\log^{(2)} s})$ for chains of size s. However, since step 1 has reduced the tree to size $n/\log^{(2)} n$, this can not lead our algorithm to a nonoptimal one. A permutation is then needed to compress the chains and a pack operation packs the tree so that less time would needed for the later iterations.

The time complexity of step 2 is $O(\frac{sG(s)}{p} + \frac{\log s}{\log^{(2)} s} \log(n/p))$. Notice that, since the size of the contracted trees form a geometric series, the term $sG(s)/p$ remains unchanged. Because $s \leq n/\log^{(2)} n$, the time complexity is $O(n/p + \log n)$.

After these two steps the size of the tree has been reduced to p. This accomplishes the first stage of our algorithm.

Theorem 10: Parallel tree contraction can be done in time $O(n/p + \log n)$ on PRAM using p shared memory cells. $\square$

The essence of the parallel tree algorithms presented in this paper is the technique of list and tree contractions. Depending on the operations defined on the elements associated with list or tree nodes, many different list and tree functions can be computed efficiently on PRAM using p shared cells. It is not difficult to construct a function which requires only $o(p)$ shared cells to achieve time $O(n/p + \log n)$. We show that $\Omega(p)$ shared cells are required for our tree algorithms to achieve time $O(n/p + \log n)$ by demonstrating that there exists a tree function which requires that many shared cells. Our proof requires that each shared cell has only $O(\log n)$ bits.

Theorem 11: There is a tree function which requires $\Omega(p)$ shared cells to be computed in time $O(n/p + \log n)$.

Proof: We use the well known technique of crossing sequence[U] to prove the lower bound. In t steps the sequence appeared in the p shared cells has $O(tp \log n)$ bits.

Suppose we are to compute the number of descendants for all tree nodes. The input tree is of the following form. The root has $n/4$ children and they are in the local memory of $n/4p$ processors. These $n/4$ tree nodes are said to be in group A. Each of the $n/4$ nodes has exactly one child. The rest $n/2 - 1$ tree nodes are grandchildren of nodes in group A and they are leaves. The nodes in group A compute the number of descendants of theirs by receiving information from shared cells. However, there are $\binom{n/2-1}{n/4}$ ways of attaching these leaves to their parents. Thus $\Theta(n \log n)$ bits are needed to distinguish between different ways of attachment. If $t = o(n/p)$ then the crossing sequence has $o(n \log n)$ bits which is not enough to distinguish between different ways of attachment. $\square$

4. Conclusions

We have presented basic techniques for the design of linked list prefix algorithms and applications of these algorithms. Due to page limit we are unable to explore many important applications and ramifications. We believe that both the techniques and the algorithms for linked list are fundamental and they will be indispensable building blocks for the design of fast and efficient parallel algorithms for a large variety of problems.

References

[AM1]. R. J. Anderson, G. L. Miller. Optimal parallel algorithms for the list ranking problem, USC-Tech. Rept. 1986.

[AM2]. R. J. Anderson, G. L. Miller. Deterministic parallel list ranking. LNCS 319 (J. H. Reif ed.), 81-90.

[BH]. R. A. Borodin and J. E. Hopcroft. Routing, merging and sorting on parallel models of computation. Proc. 14th ACM Symposium on Theory of Computing, 206-219(1986).

[CDHR]. B. S. Chlebus, K. Diks, T. Hagerup, T. Radzik. New simulations between CRCW PRAMs. LNCS 380, 95-104.

[CV]. R. Cole and U. Vishkin. Approximate and exact parallel scheduling with applications to list, tree and graph problems, Proc. 27th Symp. on Foundations of Comput.

Sci., IEEE, 478-491(1986).

[GPS]. A. V. Goldberg, S. A. Plotkin, G. E. Shannon. Parallel symmetry-breaking in sparse graphs, SIAM J. on Discrete Math., Vol. 1, No. 4, 447-471(Nov., 1988).

[H1]. Y. Han. Designing fast and efficient parallel algorithms. Ph.D. Thesis, Duke University, Durham, NC 27706, 1987.

[H2]. Y. Han. Parallel algorithms for computing linked list prefix. J. of Parallel and Distributed Computing 6, 537-557(1989).

[H3]. Y. Han. Matching partition a linked list and its optimization. Proc. ACM Symp. on Parallel Algorithms and Architectures, Sante Fe, New Mexico, 246-253(1989).

[H4]. Y. Han. An optimal linked list prefix algorithm on a local memory computer. Proc. 1989 ACM Computer Science Conf., Lousville, Kentucky, 278-286(1989).

[H5]. Y. Han. On the chromatic number of Shuffle graphs. TR. No. 140-89, Department of Computer Science, University of Kentucky, Lexington, KY 40506, USA.

[JM]. H. Jung, K. Mehlhorn. Parallel algorithms for computing maximal independent sets in trees and for updating minimum spanning trees, Information Processing Letters, Vol. 27, No. 5, 227-236(1988).

[KB]. N. C. Kalra, P. C. P. Bhatt. Parallel algorithms for tree traversals, Parallel Computing, Vol. 2, No. 2, June 1985.

[KMR]. C. P. Kruskal, T. Madej, L. Rudolph. Parallel prefix on fully connected direct connection machine. Proc. 1986 Int. Conf. on Parallel Processing, 278-284.

[KRS]. C. P. Kruskal, L. Rudolph and M. Snir. Efficient parallel algorithms for graph problems. Proc. 1986 International Conf. on Parallel Processing, 869-876.

[L]. N. Linial. Distributive graph algorithms — global solutions from local data. Proc. 1987 IEEE Symposium on Foundations of Computer Science, 331-336(1987).

[MR]. G. L. Miller, J. H. Reif. Parallel tree contraction and its application. Proc. 1985 IEEE Foundations of Computer Science, 478-489.

[P]. I. Parberry. On the time required to sum n semigroup elements on a parallel machine with simultaneous writes. LNCS, 227, 296-304.

[Reif]. J. H. Reif. An optimal parallel algorithm for integer sorting, Proc. 26th Symp. on Foundations of Computer Sci., IEEE, 291-298(1985).

[RJ]. K. W. Ryu, J. JáJá. List ranking on the hypercube. Proc. 1989 Int. Conf. on Parallel Processing. St. Charles, Illinois, (Aug. 1989).

[Sn]. M. Snir. On parallel searching. SIAM J. Comput., Vol. 14, No. 3, 688-708(Aug., 1985).

[Sp]. E. Sperner. Ein satz über untermenger endlichen menge, Math. Z. 27(1928), 544-548.

[TV]. R. E. Tarjan, U. Vishkin. Finding biconnected components and computing tree functions in logarithmic time, 25th Symp. on Foundations of Computer Sci., IEEE, 12-20.

[U]. J. D. Ullman. Computational aspects of VLSI, Computer Science Press, 1984.

[WH]. R. A. Wagner, Y. Han. Parallel algorithms for bucket sorting the data dependent prefix problem. Proc. 1986 Int. Conf. on Parallel Processing, 924-930.

[W]. J. C. Wyllie. The complexity of parallel computation. TR 79-387, Department of Computer Science, Cornell University, Ithaca, NY, 1979.

8
7
6
5
4
3
2
1
8
7
6
5
4
3
2
1
8
5
4
7
6
1
3
2
8
1
7
2
6
3
4
5

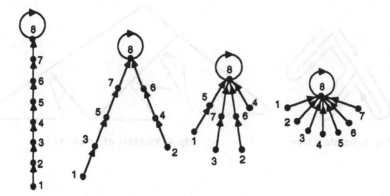

Fig. 1. The effect of pointer jumping

	0	1	2	3	4	5	6
X	x_0	x_2	x_4	x_1	x_5	x_3	x_6
NEXT	3	5	4	1	6	2	*nil*

Fig. 2. A linked list

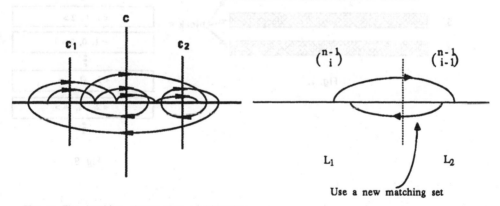

c
c1
c2

Fig. 3. The intuitive observation of bisecting

$\binom{n-1}{i}$
$\binom{n-1}{i-1}$
L1
L2
Use a new matching set

Fig. 4.

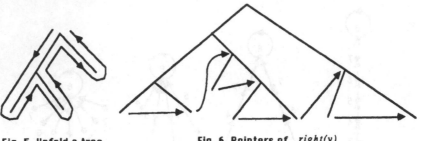

Fig. 5. Unfold a tree　　**Fig. 6. Pointers of** $right(v)$

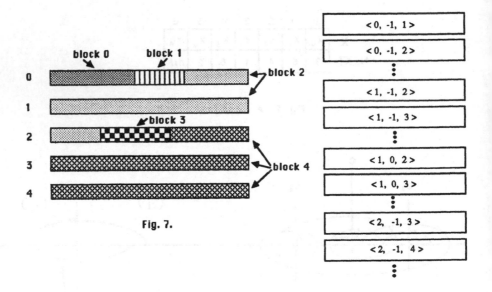

Fig. 7.

< 0, -1, 1 >

< 0, -1, 2 >

< 1, -1, 2 >

< 1, -1, 3 >

< 1, 0, 2 >

< 1, 0, 3 >

< 2, -1, 3 >

< 2, -1, 4 >

Fig. 8.

Local Tournaments and Proper Circular Arc Graphs*

Pavol HELL[†] Jørgen BANG-JENSEN[‡] and Jing HUANG[§]

Abstract

A *local tournament* is a digraph in which the out-set as well as the in-set of every vertex is a tournament. These digraphs have recently been found to share many desirable properties of tournaments. We illustrate this by giving $O(m + n \log n)$ algorithms to find a hamiltonian path and cycle in a local tournament. We mention several characterizations and recognition algorithms of graphs orientable as local tournaments. It turns out that they are precisely the graphs previously studied as proper circular arc graphs. Thus we obtain new recognition algorithms for proper circular arc graphs. We also give a more detailed structural characterization of chordal graphs that are orientable as local tournaments, i.e., that are proper circular arc graphs.

1 Local Tournaments

In [2], the second author introduced the class of locally semicomplete digraphs. For simplicity of this discussion we restrict the definition somewhat and define a *local tournament* to be an oriented digraph in which every vertex has any two out-neighbours adjacent and any two in-neighbours adjacent. In other words, the out-set as well as the in-set of every vertex is a tournament. (A digraph is *oriented* if the adjacency relation is antisymmetric. A *locally semicomplete digraph* is defined just like a local tournament, except it is not necessarily oriented, i.e., for some u, v, both arcs uv and vu may be present. Our main purpose here is with orienting undirected graphs, and it is easy to see that a graph may be oriented to be locally semicomplete if and only if it may be oriented to be a local tournament.)

There are many examples of local tournaments; they include of course all tournaments, but also all directed paths and cycles, their powers, and other classes of digraphs. The main observation of [2] has been that many nice properties of tournaments extend to this larger class of digraphs. For instance, the strong components of a local tournament can be uniquely ordered $C_1, C_2, \ldots, C_k$ so that each vertex of C_i dominates each vertex of

*Partially supported by grants from NSERC and ASI. Most of this paper was written while the first two authors were visiting the Laboratoire de Recherche en Informatique, Universite de Paris-Sud, whose hospitality is gratefully acknowledged.

†School of Computing Science, Simon Fraser University, Burnaby, B.C., Canada
‡Department of Computer Science, University of Copenhagen, Copenhagen, Denmark
§Dept. of Mathematics and Statistics, Simon Fraser University, Burnaby, B.C., Canada

C_{i+1} $(i = 1, \ldots, k-1)$ and no vertex of C_j dominates a vertex of C_i if $j > i$; moreover, if $k > 1$ then each C_i induces a tournament. On the other hand, if T is strong (i.e., if $k' = 1$) then either T is a directed cycle, or there exists a vertex u such that $T - u$ is also strong. Many similar results are discussed in [2]. In particular it is proved there that every connected local tournament has a hamiltonian path and every strong tournament has a hamiltonian cycle. (Thus a local tournament has a hamiltonian path if and only if it is connected and has a hamiltonian cycle if and only if it is strong.)

Here we present two algorithms for constructing a hamiltonian (or, if not possible, then longest) path and cycle in a local tournament. They are both based on the following lemma, showing an efficient way to insert one new vertex into an already constructed path. We measure the complexity of the next two algorithms in the number of tests performed; a test consists of determining whether, for two given vertices u, v, it is the case that u dominates v, v dominates u, or there is no arc joining u and v. Although this measure ignores other costs of concern in actual implementation of the algorithm, one can achieve the claimed complexity by storing the growing path in a suitable data structure allowing logarithmic access and insertion, such as a balanced tree, cf. [10], 6.2.3.

Lemma 1 *Let T be a local tournament, let $P = x_1, x_2, \ldots, x_r$ be a directed path of length r in T, and let x be a vertex of $T - P$ which is adjacent to a vertex x_i on P. With $O(\log r)$ tests we can find in T a directed path P' of length $r + 1$ containing the vertices $x, x_1, x_2, \ldots, x_r$.*

Proof: Suppose that x dominates x_i (the proof in the other case is analogous). Clearly, if $i = 1$ then we may let $P' = x, x_1, \ldots x_r$. Hence we may assume that $i \geq 2$. Note that either xx_{i-1} or $x_{i-1}x$ is an arc of T, because T is a local tournament. It is easy to see by the same reasoning that either there exists a j, $1 \leq j \leq i-1$, such that $x_j x$ and $x x_{j+1}$ are both arcs of T, or $x x_1$ is an arc of T (this case may be viewed as having $j = 0$). In the latter case P' is defined above; in the former case we let P' be the path $x_1, x_2, \ldots, x_j, x, x_{j+1}, \ldots, x_r$. Thus it remains to find the j efficiently. This can be done by binary search. The first test we will perform concerns the vertices x and $x_{\lceil \frac{i}{2} \rceil}$. If there is no arc between $x_{\lceil \frac{i}{2} \rceil}$ and x, or if $x_{\lceil \frac{i}{2} \rceil}$ dominates x, then j is greater than or equal to $\lceil \frac{i}{2} \rceil$. Otherwise j is smaller than $\lceil \frac{i}{2} \rceil$. Thus continuing this way we find the j with at most $O(\log r)$ tests. $\square$

Theorem 1 *Suppose the vertices of the local tournament T are given in an order $v_1, v_2, \ldots, v_n$, together with a sequence $s(2), s(3), \ldots, s(n)$ of subscripts such that each $s(i) < i$ and v_i has an arc to or from $v_{s(i)}$. Then a (directed) hamiltonian path in T can be found with $O(n \log n)$ tests.* $\square$

For practical implementation of this algorithm we must also consider the time required to obtain the ordering $v_1, \ldots, v_n$ and the corresponding sequence $s(2), \ldots, s(n)$. In general, it is not hard to see how to obtain both from a breadth first search of the underlying graph. Thus if the graph is given in a form suitable to perform a test (defined above) in time $O(1)$ and to obtain all neighbours of v in time $\deg v$ (such as some combination of adjacency matrix and adjacency lists), the above algorithm has time complexity $O(m + n \log n)$. (Here m is the number of arcs of T and $O(m)$ the time for breadth first search.) Note that if we have available an n by n matrix of zeros, the adjacency matrix can be obtained from the adjacency lists in time $O(m)$.

If the local tournament is not connected, this will be revealed during the breadth first search and a longest path will be a hamiltonian path of a largest component. We summarize this as a corollary.

Corollary 1 *There is an $O(m + n \log n)$ algorithm to find a longest (directed) path in a local tournament.* □

Another way to obtain the ordering $v_1, \ldots, v_n$ and the sequence $s(2), \ldots, s(n)$, would be to find a hamiltonian path of the underlying undirected graph. In corollary 5 we describe a situation when such a path can be found in time $O(n \log n)$, hence yielding an $O(n \log n)$ algorithm for finding a directed hamiltonian path in a local tournament.

Theorem 2 *There is an $O(m + n \log n)$ algorithm for finding a (directed) hamiltonian cycle in a strong local tournament T.*

Proof: One can use the above hamiltonian path algorithm to do this. We offer the following algorithm, which is a generalization of an algorithm for tournaments by Manoussakis [11].

Using no more than $O(m + n \log n)$ steps we find a directed hamiltonian path $x_1, x_2, \ldots, x_n$. Next we find x_i such that x_i dominates x_1 and i is as large as possible. If $i = n$ then $x_1, x_2, \ldots, x_n, x_1$ is the desired cycle. Otherwise let $C = x_1, x_2, \ldots, x_i, x_1$. Since $x_i x_{i+1}$ and $x_i x_1$ are arcs, either x_{i+1} dominates x_1, or x_1 dominates x_{i+1}. In the former case x_{i+1} may be inserted in the cycle C (between x_i and x_1). In the latter case we consider the arcs $x_1 x_{i+1}$ and $x_1 x_2$. We find that either x_{i+1} can be inserted in C to give a directed cycle of length $i + 1$, or C completely dominates x_{i+1}. Suppose C completely dominates x_{i+1}, and x_j, $j > i$, dominates a vertex x_t of C. Then we obtain a longer cycle $x_1, \ldots, x_{t-1}, x_{i+1}, x_{i+2}, \ldots, x_j, x_t, x_{t+1}, \ldots, x_i, x_1$. Continuing this way we always have a cycle $C' = x'_1, x'_2, \ldots, x'_r, x'_1$ and a hamiltonian path $x'_1, x'_2, \ldots, x'_n$. Since we have a strong digraph, it cannot happen that none of the vertices $x'_{r+1}, x'_{r+2}, \ldots, x'_n$ dominates a vertex of C. Thus this process terminates with a hamiltonian cycle C'. Since we only consider each arc once in this part of the algorithm, we will have used no more than $O(m)$ new steps. □

Corollary 2 *There exists an $O(m + n \log n)$ algorithm for finding a longest directed cycle in a local tournament.*

Proof: One can find all strong components, find a hamiltonian cycle in each, and choose the longest of them, all in time $O(m + n \log n)$. □

Note that all three of the above $O(m + n \log n)$ bounds assume that T is given in the convenient form discussed prior to 2. Also note that finding hamiltonian paths in transitive tournaments amounts to sorting and can be done with $O(n \log n)$ tests. $O(n \log n)$ algorithms for finding hamiltonian paths in *arbitrary* tournaments are given in [8], and the above algorithm further extends this to local tournaments. We also remark that (path-) merging in local tournaments can be achieved in linear time.

2   Proper Circular Arc Graphs

Here we study the class of undirected graphs that can be oriented as local tournaments. The main characterization is due to Skrien [13]. Formulated in our terminology, it says that the class in question consists precisely of proper circular arc graphs. A *circular arc graph* is the intersection graph of a family of arcs of a circle, i.e., a graph is a circular arc graph just if there exists a family of circular arcs indexed by the vertices of the graph such that two vertices are adjacent if and only if the two corresponding arcs intersect. A circular arc graph is *proper* if the family of arcs can be chosen to be inclusion-free, i.e., one is never entirely contained in another. Circular arc graphs have been found useful in a number of applications from cyclic scheduling to compiler design and psychometrics (cf. [15], [14]).

In order to describe the other characterizations we define, for an undirected graph $G = (V, E)$, the set E^* to consist of all *ordered* pairs uv where uv is an edge of G. The graph G^* has vertex set E^* and the ordered pair $uv \in E^*$ is adjacent to the ordered pair $vu \in E^*$, to any ordered pair $uw \in E^*$ with vw not an edge of G, to any ordered pair $wv \in E^*$ with uw not an edge of G, and to no other vertex of G^*.

Theorem 3 *The following statements are equivalent*

1. *Graph G is orientable as a local tournament*

2. *G is a proper circular arc graph*

3. *G does not contain as an induced subgraph the disjoint union of a cycle and an isolated vertex, or the tent (see figure 2) and an isolated vertex, or the complement of any of the graphs in figure 1*

4. *G^* is bipartite*

5. *G does not contain a sequence of edges $ab = u_1v_1, u_2v_2, \ldots, u_kv_k = ba$ such that for each $i = 1, 2, \ldots, k-1$, $u_i = u_{i+1}$ and v_i is not adjacent to v_{i+1} in G, or $v_i = v_{i+1}$ and u_i is not adjacent to u_{i+1} in G.*

Proof: (Note that the edges u_iv_i in 5. are best viewed as elements of E^*. It follows for instance that the claw, cf. figure 2, is excluded by 5.) The equivalence of the second and third statement is the result of Tucker [16]. One can derive the equivalence of the first two statements from this. In any event, this was first done by Skrien [13]. Specifically, the intersection graph of an inclusion-free family of circular arcs can be oriented by directing the edge uv from u to v if the leftmost point of the interval indexed by u preceeds, in clockwise order, the leftmost point of the (intersecting) interval indexed by v. (It is possible to choose the arcs in such a way that no two share an endpoint or cover the entire circle, [5].) It is immediate that this orientation results in a local tournament. To prove the converse, that an underlying graph of a local tournament is a proper circular arc graph, it suffices to verify that none of the graphs in 3. of theorem 3 is orientable as a local tournament. This has been done by one of us [9], but we ommit the details since we consequently discovered the reference [13]. To prove the equivalence of 1. and 4., assume that G is oriented as a local tournament T, and define a bipartition of E^*

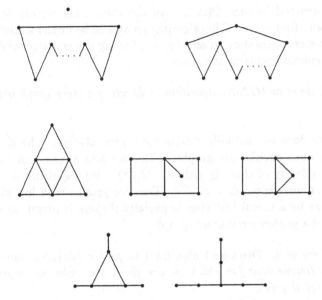

Figure 1: Subgraphs forbidden in the complement of G

into E_1, consisting of all pairs uv which are arcs of T, and E_2, consisting of all pairs uv such that vu is an arc of T. It is easy to deduce, from the definitions, that both E_1 and E_2 are independent sets of G^*. Conversely, if E_1, E_2 is a bipartition of E^* into two independent sets of G^*, then E_1 contains precisely one of the pairs uv, vu for any edge uv of G. Orienting each edge according to its representative in E_1 results in a local tournament. The equivalence of 4. and 5. is best seen if one defines a binary relation Ψ on E^* by $uv\Psi vw$ if $uv, vw \in E^*$ but $uw \notin E$, $uv\Psi zu$ if $zu, uv \in E^*$ but $zv \notin E$, $uv\Psi uv$ if $uv \in E^*$, and no other pairs being related. It is clear that $uv\Psi u'v'$ just if uv is adjacent to $v'u'$ in G^*. Thus an odd cycle in G^* is the same object as the sequence $ab = u_1v_1, u_2v_2, \ldots, u_kv_k = ba$ of 5. The transitive closure Ψ^* of Ψ is an equivalence relation, and 5. really says that no $ab\Psi^*ba$. □

It is easy to see that G has precisely two orientations as a local tournament, one converse of the other, if and only if G^* is connected, and if and only if Ψ^* has exactly two equivalence classes.

Using the above theorem we have several choices for a concise certificate of a graph being a proper circular arc graph, or orientable as a local tournament: we can use an orientation as a local tournament, or a representation as the intersection graph of an inclusion-free family of circular arcs. In the same vein, for a concise certificate of not being a proper interval graph, or orientable as a local tournament, we may take the graphs of Figure 1, or an odd cycle in G^*, or a sequence from 5. Finally, either 4. or 5. can serve as a method to recognize these graphs.

Corollary 3 *There is an $O(\Delta m)$ algorithm to decide if a given graph is orientable as a local tournament and to find such an orientation if it exists.*

Proof: The graph G^* has $O(m)$ vertices and $O(\sum_{uv \in E} degu + degv) = O(\Delta m)$ edges and can be constructed in time $O(\Delta m)$. In the same time we can test whether it is bipartite by breadth first search. This algorithm may also be implemented to work directly on G, by assigning an orientation to an edge uv, finding the equivalence class of uv under Ψ^*, etc. The complexity remains the same. □

Corollary 4 *There is an $(O\Delta m)$ algorithm to decide if a given graph is a proper circular arc graph.* □

Our algorithm does not actually produce a representation. The classical algorithm for recognizing proper circular arc graphs, [17], does find a representation if one exists. However, the complexity of that algorithm is $O(n^2)$. Our algorithm is more efficient for graphs with maximum degree $\Delta = o(\sqrt{n})$. Thus for graphs with low maximum degree, our algorithm may be a useful first step, especially if there is reason to suspect that the given graph is *not* a proper circular arc graph.

Corollary 5 *There is an $O(n \log n)$ algorithm to find a (directed) hamiltonian path in a connected local tournament for which we are given a circular arc representation of its underlying undirected graph.*

Proof: It follows from our earlier remarks that the underlying graph of a connected local tournament has an undirected hamiltonian path. Thus the algorithm of [4] will find one such path in time $O(n \log n)$ if the circular arc representation is given. The rest follows from the remarks following corollary 1. □

As a last consequence of our theorem 3 we note that the following problems can also be efficiently solved on local tournaments:

1. Maximum independent set - in time $O(n \log n)$, if a circular arc representation of the underlying graph is given, and in time $O(n + \Delta m)$ otherwise [6]

2. Maximum clique - in time $O(n^2 \log \log n)$, [1]

3. Chromatic number - in time $O(n^{3/2} \log n)$, if a representation is given, [14]

4. Minimum cover by cliques - in time $O(n^2)$, [7]

5. Minimum dominating set and the domatic number - in time $O(nm)$ and $O(n^2 \log n)$ respectively, [3].

3 Chordal Graphs

In this section we obtain a more detailed structural characterization of graphs orientable as strong tournaments, i.e., proper circular arc graphs, for the case of chordal graphs. Our starting point is the following result of Roberts ([12]):

Theorem 4 *An interval graph G has an inclusion-free representation if and only if G is claw-free.* □

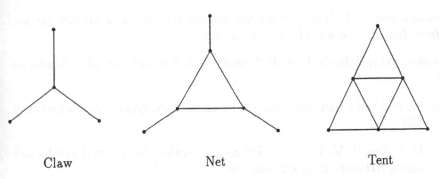

Figure 2: The claw, the net, and the tent

This theorem in effect characterizes those proper circular arc graphs, and graphs orientable as local tournaments, that are interval graphs. The only forbidden subgraph is the claw, $K_{1,3}$. We have obtained a similar characterization for chordal graphs:

Theorem 5 *A chordal graph is orientable as a local tournament, i.e., is a proper circular arc graph, if and only if it is claw-free and net-free.*

A graph is *net-free* if it does not contain the net, the six-point graph obtained from K_3 by adding a pendant edge at each vertex.

Proof: It is immediate that neither the claw nor the net are orientable as local tournaments. Thus any G containing one of them is not orientable as a local tournament. On the other hand, suppose G is a net-free and claw-free chordal graph. Wegner, cf. [5], page 195, proved that this implies that G is an interval graph or it contains the tent – the sixpoint graph obtained from K_3 by joining each edge to a new vertex. If G is an interval graph, then theorem 4 assures that it is a proper circular arc graph. Suppose therefore that G contains the tent as an induced subgraph. We have proved, by a detailed analysis to appear elsewhere, that in this case G must be obtained from the tent by replacing each vertex with a complete graph (and making completely adjacent any two complete graphs corresponding to adjacent vertices of the tent). The tent is easily seen to be orientable as a local tournament (orient all arcs clockwise). On the other hand, any digraph obtained from a local tournament by substituting a tournament at each vertex is also a local tournament. Thus it suffices to orient all edges of G within the complete subgraphs arbitrarily, and all edges between the complete subgraphs according to the local tournament orientation of the tent; the result is an orientation of G as a local tournament. □

References

[1] A. Apostolico and S. E. Hambrusch, Finding maximum cliques on circular arc graphs, *Information Processing Letters* 26 (1987) 209 - 215.

[2] J. Bang-Jensen, Locally semicomplete digraphs, *J. Graph Theory*, to appear.

[3] M. A. Bonuccelli, Dominating sets and domatic number of circular arc graphs, *Discrete Applied Math.* 12 (1985) 203 - 213.

[4] M. A. Bonuccelli and D. P. Bovet, Minimum disjoint path covering for circular arc graphs, *Information Processing Letters* 8 (1979) 159 - 161.

[5] M. C. Golumbic, **Algorithmic Graph Theory and Perfect Graphs**, Academic Press 1980.

[6] M. C. Golumbic and P. L. Hammer, Stability in circular arc graphs, *J. Algorithms* 9 (1988), 314 - 320.

[7] U. I. Gupta, D. T. Lee, J. Y.-T. Leung, Efficient algorithms for interval graphs and circular arc graphs, *Networks* 12 (1982) 459 -467.

[8] P. Hell and M. Rosenfeld, The complexity of finding generalized paths in tournaments, *J. Algorithms* 4 (1983) 303 - 309.

[9] J. Huang, A result on local tournaments, unpublished manuscript, 1989.

[10] D. Knuth, **Art of Computer Programming**, v.3, Sorting and Searching, Addison-Wesley, 1973.

[11] Y. Manoussakis, A linear algorithm for finding hamiltonian cycles in tournaments., Rapport de recherche 488, 1989, Laboratoire de Recherche en Informatique, Universite de Paris - Sud.

[12] F. S. Roberts, Indifference graphs, in **Proof techniques in Graph Theory** (F. Harary ed.), pp. 139 - 146; Academic Press, 1969.

[13] D. J. Skrien, A relationship between triangulated graphs, comparability graphs, proper interval graphs, proper circular arc graphs, and nested interval graphs, *J. Graph Theory* 6 (1982) 309-316.

[14] A. Teng and A. Tucker, An $O(qn)$ algorithm to q-color a proper family of circular arcs, *Discrete Math.* 55 (1985) 233 - 243.

[15] A. Tucker, Coloring a family of circular arcs, *SIAM J. Appl. Math.* 29 (1975) 493 - 502.

[16] A. Tucker, Structure theorems for some circular arc graphs, *Discrete Math.* 7 (1974) 167 -195.

[17] A. Tucker, Matrix characterizations of circular arc graphs, *Pacific J. Math.* 39 (1971) 535 - 545.

Fast Algorithms for the Dominating Set Problem
on Permutation Graphs

Kuo–Hui Tsai[1]   Wen–Lian Hsu[2]

[1] Department of Electrical Engineering and Computer Science, Northwestern University, Evanston, ILLINOIS 60208, U.S.A.

[2] Institute of Information Science, Academia Sinica, Taipei, Taiwan, R.O.C.

1. Introduction

Define a graph $G = (V, E)$ to be a *permutation graph* if there exists a permutation π of $\{1,2,3, \dots , |V|\}$ such that $(i,j) \in E$ if and only if $(i-j)(\pi^{-1}(i) - \pi^{-1}(j)) < 0$. It is possible [5] to determine whether a given graph G is a permutation graph, and if so, find the defining permutation π in $O(n^2)$ time, where $n = |V|$. In this paper, we assume that π is given.

A vertex i is said to *dominate* another vertex j if $(i,j) \in E$. A set S of vertices of a graph $G = (V,E)$ is a *dominating set* for G if every vertex in $V \backslash S$ is adjacent to some vertex in S. The problem of finding a *minimum dominating set* (MDS) is NP–hard in general. In this paper we consider the *minimum cardinality dominating set* (MCDS) and the *minimum weight dominating set* (MWDS) problems on permutation graphs. Based on dynamic programming, Farber and Keil [4] gave an $O(n^2)$ algorithm for the MCDS problem and an $O(n^3)$ algorithm for the MWDS problem assuming the defining permutation π is given. Another related problem of finding a minimum weight independent dominating set was solved in $O(n^2)$ time by Branstadt and Kratsch [3], and later solved in $O(n\log^2 n)$ time by Atallah, Manacher and Urrutia [1].

We present algorithms that solve the MCDS problem in $O(n\log\log n)$ time, and the MWDS problem in $O(n^2\log^2 n)$ time. Note that sorting is not necessary since π is given. Our improvement is based on the following three factors: (1) an observation on the monotone ordering among the intermediate terms in the dynamic programming, (2) a new construction formula for those intermediate terms, and (3) efficient data structures for manipulating these terms. We describe our algorithm for the cardinality case in Sections 2 and the weighted case in Section 3.

2. An $O(n\log\log n)$ algorithm for the cardinality case

Consider a permutation graph G on n vertices defined by a permutation π. For each i, j $\in$ {1,2,3, ... n}, define $V_i = \{\pi(1),\pi(2),\pi(3), ... ,\pi(i)\}$, and $V_{ij} = V_i \cap \{1,2,...,j\}$. Let S be any subset of {1,2,3, ... ,n}. Define $\max(S) = \max \{ i \mid i \in S \}$. For each i and j, Farber and Keil construct a set D_{ij} satisfying that:

(1) D_{ij} is a minimum cardinality subset of V_i dominating V_{ij}

(2) $\max(D_{ij})$ is as large as possible

Define $d_{ij} = |D_{ij}|$ and $m_{ij} = \max(D_{ij})$. Instead of dealing with the actual D_{ij} sets, Farber and Keil associate with each i and j a pair (d_{ij},m_{ij}) and update the pair in each iteration. At the end, an optimal D_{nn} can be obtained easily by backtracking. Since the calculation of each (d_{ij},m_{ij}) takes constant time. The complexity of the algorithm is $O(n^2)$. To calculate (d_{ij},m_{ij}) with i fixed and j = 1, ..., n, one only need to know the pairs $(d_{i-1,j},m_{i-1,j})$ for j = 1, ..., n. Hence, $O(n)$ space is sufficient.

In Farber and Keil's algorithm, each D_{ij} must come from one of the following three formulas: (F1) = $D_{i-1,j}$; (F2) = $D_{i-1,j} \cup \{\max(V_i)\}$; and (F3) = $D_{i-1,\pi(i)} \cup \{\pi(i)\}$. We shall rewrite the formula in terms of the changes in the (d_{ij},m_{ij})–values. Consider the change from the (i–1)–th column to the i–th column. For convenience, let $y = \pi(i)$, $\overline{d} = d_{i-1,y}$, $\overline{m} = m_{i-1,y}$, and $\overline{k} = \max(V_i)$. Then the updating rules can be rewritten as follows:

	Formula	(d,m)
(1) $\overline{m} > y$		
(1.1) $d_{i-1,j} > \overline{d} + 1$	(F3)	$(\overline{d}+1,\overline{m})$
(1.2) $d_{i-1,j} = \overline{d} + 1$		
(1.2.1) $m_{i-1,j} \leq \overline{m}$	(F3)	$(\overline{d}+1,\overline{m})$
(1.2.2) $m_{i-1,j} > \overline{m}$	(F1)	$(d_{i-1,j},m_{i-1,j})$
(1.3) $d_{i-1,j} = \overline{d}$		
(1.3.1) $m_{i-1,j} < y$	(F2)	$(\overline{d}+1,\overline{k})$
(1.3.2) $m_{i-1,j} \geq y$	(F1)	$(d_{i-1,j},m_{i-1,j})$
(1.4) $d_{i-1,j} < \overline{d}$	(F1)	$(d_{i-1,j},m_{i-1,j})$
(2) $\overline{m} \leq y$		
(2.1) $d_{i-1,j} > \overline{d} + 1$	(F3)	$(\overline{d}+1,y)$
(2.2) $d_{i-1,j} = \overline{d} + 1$		
(2.2.1) $m_{i-1,j} \leq y$	(F3)	$(\overline{d}+1,y)$
(2.2.2) $m_{i-1,j} > y$	(F1)	$(d_{i-1,j},m_{i-1,j})$
(2.3) $d_{i-1,j} = \overline{d}$		
(2.3.1) $j \geq y$	(F2)	$(\overline{d}+1,\overline{k})$

$$(2.3.2) \quad j < y \qquad \text{(F1)} \qquad (d_{i-1,j}, m_{i-1,j})$$
$$(2.4) \ d_{i-1,j} < \bar{d} \qquad \text{(F1)} \qquad (d_{i-1,j}, m_{i-1,j})$$

A pictorial description of the block changes (in case (1)) is shown in Figure 1.

Let M be the matrix whose (i,j)–th entry is (d_{ij}, m_{ij}). Farber and Keil's algorithm calculates all entries of M in a column by column fashion. Hence, $\Omega(n^2)$ time is necessary. Our algorithm keeps track of the (d,m)–pairs of each column in an efficient priority queue and updates them form one column to the next through simple insertions and deletions in the data structure. The following two lemmas provide important monotone properties of the matrix M.

Lemma 2.1. *For each i, if $k < j$, then $d_{ik} \leq d_{ij}$.*

Lemma 2.2. *For each i, if $k < j$ and $d_{ik} = d_{ij}$, then $m_{ik} \geq m_{ij}$.*

From Lemmas 2.1 and 2.2, for a fixed i, all (i,j)–entries with the same (d,m)–values must be contiguous. For a fixed column i, define a *d–block* (resp. *(d,m)–block*) to be a maximal consecutive (i,j)–entries in column i with the same d–value (respectively, (d,m)–value). Clearly, each (d,m)–block is contained in a d–block. To describe the change of entries in M, it suffices for us to describe the change of these block structures from column to column.

For each (d,m)–block in the i–th column, define the entry with the least j to be the *marker* of this block. Hence, changing the block structure from the (i−1)–th column to the i–th column is equivalent to changing the markers in the (i−1)–th iteration to the corresponding markers in the i–th iteration in the priority queue. It is relatively easy to describe an O($n\log n$) algorithm using an array to store all entries of a column. However, to obtain an O($n\log\log n$) algorithm, we need two data structures to store the markers of a column: a linked list and an efficient priority queue (as described in van Emde Boas [6]). In the latter data structure, with O(n) preprocessing time, one can execute each of the following operations in O($\log\log n$) time: insert(j), delete(j), predecessor(j), successor(j).

A two–level linked list is used to store the markers and their (d,m)–values. For each d value, we have a node representing the marker with value (d,m), where m is the smallest possible. These nodes are then linked together according to their descending d–values. This constitutes the first level linked list. In the second level list, all markers with the same d values are linked together according to their ascending m values.

In addition, we keep a priority queue for the markers that speeds up an important search operation in our algorithm. To split a block, we insert a new marker into the queue.

To merge two blocks, we delete a marker from the queue. We will show that the total number of insertion and deletion operations on the priority queue can be bounded by $O(n)$.

Algorithm /* Updating the (d,m)–values from the (i–1)–th column to the i–th column*/

1. Let $y = \pi(i)$. Find the (d,m)–block containing the (i,y)–entry by finding its marker in the queue through an insertion. Let $\bar{d} = d_{i-1,y}$, $\bar{m} = m_{i-1,y}$; $\bar{k} = \max(V_i)$;

2. If $\bar{m} > y$ then perform steps 3 and 4 (see Figure 1), otherwise perform steps 5 and 6.

3. [**divide the $\bar{d}+1$ group when $\bar{m} > y$**] Find the marker p with value ($\bar{d}+1$,m'), where m' ($\leq \bar{m}$) is as large as possible through a linear search on the list. Then merge all the (d,m)–blocks above (and including) the block of p into one new block with marker p by deleting all markers larger than p in the list and the queue. Associate with p the value ($\bar{d}+1,\bar{m}$) (case 1.1 & 1.2.1).

4. [**divide the $\bar{d}$ group when $\bar{m} > y$**] Find the marker p with value ($\bar{d}$,m'), where m' ($<$ y) is as large as possible through a linear search on the list. Then, within the $\bar{d}$–block, merge all the (d,m)–blocks above (and including) the block of p into one new block with marker p by deleting their corresponding markers in the list and the queue. Associate with p the value ($\bar{d}+1,\bar{k}$) (case 1.3.1).

5. [**divide the $\bar{d}+1$ group when $\bar{m} \leq y$**] Find the marker p with value ($\bar{d}+1$,m'), where m' ($\leq$ y) is as large as possible through a linear search on the list. Then merge all the (d,m)–blocks above (and including) the block of p into one new block with marker p by deleting all markers larger than p in the list and the queue. Associate with p the value ($\bar{d}+1$,y) (case 2.1 & 2.2.1).

6. [**divide the $\bar{d}$ group when $\bar{m} \leq y$**] Let the marker of the ($\bar{d},\bar{m}$)–block be p. Split the ($\bar{d},\bar{m}$)–block into two blocks by inserting y (a new marker) into the queue (keep the marker p). Then, within the $\bar{d}$–block, merge all the (d,m)–blocks above (and excluding) the block of y into one new block with marker y by deleting their corresponding markers in the list and the queue. Associate with y the value ($\bar{d}+1,\bar{k}$) (case 2.3.1).

Initially, we have at most $O(n)$ blocks. In each iteration, at most three new blocks are generated. Hence, the total number of blocks generated in the algorithm is $O(n)$. Since the marker of each block can be inserted and deleted at most once, the number of insertions and deletions are at most $O(n)$. Hence, the time spent on the priority queue is at most $O(n\log\log n)$. Furthermore, the time spent on the liked list can be easily shown to be $O(n)$. Therefore, the MDS problem can be solved in $O(n\log\log n)$ time.

3. An $O(n^2 log^2 n)$ algorithm for the weighted case

The MWDS problem was solved in $O(n^3)$ time by Farber and Keil using dynamic programming. They defined the term, S_{ijk}, as a minimum weight set of vertices in V_{ik} containing k and dominating V_{ij}, if it exists. Otherwise, S_{ijk} denotes a dummy set, N (standing for "nonexist"), whose weight is defined to be 1 plus the total weight of V. Initially, set $S_{ijk} = \{\pi(1)\}$ if $k = \pi(1)$ and N, otherwise; set $S_{i0k} = \{k\}$ if $\pi^{-1}(k) \leq i$ and N, otherwise. By definition, the minimum weight member of $\{S_{nn1}, S_{nn2}, S_{nn3}, \cdots , S_{nnn}\}$ is a minimum weight dominating set on V. Since the calculation of each S_{ijk} takes only constant time, they provide an $O(n^3)$ algorithm for the MWDS problem.

Rewrite the formula of S_{ijk} as follows:

(1) $k < \pi(i)$

$w(S_{ijk}) \quad = w(S_{i-1,j,k}) \qquad\qquad$ if $j < \pi(i)$

$\qquad\qquad\quad = w(N) \qquad\qquad\qquad$ if $j \geq \pi(i)$

(2) $k = \pi(i)$

$w(S_{ijk}) \quad = Min(w(S_{ij1}) : t < k) + w(\pi(i)) \qquad$ if $j < \pi(i)$

$\qquad\qquad\quad = w(S_{i,k-1,k}) \qquad\qquad\qquad$ if $j \geq \pi(i)$

(3) $k > \pi(i)$ and $\pi^{-1}(k) > i$

$w(S_{ijk}) \quad = w(N) = w(S_{i-1,j,k})$

(4) $k > \pi(i)$ and $\pi^{-1}(k) \leq i$

$w(S_{ijk}) \quad = S_{i,j-1,k} \qquad\qquad\qquad\qquad\qquad\qquad$ if $j = \pi(i)$

$\qquad\qquad\quad = min(w(S_{i-1,j,k}), w(S_{i,\pi(i)-1,k})+w(\pi(i))) \qquad$ if $j > \pi(i)$

$\qquad\qquad\quad = S_{i-1,j,k} \qquad\qquad\qquad\qquad\qquad\qquad$ if $j < \pi(i)$

We shall adopt a plane sweep technique to speed up the calculation of the $w(S_{ijk})$'s. Divide the calculation into n stages, indexed by i. In stage i, compute the matrix M_i whose (j,k)–th entry is $w(S_{ijk})$. The following lemma describes a local monotone property of M_i.

Lemma 3.1. *If $j < t$, then $w(S_{ijk}) \geq w(S_{itk})$.*

Lemma 3.1 implies that all the (j,k)–entries with the same $w(S_{ijk})$ values must be contiguous. For each column k in M_i, define an s_k–*block* (or s_k–*interval*) to be a maximal contiguous sequence of (j,k)–entries in column k such that the weight of their corresponding set equals s. Thus, we shall concentrate on changing the block structures from one matrix to the next.

We shall employ segment–tree to speed up the calculation of $min(w(S_{ijt}):t < k)$.

This data structure, originally introduced by Bentley, has many applications in computational geometry such as range tree, interval tree. A segment tree $T(1..n)$ is a binary tree witthn n leaves. The i–th leaf corresponds to the interval [i,i], and each internal node corresponds to the union of the two intervals corresponding to its two sons. An interval corresponding to some node in this tree is called a *standard interval*. Bentely proved that any interval in [1..n] can be partitioned into at most $logn$ standard intervals. Such a partition is called a *standard partition* and is represented by a set of nodes in $T[1..n]$. Although the number of possible intervals with endpoints in $\{1,...,n\}$ is $O(n^2)$, the number of standard intervals is only $O(n)$. Furthermore, the number of possible intervals containing a specific integer k is $O((n-k) \cdot k)$, but there can be at most $O(log\ n)$ such standard intervals (all of them lie on the path from the root to the leave [k,k] in the segment tree). This is an important fact used in our algorithm..

Our data structure can be described as **two–level segment trees**. The *primary segment tree*, denoted by T, is used to store standard intervals (or blocks) for composing the s_k–intervals. Each leaf in T corresponds to a distinct index j. Associate with each node v in T a *secondary segment tree* T_v: each leaf of T_v corresponds to a distinct index k. Each node in T_v contains an s–value such that the s–value of each internal node v is the minimum of the s–values of its two sons.

Based on the matrix M_{i-1}, we calculate the matrix M_i in a column by column fashion. To create a new s_k–interval in column k, we first find the standard partition P of the s_k–interval. Then for every standard interval v in P, insert the s–value of the interval into the k–th leaf in its associated secondary tree T_v, and update the the the s–values of internal nodes (along the path from the k–th leaf to the root) in T_v. The complexity of these operations are analyzed in the following lemmas.

Lemma 3.2. *Both insertion and deletion of an s_k–interval can be executed in $O(log^2\ n)$ time.*

Lemma 3.3. *For fixed j and k, min $(w(S_{ijt}):t < k)$ can be found in $O(log^2\ n)$ time.*

Besides the two–level tree structure, we maintain a linked list structure which allows us to trace all the s_k–intervals of the same k value from the greatest j to the least j. A pictorial description of the change from M_{i-1} to M_i is shown in Figure 2.

Algorithm UPDATING M_i FOR THE WEIGHTED CASE
For k = 1 to n do
1.　　 k < π(i)
　　 Delete all the s_k–intervals which is contained in $(\pi(i), n]$. Let r = [low,high] be a

s_k–interval such that low $\leq \pi(i) \leq$ high. Replace r by the s_k–interval, $[\text{low},\pi(i)-1]$ (could be empty) with the same s–value. Then, create a new s_k–interval, $[\pi(i),n]$, whose s–value is $w(N)$.

2. $k = \pi(i)$

For each $j < \pi(i)$, create a s_k–interval, $[j,j]$, whose s–value is the minimum of $(w(S_{ijt}):t < k) + w(\pi(i))$. Create a new s_k–interval, $[\pi(i),n]$, whose s–value is $w(S_{ik-1k})$.

3. $k > \pi(i)$ and $\pi^{-1}(k) < i$

Followed the linked list structure, we can find a smallest j^*, such that the s–values of all the s_k–intervals contained in $[j^*,n]$ is greater than $w(S_{i\pi(i)-1k}) + w(\pi(i))$. Let $j_{max} = \max(j^*,\pi(i))$. Then we merge all the s_k–intervals contained in $[j_{max},n]$ into a new s_k–interval whose s–value is $w(S_{i\pi(i)-1k}) + w(\pi(i))$.

End_Of_For

Theorem 3.4. *The total number of s_k–intervals generated in the algorithm is bounded by $O(n^2)$.*

By 3.2, 3.3 and 3.4, the time complexity of our algorithm is $O(n^2 log^2 n)$.

References

1.   M. J. Atallah, G. K. Manacher and J. Urrutia, *"Finding a minimum independent dominating set in a permutation graph"*, **Discrete Applied Math.** 21(3), 1988,177–183.

2.   J. L. Bently, *"Algorithms for Klee's rectangle problems"*, Carnegie–Mellon University, Pittsburg, Penn., Department of Computer Science, uunpublished notes, 1977.

3.   A. Brandstadt and D. Kratsch, *"On domination problems for permutation and other graphs"*, **Theoretical Computer Science 54**, 1987, 181–198.

4.   M. Farber and J. M. Keil, *"Domination in permutation graphs"*, **J. Algorithm 6** 1985, 309–321.

5.   J. Spinrad, *"Transitive orientation in $O(n^2)$ time"*, Proc. 15th Annual ACM **Symp. on Theory of Computing**, 1983, 457–466.

6.   P. Van Emde Boas, *"Preserving order in a forest in less than logarithmic time and linear space"*, **Information Processing Letter 6**(3), 1977, 80–82.

116

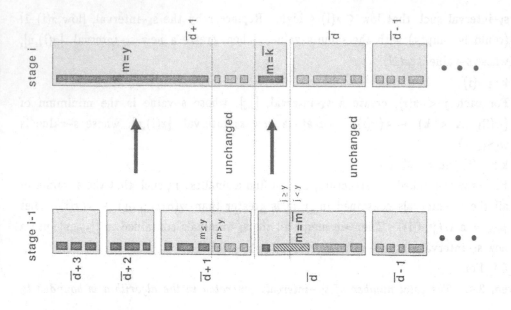

Figure 1 (b) Updating (d, m) in case of $\overline{m} \le y$

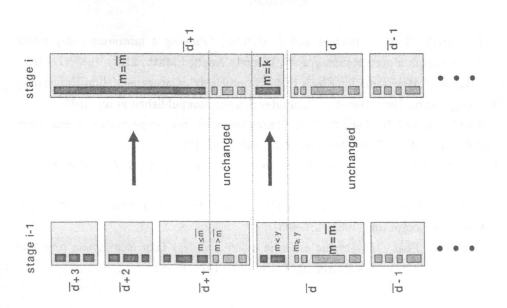

Figure 1 (a) Updating (d, m) in case of $\overline{m} > y$

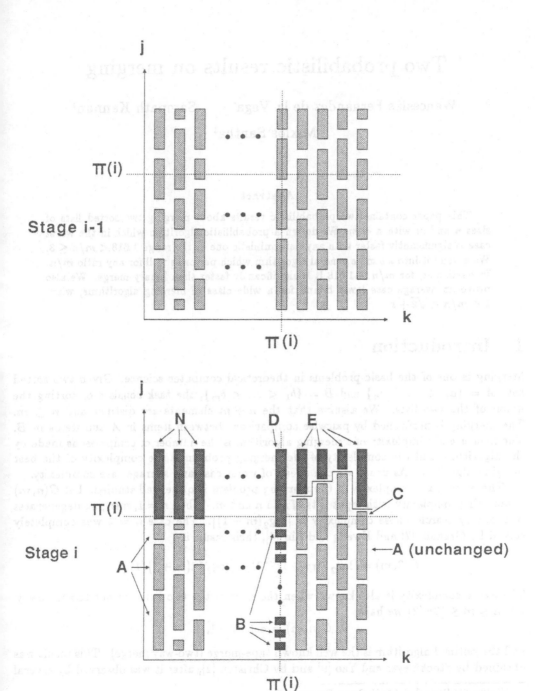

Figure 2. Updating $w(S_{ijk})$ in the weighted case

Two probabilistic results on merging

Wenceslas Fernandez de la Vega*   Sampath Kannan†

Miklos Santha‡

Abstract

This paper contains two probabilistic results about merging two sorted lists of sizes n and m with $n < m$. We design a probabilistic algorithm which in the worst case is significantly faster than any deterministic one in the range $1.618 < m/n \leq 3$. We extend it into a simple general algorithm which performs well for any ratio m/n. In particular, for $m/n > 1.618$ it is significantly faster than binary merge. We also prove an average case lower bound for a wide class of merging algorithms, when $1 < m/n < \sqrt{2} + 1$.

1   Introduction

Merging is one of the basic problems in theoretical computer science. Given two sorted lists $A = \{a_1 < \ldots < a_n\}$ and $B = \{b_1 < \ldots < b_m\}$, the task consists of sorting the union of the two lists. We assume that the $n + m$ elements are distinct and $n \leq m$. The merging is performed by pairwise comparisons between items in A and items in B. The measure of complexity of a merging algorithm is the number of comparisons made by the algorithm, and the complexity of the merging problem is the complexity of the best merging algorithm. As usual, we can speak of worst case and average case complexity.

The worst case complexity of the merging problem is quite well studied. Let $C(n, m)$ denote this complexity with input lists of size n and m. When $n = 1$, merging degenerates into binary search whose complexity is $\lceil \log_2(m + 1) \rceil$. The case $n = 2$ was completely solved by Graham [7] and Hwang and Lin [4], their result is

$$C(2, m) = \lceil \log_2 7(m + 1)/12 \rceil + \lceil \log_2 14(m + 1)/17 \rceil.$$

The exact complexity is also known when the sizes of the two lists are not too far away. For $n \leq m \leq \lfloor 3n/2 \rfloor$ we have

$$C(n, m) = n + m - 1,$$

and the optimal algorithm is the well known tape-merge (two-way merge). This result was obtained by Stockmeyer and Yao [9] and by Christen [2], after it was observed by several

---

*Université Paris-Sud, 91405 Orsay, France

†University of California, Berkeley CA 94720, Supported by NSF Grant CCR85-13926

‡Université Paris-Sud, 91405 Orsay, France

people for $n \leq m \leq n + 4$. The best known algorithm — binary merge — which gives satisfying result for any values of n and m is due to Hwang and Lin [5]. Let $\mathcal{N}$ denote the set of non-negative integers. Set $m/n = 2^t x$, where $t \in \mathcal{N}$ and $1 < x \leq 2$ are uniquely determined, and let $BM(n, m)$ denote the complexity of binary merge. Then

$$BM(n, m) = \lfloor (t + x + 1)n \rfloor - 1,$$

and Hwang and Lin have also shown that

$$BM(n, m) \leq L(n, m) + n,$$

where $L(n, m) = \log_2 \binom{n+m}{n}$ is the lower bound coming from information theory. This means that if the ratio m/n is high, then the relative overwork done by binary merge is certainly small. As the relative overwork might be significant for small ratios of m/n, it is important to look for improvements in this case.

Several algorithms were proposed which in some range for m/n perform better than binary merge. Let us say that merging algorithm A_1 with running time $T_1(n, m)$ is *significantly* faster for some fixed ratio m/n than merging algorithm A_2 with running time $T_2(n, m)$, if $T_2(n, m) - T_1(n, m) = \Omega(n)$. The merging algorithm of Hwang and Deutsch [6] is better than binary merge for small values of n, but is not significantly faster. The first significant improvement over binary merge was proposed by Manacher [8], he improved it for $m \geq 8n$ by $31n/336$ comparisons. It was further improved by Christen [1] who designed an ingenious but quite involved merging algorithm. It is better than binary merge if $m > 3n$, it uses at least $n/4$ comparisons less if $m \geq 4n$, and asymptotically it uses at least $n/3 - o(n)$ comparisons less when m/n goes to infinity. On the other hand, this algorithm is worst than binary merge when $m < 3n$.

In the first part of this paper we propose a probabilistic algorithm for merging. The algorithm will at some points flip a (biased) coin, and its next step will depend on the result of the coin toss. The algorithm is quite simple, and works well for every values of n and m. It is significantly faster then binary merge for $m/n > (\sqrt{5} + 1)/2 \approx 1.618$. To the best of our knowledge binary merge is the best algorithm known for $m/n \leq 3$, thus our probabilistic algorithm is significantly faster than any deterministic one in this range.

In the second part of the paper we prove a non-trivial average case lower bound for a wide class of merging algorithms. We say that a merging algorithm is *insertive* if for each element of the smaller list the comparisons which involve it are all made one after the other. In other words in the binary decision tree associated with the algorithm, along any path, for every element a of the smaller list, the nodes which represent comparisons involving the element a appear in consecutive positions. Apart from this constraint the decision tree can be arbitrary. Many merging algorithms such as repeated binary search, tape merge and binary merge are insertive. We will prove that if $(1 + \eta) \leq m/n \leq (\sqrt{2} + 1 - \eta)$ for some constant $\eta > 0$, then every insertive merging algorithm makes on the average at least a constant factor more comparisons than the information theoretical lower bound.

The rest of the paper is organized as follows: In Section 2.1 we describe our algorithm. The analysis of its complexity will be done in Section 2.2 and we also prove there that the particular bias of the coin in the algorithm was optimally chosen. In Section 3 we prove the average case lower bound result for insertive algorithms. Finally, in Section 4 we mention some open problems. Throughout the paper the logarithm function in base 2 will be denoted log.

2 The probabilistic algorithm

2.1 Description of the algorithm

For this section let $s = (\sqrt{5}-1)/2 \approx 0.618$, and $r = (\sqrt{2}-1+\sqrt{2}s)^2 \approx 1.659$, these number play a considerable role in the algorithm and in its analysis. The heart of the algorithm MERGE is the probabilistic procedure PROBMERGE which merges two already sorted lists, where the longer list contains more than $(1+s)$-times, but at most $2r$-times as many elements as the shorter one. The intuition underlying PROBMERGE is described below. We know that if $m \le \lfloor 3n/2 \rfloor$ then the tape merge algorithm is best possible. The tape merge algorithm can be thought of as inserting the a's in order in list B. At each stage the next a to be inserted is compared with the first 'eligible' element in B. If however B is, for example, a list of length $2n$ then since on the average between every two consecutive a's there are two b's, it might be better to compare the next a to be inserted with the second eligible b. However a deterministic algorithm which tries to do this turns out to be no better than tape merge. Its worst case complexity is also $3n - 1$. A natural idea is to try to compare the next a to be inserted with either the first or the second eligible element in B, the decision being made probabilistically. In fact this is our algorithm.

We will use sentinel elements in the algorithm. They make the description simpler for the price of a little loss of efficiency.

Procedure PROBMERGE

Input: $A = \{a_1 < \ldots < a_n\}$ and $B = \{b_1 < \ldots < b_m\}$, where $1 + s < m/n \le 2r$, and sentinel elements $b_0 = -\infty$, $b_{m+1} = b_{m+2} = \infty$.

Output: The merged list.

- $p := \begin{cases} s & \text{if } 1+s < m/n \le 2+s \\ \sqrt{m/n} - 1 & \text{if } 2+s < m/n \le 2r \end{cases}$ $\left\{ \begin{array}{l} \textit{the choice of the} \\ \textit{probability value} \end{array} \right\}$

- $i := 1; j := 1$ $\{i$ indexes A, j indexes $B\}$

- **while** $i \le n$ **do**

  - with probability $1 - p$ compare a_i with b_j
    - if $a_i < b_j$ then $i := i + 1$ $\{b_{j-1} < a_i < b_j\}$
    - else $j := j + 1$
  - with probability p compare a_i with b_{j+1}
    - if $a_i < b_{j+1}$ then
      - compare a_i with b_j
      - if $a_i < b_j$ then $i := i + 1$ $\{b_{j-1} < a_i < b_j\}$
      - else $i := i + 1; j := j + 1$ $\{b_j < a_i < b_{j+1}\}$

        ∗ **else** $j := j + 2$

    • **end**

The general algorithm MERGE uses PROBMERGE as a subroutine. Given lists A and B of size n and $m > (1 + s)n$ respectively, the algorithm calls directly PROBMERGE if $m \le rn$. Otherwise it picks a uniformly spaced sublist C of B of cardinality between rn and $2rn$. More precisely, the spacing between the elements of C is 2^t for t a non-negative integer. The algorithm first merges A with C probabilistically. This determines for each a in A a list of $2^t - 1$ b's in which a should be inserted. The insertion is done deterministically by binary search and takes t steps. Let us observe that MERGE actually coincides with the procedure PROBMERGE while $m \le 2rn$.

Algorithm MERGE

Input: $A = \{a_1 < \ldots < a_n\}$ and $B = \{b_1 < \ldots < b_m\}$, where $(1 + s) < m/n$. For $r < m/n$ set $m = 2^t x n$, $t \in \mathcal{N}$, $r < x \le 2r$.

Output: The merged list.

1. If $m/n \le r$ then PROBMERGE(A, B).

2. Let $C = \{c_1 < \ldots < c_{\lceil xn \rceil}\}$ be the sublist of B, where $c_k = b_{(k-1)2^t + 1}$ for $k = 1, \ldots, \lceil xn \rceil$, and define sentinel elements $c_0 = -\infty$, $c_{\lceil xn \rceil + 1} = \infty$.

3. PROBMERGE(A, C).

4. For $i = 1, \ldots, n$ let $0 \le j_i \le \lceil xn \rceil$ be such that $c_{j_i} < a_i < c_{j_i + 1}$.

5. For $i = 1, \ldots, n$ insert a_i by binary search into the list $\{b_{(j_i - 1)2^t + 2} < \cdots < b_{j_i 2^t}\}$.

2.2 Analysis of the algorithm

We would like to analyse the expected number of comparisons made by the algorithm PROBMERGE. Let $T(n, m)$ be this number when the input of the procedure is two lists of size respectively n and m, $1 + s < m/n \le 2r$, and let p be the probability value defined inside the procedure.

Theorem 1 *We have*

$$T(n, m) \le \begin{cases} sm + (1 + s)n & \text{if } 1 + s < m/n \le 2 + s \\ 2\sqrt{mn} & \text{if } 2 + s < m/n \le 2r. \end{cases}$$

Proof The outcome of the procedure uniquely determines n non-negative integers $k_1, \ldots, k_n$, where k_i is the number of b's which are between a_{i-1} and a_i. As the list B contains m elements, $\sum_{i=1}^{n} k_i \le m$. The procedure puts the a's into B one by one. Let f_k be the expected number of comparisons to put an a into B, when the number of b's between a

and its predecessor in A is k. f_k doesn't depend on the index of the element being inserted. Then

$$T(n,m) = \max_{\substack{k_1,\ldots,k_n \\ \sum_{i=1}^{n} k_i \leq m}} \sum_{i=1}^{n} f_{k_i}.$$

Thus we are interested in the values f_k. If $k \geq 2$, then after the first comparison the element a jumps over two b's with probability p, and jumps over one b with probability $1 - p$. This means that we get the following recurrence relation for f_k:

$$f_k = (1-p)f_{k-1} + pf_{k-2} + 1,$$

with the initial conditions

$$f_0 = 2p + 1(1-p) = p + 1,$$

$$f_1 = 3p(1-p) + 2((1-p)^2 + p) = -p^2 + p + 2.$$

By standard technique, the solution of this linear recurrence is

$$f_k = \frac{k}{p+1} + \frac{p^3 + p^2 - p}{(p+1)^2}(-p)^k + \frac{2p^2 + 4p + 1}{(p+1)^2}.$$

This tells us that

$$\sum_{i=1}^{n} f_{k_i} = \sum_{i=1}^{n} \frac{k_i}{p+1} + \frac{p^3 + p^2 - p}{(p+1)^2} \sum_{i=1}^{n}(-p)^{k_i} + \frac{2p^2 + 4p + 1}{(p+1)^2}n.$$

The term $\sum_{i=1}^{n} \frac{k_i}{p+1}$ is always bounded from above by $\frac{m}{p+1}$. If $p = s$, we have $p^3 + p^2 - p = 0$, and the result follows. If $p > s$ (when $2 + s < m/n$), we have $p^3 + p^2 - p > 0$, and $\sum_{i=1}^{n}(-p)^{k_i}$ is maximized by choosing $k_i = 0$. Simple arithmetic gives the result also in this case. $\quad\square$

The global algorithm MERGE calls PROBMERGE with two lists of size n and $\lceil nx \rceil$ and then makes n binary searches, each in a list of size $2^t - 1$. Applying Theorem 1, we immediately get the following result:

Theorem 2 *Let $E(n,m)$ denote the expected number of comparisons made by MERGE. For $m/n \leq r$ set $m/n = x$, and for $m/n > r$ set $m/n = 2^t x$, $t \in \mathcal{N}$, $r < x \leq 2r$. Then*

$$E(n,m) \leq \begin{cases} (t + sx + 1 + s)n + 1 & \text{if } r < x \leq 2 + s \\ (t + 2\sqrt{x})n + 1 & \text{if } 2 + s < x \leq 2r. \end{cases}$$

Corollary 1 *The algorithm MERGE is significantly faster than binary merge for any fixed ratio $m/n > 1 + s$.*

Proof With the notation of Theorem 2 we have

$$BM(n,m) - E(n,m) \geq \begin{cases} (x - sx - s)n - 3 & \text{if } (1+s) < x \leq 2 \\ ((1-s) - x(s - 1/2))n - 3 & \text{if } 2 < x \leq 2 + s \\ (2 - \sqrt{x})^2 n/2 - 3 & \text{if } 2 + s < x \leq 2r. \end{cases}$$

This difference is $\Omega(n)$. □

Let us point out an interesting special case of the above result. When $m = 2n$, PROB-MERGE merges the two lists with less than $2.855n$ expected comparisons, whereas the best known deterministic algorithm [5] performs $3n - 2$ comparisons. A simple calculation also yields the following relation between the expected running time of MERGE and the information theoretic lower bound. The maximum difference is attained for values of $m/n \to \infty$ with $x = 2r$.

Corollary 2 $E(n, m) - L(n, m) \leq 0.471n$

We also claim that the probability value p was optimally chosen in the procedure PROBMERGE. Let $T_p(n, m)$ be the expected number of comparisons made by PROB-MERGE when the probability $0 \leq p \leq 1$ is taken as a variable. Let us observe that we have just analyzed $T_s(n, m)$ and $T_{\sqrt{m/n}-1}(n, m)$ in Theorem 1. For the purpose of this analysis let us permit any input lists such that the ratio m/n is at least 1. The following theorem which we state here without proof can be obtained by doing some calculus.

Theorem 3 *For every large enough n and m, for every p we have*

$$T_p(n, m) \geq \begin{cases} T_0(n, m) & \text{if } 1 \leq m/n \leq 1 + s, \\ T_s(n, m) & \text{if } 1 + s \leq m/n \leq 2 + s \\ T_{\sqrt{m/n}-1}(n, m) & \text{if } 2 + s \leq m/n \leq 4, \\ T_1(n, m) & \text{if } 4 \leq m/n. \end{cases}$$

Thus while $m/n \leq (1 + s)$, our probabilistic procedure gives out deterministic tape-merge as a special case. In the range $1 + s \leq m/n \leq 2 + s$ the best choice is the constant $p = s$. For $m/n > 2 + s$, the best choice grows with m/n until 1, when the procedure (obviously) degenerates into a deterministic one. There is no reason to use PROBMERGE for large values of m/n, MERGE is significantly faster when $m/n > 2r$. In fact that is how r is chosen.

3 The lower bound

In this section we will prove the following theorem:

Theorem 4 *For every $\eta > 0$ there exists $\epsilon > 0$ such that for every large enough n and m with $(1 + \eta)n \leq m \leq (\sqrt{2} + 1 - \eta)n$, every insertive merging algorithm makes at least $(1 + \epsilon) \log L(n, m)$ comparisons on the average.*

Proof Here we will prove the theorem in the special case $m = 2n$ wich already contains the main ideas for the general result. The proof depends on the following Unbalance Lemma:

Unbalance Lemma *Let T denote a binary decision tree with k leaves, let the set of internal nodes be $\{D_1...., D_{k-1}\}$, and let d_j be the number of leaves of the subtree with root D_j. Let us suppose that for some $J \subseteq \{1, 2, \ldots, k - 1\}$, the subset of nodes $\{D_j : j \in J\}$ satisfies:*

i) There exists $\epsilon_1 > 0$ such that $\sum_{j \in J} d_j \geq \epsilon_1 k \log_2 k$.

ii) There exists $\epsilon_2 > 0$ such that for every $j \in J$, the answer probabilities (fraction of leaves in left and right subtrees) at the node D_j lie outside the interval $(1/2 - \epsilon_2, 1/2 + \epsilon_2)$.

Then, the average path length of T is at least

$$(1 + \epsilon_1 \epsilon_2^2) \log_2 k. \quad \square$$

Let us fix some insertive algorithm, and let T be the binary decision tree associated with this algorithm. Let the internal nodes of T be $D_1, \ldots, D_{k-1}$, where $k = \binom{n+m}{n}$. Since the average running time of the algorithm is the average path length of T, it will suffice to prove that the assumptions of the Unbalance Lemma are fulfilled for some subset of the nodes and for some proper ϵ_1 and ϵ_2. To begin, notice that there is a bijection between the outcomes of a merging algorithm and the set Z of words containing precisely n letters, a and m letters, b. For any word $w \in \{a, b\}^*$, let $|w|_a$ and $|w|_b$ denote respectively the number of occurrences of a and b in w. Let $c_3 > 0$ be an appropriate constant which will be specified later. For any $n \leq i \leq m$ and $w \in Z$, let $decomp(w, i)$ denote the decomposition $uvxy = w$, where the factors u, v, x, y have the following lengths:

- $|u| = i - 3c_3$,

- $|v| = |x| = 3c_3$,

- $|y| = 3n - i - 3c_3$.

Let us denote by Z_i the set of words $w \in Z$ such that $decomp(w, i)$ satisfies the following conditions:

1. $vx = (bba)^{c_3}(abb)^{c_3}$,

2. for every decomposition $y = y'y''$, where $|y'| = l$, and for every decomposition $u = u''u'$ where $|u'| = l$, we have

$$(2/3 - 1/30)l - c_2 \leq |y'|_b, |u'|_b \leq (2/3 + 1/30)l + c_2,$$

where $c_2 > 0$ is an appropriate constant also to be specified later. The value $1/30$ in the second condition is arbitrary, any constant $\theta > 0$ such that $2/3 - \theta > 1/2$ and $2/3 + \theta < 1/\sqrt{2}$ will suffice. We will define a subset of nodes satisfying the assumptions of the Unbalance Lemma using the sets Z_i. First we prove two claims about these sets.

Claim 1 *There exists a constant $\epsilon_3 > 0$ such that for all $n \leq i \leq m$ we have:*

$$|Z_i| \geq \epsilon_3 |Z|.$$

Proof Let w be a random element of Z, and let $n \leq i \leq m$. The first requirement is obviously satisfied by w with constant probability. We will show that under this hypothesis, the conditional probability that $|y'|_b$ falls outside the interval $(19l/30 - c_2, 7l/10 + c_2)$ is less then 0.48 . As the second requirement is symetrical in $|y'|_b$ and $|u'|_b$, this will imply the claim. We shall (rather crudely) bound this probability by the sum $\sum_{1 \leq l \leq |y|} s(l)$, where $s(l)$ denotes the probability that the left factor of y of length l violates the second requirement. Clearly $s(l) = q(l) + r(l)$, where $q(l)$ is the probability that the left factor of y of length l contains less than $19l/30 - c_2$ occurrences of b's, and $r(l)$ is the probability that the same left factor contains more than $7l/10 + c_2$ occurrences of b's. Hoeffding's bound [3] about sampling without replacement gives us for every l,

$$q(l), r(l) < 0.9992^l.$$

Let us choose c_1 such that $\sum_{l=c_1+1}^{\infty} 0.9992^l < 0.24$. We can take for example $c_1 = 10800$ (this is not the smallest possible choice). Now, in function of c_1 we can choose c_2 such that $q(l) = r(l) = 0$ whenever $l \leq c_1$. If we take $c_2 = 19c_1/30 = 6840$, then $19l/30 - c_2 \leq 0$, and clearly $q(l) = 0$. It can be checked that this choice is good also for $r(l)$. Hence

$$\sum_{1 \leq l \leq |y|} s(l) < 2 \sum_{l=10800+1}^{\infty} 0.9992^l < 0.48 . \quad \square$$

We are now ready to choose the constant c_3 in function of c_2 such that the following Claim becomes true. Let $c_3 = 10c_2 = 68400$.

Claim 2 *Let $w \in Z_i$ for some $n \leq i \leq m$, and let $uvxy = decomp(w,i)$. Let z be a non-empty prefix of xy (a suffix of uv) such that either the letter which follows (precedes) z in w is an a, or there is no more a in w after (before) z. Then we have*

$$19/30 \leq |z|_b/|z| \leq 7/10.$$

Proof We show the upper bound. If z is a prefix of x, than $|z|_b/|z| = 2/3$. Otherwise this fraction is maximized when the corresponding prefix of y contains the most possible b's. If this prefix is of length l, we have

$$|z|_b/|z| \leq \frac{2c_3 + 7l/10 + c_2}{3c_3 + l} = 7/10. \quad \square$$

We can now define a subset of the nodes of T which satisfies the conditions of the Unbalance Lemma. For every couple (w,i) such that $n \leq i \leq m$ and $w \in Z_i$, we first define an internal node $V(i,w)$ of T. Let $uvxy = decomp(w,i)$, and let $h = h(w,i)$ be such that a_h is the first instnce of a in x (thus $h-1$ is the number of instances of a in uv). If on the path of T corresponding to w the comparisons involving a_{h-1} preceede those involving a_h, then let $V(w,i)$ be the node of the decision tree at which the first comparison involving a_h takes place. Otherwise let $V(w,i)$ be the node at which the first comparison involving a_{h-1} takes place. Finally let

$$J = \{j : D_j = V(w,i) \text{ for some } w \in Z_i\}.$$

We now show that the two conditions of the Unbalance Lemma are met by the nodes whose indices are in J. First we will show that

$$\sum_{j \in J} d_j \geq \sum_{n \leq i \leq m} |Z_i|.$$

On the right hand side every word $w \in \bigcup_{n \leq i \leq m} Z_i$ is counted with multiplicity t_w, where

$$t_w = |\{i : w \in Z_i\}|.$$

If $i \neq i'$, then $V(w, i) \neq V(w, i')$, because $|h(w, i) - h(w, i')| > 1$. Thus the leaf corresponding to w is counted on the left hand side at least t_w times. Therefore by using Claim 1 we get

$$\sum_{j \in J} d_j \geq \epsilon_3 n \binom{n + m}{n} = \epsilon_1 k \log k.$$

We now turn to the second condition of the Unbalance Lemma. Let $D_j = V(w, i)$ for some $w \in Z_i$, and let's suppose without loss of generality that the comparisons involving a_{h-1} preceede those involving a_h. Let a_{h+r} denote the leftmost right neighbour of a_h which has already been treated when the comparisons involving a_h begin at D_j. (If there is no such element we set $r = n - h + 1$). Let us further suppose that a_{h-1} and a_{h+r} are separated by s occurrences of b for some $s \geq 1$, which we denote by $b_e, \ldots, b_{e+s-1}$. Let us observe that $s/(r + s) = |z|_b/|z|$ for some initial segment z of xy, where z satisfies the conditions of Claim 2. Therefore we have

$$19/30 \leq s/(r + s) \leq 7/10.$$

The comparison at D_j is $a_h : b_{e+l-1}$ for some $1 \leq l \leq s$. Let $p(r, s, l) = Pr[a_h > b_{e+l-1}]$ denote the answer probability at D_j. Among the leaves of $V(w, i)$, the relative rankings of the sets $\{a_h, \ldots, a_{h+r-1}\}$ and $\{b_e, \ldots, b_{e+s-1}\}$ are all equally represented. For any word belonging to the leaves of D_j, the relative rank of a_h whithin the pooled set is at least l if and only if $a_h > b_{e+l-1}$. Thus we have

$$p(r, s, l) = \frac{s(s - 1) \ldots (s - l + 1)}{(r + s)(r + s - 1) \ldots (r + s - l + 1)}.$$

We claim that for every l, the answer probability $p(r, s, l)$ falls outside the interval $(0.49, 0.51)$, thus we can choose for example $\epsilon_2 = 0.01$. We prove this in two cases according to the value of l. If $l = 1$ then $p(r, s, 1) = s/(r + s) \geq 19/30$. If $l \geq 2$ then we have

$$p(r, s, l) \leq \left(\frac{s}{r + s}\right)^2 \leq 0.49,$$

and the result follows. $\square$

4 Open Problems

We have shown how to significantly improve upon the tape merge algorithm when the ratio m/n is at least the golden ratio. An interesting open question is the value of the smallest ratio m/n where an improvement — probabilistic respectively deterministic — can be obtained. Although the lower bounds in [2,9] hold only for deterministic algorithms, we conjecture that not even a probabilistic algorithm can achieve improvements over tape merge for ratios less than the golden ratio.

It would be interesting to design other more complex probabilistic algorithms. Generalizing our average case lower bound for arbitrary merging algorithms also remains open.

Acknowledgements

Thanks are due to Charles Delorme for his most valuable assistance.

References

[1] C. Christen (1978) *Improving the bound on optimal merging*, Proceedings of 19th FOCS, pp. 259-266.

[2] C. Christen (1978) *On the optimality of the straight merging algorithm*, Publication 296, Dép. d'Info. et de Rech. Op., Université de Montréal.

[3] V. Chvatal (1984) *Probabilistic methods in graph theory*, Annals of Operations Research 1, pp. 171-182.

[4] F. K. Hwang and S. Lin (1971), *Optimal merging of 2 elements with n elements*, Acta Informatica 1, pp. 145-158.

[5] F. K. Hwang and S. Lin (1972), *A simple algorithm for merging two disjoint linearly ordered lists*, SIAM J. of Comput. 1, pp. 31-39.

[6] F. K. Hwang and D. N. Deutsch (1973) *A class of merging algorithms*, JACM, Vol 20, No. 1, pp. 148-159.

[7] D. E. Knuth (1973), The Art of Computer Programming, Volume 3: Sorting and Searching, Addison-Wesley.

[8] G. K. Manacher (1979), *Significant improvements to the Hwang-Ling merging algorithm*, JACM, Vol. 26, No. 3, pp. 434-440.

[9] P. K. Stockmeyer and F. F. Yao (1980) *On the optimality of linear merge*, SIAM J of Comput. 9, pp. 85-90.

[10] A. Yao (1977), *Probabilistic computations: Towards a unified mesure of complexity*, Proc. 18th FOCS, pp. 222-227.

Randomized Broadcast in Networks

Uriel Feige[*]   David Peleg[*†]   Prabhakar Raghavan[‡]

Eli Upfal[§]

Abstract

In this paper we study the rate at which a rumor spreads through an undirected graph. This study has two important applications in distributed computation: (1) in simple, robust and efficient broadcast protocols; (2) in the maintenance of replicated databases.

1. Introduction

Let $G = (V, E)$ be a connected, undirected graph on N vertices. One vertex initially knows of a "rumor" that has to be conveyed to every other vertex in V. The rumor is propagated as follows: at each step, every vertex that knows of the rumor chooses one of its neighbors in G uniformly at random, and informs it of the rumor. How many steps elapse before every vertex knows the rumor? The answer clearly depends on the nature of G; for instance, if G were the complete graph on N vertices, K_N, it is well known [FG85, Pit87] that $O(\log N)$ steps suffice almost surely.

Consider the standard model of point to point communication networks, described by a connected, undirected graph: the vertices represent the processors of the network and the edges represent bidirectional communication channels between the vertices. The study of rumor propagation has at least two applications to distributed computing in such networks. The first is in algorithms for *broadcasting*: a single processor wishes to broadcast a piece of information to all the other processors in the network (see [HHL88] for a comprehensive survey).

---

[*]The Weizmann Institute of Science, Rehovot, Israel.

[†]Supported in part by an Allon Fellowship.

[‡]IBM T.J. Watson Research Center, Yorktown Heights, NY. A portion of this work was done while the author was visiting the Weizmann Institute of Science, Rehovot, Israel.

[§]IBM Almaden Research Center, San Jose, CA, and Department of Applied Mathematics, The Weizmann Institute of Science, Rehovot, Israel. Work at the Weizmann Institute supported in part by a Bat-Sheva de Rothschild Award and by a Revson Career Development Award.

There are at least three advantages to a randomized broadcast algorithm such as the one we have described:

(1) Simplicity — the entire algorithm is simple and local, with no need to know the current global topology and no need to remember whether the rumor has already been sent to a particular neighbor. Despite this simplicity, the algorithm achieves fast broadcast, as we shall see.

(2) Scalability — the algorithm is independent of the size of the network; it does not grow more complex as the network grows.

(3) Robustness — randomized broadcast is robust in that it works well even in the face of link/node failures in the network. A simple example might help here. Let us again consider broadcast on the complete graph K_N; it is possible to devise deterministic broadcast algorithms that achieve $O(\log N)$-step broadcast in the absence of faults, just as our random broadcast did. Now let $t < cN$ be a positive integer for a constant $c < 1$; suppose that an adversary were allowed to choose t links in the network to "break". It is easy to show now (details omitted in this abstract) that for any deterministic algorithm an adversary can select the link failures such that the algorithm takes $\Omega(t + \log N)$ steps, whereas randomized broadcast achieves $O(\log N)$ steps almost surely no matter how the adversary chooses the link failures.

A second application comes from the maintenance of replicated databases, for instance in name servers in a large corporate network [DGH+87]. There are updates injected at various nodes, and these updates must propagate to all the nodes in the network. At each step, a processor and its neighbor check whether their copies of the database agree,[1] and if not make the necessary reconciliation. The goal is that eventually all copies of the database converge to the same contents. Thus here we have a situation with multiple rumors propagating in the network. However, the basic analysis of this case can be reduced to the case of the analysis of a singe rumor; details are omitted here for brevity and will be given in the final version.

The spread of rumors in a network bears a superficial resemblance to another stochastic process that has been studied much by computer scientists: the random walk [AKL+79]. In rumor propagation every node that has already seen the rumor is broadcasting it at every subsequent step, whereas in the random walk propagation occurs from exactly one node at each step. The tools used in random walks, such as algebraic methods [BK88] or physical analogies [CRR+89] seem to be inappropriate to our study here. Also related to our work is the mathematical theory of epidemics [GN64, LR53], although that theory deals with variants of rumor broadcast on the complete graph (rather than general graphs as we consider here).

In Section 2 we give some fundamental theorems governing the rate of prop-

---

[1] It suffices for the nodes to compare checksums or "fingerprints" rather than the entire databases.

agation of rumors in graphs. While these theorems yield tight results for many classes of graphs, it is sometimes necessary to examine the structure of the graph more closely to precisely determine the rate of rumor propagation. We illustrate this in Section 3, where we show that a rumor reaches all vertices of a hypercube in $\Theta(\log N)$ steps almost surely; elementary analysis only yields $O(\log^2 N)$ steps. The hypercube also illustrates the difficulties introduced in the analysis due to the statistical correlation of several copies of a rumor repeatedly running into each other.

2.   Definitions and General Results

Let $deg(v)$ denote the degree of a vertex v. The length of the shortest path connecting two vertices v and u is denoted by $dist(v, u)$, the length of the shortest path connecting a vertex v to some member of a set of vertices C is denoted by $dist(u, C)$. The diameter of a network G is denoted by $diam(G)$. All logarithms in the paper are to the base 2.

In this work we analyze the following random broadcast procedure; initially, at least one vertex in V starts off knowing the rumor.

> **Procedure $\mathcal{RB}$**
> **repeat**
>
>> **for all $v \in V$, in parallel do**
>>> **if** v has already received the rumor **then**
>>>> v randomly chooses a neighbor with equal probability, and sends it the rumor;
>
>> **end**

Given a network G, the number of iterations of the procedure $\mathcal{RB}$ until the rumor reaches all the vertices of G is a random variable that depends on the topology of G. We say that $T(G)$ is the *almost sure coverage time* of a network G, if after $T(G)$ iterations of the procedure all vertices in G receive the rumor with probability $1 - 1/N$.

Theorem 2.1: For a general network G, with N vertices,
(1) $\log N \leq T(G) \leq N \log N$,
(2) There are networks G for which $T(G) = \Omega(N \log N)$,
(3) There are networks G for which $T(G) = O(\log N)$.

Proof: Each iteration of the procedure $\mathcal{RB}$ can at most double the number of vertices that received the rumor; thus $T(G) \geq \log N$.

Let $v = x_0, x_1, ..., x_\ell = u$ denote a shortest path connecting v to u in G. A vertex w is connected to two vertices x_i and x_j on the path only if they are at most one apart on the path (else there is a shorter path). Thus, $\sum_{j=0}^{\ell} deg(x_j) \leq 3N$. The expected number of iterations between the time x_j receive a rumor until the time x_j sends the rumor to x_{j+1} is bounded by $deg(x_j)$. Thus, the expected number of iterations until u receives the rumor is $3N$, and after $6N$ iterations the probability that u does not receive the rumor is bounded by $1/2$. After $12N \log N$ iterations of the procedure, the probability that there exists a vertex that did not receive the rumor is bounded by $N2^{-2\log N} \leq 1/N$.

The bounds are tight within a constant factor since $O(\log N)$ iteration suffice for the complete graph [FG85], and a star graph clearly requires $\Omega(N \log N)$ iterations. □

Theorem 2.2: For every network G

$$T(G) = O\left(deg(G)(diam(G) + \log N)\right).$$

Proof: Given a shortest path between two vertices in G, the probability that the rumor fails to traverse that path in $3deg(G)(diam(G) + 2\log N)$ iterations is bounded by the Chernoff bound [Che52] by $1/N^2$. Thus, the probability that a rumor does not reach all vertices of the graph within this number of steps is bounded by $1/N$. □

Corollary 2.3: For a bounded degree graph G, $T(G) \leq O(diam(G))$.

We also show that, in a certain sense, "almost all" connected graphs have $T(G)$ that is $O(\log N)$ (the minimum possible). For $p \in [0, 1]$, we say that G is drawn from $\mathcal{G}_{N,p}$ if G is an N-vertex graph each edge of which is present independently with probability p. (Thus $\mathcal{G}_{N,1/2}$ is the space of all N-vertex graphs chosen equiprobably.) Similarly, $\mathcal{G}_{N,M}$ denotes the space of graphs on N vertices with M randomly placed edges. The proof of the following theorem is left to the full paper.

Theorem 2.4: For almost all $G \in \mathcal{G}_{N,p}$ ($G \in \mathcal{G}_{N,M}$),
(1) If $p \leq (\log N - \omega(N))/N$ (resp. $M \leq (\log N - \omega(N))N/2$), where $\omega(N) \to \infty$, then $T(G) = \infty$.
(2) If $p = (\log N + \omega(N))/N$ (resp. $M = (\log N + \omega(N))N/2$), where $\omega(N) \to \infty$, $\omega(N) = O(\log \log N)$, then $\mathcal{RB}(G) = \Theta(\log^2 N)$.
(3) If $p \geq (1 + \epsilon)(\log N)/N$ (resp. $M = ((1 + \epsilon)N \log N)N/2$), for some fixed $\epsilon > 0$, then $\mathcal{RB}(G) = \Theta(\log N)$.

3. Random Broadcast on the Hypercube

We now focus on the hypercube, an important network for parallel computation for which the general theorems do not give the correct upper bound. For instance, Theorem 2.2 only gives an upper bound of $O(\log^2 N)$ on the rumour cover time. Here we determine the rumor cover time of the hypercube, illustrating some of the difficulties and techniques in proving such a bound. Let $H_n = (V_n, E_n)$ denote the n-dimensional hypercube, where $V_n = \{0, 1\}^n$ and

$$E_n = \{(x, y) \mid x, y \in V_n, \ x \text{ and } y \text{ differ in exactly one bit}\}.$$

The network has $|V_n| = 2^n = N$ nodes, $|E_n| = n \cdot N/2$ edges, and diameter $n = \log N$.

Theorem 3.1: For the n-dimensional hypercube H_n, $T(H_n) = \Theta(n) = \Theta(\log N)$.

Proof: The diameter of the hypercube is n, but the expected number of steps to traverse any single path of length $O(n)$ on the hypercube is $\Theta(n^2)$. To prove the theorem we need to analyze the progress of the rumor along many paths in parallel. The main difficulty in this analysis is that the paths are not disjoint, and thus introduce dependencies to the analysis. We overcome this difficulty by analyzing the progress of the procedure in two phases. The first phase brings the rumor to within a distance of αn from all vertices of the hypercube. The second phase completes the distribution.

Definition 1: A set of vertices C, α-*approximates* a vertex t if $dist(t, C) \leq \alpha n$. C is an α-*cover* of the hypercube if it α-approximates all vertices of V_n.

Lemma 3.2: For any $0 < \alpha < 1$, after $3n/\alpha$ iterations of $\mathcal{RB}$, with probability $1 - 2^{-2n}$, the rumor has reached every vertex of some α-cover C.

Proof: In analyzing the process of α-approximating a vertex t, we concentrate on one path $s = x_0, x_1, \ldots$ generated by $\mathcal{RB}$ as follows:
(1) s is the vertex that initiated the rumor.
(2) x_{i+1} is the first vertex satisfying $dist(x_{i+1}, t) < dist(x_i, t)$ that received the rumor from x_i.

The probability that in $3n/\alpha$ iterations of $\mathcal{RB}$ the path does not α-approximate the vertex t is bounded by the probability that there is a set of αn coordinates that were not hit in $3n/\alpha$ successive trials. This probability is

$$\leq \binom{n}{\alpha n} \left(1 - \alpha\right)^{3n/\alpha} < 2^n e^{-3n} \leq 2^{-2n-1}.$$

Thus, the probability that there exists a vertex that is not α-approximated is bounded by $1 - 2^{-n-1}$. ☐

Definition 2: For any vertex v and integer $1 \leq h \leq n$, the *band* $B(v, h)$ is the set of vertices at distance exactly h from v, i.e.,

$$B(v, h) = \{u | dist(u, v) = h\}.$$

Lemma 3.3: Given a 1/64-cover C, for any vertex t there exist an integer h and a set of vertices $S_h(t)$ such that the following properties hold.
(1) $31n/64 < h < 33n/64$,
(2) $S_h(t) \subseteq B(t, h) \cap C$,
(3) $|S_h(t)| \geq 2^{n/4}$, and
(4) the distance between every two vertices of $S_h(t)$ is at least $n/8$.

Proof: Let $S' \subseteq C$ be some minimal collection of vertices of C that 1/64-approximate the vertices of $B(t, n/2)$. The number of vertices in this band

$$|B(t, \tfrac{n}{2})| = \binom{n}{n/2} \geq \frac{2^n}{\sqrt{2\pi n}}.$$

A vertex may serve as an α-approximation only for vertices at distance αn or less from it. In the full version we prove that the number of vertices that can have the same 1/64-approximation is bounded by,

$$Y = \sum_{i=1}^{n/64} \binom{n}{i} \leq \frac{n}{64}\binom{n}{n/64} \leq n(4(64-1))^{n/64} \leq n \cdot 252^{n/64},$$

and

$$|S'| \geq \frac{|B(t, \tfrac{n}{2})|}{|Y|}.$$

By the definition of 1/64-approximation, S' contains vertices at distances $n/2 \pm n/64$ from t. Therefore, there is a band $B(t, h)$ for some h in this range such that the set $S'' = S' \cap B(t, h)$ satisfies

$$|S''| = |S' \cap B(t, h)| \geq \frac{64|B(t, n/2)|}{n|Y|}.$$

Using the greedy method, we now pick vertices from S'' into our final set $S_h(t)$, each at distance $n/8$ or more from all previously selected vertices. The number of vertices that each new pick rules out is at most

$$Z = \sum_{i=1}^{n/8} \binom{n}{i} \leq \frac{n}{8}\binom{n}{n/8} \leq n(4(8-1))^{n/8} \leq n \cdot 28^{n/8}.$$

Thus the total number of vertices we can pick into $S_h(t)$ is at least

$$\frac{|S''|}{|Z|} \geq \frac{64|B(t, \frac{n}{2})|}{n|Y||Z|} \geq \frac{2^{n+6}}{252^{n/64}28^{n/8}n^3\sqrt{2\pi n}} \geq 2^{n/4}$$

for sufficiently large n. □

In analyzing the process of reaching a vertex t in the second phase of distributing the rumor, we concentrate only on paths that start at vertices in $S_h(t)$. In order to define a sequence of independent events we further restrict our discussion to paths that were built at a pre-defined set of time intervals. Formally, consider a set of paths $L(t, \tau, h, k)$, generated by a sub-process of the procedure $\mathcal{RB}$ in the following way:

1. The process starts at time τ after all vertices in $S_h(t)$ have already received the rumor.

2. Each $s \in S_h(t)$ is a start-point of a path in $L(t, \tau, h, k)$.

3. Let $x_0, x_1, ...$ denote a path in $L(t, \tau, h, k)$. For every i, $x_i \in B(t, h-i)$, and x_{i+1} is the first vertex in $B(t, h-i-1)$ that received the rumor from x_i in the time interval $[\tau + ik, ..., \tau + (i+1)k - 1]$. If no vertex in $B(t, h-i-1)$ received the rumor from x_i in that time interval, x_i is the last vertex in this path.

Note that the set of paths $L(t, \tau, h, k)$ are independent from the set of paths $L(t, \tau + k, h, k)$, since for every i, the i-vertices of the paths in the first set, and the i-vertices of paths in the second set, were chosen at different time intervals.

Lemma 3.4: Assume that all vertices in $S_h(t)$ have received the rumor at time τ. The probability that at least one of the paths in $L(t, \tau, h, 11)$ reaches the vertex t is at least $2/3$.

Proof: Let $m = |S_h(t)|$, and let $s_1, ..., s_m$ denote the vertices of $S_h(t)$. Denote by p_d the probability that a path from $s \in S_h(t)$ reaches the band $B(t, d)$. By symmetry, this probability is equal for all $s \in S_h(t)$. Let x_i be the random variable defined by

$$x_i = \begin{cases} 1 & \text{if the path from } s_i \text{ reaches } t, \\ 0 & \text{otherwise.} \end{cases}$$

Let $p = p_0 = Prob(x_i = 1)$, and let $X = \sum_{i=1}^{m} x_i$.

Assuming that the path from s_i successfully reaches distance d from t, the probability that it reaches distance $d - 1$ is $(1 - (1 - d/n)^{11})$. In other words,

$$p_d = p_{d+1} \cdot (1 - (1 - \frac{d+1}{n})^{11}).$$

In the full version we show that for sufficiently large n,

$$p \geq \prod_{d=33n/64}^{1} \left(1 - \left(1 - \frac{d}{n}\right)^{11}\right) \geq 2^{-\theta n/11} > 2^{-n/4 + 4}.$$

By lemma 3.3, $m > 2^{n/4}$, hence the expected number of successful paths that reach t satisfies $E[X] = mp > 12$.

To bound the probability that at least one of the m paths reaches t we need a bound on $Var[X] = \sum_{i,j} E[x_i x_j] - (E[X])^2$.

Let $\mathcal{E}_{i,j}^1$ denote the event that s_j reaches t without intersecting s_i's path, and let $\mathcal{E}_{i,j}^2$ denote the event that s_j intersects s_i's path (and from then on the two paths merge). Then

$$Prob(x_i x_j = 1) = Prob(x_i = 1) \cdot \left(Prob(\mathcal{E}_{i,j}^1) + Prob(\mathcal{E}_{i,j}^2)\right).$$

Clearly, $Prob(\mathcal{E}_{i,j}^1) \leq p$. It remains to bound $Prob(\mathcal{E}_{i,j}^2)$. Let $\mathcal{E}_{i,j}^2(d)$ denote the event: s_j's path intersects s_i's path in band $B(t,d)$ and they did not intersect before. Clearly, $\mathcal{E}_{i,j}^2 = \cup_{d>0} \mathcal{E}_{i,j}^2(d)$.

The path from s_j can meet the path from s_i only in vertices v satisfying

$$v \in B(t,d) \cap B(s_j, h-d) \cap B(s_i, h-d).$$

Furthermore, there is only one set of $h-d$ coordinates that the path for s_2 can cross in its first $h-d$ transitions in order to meet the other path in $B(t,d)$.

The two origins s_i and s_j are at a distance of at least $n/8$ apart, and thus there exists some $d' \leq h - n/16$ such that

$$Prob(\mathcal{E}_{i,j}^2(d)) \leq \begin{cases} 0 & d > d', \\ p_d / \binom{h}{d} & d \leq d'. \end{cases}$$

Thus,

$$\sum_{d=n/16}^{h} Prob(\mathcal{E}_{i,j}^2(d)) \leq \frac{33n}{64} Prob(\mathcal{E}_{i,j}^2(h - \frac{n}{16})) \leq \frac{n}{\binom{31n/64}{n/16}} \leq n 2^{1-n/16} \leq \frac{p}{100}$$

for sufficiently large n.

For the last segment of the path we need a different analysis. Let R_d denote the ratio between the probabilities of intersection at band d and at band $d+1$. Using (1),

$$R_d = \frac{Prob(\mathcal{E}_{i,j}^2(d))}{Prob(\mathcal{E}_{i,j}^2(d+1))} = \frac{p_d \binom{h}{d+1}}{p_{d+1} \binom{h}{d}} = \left(1 - \left(1 - \frac{d+1}{n}\right)^{11}\right) \frac{h-d}{d+1}.$$

We are interested in R_d only for $0 \le d \le n/16$. In this range $R_d > 2$. Furthermore, for constant d and sufficiently large n,

$$R_d \ge \frac{11(d+1)}{n} \cdot \frac{h}{d+1} \ge \frac{11(d+1)}{n} \cdot \frac{31n}{64(d+1)} > 5 .$$

Recalling that the probability of intersection at $d = 0$ is at most p,

$$Prob(\mathcal{E}_{i,j}^2) < p \sum_{d=1}^{8} 5^{-d} + \frac{p}{5^8} \sum_{d=1}^{n/16-8} 2^{-d} + \frac{p}{100} < \frac{p}{4} .$$

Thus, $Ex_i x_j \le p^2(1 + 1/4)$ when $i \ne j$ and p otherwise.

$$Var[X] = \sum_{i,j} Ex_i x_j - (E[X])^2 \le m^2 p^2 (1 + \frac{1}{4}) + mp - (mp)^2 \le \frac{p^2 m^2}{3}$$

for sufficiently large n (recalling that $m \ge 2^{n/4}$ by Lemma 3.3).

Using Chebyshev's inequality,

$$Prob\{X = 0\} \le \frac{Var[X]}{(E[X])^2} \le 1/3.$$

$\square$

Lemma 3.5: Let C be a 1/64-cover, and assume that at time τ_0 all the vertices of C already received the rumor. There exists a constant $c > 0$ such that after cn additional iterations of the Procedure $\mathcal{RB}$ all vertices receive the rumor with probability $1 - 2^{-2n}$.

Proof: Let $\mathcal{L}(t, \tau, h, k)$ denote the event: at least one of the paths in $L(t, \tau, h, k)$ reached the vertex t. Let $k = 11$. Consider the following sequence of ℓ events:
$\mathcal{L}(t, \tau_0, h, k), \mathcal{L}(t, \tau_0 + k, h, k), \mathcal{L}(t, \tau_0 + 2k, h, k), ..., \mathcal{L}(t, \tau_0 + (2n-1)k, h, k),$

The ℓ events are independent, since each event consider the performance of vertices in each band at a different interval of k iterations.

By Lemma 3.4 each event holds with probability greater than $2/3$, thus, at time $\tau_0 + 2nk + kh$ the probability that t did not receive the rumor is bounded by $(1/3)^{2n} n \le 2^{-2n-1}$, and the probability that any vertex did not receive the rumor is bounded by 2^{-n-1}. $\square$

We are now ready to complete the proof of the main theorem. By Lemma 3.2 after $\tau_0 = 3n/\alpha$ iterations, with probability $1 - 2^{-n-1}$ there exists an α-cover, for $\alpha = 1/64$, in which all vertices received the rumor. By Lemma 3.4 an additional cn iterations guarantee that the rumor reaches every vertex with probability $1 - 2^{-n-1}$, thus $O(n) = O(\log N)$ iterations are sufficient for distributing the rumor among all vertices of the hypercube with probability $1 - 1/N$. $\square$

References

[AKL⁺79] R. Aleliunas, R. M. Karp, R. J. Lipton, L. Lovász, and C. Rackoff. Random walks, universal traversal sequences, and the complexity of maze problems. In *20th Annual Symposium on Foundations of Computer Science*, pages 218–223, San Juan, Puerto Rico, October 1979.

[BK88] A.Z. Broder and A.R. Karlin. Bounds on covering times. In *29th Annual Symposium on Foundations of Computer Science*, pages 479–487, White Plains, NY, October 1988.

[Che52] H. Chernoff. A measure of asymptotic efficiency for tests of a hypothesis based on the sum of observations. *Annals of Math. Stat.*, 23:493–509, 1952.

[CRR⁺89] A. K. Chandra, P. Raghavan, W.L. Ruzzo, R. Smolensky, and P. Tiwari. The electrical resistance of a graph captures its commute and cover times. In *Proceedings of the 21st Annual ACM Symposium on Theory of Computing*, pages 574–586, Seattle, May 1989.

[DGH⁺87] A. Demers, D. Greene, C. Hauser, W. Irish, J. Larson, S. Shenker, H. Sturgis, D. Swinehart, and D. Terry. Epidemic algorithms for replicated database management. In *6th ACM Symp. on Principles of Distributed Computing*, pages 1–12, 1987.

[FG85] A.M. Frieze and G.R. Grimmett. The shortest-path problem for graphs with random arc-lengths. *Discrete Applied Math.*, 10:57–77, 1985.

[GN64] W. Goffman and V.A. Newill. Generalization of epidemic theory — an application to the transmission of ideas. *Nature*, 204:225–228, 1964.

[HHL88] S.M. Hedetniemi, S.T. Hedetniemi, and A.L. Liestman. A survey of gossiping and broadcasting in communication networks. *Networks*, 18:319–349, 1988.

[LR53] H.G. Landau and A. Rapoport. Contribution to the mathematical theory of contagion and spread of information: I. spread through a thoroughly mixed population. *Bull. Math. Biophys.*, 15:173–183, 1953.

[Pit87] B.M. Pittel. On spreading a rumour. *SIAM J. Applied Math.*, 47:213–223, 1987.

On the Construction of Abstract Voronoi Diagrams, II [1]

by

R. Klein[2]

K. Mehlhorn[3]

S. Meiser[3]

Abstract: Abstract Voronoi Diagrams are defined by a system of bisecting curves in the plane, rather than by the concept of distance [K88a,b]. Mehlhorn, Meiser, Ó' Dúnlaing [MMO] showed how to construct such diagrams in time $O(n \log n)$ by a randomized algorithm if the bisecting curves are in general position. In this paper we drop the general position assumption. Moreover, we show that the only geometric operation in the algorithm is the construction of a Voronoi diagram for five sites. Using this operation, abstract Voronoi diagrams can be constructed in a purely combinatorial manner. This has the following advantages: On the one hand, the construction of a five-site-diagram is the only operation depending on the particular type of bisecting curves and we can therefore apply the algorithm to all concrete diagrams by simply replacing this operation. On the other hand, this is the only operation computing intersection points; thus, problems arising from instable numerical computations can occur only there.

Key words: Voronoi diagrams, randomized algorithms

---

[1] This work was supported partially by the DFG, grant SPP Me 620/6, and partially by the ESPRIT II Basic Research Actions Program of the EC under contract No. 3075 (project ALCOM)

[2] FB Mathematik/Praktische softwareorientierte Informatik, Universität–GHS–Essen, D-4300 Essen 1, West Germany

[3] FB Informatik, Universität des Saarlandes, D-6600 Saarbrücken, West Germany

I. Introduction

The Voronoi diagram of a set of sites S in the plane partitions the plane into regions, called Voronoi regions, one for each site. The region of site p contains all points of the plane that are closer to p than to any other site in S.

Such partitions play an important role in different areas of science. In computer science, the Voronoi diagram belongs to the most useful data structures. Its structural properties depend on the underlying distance measure and the type of sites considered. Surveys on the variety of Voronoi diagrams that have been heretofore investigated in publications are given in [LS] and in [A].

In order to provide a unifying concept for the study of Voronoi diagrams, *abstract Voronoi diagrams* were introduced [K88a,b]. They are not based on the notion of distance but on systems of bisecting curves as primary objects. For any two sites, p and q, in S, let $J(p,q)$ denote a curve that is homeomorphic to the line and divides the plane into a p-region and a q-region. Then the Voronoi region of p with respect to S is the intersection of all p-regions, as q varies in $S - \{p\}$. The family of curves $J(p,q)$, $p \neq q$, $p,q \in S$, is called *admissible* if for each subset of sites $S' \subseteq S$ the Voronoi regions w.r.t. S' are path-connected and form a partition of the plane.

The algorithm of [MMO] is an instance of randomized incremental construction [CS], i.e., the Voronoi diagram is constructed incrementally by adding the sites one by one in random order. It constructs abstract Voronoi diagrams for a restricted class of bisecting curves, namely curves in general position. In particular, two different bisecting curves intersecting at a common point x must cross at x and no four curves may intersect in a common point. The first restriction forbids, for example, diagrams of point sites under the L_1-metric (Manhattan-metric), as Figure 1 shows, cf. [L].

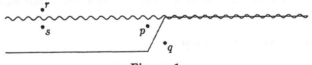

Figure 1.

In this paper, we drop the general position assumption. We only forbid some very special arrangements of bisecting curves. These forbidden arrangements do not occur for bisecting curves induced by *nice* metrics, including all norms, cf. [K88c]. The shapes of resulting Voronoi diagrams are described in section II. We show that Voronoi regions are homeomorphic to discs and that vertices and edges of the diagram can be described by only 3 and 4 sites, respectively. Next we show that when adding a new site to the diagram, all information necessary to compute the region of the new site can be obtained from the diagrams of the four characteristic sites of each intersected edge. This implies that the only non-combinatorial operation in the algorithm is the construction of a Voronoi diagram for 5 sites. All numerical operations take place in this particular operation. The algorithm of Guibas/Stolfi [GS] for Euclidean diagrams also has this property but neither the Plane-Sweep- nor the Divide & Conquer-algorithm do. The latter algorithms need to sort the sites by x-coordinates. Moreover, the Plane-Sweep-algorithm sorts the computed events by x-coordinates; the Divide & Conquer-algorithm sorts the nodes of the diagram by y-coordinates in its merge step. In both cases, objects are compared to each other that are not at all related in their topology. Therefore, it may be difficult to make geometric decisions in a consistent manner.

The analysis of our algorithm makes a novel use of perturbation. Clarkson and Shor's analysis [CS] of randomized incremental construction relies on the input being in general position. Although Klein [K89] has shown how to perturb abstract Voronoi diagrams so as to generate bisecting curves in general position, this perturbation is much too costly to be used in an algorithm. We use it in the analysis by showing that the running time on the perturbed system dominates the running time on the original system and applying [CS] to bound the running time on the perturbed system. Thus perturbation is only used in the analysis; the algorithm itself does not know about it and works on the original system.

Throughout the paper, we use the following notation: For a subset $X \subseteq \mathbb{R}^2$ the closure, boundary and interior of X are denoted by $\overline{X}$, $bd\ X$ and $int\ X$, respectively.

II. Abstract Voronoi Diagrams

Let $n \in \mathbb{N}$, and for every pair of integers p, q such that $1 \le p \ne q < n$ let $D(p, q)$ be either empty or an open unbounded subset of $\mathbb{R}^2$ and let $J(p, q)$ be the boundary of $D(p, q)$. We postulate:

1) $J(p, q) = J(q, p)$ and for each p, q such that $p \ne q$ the regions $D(p, q)$, $J(p, q)$ and $D(q, p)$ form a partition of $\mathbb{R}^2$ into three disjoint sets.

2) If $\emptyset \ne D(p, q) \ne \mathbb{R}^2$ then $J(p, q)$ is homeomorphic to the open interval $(0, 1)$.

We call $J(p, q)$ the bisecting curve for sites p and q. The abstract Voronoi diagram is now defined as follows:

Definition 1: Let $S = \{1, \ldots, n - 1\}$ and

$$R(p, q) := \begin{cases} D(p, q) \cup J(p, q) & \text{if } p < q \\ D(p, q) & \text{if } p > q \end{cases}$$

$$VR(p, S) := \bigcap_{\substack{q \in S \\ q \ne p}} R(p, q)$$

$$V(S) := \bigcup_{p \in S} bd\ VR(p, S)$$

$VR(p, S)$ is called the Voronoi region of p w.r.t. S and $V(S)$ is called the Voronoi diagram of S. □

We postulate that the Voronoi regions and the bisecting curves satisfy the following three conditions:

1) Any two bisecting curves intersect in only a finite number of connected components.
2) For any non-empty subset S' of S
   A) if $VR(p, S')$ is non-empty then $VR(p, S')$ is path-connected and has non-empty interior for each $p \in S'$,
   B) $\mathbb{R}^2 = \bigcup_{p \in S'} VR(p, S')$ (disjoint)
3) For any two bisecting curves $J(p, q)$ and $J(p, r)$ and for all points $v \in J(p, q) \cap J(p, r)$, if for any neighborhood $U(v)$ of v
   a) $J(p, r)|_{U(v)} \cap D(q, p)|_{U(v)} = \emptyset$ then $D(p, r)|_{U(v)} \subseteq D(p, q)|_{U(v)}$
   or
   b) $J(p, r)|_{U(v)} \cap D(p, q)|_{U(v)} = \emptyset$ then $D(r, p)|_{U(v)} \subseteq D(q, p)|_{U(v)}$.

Informally speaking, Condition 3 says that any two bisecting curves $J(p, q)$ and $J(p, r)$ intersecting in a point v without crossing, i.e., only touching at v, must have the p-region on the same side. Note that both bisecting curves belong to the same site p.

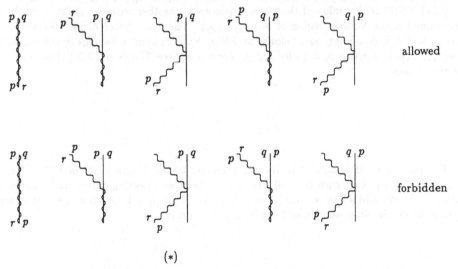

(*)

Figure 2. The upper row shows situations allowed by Condition 3, the lower row shows the corresponding forbidden situations

Condition 3 excludes the degenerate shapes of Voronoi regions shown in Figure 3. As a

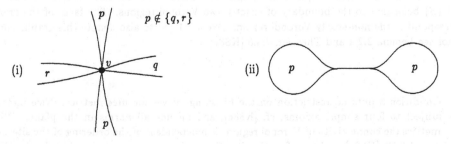

Figure 3.

consequence of Theorem 2.3.5 [K88a] the point v in Figure 3 (i) must belong to the region of site p and both p-sectors are connected via v. This implies that these sectors cannot be separated by $J(p,q)$ and $J(p,r)$. Since both bisecting curves must pass through v, they are arranged as in Figure 2 (*). This is a contradiction to Condition 3. Situation (ii) in Figure 3 is impossible for the same reason; on both sides of the thin part of p-region there must be regions of different sites r, q (Theorem 2.3.5, [K88a]). In general we have:

Fact 1: The following holds for each point $v \in V(S)$: Arbitrarily small neighborhoods $U(v)$ of v exist having the following properties. Let $VR(p_1, S), VR(p_2, S), \ldots, VR(p_k, S)$ be the sequence of Voronoi regions traversed on a counterclockwise march around the boundary of $U(v)$ and let $I_1, I_2, \ldots, I_k$ denote the corresponding intervals of $\partial U(v)$, where $I_j = \langle w_j, w_{j+1} \rangle \subseteq VR(p_j, S)$ for $1 \leq j \leq k$ (indices must be read mod k). The intervals may be open, half-open or closed. We have $w_j \neq w_{j+1}$ for $1 \leq i \leq k$. The common boundary of $VR(p_{j-1}, S)$ and $VR(p_j, S)$ defines a curve segment $\beta_j \subseteq J(p_{j-1}, p_j)$ connecting v and w_j. $V(S) \cap U(v)$ is the union of the curve segments β_j together with the point v. Each β_j is contained in the Voronoi region of $min\{p_{j-1}, p_j\}$. The open "piece of pie" bordered by β_j, β_{j+1} and I_j is non-empty and belongs to $VR(p_j, S)$. The point v belongs to the region of $min\{p_1, \ldots, p_k\}$. Finally, $p_i \neq p_j$ for $i \neq j$. For a proof see Theorem 2.3.5 [K88a] and the discussion above. $\qquad \square$

For the sequel, it is helpful to restrict attention to the "finite part" of $V(S)$. Let Γ be a simple closed curve such that all intersections between bisecting curves lie in the inner domain of Γ. We add a site ∞ to S, define $J(p, \infty) = J(\infty, p) = \Gamma$ for all p, $1 \leq p < n$, and $D(\infty, p)$ to be the outer domain of Γ for each p, $1 \leq p < n$.

Fact 2: Each Voronoi region is simply-connected and its boundary is a simple closed curve. Moreover, the closure of each non-empty Voronoi region $VR(p, S)$, $p \neq \infty$, is homeomorphic to a closed disc. A Voronoi diagram can be represented as a planar graph in a natural way. The vertices of the graph are the points of $V(S)$ which belong to the boundary of three or more Voronoi regions; the edges of the graph correspond to the maximal connected subsets of $V(S)$ belonging to the boundary of exactly two Voronoi regions. The faces of the graph correspond to the non-empty Voronoi regions. We use $V(S)$ to also denote this graph. For a proof see Lemma 2.2.4 and Theorem 2.3.5 [K88a]. $\qquad \square$

Condition 3 puts no restriction on the bisecting curves for *nice* metrics. Nice metrics are subject to four simple axioms, cf. [K88c], and include all norms in the plane. With *nice* metrics the connectivity of Voronoi regions is independent of the ordering of the sites, cf. Theorem 1.2.13, [K88a], and therefore the situation of Figure 4 cannot arise. In this situation,

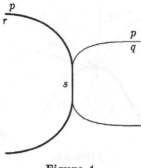

Figure 4.

p-region would not be connected in $V(\{p, q, r\})$ if the sites are ordered $r < p < q$. Segment s would belong to r-region.

R. Klein showed recently [K89] that the following conditions are equivalent to our conditions (2A) and (2B). These conditions can be tested more easily.

Lemma 1: In any system of bisecting curves Conditions (2A) and (2B) are fulfilled iff the following assertions are fulfilled.

1) $R(p, q) \cap R(q, r) \subseteq R(p, r)$ holds for any three sites $p, q, r \in S$. (Transitivity)
2) Let $J(p, q)$ and $J(p, r)$ be such oriented that the p-region lies on their left hand side. Then the two curves cross at most twice and do not constitute a clockwise cycle in the plane. $\qquad\square$

We next take a closer look at the elements of $V(S) = \bigcup_{r \in S} bd\ VR(r, S)$. Let x be a point of $V(S)$. Since Voronoi regions partition the plane, x lies on the boundary of at least two Voronoi regions.

Case 1: x lies on the boundary of exactly two Voronoi regions, say p-region and q-region.

By Fact 1, Figure 5 shows $V(S)$ in a sufficient small neighborhood $U(x)$ of x. Thus x is a

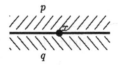

Figure 5.

point on an edge of the Voronoi diagram. The bisecting curve defining the edge is determined by the sites p and q. Note that the endpoints of an edge do not belong to the edge.

Case 2: x lies on the boundary of $k \geq 3$ Voronoi regions.

In this case a neighborhood of x consists of k sectors belonging to different Voronoi regions, and x is a common endpoint of k edges of $V(S)$, i.e., a vertex of $V(S)$. Let $p, q, r_1, r_2, \ldots, r_{k-2}$

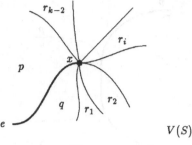

Figure 6.

be the Voronoi regions involved, enumerated in counterclockwise order. Then, in particular, x is the endpoint of an edge e separating p-region from q-region.

The following lemma shows that every site $r_1, \ldots, r_{k-2}$ takes equal responsibility in the definition of x as an endpoint of e.

Lemma 2: Let $L \subseteq S$ and $\{p, q, r_i\} \subseteq L$ for some i, $1 \leq i \leq k-2$. Then x is also an endpoint of e in $V(L)$.

Proof: The regions of p, q, r_i cannot become smaller in $V(L)$ compared to $V(S)$. Thus in a neighborhood of x in $V(L)$, at least the regions of p, q and r_i are represented. Since each region can be represented only by one sector (Fact 1), the edge e must end in the point x in $V(L)$ as well. □

Besides x the site r_i can define at most one more endpoint x' of an edge separating the p- and q-regions, because the Voronoi diagramm of the three sites p, q and r_i can have at most two vertices in its "finite part", cf. Figure 7 (for a proof see Lemma 3.5.2.5, [K88a]). The points x and x' must necessarily be endpoints of the same edge. We can distinguish

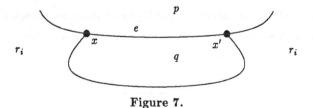

Figure 7.

between x and x' by looking at the counterclockwise orders p, r_i, q resp. p, q, r_i of regions around the vertex.

Lemma 3: In $V(S)$ at most one vertex can exist with the regions of p, q and r appearing in counterclockwise order in its neighborhood. Moreover, this vertex also exists in any diagram $V(L)$ where $\{p, q, r\} \subseteq L \subseteq S$.

Proof: Otherwise, consider two occurences v_1, v_2 of such vertices in $V(\{p, q, r\})$, as in Figure 8. Since the p- and q-regions are each connected, two paths P and Q exist, connecting

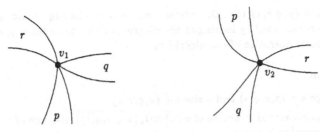

Figure 8.

v_1 and v_2 but not containing them: P is contained in p-region and Q is contained in q-region. Now consider a tour on the cycle $v_1 \circ P \circ v_2 \circ Q$. In v_1, we have r-land to the right; in v_2, we have r-land to the left; i.e., r-land lies inside the circle as well as outside. But since every site can be represented only once at a vertex, r-land can cross the circle neither at v_1 nor at v_2 from the inside to the outside. $\qquad\square$

Corollary 1: Any edge e of $V(S)$ is determined uniquely by at most four sites and their counterclockwise order, say (p, r, q, s). The same edge also exists in any diagram $V(L)$ where $\{p, q, r, s\} \subseteq L \subseteq S$.

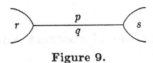

Figure 9.

Proof: This follows from Lemma 3 and the fact that $VR(o, S) \subseteq VR(o, L)$ for $o \in \{p, q\}$. $\quad\square$

As an important instance of the corollary the case $L = \{p, q, r, s\}$ tells us that the edge e exists even in $V(\{p, q, r, s\})$. This fact is important for the algorithm of [MMO]. This algorithm is incremental, i.e., at time t the Voronoi diagram $V(R)$ for a certain subset $R \subseteq S$ is already constructed and we add a new site not already in R to the Voronoi diagram. Some edges in $V(R)$ are thereby shortened and others are erased if they are covered by the new region. In order to construct the new region, we have to determine these intersections; this implies examining intersections between bisecting curves. Theorem 1 characterizes which bisecting curves we need to consider.

Theorem 1: Let $e = (p, r, q, s)$ be an edge of the Voronoi diagram $V(R)$ for sites $p \neq q \in R \subseteq S$, $r, s \in R - \{p, q\}$. Then for any set $L \subseteq R$ with $p, q, r, s \in L$, the edge e is also an edge of $V(L)$ and we have $e \cap \overline{VR(t, R \cup \{t\})} = e \cap \overline{VR(t, L \cup \{t\})}$ for any site $t \notin R$.

Proof: By corollary 1, edge e also exists in $V(L)$. Now consider any point $x \in e$. We show that a suffiently small neighborhood $U(x)$ looks the same in $V(L \cup \{t\})$ as in $V(R \cup \{t\})$. Choose $U(x)$ such that $y \in VR(p, R) \cup VR(q, R)$ for all points $y \in U(x)$. Let $o \in \{p, q, t\}$ be such that $y \in VR(o, R \cup \{t\})$ for a point $y \in U(x)$. From $VR(o, R \cup \{t\}) \subseteq VR(o, L \cup \{t\})$ we conclude $y \in VR(o, L \cup \{t\})$. The theorem follows since the Voronoi regions form a partition of $U(x)$. $\qquad\square$

If we take $L = \{p, q, r, s\}$ then the intersection between the edge e and the region of a new site t can be constructed by looking at the diagram of only five sites p, q, r, s and t. We take this as the basic operation of our algorithm.

Basic Operation

Input: An edge $e = (p, r, q, s)$ and a site $t \notin \{p, q, r, s\}$

Output: The combinatorial structure of $e \cap \overline{VR(t, \{p, q, r, s, t\})}$, i.e., one of the following:
   1) intersection is empty
   2) intersection is non-empty and consists of a single component:
      a) e itself
      b) a segment of e adjacent to the (p, r, q) endpoint
      c) a segment of e adjacent to the (p, q, s) endpoint
      d) a segment not adjacent to any endpoint of e
   3) intersection is non-empty and consists of exactly two components
   4) neither of the above

Remark: We show in Lemma 6, section III, that case 4 never arises and that the two components in case 3 are adjacent to one endpoint of e each.

III. Incremental construction of abstract Voronoi diagrams — the general case

In this section, we describe the incremental construction algorithm. We start with three sites ∞, p, q where p and q are chosen at random and then add the remaining sites in random order. At the general step, we have to consider a set $R \subseteq S$ of sites with $\infty \in R$ and $|R| \geq 3$. We maintain the following data structures.

1) The Voronoi diagram $V(R)$: It is stored as a planar graph as described in the previous section.

2) The conflict graph $G(R)$: The conflict graph is a bipartite graph. The nodes on one side are the vertices and edges of the Voronoi diagram $V(R)$ and on the other side the sites in $S - R$. Correspondingly, there are two types of edges in the conflict graph. An edge of the first type connects edge e of $V(R)$ and site $s \in S - R$ iff $e \cap \overline{VR(s, R \cup \{s\})} \neq \emptyset$. An edge of the second type connects a vertex $v \in V(R)$ and a site $s \in S - R$ iff $v \in \overline{VR(s, R \cup \{s\})}$. We say: the site s is in conflict with the edge e (the vertex v).

In the sequel, we proceed as in the argumentation of [MMO]. Because we drop the general position assumption, the boundary of a new region can now contain vertices or even edges of the Voronoi region constructed so far. Therefore we will go through all proofs of [MMO] again and take care of the boundary of Voronoi regions.

We assume that the Voronoi diagram $V(R)$ and the conflict graph $G(R)$ for $R \subseteq S$ are already constructed and discuss how to adjust $V(R)$ and $G(R)$ when adding a new site $s \in S - R$ to R. We first concentrate on $V(R)$. From now on, let $S := VR(s, R \cup \{s\})$.

Lemma 4: $\overline{S} = \emptyset \iff deg_{G(R)}(s) = 0.$

Proof: If $\overline{S} = \emptyset$ then clearly $deg_{G(R)}(s) = 0$. Thus let $deg_{G(R)}(s) = 0$. Then $\overline{S} \cap V(R) = \emptyset$ and $\overline{S} \subseteq int\ VR(p, R)$ for a site $p \in R$, since $V(R) = \bigcup_{r \in R} bd\ VR(r, R)$. Thus $VR(p, R \cup \{s\})$ is not simply-connected. $\qquad\square$

We are interested in the part of the diagram that is covered by S. Thus let $I := V(R) \cap \overline{S}$.

Lemma 5: I is a connected set and intersects $bd\ S$

Proof: Let $I_1, I_2, \ldots, I_k$ be the connected components of I for some k. No I_j can be contained entirely in the interior $int\ S$ of S. Otherwise, a simple closed curve $C \subseteq intS$ would exist, such that I_j is contained in the inner domain of C and C does not intersect $V(R)$. Thus $C \subseteq VR(r, R)$ for some $r \in R$. Since Voronoi regions are simply-connected, C and its interior must belong to $VR(r, R)$ and hence, C cannot contain I_j in its interior.

Thus let $k \geq 2$. Then a path $P \subseteq \overline{S} - I$ exists, connecting two points x and y on the boundary of S and separating I_1 from I_2, cf. Figure 10. From $P \cap I = \emptyset$ we have $P \cap V(R) = \emptyset$

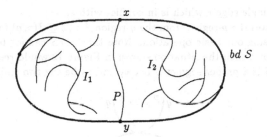

Figure 10.

and thus $P \subseteq int\ VR(r, R)$ for a site $r \in R$. Since $x, y \in P$ all sufficiently small neighborhoods $U(x)$ and $U(y)$ are entirely contained in $VR(r, R)$. The points in the intersection of these neighborhoods with the complement of S thus lie in $VR(r, R \cup \{s\})$ and can be connected by a path $Q \subseteq int\ VR(r, R \cup \{s\}) \subseteq int\ VR(r, R)$. The cycle $P \circ Q$ is therefore entirely contained in $int\ VR(r, R)$ and contains I_1 or I_2 in its interior. This is a contradiction. $\qquad\square$

Lemma 6: Let e be an edge of $V(R)$. If $e \cap \overline{S} \neq \emptyset$, then either $e \cap \overline{S} = V(R) \cap \overline{S}$ and $e \cap \overline{S}$ is a single component or $e - \overline{S}$ is a single component.

Proof: Assume first that $e \cap \overline{S} = V(R) \cap \overline{S}$. Because $V(R) \cap \overline{S}$ is connected by Lemma 5, we conclude that $e \cap \overline{S}$ is connected. Assume next that $e \cap \overline{S} \neq V(R) \cap \overline{S}$. Then with every point $x \in e \cap \overline{S}$ one of the subpaths of e connecting x to an endpoint of e must be contained in $\overline{S}$. Hence $e - \overline{S}$ is a single component. $\qquad\square$

We need to construct the boundary of S. From the discussion in section II we know that the boundary of S is a simple closed curve. Therefore $bd\ S \cap V(R)$ is a cyclic sequence of

- proper crossings between edges of $V(R)$ and $bd\ S$
- vertices of $V(R)$
- parts of edges of $V(R)$.

In the case of general position, $bd\,S \cap V(R)$ contains only intersection points of the first type. Also, the cyclic sequence of these intersection points and, thus, the new edges on $bd\,S$ are easy to find by simply walking along the boundary edges of the splitted regions, cf. Figure 11. The correctness of this procedure follows from the connectivity of $\overline{S} \cap V(R)$. We will now

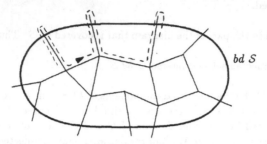

Figure 11.

discuss how to extend this strategy to degenerate cases. We distinguish two major cases.

Case 1: There is a single edge e which is in conflict with s.
Assume that e separates the p-region from the q-region, i.e., $e \subseteq J(p,q)$ for two sites $p, q \in S$. We compute $e \cap \overline{S}$ using our basic operation. Note that $e \cap \overline{S}$ cannot be a single point. If $e \cap \overline{S}$ were a single point then the situation shown in Figure 12 would result, a contradiction to Fact 1. Thus $e \cap \overline{S}$ is a proper segment. Let v_1 and v_2 be the two endpoints of $e \cap \overline{S}$. We

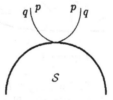

Figure 12.

delete $e \cap \overline{S}$ from the diagram, add v_1 and v_2 as vertices and two edges with endpoints v_1 and v_2. The two edges are part of $J(p,s)$ and $J(q,s)$, respectively.

Case 2: There are at least two edges which are in conflict with s.
Then $e - \overline{S}$ is a single component for each edge e of $V(R)$. The conflict graph gives us all edges in conflict with s. Let $e \subseteq J(p,q)$ be such an edge, i.e., $e \cap \overline{S} \neq \emptyset$. If $e \subseteq \overline{S}$ then e disappears completely. If $\emptyset \neq e - \overline{S} \neq e$ then some part of e survives. Let e_1 be one of the components of $e \cap \overline{S}$; e_1 can be computed by our basic operation. Then $\overline{e_1}$ contains exactly one endpoint of e. Let w be the other endpoint of $\overline{e_1}$. Then w lies on $J(p,q)$ and also on the boundary of S. Thus in a small neighborhood $U(w)$ of w we have p-, q- and s-region, i.e., w is a vertex of $V(R \cup \{s\})$. Also, in $V(R \cup \{s\})$ exactly three edges are incident to w; they are parts of $J(p,q)$, $J(q,s)$ and $J(p,s)$, respectively.

Consider the vertices of $V(R)$ next. If no edge e of $V(R)$ has v as an endpoint of $e - \overline{S}$ then v disappears from the diagram. So let us assume that at least one edge e of $V(R)$ has v as an endpoint of $e - \overline{S}$, i.e., v does not lie in the interior of S. The vertex v has degree at least three in $V(R)$. Also, either v lies outside $\overline{S}$ or at least one edge incident to v must intersect $\overline{S}$ in all neighborhoods $U(v)$ because otherwise the situation shown in Figure 13 would result, contradicting Fact 1. If v lies outside $\overline{S}$ then nothing changes in the vicinity

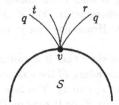

Figure 13.

of v. So let us assume that v lies on the boundary of S. Since $bd\ S$ is a simple closed curve which passes through v the edges incident to v split into two groups, those which intersect $\overline{S}$ in all neighborhoods $U(v)$ of v and those which do not. Both groups form a contiguous subsequence of the cyclic adjacency list of v, cf. Figure 14. Let f' and f'' be the first and

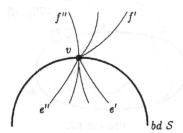

Figure 14.

the last edge of the subsequence of edges which run outside $\overline{S} \cap U(v)$ for a sufficiently small neighborhood $U(v)$ of v; $f' = f''$ is possible. Let e'' and e' be the first and last edge in counterclockwise order of the subsequence of edges which run inside $\overline{S} \cap U(v)$ for all small neighborhoods $U(v)$ of v. Let f' and e' border p-region and let f'' and e'' border q-region. Then we delete the edges between e'' and e' and add two new edges incident to v. These edges are part of $J(q,s)$ and $J(p,s)$, respectively.

At this point we have constructed all vertices of $V(R \cap \{s\})$ and their cyclic adjacency lists. We have not yet linked the two occurrences of each edge. Let $I := V(R) \cap \overline{S}$. We know an embedding of I into the plane. The boundary of the outer face is a closed curve since I is connected. Also, all the vertices on $bd\ S$ lie on I and hence a clockwise traversal of the boundary of the outer face yields the cyclic ordering of the vertices of $bd\ S$. This allows us to link the two occurrences of each edge. We conclude in

Lemma 7: $V(R \cup \{s\})$ can be constructed from $V(R)$ and $G(R)$ in $O\big(deg_{G(R)}(s) + 1\big)$ steps. Here our basic operation is considered as *one* step. $\qquad \square$

Let us now turn to the computation of $G(R \cup \{s\})$. We distinguish three cases: edges of $V(R \cup \{s\})$ which already were edges of $V(R)$, edges which are part of edges of $V(R)$, and edges which are completely new.

For any edge e of $V(R)$ not in conflict with s, the conflict information remains the same. Since $e \cap \overline{VR(s, R \cup \{s, t\})} \subseteq e \cap \overline{VR(s, R \cup \{s\})} = \emptyset$ the intersection $e \cap \overline{VR(t, R \cup \{t\})}$ remains unchanged when we add s to R for any $t \in S - R - \{s\}$.

If e is in conflict with s but $e \not\subseteq \overline{S}$ then $e - \overline{S}$ consists of at most two subsegments by Lemma 6. Let e' be one of those subsegments. For e' the same argumentation applies as above since $e' \cap \overline{VR(s, R \cup \{s\})} = \emptyset$. Hence any site in conflict with e' must be in conflict with e.

For a newly constructed edge, we show that it is sufficient to gather its conflict information from edges on that part of the boundary of the splitted region that is covered by S. Let e_{12} be such a new edge with endpoints x_1 and x_2. The new edge e_{12} either intersects the interior of exactly one Voronoi region $VR(p, R)$ or it runs on the boundary of a region $VR(p, R)$ such that $VR(p, R)$ and S are on different sides of e_{12}. This follows from the discussion in Case 1 and Case 2 above.

Let P be the part of the boundary of $VR(p, R)$ between x_1 and x_2 such that in all sufficiently small neighborhoods of x_1 and x_2 the path P is contained in $\overline{S}$.

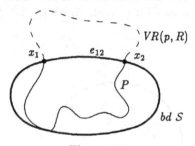

Figure 15.

Claim: $P \subseteq \overline{S}$

Proof: If P crosses the boundary $bd\ S$ of S then $V(R) \cap \overline{S}$ becomes unconnected, contradicting Lemma 5. $\qquad \square$

Lemma 8: Let $t \in S - R - \{s\}$ be in conflict with e_{12} in $V(R \cup \{s\})$. Then t is in conflict with P in $V(R)$.

Proof: Consider $VR(p, R)$. By the definition of conflict, a point $x \in e_{12}$ exists such that $x \in$

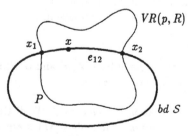

Figure 16.

$\overline{VR(t, R \cup \{s, t\})} \subseteq \overline{VR(t, R \cup \{t\})}$. Since we claim a contradiction we assume that t is not in conflict with P in $V(R)$, i.e., $\overline{VR(t, R \cup \{t\})} \cap P = \emptyset$. If x is on P the contradiction is evident. Thus, assume $x \notin P$. By assumption, $VR(t, R \cup \{s, t\}) \cap U(x_1) \subseteq VR(t, R \cup \{t\}) \cap U(x_1) = \emptyset$ for any sufficiently small neighborhood $U(x_1)$ of x_1. Now consider in any such neighborhood of x_1 the wedge spanned by e_{12} and the part of bd $VR(p, R)$ outside S. The points in this wedge all belong to $VR(p, R \cup \{s, t\})$. The same is true for any sufficiently small neighborhood of x_2. Since $VR(p, R \cup \{s, t\})$ is connected, there is a path Q from x_1 to x_2 running completely inside $VR(p, R \cup \{s, t\}) \subseteq VR(p, R \cup \{t\})$ except at the endpoints.

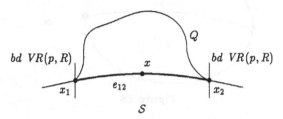

Figure 17.

We can assume that Q does not touch the boundary bd S of S (and therefore not includes x). By definition of P and Q the Voronoi region $VR(t, R \cup \{t\})$ cannot intersect these two paths. Moreover, x lies in the interior of the cycle $x_1 \circ P \circ x_2 \circ Q$; otherwise $VR(p, R)$ would not be simply connected. From $P \cap \overline{VR(t, R \cup \{t\})} = \emptyset$ and $x \in \overline{VR(t, R \cup \{t\})}$ we conclude that $VR(t, R \cup \{t\})$ lies in the interior of the cycle. Since bd $VR(t, R \cup \{t\}) \cap x_1 \circ P \circ x_2 = \emptyset$ this is a contradiction to the fact that $VR(p, R \cup \{t\})$ is simply connected. $\square$

Now we know all candidate-sites possibly in conflict with e_{12}, namely the sites in conflict with an edge or a vertex on P. Using our basic operation, we can test whether a conflict exists with these candidate sites or not. The conflict information of a newly constructed vertex $v = e \cap bd$ S is obtained from the conflict information of edge e in the same way. Note that we visited only conflicting edges and that each such edge can belong at most 2 times to a path P by planarity. We conclude in

Lemma 9: $G(R \cup \{s\})$ can be constructed from $V(R)$ and $G(R)$ in time

$$O\Big(\sum_{\{f, s\} \in G(R)} deg_{G(R)}(f) \Big).$$

Theorem 2: a) Let $s \in S - R$. Then the data structures $G(R \cup \{s\})$ and $V(R \cup \{s\})$ can be obtained from $G(R)$ and $V(R)$ in time

$$O\Big(\sum_{\{f, s\} \in G(R)} deg_{G(R)}(f) \Big).$$

b) For $R \subseteq S$, $|R| = 3$ and $\infty \in R$, the data structures $V(R)$ and $G(R)$ can be set up in time $O(n)$ where $n = |S|$.

Proof: a) This point summarizes Lemmas 7 and 9.

b) The Voronoi diagram $V(R)$ for three sites ∞, p and q has the structure shown in Figure 18 and can certainly be set up in time $O(1)$. Also for each of the edges e of $V(R)$ and each of the $n-3$ sites in $S-R$ one can test $e \cap \overline{VR(t, R \cup \{t\})} \neq \emptyset$ by our basic operation in time $O(1)$. This proves (b). $\qquad\Box$

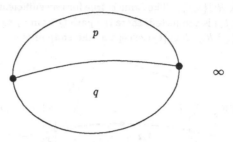

Figure 18.

We next turn to the analysis of the algorithm. Our goal is to analyze the running time of our algorithm by an appeal to Clarkson and Shor's results about the running time of randomized incremental construction. However, Clarkson and Shor's analysis relies heavily on their general position assumption. It states that each region (here edges and vertices) is *uniquely* determined by a bounded number of objects (here sites). Now observe that any vertex that has $r_1, \ldots, r_l$-regions, $l > 3$, incident to it is defined by any 3-tuple (r_i, r_j, r_k), $1 \leq i < j < k \leq l$, by Lemma 3. A similar observation holds for any edge with endpoints of degree greater than 3. This shows that the general position assumption of [CS] is generally not satisfied, when the bisecting curves are not in general position. However, the general position assumption of [CS] is satisfied if the bisecting curves are in general position as defined in [MMO], i.e., any intersection is a proper crossing and no four curves have a common point.

The idea behind our proof is now as follows. We take our system of bisecting curves and conceptually perturb it so as to obtain a system in general position. The perturbation is carried out in such a way that no conflicts are lost. This allows us to show that the cost of one iteration step of the algorithm on the original system is bounded by a constant times the cost of the corresponding step on the perturbed system; the asymptotic time bound thus holds for both. The perturbed system is in general position and hence Clarkson and Shor's analysis applies.

In [K89] Klein proves the following

Theorem 3: Let $\mathcal{J} = \{J(p, q); p, q \in S, p \neq q\}$ be a system of bisecting curves as defined in section II. Then $\mathcal{J}$ can be deformed in such a way that the resulting system $\tilde{\mathcal{J}}$ has the following properties:
1) No other intersections than cross-points exist between the curves in $\tilde{\mathcal{J}}$. At most 3 curves cross at a point (general position)
2) For each subset $R \subseteq S$: After contracting some edges, the perturbed Voronoi diagram $\hat{V}(R)$ differs from the original diagram $V(R)$ only by a deformation.

A perturbation is neccessary for those curves that touch each other or even share a common segment and for more than 3 curves intersecting in a common point. Curves sharing a common segment are spread out like wires of a cable. A point common to more than 3 curves or a touching point of two curves is split into a graph of newly constructed intersection points and curves segments connecting them. These segments give rise to the edges in $\tilde{V}(R)$ that are contracted in Theorem 3.

We will now show that no conflict information is lost by the perturbation. First consider an edge e in $V(R)$ that is in conflict with some site t in $S - R$. Edge e also exists in $\tilde{V}(R)$. Moreover, Theorem 3 implies that e exists in $\tilde{V}(R \cup \{t\})$ iff e exists in $V(R \cup \{t\})$. Thus, each edge of $V(R)$ is also represented in $\tilde{V}(R)$ and their conflict information is the same.

Next consider a vertex $v \in V(R)$ that is replaced by a graph $\mathcal{G}(v)$ in $\tilde{V}(R)$. Analogously, Theorem 3 implies that v exists in $V(R \cup \{t\})$ iff $\mathcal{G}(v)$ exists in $\tilde{V}(R \cup \{t\})$. Hence the vertices of $\mathcal{G}(v)$ all have the same conflict information as v.

Let us now compare the running time of our algorithm on both systems of bisecting curves $\mathcal{J}$ and $\tilde{\mathcal{J}}$, respectively. Addition of a new site $s \in S - R$ to $V(R)$ and $\tilde{V}(R)$ takes time proportional to

$$\sum_{\{f,s\} \in G(R)} deg_{G(R)}(f) \quad \text{and} \quad \sum_{\{f,s\} \in \tilde{G}(R)} deg_{\tilde{G}(R)}(f), \quad \text{respectively.}$$

As discussed above we have $deg_{G(R)}(e) = deg_{\tilde{G}(R)}(e)$ for any edge $e \in V(R)$ and $deg_{G(R)}(v) = deg_{\tilde{G}(R)}(v')$ for any vertices $v \in V(R), v' \in \mathcal{G}(v) \subseteq \tilde{V}(R)$; thus $deg_{G(R)}(s) \leq deg_{\tilde{G}(R)}(s)$. Hence, any iteration step of the algorithm on the original curve system costs at most a constant factor more than the corresponding step for the perturbed system.

Theorem 4: The abstract Voronoi diagram $V(S)$ of n sites can be constructed by a randomized algorithm in time $O(n \log n)$.

Proof: In [CS] Clarkson and Shor proved that randomized incremental construction has expected running time

$$O\left(m(n) + n \cdot \sum_{1 \leq r \leq n/2} m(r)/r^2 + n\right)$$

provided that their general position assumption holds, initialization takes time $O(n)$ and addition of an object (here site) s to the set R takes time proportional to

$$\sum_{\{f,s\} \in G(R)} deg_{G(R)}(f);$$

the summation is over all regions (here edges and vertices) of the current structure (here Voronoi diagram $V(R)$) which conflict with site s. Also, $m(r)$ is the expected size of the structure for a random subset $R \subseteq S$ of r elements. In our case we have $m(r) = O(r)$ since the Voronoi diagram is a planar graph with r regions and therefore has less than $3r - 6$ edges and $2r - 4$ vertices by Euler's relation, initialization takes time $O(n)$ by Theorem 2b and the time bound for the update step holds by Theorem 2a. Thus the running time of our algorithm on the perturbed system is $O(n \log n)$. From the considerations above we conclude that the same bound holds for the original system. □

Remark: In our algorithm ∞ is always a member of R. An inspection of Clarkson's argument shows that this minor deviation from randomness does not change the time bound.

IV. Conclusions and Open Problems

We showed under very general assumptions on the bisecting curves that the construction of abstract Voronoi diagrams can be transformed efficiently and purely combinatorially into the construction of abstract Voronoi diagrams for five sites. Further research will be done on the question of how the construction of the 5-site-diagram can be carried out efficiently dependent on the type of bisecting curves. Our goal is a "universal" algorithm for Voronoi diagrams. Nevertheless, we still have open problems to solve: Can the concept of abstract Voronoi diagrams be generalized to higher dimensions? What can be done in two dimensions without the assumption that bisectors are non-closed curves?

References:

[A]   F. Aurenhammer (1988): *Voronoi Diagrams — A survey*, tech. report 263, Institutes for Information Processing, Graz Technical University, Austria

[CS]   K. L. Clarkson, P. W. Shor (1988): *Algorithms for Diametral Pairs and Convex Hulls that are Optimal, Randomized and Incremental*, Proc. 4th ACM Symposium on Computational Geometry, pp. 12–17

[GS]   L. Guibas, J. Stolfi (1985): *Primitives for the Manipulation of General Subdivisions and the Computation of Voronoi Diagrams*, ACM Transactions on Graphics 4, pp. 74–123

[K88a] R. Klein (1988): *Abstract Voronoi Diagrams*, Habilitationsschrift, Mathematics Faculty of University of Freiburg i. Br., LNCS 400

[K88b] R. Klein (1988): *Abstract Voronoi Diagrams and their Applications (extended abstract)*, in: H. Noltemeier (Ed.), Computational Geometry and its Applications (CG '88), LNCS 333, pp. 148–157

[K88c] R. Klein (1988): *Voronoi Diagrams in the Moscow Metric (extended abstract)*, in: J. van Leeuwen, (Ed.), Graphtheoretic Concepts in Computer Science (WG '88), LNCS 344, pp. 434–441

[K89]   R. Klein (1989): *Combinatorial Properties of Abstract Voronoi Diagrams*, WG 89, Rolduc, to appear in LNCS

[L]   D. T. Lee (1980): *Two-Dimensional Voronoi Diagrams in the L_p-Metric*, J. ACM 27, pp. 604–618

[LS]   D. Leven, M. Sharir (1986): *Intersection and Proximity Problems and Voronoi Diagrams*, in: J. Schwartz and C. K. Yap (eds.), Advances in Robotics, Vol. 1, Lawrence Erlbaum, pp. 187–228

[MMO]   K. Mehlhorn, S. Meiser, C. Ó' Dúnlaing (1989): *On the Construction of Abstract Voronoi Diagrams*, Journal of Discrete & Computational Geometry, to appear

Searching in Higher Dimension

BERNARD CHAZELLE

Department of Computer Science
Princeton University
Princeton, NJ 08544

Abstract:

Multidimensional searching addresses search problems where the underlying universe lacks a total order. Hit detection in computer graphics, range searching in databases, point location in geographical maps are well-known instances of such problems. We will review some recent developments in that area. Point location in a nonlinear universe involves preprocessing a collection of real-algebraic varieties to facilitate the location of a query point. By using probabilistic data structuring techniques we can reduce the problem to the general question of stratifying semi-algebraic sets into simple smooth manifolds. Another subject of current interest is the manipulation of lines and segments in Euclidean 3-space. Given a set of lines one may wish to compute their upper envelope or be ready to answer questions of the form: given a new line (or a set of new lines), does it (or do they) all lie above the lines in the database? This problem has applications to ray-tracing in a polyhedral environment. Another subject with interesting recent developments is *polytope range searching*: the problem involves preprocessing a collection of points in d-space so that, given a query simplex q, the number of points inside q can be evaluated efficiently. We wil report on practical solutions which offer a whole spectrum of quasi-optimal tradeoffs between storage and query time.

Finding Extrema With Unary Predicates

David G. Kirkpatrick [*]

and

Feng Gao [†]

Computer Science Department

University of British Columbia

Abstract

We consider the problem of determining the maximum and minimum elements of a set $X = x_1, \ldots, x_n$ of integers, drawn from some universe $\mathcal{U}$, using only unary predicates of the inputs. It is shown that $\Theta(n + log|\mathcal{U}|)$ unary predicate evaluations are necessary and sufficient, in the worst case. Results are applied to i) the problem of determining approximate extrema of a set of real numbers, in the same model, and ii) the multiparty broadcast communication complexity of determining the extrema of a set of integers held by distinct processors.

1   Introduction

A familiar guessing game has one player A choose a number x while a second player B attempts to determine x by a short sequence of yes-no questions. It is widely appreciated (even by those who are not computer scientists) that a guessing strategy exists for B which is guaranteed to identify the chosen number in at most $\log_2 |\mathcal{U}|$ questions, where $\mathcal{U}$ denotes the universe of possible numbers. The optimality of this strategy is easily established by an elementary adversary argument (*cf.* [1]); in effect A tries to postpone her specific choice as long as possible.

We are interested in a generalization of this game in which some number $n \geq 1$ players $A_1, \cdots, A_n$ choose numbers $x_1, \cdots, x_n$ respectively while a last player B attempts to determine some function $f(X)$ of $X = (x_1, x_2, \cdots, x_n)$ by means of yes-no questions directed to $A_1, \cdots, A_n$ individually. It follows from the conventional game that B can determine all of the values $x_1, \cdots, x_n$ (and thereafter, $f(X)$) in at most $n \log |\mathcal{U}|$ questions. However, evaluation of f many not require exact knowledge of all of its arguments and hence more efficient guessing strategies may exist for specific functions of interest. In this paper we show that this is the case for the functions max and min; their complexities in this yes-no question model are shown to be $\Theta(n + \log |\mathcal{U}|)$.

[*]Supported in part by Natural Sciences and Engineering Research Council Grant A3583 and a BCASI Fellowship.

[†]Supported in part by Natural Sciences and Engineering Research Council Grant A0482

Before setting out our specific results in more detail it is worth describing some of the motivation for our study. The multiplayer guessing game models in a natural way the evaluation of functions in two realistic computational settings. The first involves a situation in which the function arguments are provided by independent black boxes. Since the arguments are not available explicitly they can not be directly compared, for example. Such would be the case if, for example, each x_i were defined to be the zero-crossing of some monotone function on U. In this situation, it is reasonable to restrict algorithms to the use of (possibly constrained) unary predicates of the x_i's.

The second setting captured by the multiplayer guessing game involves cooperative distributed computation. In this case, the arguments $x_1, \cdots, x_n$ are known individually and explicitly to n distinct processors $P_1, \cdots, P_n$ respectively, and the objective is to collectively determine $f(X)$ by using the least amount of broadcast (one-to-all) communication. In this case individual processors can communicate information based on their evaluation of arbitrary unary functions of their own input, together with earlier communications. Each bit of communication of processor P_i can be expressed as a unary predicate of x_i. Our model for multiparty communication complexity is a natural extension of the two-party model of Yao [6]. It closely resembles the multiparty models of [2] and [3] in which parties communicate through a shared "blackboard". Our model differs from these in that we assume no sharing of input values.

Hereafter we will restrict our attention to the cases where f is either max or min, the complexity of which are well understood on a binary comparison model [4]. The possibility for improvement over the naive $O(n \log |U|)$ algorithm suggests itself almost immediately in these cases. Obviously, if one has determined the value of x_1 (or more generally $\max\{x_1, \cdots, x_i\}$) then the universe of interest for all subsequent values has been reduced. (If x_{i+1} is determined to be no larger than $\max\{x_1, \cdots, x_i\}$ then it can be ignored without being known explicitly.) However, this observation alone does not suffice to produce an asymptotic improvement, in the worst case.

The naive algorithm (and, presumably any other approach that has suggested itself to the reader) exploits the total order of U and uses only *threshold predicates*, that is questions of the form "is x_i greater than or equal to c", for fixed constants c. We have intentionally included non-threshold predicates in our formulation of problem since their use is another potential source of improvement over the naive schemes. As it turns out, for the computation of max and min, threshold predicates suffice to realize (to within small constant factors) the lower bounds required of arbitrary unary predicates.

A (temporary) restriction to threshold predicates does serve to simplify the expression of both lower bound arguments and algorithms. Under this restriction our problems can be interpreted as natural matrix searching problems. This matrix problem together with the associated lower bound results are described in section 2. The corresponding algorithms (upper bounds) are presented in section 3. Section 4 interprets the algorithmic results in the communication complexity setting and describes the more general (non-threshold predicates) lower bounds.

Our principal results for the computation of $x_{\max} = \max\{x_1, \cdots, x_n\}$ can be summarized as follows:

i. Determining $x_{\max}$ requires $\Omega(n + \log |U|)$ unary predicate evaluations.

ii. If U is the set of all integers $x_{\max}$ can be determined in $O(n + \log |x_{\max}|)$ threshold predicate evaluations.

iii. If $\mathcal{U}$ is the set of reals, then $x_{\max}$ can be determined to within absolute error ϵ in $O(n + \log x_{\max} + \log 1/\epsilon)$ threshold predicate evaluations. If $\mathcal{U}$ is the set of reals in the range $(0, 1]$, then $x_{\max}$ can be determined to within relative error $\epsilon \leq 1$ in $O(n + \log \log 1/x_{\max} + \log 1/\epsilon)$ threshold predicate evaluations.

iv. The communication complexity of determining x_{max} in the multiparty broadcast model is $\Omega(n)$ messages and $\Omega(n + \log |\mathcal{U}|)$ bits and these bounds are simultaneously realizable.

2 Lower bounds for threshold predicates

Suppose that $\mathcal{U}$ is the set of integers in the range $[0, m]$, where $m \geq 1$. If we restrict our attention to algorithms that use only threshold predicates then the problem of determining $x_{\max} = \max\{x_1, \cdots, x_n\}$ can be recast as a matrix searching problem. Consider the binary matrix $M[1 : n, 1 : m]$ defined by $M[i, j] = 1$, if $x_i \geq j$ (and $M[i, j] = 0$ otherwise). The evaluation of a threshold predicate on input x_i amounts to a single probe into the matrix M. Determining $x_{\max}$ corresponds to locating the rightmost (highest indexed) column of M containing a 1. Since the rows of M are monotonically decreasing it is clear that exhausive search of M is unnecessary. Indeed, the naive algorithm sketched in the introduction can be viewed as the (binary) search in successive rows for the rightmost one in that row. The question is, is there a significantly better probing strategy? Note that the constraints implicit in the structure of M make the problem fundamentally different from that associated with the detection of graph properties (*cf.* [5]).

We start by formulating a simple adversary-based lower bound on the number of probes required to determine $x_{\max}$.

Lemma 2.1. Finding $x_{\max}$ requires $\Omega(n + \log m)$ probes of M even if it is known that $x_{\max} - 1 \leq x_i \leq x_{\max}$, for $1 \leq i \leq n$.

Proof. (sketch) The adversary responds to successive probes in such a way as to maximize the range of possible values that $x_{\max}$ could assume. When this range has been reduced to two elements $\{j, j + 1\}$, the adversary responds $M[i, j] = 1$ and $M[i, j + 1] = 0$ for all subsequent probes. Since $\log m$ probes are required to reduce the range to size two and thereafter n probes are required to confirm that $x_{\max} = j$ (and not $j + 1$), the result follows. $\square$

Remark. Essentially the same strategy shows that $\Omega(n + \log m)$ probes are required to determine $x_{\min}$ as well.

Corollary 2.1. If $\mathcal{U}$ is the set of integers in the range $[m_L, m_U]$, where $m_L < m_U$, then determining $x_{\max}$ or $x_{\min}$ requires $\Omega(n + \log(m_U - m_L))$ threshold predicate evaluations.

It is clear from corollary 2.1 that the worst case complexity of finding extrema in an arbitrary collection of integers can be arbitrarily high. A second corollary suggests that it may be more natural to attribute this complexity to the size of the output.

Corollary 2.2. If $\mathcal{U}$ is the set of integers in the range $[2^k + 1, 2^{k+1}]$, where $k \geq 1$, then determining $x_{\max}$ (or $x_{\min}$) requires $\Omega(n + \log x_{\max})$ threshold predicate evaluations.

Despite the assumptions stated in Lemma 2.1 (which amount to significant constraints on the adversary) the lower bounds of this section can be realized (to within small constant factors) by a fairly straightforward algorithm. What the lower bounds suggest is that the algorithm must somehow globally reduce the range of candidates for $x_{\max}$ without investing too many probes in any particular row of M.

3   Algorithms using threshold predicates

We continue with the matrix formulation of the extrema-finding problems introduced in section 2. For convenience we extend the matrix M by two columns and define $M[i, 0] = 1$ and $M[i, m + 1] = 0$, for $1 \leq i \leq n$. (Note this does not alter the monotonicity of the rows of M).

As the algorithm proceeds the state of knowledge about x_i is captured entirely by the positions of the rightmost 1 and the leftmost 0 discovered (so far) in row i. We denote these quantities by $\rho(i)$ (rightmost 1) and $\lambda(i)$ (leftmost 0), respectively. (Initially, $\rho(i) = 0$ and $\lambda(i) = m + 1$). We denote by $\rho_{\max}$ the value $\max\{\rho(i) : 1 \leq i \leq n\}$, the rightmost 1 discovered so far. At any point in time row i is said to be *active* if $\lambda(i) > \rho_{\max} + 1$, that is it remains a possibility that $x_i > \rho_{\max}$.

The idea of the algorithm is to maintain a set I including the indices of all active rows. At any time $\rho_{\max} = \max\{x_i : i \notin I\}$. The algorithm continues until I has been reduced to the empty set, at which point $\rho_{\max} = x_{\max}$. More specifically, I is maintained as a collection of disjoint sets $B_s, B_{s+1}, \cdots, B_t$ (called *blocks*), where $|B_k| = \alpha_k$ and $\lambda(i) = \beta_k$, for all $i \in B_k$ (i.e. indices in the same block share the same λ value). Figure 1 gives a snapshot of the known contents of M at an arbitrary point during the execution of the algorithm. The γ values (gaps between blocks) play a role in describing invariant properties of the algorithm. Note that if $\rho_{\max} < x_{\max}$ (that is the rightmost 1 has not yet been discovered) then any rightmost 1 must lie in the unshaded portion of the matrix.

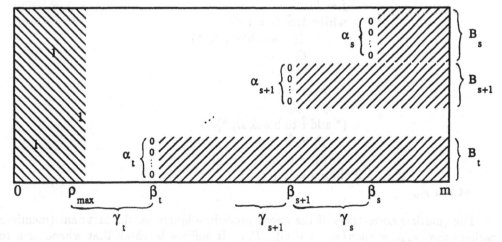

Figure 1

We are now in a position to present the algorithm in pseudo-code. We use the notation $s \nwarrow S$ (respectively $S \nwarrow s$) to denote the operation of removing an element of set S and assigning it to the variable s (respectively adding the element denoted by s to the set S).

procedure find_max (M, n, m)
    (* initialize *)
        $s \leftarrow t \leftarrow 0$
        $B_0 \leftarrow \{1, \cdots, n\}; \ \alpha \leftarrow n; \ \beta_0 \leftarrow 2^{\lceil \log(m+1) \rceil}$
        $\rho_{\max} \leftarrow 0$
    while $t \geq s$ **do**
        if $\beta_t = \rho_{\max} + 1$
            then (* eliminate block B_t *)
                $B_t \leftarrow \phi$
                $t \leftarrow t - 1$
            else
                (* select next active row *)
                $\hat{i} \nwarrow B_s$
                $\alpha_s \leftarrow \alpha_s - 1$
                $\hat{j} \leftarrow (\rho_{\max} + \beta_t)/2$
                if $M[\hat{i}, \hat{j}] = 0$
                    then
                        (* start a new block *)
                        $\alpha_{t+1} \leftarrow 1$
                        $\beta_{t+1} \leftarrow \hat{j}$
                        $B_{t+1} \nwarrow \hat{i}$
                        $t \leftarrow t + 1$
                  else
                      (* find highest indexed old block to assign $\hat{i}$ to *)
                        $\rho_{\max} \leftarrow \hat{j}$
                        $\hat{j} \leftarrow \beta_t$
                        while $M[\hat{i}, \hat{j}] = 1$ **do**
                            (* eliminate block B_t *)
                                $B_t \leftarrow \phi$
                                $t \leftarrow t - 1$
                          $\rho_{\max} \leftarrow \hat{j}$
                          $\hat{j} \leftarrow \beta_t$
                      (* add $\hat{i}$ to block B_t *)
                        $B_t \nwarrow \hat{i}$
                        $\alpha_t \leftarrow \alpha_t + 1$
            if $\alpha_s = 0$ **then** $s \leftarrow s + 1$
    return $\rho_{\max}$

The (partial) correctness of the above procedure hinges on the invariant (mentioned earlier) that $\rho_{\max} = \max\{x_i : i \notin \cup_{k=s}^{t} B_k\}$. It suffices to check that whenever a row index i is eliminated (no longer belongs to $\cup_{k=s}^{t} B_k$), $\lambda(i) \leq \rho_{\max} + 1$ (that is row i is no longer active). The termination (and hence correctness) of the procedure *find_max* is an immediate consequence of the following complexity analysis.

To analyse the complexity of our procedure we need to establish some additional invariant properties. Specifically, if we define $\gamma_t = \beta_t - \rho_{max}$, $\gamma_k = \beta_k - \beta_{k+1}$, for $s \le k < t$, and $\hat{\alpha} = \sum_{k=1}^{t} \alpha_k$, then we have the following easily established properties:

outer loop invariants

$$\gamma_t = 2^{\lceil \log(m+1) \rceil - \hat{\alpha}}$$
$$\gamma_{t-1} = \gamma_t \cdot 2^{\alpha_t - 1}, \qquad \text{provided } t > 2$$
$$\text{and} \quad \gamma_{k-1} = \gamma_k \cdot 2^{\alpha_k}, \qquad \text{for } s < j \le t-1$$

inner loop invariants

$$\gamma_t = 2^{\lceil \log(m+1) \rceil - \hat{\alpha} - 1}$$
$$\text{and} \quad \gamma_{k-1} = \gamma_k \cdot 2^{\alpha_k}, \text{ for } s < k \le t$$

Combining the above invariant properties it is now straightforward to establish a relationship between the number of probes made at any point in time, denoted #probes, and the variables γ_t and α_k, $0 \le k \le t$.

Specifically, for the outer loop we have

$$\#probes = 2n + 2\lceil \log(m+1) \rceil - 2 \log \gamma_t$$
$$- 2\sum_{k=s}^{t} \alpha_k - t$$

and for the inner loop we have

$$\#probes = 2n - 1 + 2\lceil \log(m+1) \rceil - 2 \log \gamma_t$$
$$- 2\sum_{k=s}^{t} \alpha_k - t$$

Since the variables $\gamma_t, \alpha_k, 0 \le k \le t$, and t are all non-negative it follows immediately that the total number of probes is bounded above by $2(n + \lceil \log(m+1) \rceil)$. Hence we have established the following:

Lemma 3.1. Procedure *find_max* determines the index x_{max} the rightmost column of matrix $M[1:n, 1:m]$ containing a 1, using at most $2(n + \lceil \log(m+1) \rceil)$ probes.

We now turn our attention to some direct applications of lemma 3.1.

Theorem 3.1. If $\mathcal{U}$ is any set of integers then x_{max} and x_{min} can be determined in $O(n + \log |\mathcal{U}|)$ threshold predicate evaluations.

Proof. Exploit the natural (order preserving) bijection from $\mathcal{U}$ to $\{0, \cdots, |\mathcal{U}| - \infty\}$. To compute x_{min} note that $\min\{x_1, \cdots, x_n\} = m - \max\{m - x_1, \cdots, m - x_n\}$. $\square$

Theorem 3.2. If $\mathcal{U}$ is the set of all integers then x_{max} and x_{min} can be determined in $O(n + \log(|x_{max}| + 1))$ and $O(n + \log(|x_{min}| + 1))$ threshold predicate evaluations respectively.

Proof. It suffices to compute suitable upper bounds on $\mid x_{\max} \mid$ and $\mid x_{\min} \mid$ efficiently. For example, if $x_{\max}$ is positive the following fragment determines a bound m satisfying $m/2 \le x_{\max} < m$ in $O(n + \log(x_{\max} + 1))$ probes.

$\hat{i} \leftarrow 1, m \leftarrow 1$
while $\hat{i} \le n$ do
  if $M[\hat{i}, m] = 1$
      then $m \leftarrow 2m$
      else $\hat{i} \leftarrow \hat{i} + 1$

Similarly, assuming $x_{\min}$ is positive the following fragment determines a bound m satisfying $m/2^n \le x_{\min} < m$ in $O(n + \log(x_{\min} + 1))$ probes.

$\hat{i} \leftarrow 0, m \leftarrow 1$
while $M[\hat{i} + 1, m] = 1$ do
  $m \leftarrow 2m$
  $\hat{i} \leftarrow (\hat{i} + 1) \bmod n$

The cases where $x_{\max}$ and $x_{\min}$ are negative are similar. □

Theorem 3.3. If $\mathcal{U}$ is the set of all positive reals then $x_{\max}$ (respectively $x_{\min}$ can be determined to within absolute error $\epsilon > 0$ in $O(n + \log x_{\max} + \log 1/\epsilon)$ (respectively $O(n + \log x_{\min} + \log 1/\epsilon)$) threshold predicate evaluations.

Proof. It suffices to apply Theorem 3.2 to the set $\{y_1, \cdots, y_n\}$ where $y_i = \lceil x_i/\epsilon \rceil$. □

Theorem 3.4. If $\mathcal{U}$ is the set of all positive reals in $(0, 1]$ then $x_{\max}$ (respectively, $x_{\min}$) can be determined to within relative error ϵ, where $0 < \epsilon \le 1$, in $O(n + \log \log(1/x_{\max}) + \log 1/\epsilon)$ (respectively, $O(n + \log \log(1/x_{\min}) + \log 1/\epsilon)$) threshold predicate evaluations.

Proof. In this case we apply Theorem 3.3 to the set $\{z_1, \cdots, z_n\}$ where $z_i = \log_{1+\epsilon} 1/x_i$ and observe that an absolute error of 1 in determining $\min\{z_1, \cdots, z_n\}$ corresponds to a relative error of ϵ in determining $\max\{x_1, \cdots, x_n\}$. Since $\log z_i = \log \log 1/x_i - \log \log 1 + \epsilon$ and $\log 1 + \epsilon \ge \epsilon$ for $\epsilon \in (0, 1]$, the result follows. □

4 Communication complexity of extrema finding

As we noted in the introduction, the determination, using unary predicates, of $\max\{x_1, \cdots, x_n\}$ corresponds in a natural way to the distributed evaluation of $\max\{x_1, \cdots, x_n\}$ on a multiparty broadcast model of distributed computation. In fact a slight variant of our algorithm *find_max* presented in section 3 translates to a optimal protocol for the cooperative evaluation of $\max\{x_1, \cdots, x_n\}$. In this protocol processor P_i holds x_i and broadcasts (in its turn) bits from specific positions of row i. If we ignore the rows that have ceased to be active (the corresponding processors become passive in the protocol) a natural implementation of procedure *find_max* processes rows of M in a round-robin fashion. When its row is selected as the next active row, processor P_i simulates the sequence of probes of the ith row of M specified by the procedure *find_max*, and packages the corresponding bit sequence into a single message for broadcast to the other processors. This information suffices for all of the other processors to update their ongoing simulations of procedure *find_max*.

As it has been described, each probe made by procedure *find_max* corresponds to a single bit of communication in the protocol. Thus Theorem 3.1 can be reinterpreted as asserting that x_{max} and x_{min} can be computed using $O(n + \log |\mathcal{U}|)$ bits of (broadcast) communication in a distributed computation model. Note that the protocol may use asymptotically as many messages as it does bits (since the procedure *find_max* may make only $O(1)$ probes for each repetition of its outer loop). However a very simple modification of procedure *find_max* ensures that the corresponding communication protocol uses only $O(n)$ messages while retaining the same asymptotic bound on the number of bits. The idea is to replicate rows $\lceil (\log |\mathcal{U}|)/n \rceil$ times. The resulting matrix has $n' = O(n + \log |\mathcal{U}|)$ rows (with, of course, the same value of x_{max}) and hence requires (asymptotically) the same number of probes. Assuming that bits from replicated rows are bundled into the same message, it follows that each message contains at least $\lceil (\log |\mathcal{U}|)/n \rceil$ bits and hence there are $O(n)$ messages in total. We summarize the above construction in the following:

Theorem 4.1. There exists a multiparty communication protocol that determines $\max\{x_1, \cdots, x_n\}$ in $O(n)$ (broadcast) messages with a total of $O(n + \log |\mathcal{U}|)$ bits.

We conclude this section by outlining our argument that the protocol described in Theorem 4.1 is asymptotically optimal. (A more detailed proof, including a more formal description of our communication model, will appear elsewhere.) Specifically:

Theorem 4.2. The communication complexity of determining $\max\{x_1, \cdots, x_n\}$ in the multiparty broadcast model is $\Omega(n)$ messages and $\Omega(n + \log |\mathcal{U}|)$ bits.

Proof. The message lower bound is clear; in the worst case each participant must communicate something (here we exploit asynchrony) or else the value of $\max\{x_1, \cdots, x_n\}$ can not be known to all processors.

The bit lower bound is a little less obvious and requires, in part, a modest generalization of Lemma 2.1. First, we note that our translation of an algorithm for determining x_{max} with unary predicates to a (broadcast) protocol for computing x_{max}, can be reversed. If we are only interest in bit complexity any protocol can be trivially modified to use only 1-bit messages. Since each such message can be interpreted as the evaluation of some unary predicate on one of the inputs, a protocol specifies an algorithm for determining x_{max} using unary (but not necessarily threshold) predicates.

At any point in time there is partial information available to all of the processors about each of the numbers x_i. If a specific number is consistent with this partial information we say that it is an *i-candidate*. If there exist candidates for all of the other x_j that make a given i-candidate the largest, then this i-candidate is said to be *viable*. The adversary maintains the subset V of $\mathcal{U}$ consisting of numbers that are viable i-candidates, for all i. With each successive unary predicate the adversary responds in such a way that the size of V is decreased by at most 2. (Note that every element of V must remain viable for at least one of the two outcomes of the predicate.) Since $V = \mathcal{U}$ initially, the adversary can force at least $\log |\mathcal{U}|$ predicate evaluations before x_{max} is determined. $\square$

5 Concluding remarks

We have identified the complexity of extrema finding using unary predicates to within a factor of 2. It will be clear that the upper bound can be improved somewhat by simply avoiding questions (probes) whose results are implied by earlier questions. (This is most clearly evident by studying the algorithm for the case $n = 1$.) It remains to specify exactly the complexity of extrema finding in this model.

It is of interest to study the complexity of other basic functions in this same model. Similarly, our results motivate the investigation of the multiparty communication complexity of other basic functions.

References

[1] Aho, A.V., Hopcroft, J.E. and Ullman, J.D., *The Design and Analysis of Computer Algorithms*, Addison-Wesley, 1975.

[2] Babai, L., Nisan, N. and Szegedy, M., Multiparty protocols and logspace-hard pseudorandom sequences, Proc. 21st ACM STOC (1989), 1-11.

[3] Chandra, A., Furst, M. and Lipton, R., Multiparty protocols, Proc. 15th ACM STOC (1983), 94-99.

[4] Knuth, D.E., *The Art of Computer Preogramming, Vol.3*, Addison-Wesley, 1973.

[5] Rivest, R. and Vuillemin, J., On recognizing graph properties from adjacency matrices, *Theoretical Comuter Science* 3 (1978), 371-384.

[6] Yao, A.C., Some complexity questions related to distributed computing, Proc. 11th ACM STOC (1979), 209-213.

Implicitly Searching Convolutions and Computing Depth of Collision

David Dobkin*   John Hershberger†   David Kirkpatrick‡   Subhash Suri§

Abstract

Given two intersecting polyhedra P, Q and a direction d, find the smallest translation of Q along d that renders the interiors of P and Q disjoint. The same question can also be asked without specifying the direction, in which case the minimum translation over all directions is sought. These are fundamental problems that arise in robotics and computer vision. We develop techniques for implicitly building and searching convolutions and apply them to derive efficient algorithms for these problems.

1 Introduction

The computation of spatial relationships among geometric objects is a fundamental problem in such areas as robotics, computer-aided design, VLSI layout, and computer graphics. In a dynamic environment where objects are mobile, intersection or proximity among objects has obvious applications. Consider, for instance, the problem of collision detection in robot motion planning. The Euclidean distance is a commonly used measure in these areas. Numerous efficient algorithms are known for computing the minimum distance between two polyhedra in two and three dimensions. Whenever two objects intersect, this distance measure is zero. Thus, it fails to provide any information about the *extent of penetration*. The notion of *negative distance* has been proposed by Buckley and Leifer [5] and Cameron and Culley [6] to rectify this discrepancy. We follow Keerthi and Sridharan [17] and define the following measure of negative distance.

Given two intersecting polyhedra P and Q and a direction d, let $\sigma_d(P,Q)$ denote the minimum distance by which Q must be translated along d to make the interiors of P and Q disjoint. We call $\sigma_d(P,Q)$ the *depth of collision* between P and Q in direction d. The *minimum depth of collision* between P and Q, denoted $\sigma(P,Q)$, is defined as the minimum of $\sigma_d(P,Q)$ over all directions d. The definition of $\sigma(P,Q)$ agrees with the measure proposed in [6].

The problem of computing the depth of collision between two intersecting objects is addressed by Cameron and Culley [6], however, the complexity of their algorithm is not analyzed. Keerthi

---

*Department of Computer Science, Princeton University, NJ 08544, U.S.A.

†DEC Systems Research Center, 130 Lytton Street, Palo Alto, CA 94301, U.S.A.

‡Department of Computer Science, University of British Columbia, Vancouver, BC V6T 1W5, Canada.

§Bell Communications Research, 445 South Street, Morristown, NJ 07960, U.S.A.

and Sridharan [17] consider the problem for two convex polygons. Let P and Q be two intersecting convex polygons, with n and m vertices, respectively. Keerthi and Sridharan [17] present an $O(\log n \log m)$ time algorithm for computing the directional depth of collision, $\sigma_d(P, Q)$. For the minimum depth of collision, $\sigma(P, Q)$, they achieve $O(n + m)$ time. In this paper, we derive the following improvements and extensions of these results.

Given two intersecting convex polygons P and Q, where $|P| = n$ and $|Q| = m$, we can determine $\sigma_d(P, Q)$ for any direction d in optimal time $O(\log n + \log m)$. In addition, after linear-time preprocessing, we can in $O(\log(n+m))$ time compute the minimum depth of collision, $\sigma(P, Q)$, for any placement of Q.

Next, we relax the convexity assumption on one of the polygons. Given a simple nonconvex polygon P and a convex polygon Q, we can build a data structure in $O(n \log^2(n + m) + m)$ time such that both directional depth and minimum depth of collision queries can be answered for any placement of Q in time $O((n + m) \log(n + m))$. The data structure can be extended to support directional depth of collision queries in $O(n^{2/3+\delta})$ time. The data structure requires $O(n + m)$ space. The query time can be reduced to $O(\sqrt{n} \log^2 n)$, by increasing the preprocessing time and space.

We then consider the directional depth of collision problem for two convex polytopes in three dimensions. For suitably preprocessed polyhedra P and Q, we can determine $\sigma_d(P, Q)$ in time $O(\log n \log m)$, for any direction d. The representation assumed for the polyhedra is the hierarchical representation, originally proposed by Dobkin and Kirkpatrick [8].

The unifying idea in all our algorithms is implicitly searching the convolution of two polyhedra. It is well-known that if two polytopes P and Q intersect, then the difference of their reference vectors lies in their convolution. Thus the problem of computing depth of collision reduces to searching a convolution. In all interesting cases, however, an explicit construction of the convolution can be rather expensive; in two dimensions, the convolution can have quadratic size if one of the polygons is nonconvex, and in three dimensions, the size can be quadratic even for two convex polyhedra. We develop methods for building and searching the convolutions implicitly, and apply them to solve the depth of collision problems in time sublinear in the size of the convolution.

2 Minkowski Sum

Given two sets A and B of vectors, their *Minkowski sum* or *convolution* is the set

$$A \oplus B = \{a + b \mid a \in A \text{ and } b \in B\}.$$

In all our applications, the ambient space is two or three-dimensional Euclidean space, the sets are polyhedra, and the elements are points taken as vectors. We will need to distinguish between various placements of the same shape (or solid) in the space, and so each polyhedron is endowed with a *reference point*. Since only the translates of a polyhedron are of interest, the placement of a polyhedron is completely specified by the coordinates of its reference point. The polyhedron P translated by a vector z is denoted P^z. Symbols without the superscript denote the "default" placement of the associated polyhedron, where the reference point is at the origin. Finally, the polyhedron P rotated $180°$ about the origin is denoted P^{-1}:

$$P^{-1} = \{-a \ : \ a \in P\}.$$

Lemma 2.1 *Let P and Q be two polyhedra, and let x and y be vectors. Then, $P^x \cap Q^y \neq \emptyset \iff y - x \in P \oplus Q^{-1}$.*

Let P and Q be two intersecting polyhedra. Without loss of generality, assume that the reference point of P coincides with the origin of our coordinate system, and that the reference point of Q is given by the vector q. Their initial placements satisfy $P \cap Q^q \neq \emptyset$, which by the preceding lemma implies that $q \in P \oplus Q^{-1}$. A displacement of Q in direction d is denoted by Q^{q+td}, for some $t \geq 0$. Thus, the problem of computing $\sigma_d(P, Q^q)$ is equivalent to finding the *minimum* t such that $P \cap Q^{(q+td)} = \emptyset$. By Lemma 2.1, this is equivalent to finding the *minimum* t such that

$$q + td \ \notin \ P \oplus Q^{-1}. \tag{1}$$

Since $q + td$ is the parametric equation of a ray, whose origin is q and direction is d, the minimum t for which (1) holds is determined by the (first) intersection of the ray $q + td$ with the boundary of $P \oplus Q^{-1}$. Thus finding the depth of collision in a fixed direction reduces to the problem of intersecting a ray with a Minkowski-sum polyhedron.

The problem of finding $\sigma(P, Q)$, the smallest translation in *any* direction that would separate P and Q, is equivalent to finding the minimum distance between q and the boundary of $P \oplus Q^{-1}$:

$$\sigma(P, Q^q) \ = \ \min(t \geq 0 \mid \exists d \text{ with } |d| = 1 \text{ and } q + td \notin P \oplus Q^{-1}). \tag{2}$$

3 Convex Polygons

It is well-known that the Minkowski sum of two convex polyhedra is a convex polyhedron. In two dimensions, if P and Q are convex polygons of n and m vertices, respectively, then $P \oplus Q$ is a convex polygon of $n + m$ vertices; an exception occurs if P and Q have parallel edges; in that case the convolved polygon has fewer than $n + m$ vertices. In the plane, $P \oplus Q$ has a particularly simple characterization: its edges are those of P and Q, merged in slope order [13]. This fact allows one to compute $P \oplus Q$ in linear time.

Theorem 3.1 ([13]) *Given two convex polygons P and Q, with n and m vertices, respectively, we can compute their convolution $P \oplus Q$ in time $O(n + m)$.*

The directional depth of collision between P and Q can be computed by implicitly building that portion of $P \oplus Q^{-1}$ that intersect the ray from q in direction d. Guibas and Stolfi [14] use a similar method to detect whether two given convex polygons intersect. Observe that testing whether P intersects Q^q is equivalent to checking if $q \in P \oplus Q^{-1}$. Guibas and Stolfi achieve $O(\log n)$ time by determining the position of those edges of the convolution that intersect the horizontal line through q. The containment of q in the convolution can be determined by testing against these edges. The method of Guibas and Stolfi can be easily modified to determine, in $O(\log n)$ time, the point of intersection between a ray originating from an interior point q and the convolution $P \oplus Q^{-1}$.

We can also perform a binary search directly in the *object space*, rather than the convolution (i.e., the *configuration space*) to compute $\sigma_d(P, Q)$. This solution is presented in the Appendix.

Next, to solve the problem of finding the minimum depth of collision over all directions, we first compute the convolution $P \oplus Q^{-1}$, in $O(n)$ time; we assume that each of the polygons has at most n vertices. The minimum depth is found by determining the point closest to q on the boundary of $P \oplus Q^{-1}$. The latter task can be accomplished easily in linear time by checking all the edges of the convolution.

Once $P \oplus Q^{-1}$ is computed, we can also answer the following query efficiently: given Q^y, find the minimum translation needed in any direction to separate Q from P. To handle these queries, we compute the medial axis of the polygon $P \oplus Q^{-1}$ and build a point-location structure for it, using $O(n)$ time altogether [2, 10, 18]. Given a query placement Q^y, we can find the closest point to y on the boundary of $P \oplus Q^{-1}$ in $O(\log n)$ time by point location in the medial-axis diagram. This establishes the following theorem.

Theorem 3.2 *Let P and Q be two convex polygons, with a total of n vertices. After $O(n)$ time preprocessing, given a placement Q^y, we can compute $\sigma(P, Q^y)$ in time $O(\log n)$.*

In the following sections, we extend these ideas to solve similar problems for non-convex polygons and three-dimensional convex polyhedra. The main difficulty in these generalizations is that the Minkowski sum of non-convex polygons or 3D polyhedra may have quadratic complexity, which makes their explicit construction too costly. Instead, we develop methods for implicitly constructing and searching these structures.

4   One Convex and One Nonconvex Polygon

Let P be a simple nonconvex polygon of n vertices and let Q be a convex polygon of m vertices. Given a placement of Q, we want to compute the depth of collision, either in a fixed direction or the minimum over all directions. The Minkowski sum of P and Q is again a polygon; however, unlike in the convex case, $P \oplus Q$ is in general multiply connected, and may have $\Omega(nm)$ vertices. Therefore, an explicit construction of the convolution may be too expensive. Our main result in this section is a method for constructing and searching an implicit representation of $P \oplus Q$. The representation can be constructed in $O(n \log^2(n+m) + m)$ time, stored in $O(n+m)$ space, and searched efficiently for the depth queries. At the top-level, our method consists of the following steps.

Construct Convolution

Step 1. Triangulate P. Let $T_1, T_2, \ldots, T_k$ be the triangles in this triangulation.

Step 2. Compute an implicit representation of the convolutions $R_i = T_i \oplus Q^{-1}$, for $i = 1, 2, \ldots, k$.

Step 3. Compute an implicit representation of the boundary of $R = \cup R_i$.

4.1 Details of the Construction

We fix a Cartesian coordinate system in the plane and, without loss of generality, assume that the origin lies in Q; we take the origin as the reference point of Q.

Step 1 is easily accomplished in $O(n \log n)$ time, using any of the well-known triangulation algorithms [11, 15]; the time complexity can be reduced to $O(n \log^* n)$ using a recent algorithm of Chazelle [7]. Pick a reference point r_i for each of the triangles T_i; for instance, we may pick the vertex closest to the origin.

Step 2 computes an implicit representation of the convolutions of these triangles with the convex polygon Q^{-1}. The key to this step is the observation that the convolution of two convex polygons has a particularly simple form, which in the case of a triangle T_i and a convex polygon Q can be determined implicitly in $O(\log m)$ time. Recall that if A and B are two convex polygons, then the boundary of $A \oplus B$ consists of edges of A and B merged in slope order [13]. To compute $T_i \oplus Q^{-1}$, therefore, we only need to locate the slopes of each of the three edges of T_i among the slopes of the edges of Q^{-1}, which takes three binary searches and $O(\log m)$ time. To represent $R_i = T_i \oplus Q^{-1}$ implicitly, we just record the three places in the array storing Q^{-1} where the edges of T_i are inserted. We call the convolution R_i a *convolved triangle*; see Figure 1 for an illustration.

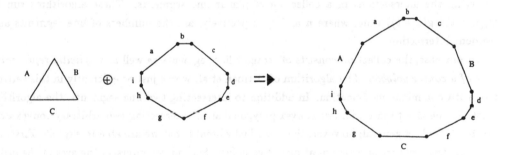

Figure 1. A convolved triangle

Next, we discuss the details of Step 3, in which we compute the boundary of the union of these convolved triangles.

Union of Convolved Triangles

We want to compute $R = \cup_{i=1}^{k} R_i$, where each of the R_i is a convex polygon with $m+3$ vertices. An explicit description of R, of course, will require $\Omega(nm)$ space in the worst case. We therefore need to exploit the fact that each R_i is (implicitly) representable as a six-sided figure: three straight lines and three "arcs" of a convex polygon. This alone still does not suffice since, in general, even the boundary of k triangles can have quadratic size. Fortunately, a result of Kedem et al. [16] shows that a collection of convolved triangles is "well-behaved"; we use their result to construct a linear-space implicit representation of the boundary of R. We use the following two facts proved in Kedem et al. [16].

Fact 4.1 ([16], p. 66) *Let A_1 and A_2 be two convex polygons with disjoint interiors, and let B be another convex polygon. Then the boundaries of $A_1 \oplus B$ and $A_2 \oplus B$ intersect in at most two points.*

Fact 4.2 ([16], p. 61) *Let A_i, $i = 1, 2, \ldots, k$, be a collection of convex polygons with disjoint interiors, and let B be another convex set. Then the boundary of $R = \cup_{i=1}^{k}(A_i \oplus B)$ has at most $6k - 12$ points of local nonconvexity.*

(Given a set S, a point $p \in S$ is called a point of *local nonconvexity* if each neighborhood of p contains two points $x, y \in S$ such that the segments $px, py \in S$ but the segment xy is not contained in S.)

A Sweep-line Algorithm

We use these facts to get an upper bound on the running time of a sweep-line based algorithm for computing the boundary of R. The algorithm we use is due to Ottman, Widmeyer and Wood [19], which in turn is based on a technique of Bentley and Ottman [4] for counting and reporting the intersections in a collection of planar line segments. These algorithms run in $O((n + t)\log(n + t))$ time, where n and t, respectively, are the numbers of line segments and segment intersections.

In our case, the collection consists of straight-line segments as well as *implicitly represented arcs of a convex polygon*. The algorithm of Ottman et al. works just as efficiently in this setting too, with one minor modification. In addition to intersecting two line segments, the algorithm must also be able to intersect two convex polygonal arcs. Intersecting two arbitrary convex arcs can be difficult in general; however, there are lucky breaks that we are able to exploit. First, by Fact 4.1, any two arcs intersect in at most two points. Second, we intersect the arcs in the order the sweep line encounters them. Thus, when our algorithm tries to compute the intersections of two arcs, their leftmost points lie on the sweep line. Since each of the arcs is given as an ordered list of vertices in an array, we can intersect them in time $O(\log m)$ by a binary search. (The details of this search are quite straightforward and can easily be worked out by the reader.) The worst-case running time of the sweep line algorithm is still $O((n + t)\log(n + t))$ since the overhead cost per intersection point is $O(\log n)$. The following is a high-level description of our algorithm.

Algorithm to Compute $R = \cup R_i$.

1. Divide the input into two halves: $R_1, \ldots, R_{\lfloor k/2 \rfloor}$ and $R_{\lfloor k/2 \rfloor+1}, \ldots, R_k$.

2. Recursively compute the union of the two halves, $U = \cup_{i=1}^{\lfloor k/2 \rfloor} R_i$, and $V = \cup_{i=\lfloor k/2 \rfloor+1}^{k} R_i$.

3. Merge the two contours U and V using the (modified) Ottman-Widmeyer-Wood algorithm.

Fact 4.2 implies that the total number of intersections t is at most $6k - 12$, which is $O(n)$, and hence the merge phase of our algorithm (step 3) takes $O(n\log(n+m))$. The entire algorithm takes $O(n\log^2(n + m) + m)$ time.

The boundary of R has $O(n)$ total vertices, and each pair of adjacent vertices is joined either by a straight line segment (corresponding to an edge of P) or an implicitly described convex arc (corresponding to some portion of the boundary of Q^{-1}). We can store this boundary in $O(n + m)$ space.

4.2 Computing the Depth of Collision using R

Having computed an implicit representation of $R = \cup R_i$, we now discuss how to use it to solve the depth of collision problem.

In the case of directional depth of collision, we are given a placement of Q, say, Q^q, and want to compute $\sigma_d(P, Q^q)$. Recalling the relation (1), this distance is equal to the distance between q and its closest point on the boundary of R in direction d. Equivalently, we need to find the first intersection of R with the ray from q in direction d. We intersect the ray with each of $O(n)$ bounding edges and arcs of R, and choose the point closest to q. We can intersect the ray with a line segment in $O(1)$ time, and find its first intersection with a convex arc of m vertices in time $O(\log m)$. Thus, the directional depth of collision can be determined from an implicit representation of R in $O(n \log m)$ time.

If the minimum depth of collision over all directions, namely $\sigma(P, Q^q)$, is desired, we need to do a little more work. We start by computing the minimum distance from q to the *explicitly stored boundary of* R; that is, we ignore the convex arcs on the boundary and find the minimum distance to the remaining edges and vertices of R. If this distance is Δ, let D be the disk of radius Δ centered on q. Let $V(q, \Delta) \subset R$ be the set of edges and vertices visible from q whose distance from q is no more than Δ; alternatively, $V(q, \Delta)$ is the intersection of the visibility polygon of q with the disk D.

Lemma 4.3 $|V(q, \Delta)| = O(n + m)$.

The following lemma provides an algorithm for computing $V(q, \Delta)$.

Lemma 4.4 *Given the implicit representation of R, we can compute $V(q, \Delta)$ in $O(n \log(n + m) + m)$ time.*

Proof: We start by computing an implicit representation of the visibility polygon V of q with respect to the boundary of R. This takes $O(n \log(n + m))$ time using a variant of a standard visibility polygon algorithm [3, 20], which we now describe. On each arc of Q in the implicit representation of R, there are at most two points z such that the ray $\overline{qz}$ is tangent to the arc.

We find these points in $O(n \log m)$ time altogether. If we cut the boundary of R at these tangent points and at all the explicitly represented vertices, we get a collection of $O(n)$ explicit segments and implicit arcs, no two intersecting. A ray from q intersects a segment or an arc in the collection in at most one point. We use an angular-sweep method to find the visibility polygon of q. We first find the intersections of all arcs and segments with the vertical ray extending up from q. The closest intersection to q belongs to V. Now we rotate the ray through 360° while maintaining the segment or arc closest to q as the minimum element of a priority queue. Because the segments and arcs are disjoint, the closest segment or arc changes only when the sweeping ray passes over an explicit vertex or a tangent point. The result of the sweep is a partition of the directions around q into $O(n)$ angular ranges such that within each range, exactly one segment or arc is visible from q. The algorithm takes $O(n \log(n+m))$ time altogether for tangent-finding, intersection-finding, and priority queue operations.

Once we have found V implicitly, we must refine it to get $V(q, \Delta)$. We compute Δ in $O(n)$ time. We wish to find any implicit arcs that cross D, but do not want to examine all the $O(nm)$ edges of V. Suppose that some implicit convex chain c is visible from q in an angular range (α, β). A geometric argument shows that any edge of c that intersects the interior of D must have its normal in the range (α, β), that is, must have the same slope as a tangent to D between α and β. Thus to find any portion of c inside D, we find the extreme points of c in directions α and β, then examine the edges between to see if any intersects D. This takes $O(\log m)$ time, plus a cost proportional to the number of edges examined. There is no overlap between the edges of Q examined in any two angular ranges, and so the total cost of computing $V(q, \Delta)$ is $O(n \log(n + m) + m)$. ∎

Once $V(q, \Delta)$ is known, the minimum distance between q and $V(q, \Delta)$ can be computed in $O(n + m)$ time. By equation (2), this distance equals the minimum depth of collision $\sigma(P, Q^q)$. We summarize the results of this section:

Theorem 4.5 *Let P be a simple nonconvex polygon and Q a convex polygon, where $|P| = n$ and $|Q| = m$. In $O(n \log^2(n + m) + m)$ time, we can build a linear-space data structure that can answer the following queries in $O((n + m) \log(n + m))$ time: (1) given a placement Q^q and a direction d, determine the directional depth of collision $\sigma_d(P, Q^q)$, and (2) given a placement Q^q, determine the minimum depth of collision $\sigma(P, Q^q)$.*

4.3 Query Answering in Sublinear Time

In this section, we show how to preprocess R so that minimum depth of collision queries can be answered in sublinear time, for an arbitrary placement of Q. Finding $\sigma_d(P, Q^q)$ is a ray-shooting problem: we want to find the first point where the ray $q + td$ hits the boundary of R. After some preprocessing, we can apply the ray-shooting techniques of Guibas, Overmars and Sharir [12] and Agarwal [1]. The preprocessing takes $O(n \log n + m)$ time and $O(n + m)$ space; queries take $O(n^{2/3+\delta} + \log m)$ time, for $\delta > 0$ and arbitrarily small.

For any implicit arc c, we know that the segment connecting the arc endpoints lies inside R (it lies inside the placement of Q that produced c). Furthermore, the segments for any two distinct arcs are disjoint. Each arc and its associated segment form an "ear" of the region R. These ears play a critical rôle in ray-shooting: any ray that enters an ear by crossing its segment will intersect the ear's arc before hitting any other boundary point of R. The point of intersection can be found in $O(\log m)$ time.

To prepare R for ray-shooting, we cut off all its ears. This leaves a polygonal region bounded by $O(n)$ line segments. We preprocess this region for ray-shooting [1, 12], which takes $O(n \log n)$ time and $O(n)$ space. We also preprocess the ears for point location; the preprocessing takes $O(n \log(n + m))$, after which a point location query can be answered in $O(\log(n + m))$ time [10].

At query time, we try to locate q in the ears. If q lies in an ear, we check whether $q + td$ hits the arc or the segment of the ear. If it hits the arc, we are done. If it hits the segment, we proceed as if q lay outside the ear at the intersection of the ray and the ear's segment. In this case, or if q lies outside all ears, we find the first explicit segment or ear segment that the ray hits, which takes $O(n^{2/3+\delta})$ time. If the ray hits an ear segment, we find the intersection of the ray with the ear's arc in $O(\log m)$ additional time. If we are willing to spend $O(n^{3/2} \log^5 n)$ time for preprocessing and $O(n \log^3 n)$ storage space, then the query time can be reduced to $O(\sqrt{n} \log^2 n)$ [1]. This solves the directional depth of collision problem.

Theorem 4.6 *Let P be a simple polygon and let Q be a convex polygon with a total of n vertices. In $O(n \log n)$ time we can build an $O(n)$ space data structure that can compute $\sigma_d(P, Q^q)$, for an arbitrary direction d and an arbitrary placement of Q, in time $O(n^{2/3+\delta})$, for an arbitrarily small but positive constant δ. The query time can be reduced to $O(\sqrt{n} \log^2 n)$ at the expense of $O(n^{3/2} \log^5 n)$ time preprocessing and $O(n \log^3 n)$ space.*

5 Convex Polyhedra in Three Dimensions

5.1 Representation of Polytopes

Let P and Q be two convex polyhedra, with a total of n vertices. We show that if P and Q are suitably preprocessed, then their directional depth of collision can be determined in $O(\log^2 n)$ time. The representation of each polyhedron consists of an inner polyhedral hierarchy and an outer polyhedral hierarchy. These hierarchical representations were introduced by Dobkin and Kirkpatrick [8], who used them for determining the separation of two convex polyhedra. To describe these data structures, consider a polytope P with the vertex set $V(P)$.[1] A sequence of polytopes $P_1, P_2, \ldots, P_k$ is called an *inner polyhedral hierarchy* of P if, for $1 \le i < k$,

(i) $P_1 = P$, and P_k is a simplex,

(ii) $P_{i+1} \subset P_i$ and $V(P_{i+1}) \subset V(P_i)$,

(iii) $V(P_i) \setminus V(P_{i+1})$ forms an independent set in P_i.

---

[1] We limit our discussion to bounded polytopes, although all our results hold for unbounded polyhedra as well, at the expense of some minor technical complications.

The *outer polyhedral hierarchy* is similar, except at each iteration we remove an independent set of faces. In particular, let $H(P)$ be the set of planes bounding the faces of P. Then the outer hierarchy of P is a sequence of polytopes $P_1, P_2, \ldots, P_k$ such that

(i) $P_1 = P$, and P_k is a simplex,

(ii) $P_{i+1} \supset P_i$ and $H(P_{i+1}) \subset H(P_i)$,

(iii) $H(P_i) \setminus H(P_{i+1})$ bounds an independent set of faces in P_i.

The *degree* of an inner polyhedral hierarchy is the maximum degree, over all i, of a vertex in $V(P_i) \setminus V(P_{i+1})$. The degree of an outer polyhedral hierarchy is analogously defined, with faces in place of vertices. The following fact is based on the property that the facial graph of a convex polytope is a planar graph, and planar graphs admit a large independent set of small degree.

Fact 5.1 ([8]) *There is an inner (resp. outer) polyhedral hierarchy of P with height $O(\log|P|)$, size $O(|P|)$, and degree $O(1)$. Furthermore, given a standard representation of the facial graph of P (cyclic orderings of edges around the vertices and the faces), one can construct these hierarchies in linear time.*

In our discussion, we assume that each polyhedron is represented by its twin polyhedral hierarchies. In particular, a convex polytope P is represented by its inner polyhedral hierarchy $P_1, P_2, \ldots, P_k$, where each P_i is represented by an outer polyhedral hierarchy. Fact 5.1 implies that this representation requires $O(n)$ space.

Using the hierarchical representation of a polytope, we can efficiently answer various extremal queries involving points, lines or planes. We state these results in the following lemma without proof; a complete proof can be found in the full version of our paper [9].

Lemma 5.2 *For any polytope P_i in the hierarchical representation of P, for $1 \le i \le k$, we can answer the following queries in $O(\log n)$ time: (1) find the points of intersection between a line l and P_i, (2) find the first point of contact between P_i and a line l moving from infinity in some direction d, (3) find the first point of contact with P_i and a plane H moving from infinity in some direction d.*

5.2 Computing the Depth of Collision

We begin with the observation that if A is a convex polytope of constant complexity, then the depth of collision $\sigma_d(A, P)$ can be computed in time $O(\log n)$. This follows because the depth of collision $\sigma_d(x, P)$ for each element (vertex, edge or face) $x \in A$ can be computed in time $O(\log n)$ using Lemma 5.2. We therefore have the following corollary of the results of the previous subsection.

Corollary 5.3 *Let A be a convex polytope of constant complexity and let P be a convex polytope of n vertices, given by its twin hierarchical representation. Then for any direction d, we can compute the depth of collision $\sigma_d(A, P)$ in time $O(\log n)$.*

We now describe our algorithm for computing the directional depth of collision between P and Q. For ease of description, let us assume that the inner hierarchies of both P and Q have k levels.

Algorithm Compute Depth 3D

Initialization. Compute the intersection between the simplices P_k and Q_k. If $P_k \cap Q_k = \emptyset$, find a separating plane H_k and go to **Phase 1**.

Otherwise (i.e., $P_k \cap Q_k \neq \emptyset$), compute $\sigma_d(P_k, Q_k)$; translate Q_k by $\sigma(P_k, Q_k)$ along direction d so that P_k and Q_k are in contact; find a plane H_k that passes through the common tangency of P_k and Q_k and separates their interiors; and go to **Phase 2**.

Phase 1. *(This phase maintains the invariant that the polytopes P_i and Q_i are nonintersecting and we know a witness plane H_i such that $P_i \subset H_i^+$ and $Q_i \subset H_i^-$.)*

(1A) Compute the polytopes $P'_{i-1} = P_{i-1} \cap H_i^-$ and $Q'_{i-1} = Q_{i-1} \cap H_i^+$.

(1B) Determine whether the following intersections are nonempty: $P_i \cap Q'_{i-1}$, $P'_{i-1} \cap Q_i$, and $P'_{i-1} \cap Q'_{i-1}$. If all three intersections are empty, go to **Step 1C**; otherwise go to **Step 1D**.

(1C) Find a plane H_{i-1} such that $P_{i-1} \subset H_{i-1}^+$ and $Q_{i-1} \subset H_{i-1}^-$. This plane can be obtained as a plane that separates P'_{i-1} and Q'_{i-1} and is tangent to both. If $i > 1$, decrement i by 1 and go to **Step 1A**; otherwise, output "$\sigma_d(P,Q) = 0$", and stop.

(1D) Compute $\sigma_d(P_{i-1}, Q_{i-1}) = \max \left\{ \sigma_d(P'_{i-1}, Q'_{i-1}),\ \sigma_d(P'_{i-1}, Q_i),\ \sigma_d(P_i, Q'_{i-1}) \right\}$.
Translate Q_{i-1} by $\sigma_d(P_{i-1}, Q_{i-1})$. Find a plane H_{i-1} such that $P_{i-1} \subset H_{i-1}^+$ and $Q_{i-1} \subset H_{i-1}^-$. Go to **Phase 2**.

Phase 2. *(This phase maintains the invariant that the polytopes P_i and Q_i are tangent to each other and we know a plane H_i such that $P_i \subset H_i^+$ and $Q_i \subset H_i^-$.)*

(2A) Compute the polytopes $P'_{i-1} = P_{i-1} \cap H_i^-$ and $Q'_{i-1} = Q_{i-1} \cap H_i^+$.

(2B) Compute $\Delta\sigma_d(P_{i-1}, Q_{i-1}) = \max \left\{ \sigma_d(P'_{i-1}, Q'_{i-1}),\ \sigma_d(P'_{i-1}, Q_i),\ \sigma_d(P_i, Q'_{i-1}) \right\}$.

(2C) Translate Q_{i-1} by $\Delta\sigma_d(P_{i-1}, Q_{i-1})$. Update

$$\sigma_d(P_{i-1}, Q_{i-1}) = \sigma_d(P_i, Q_i) + \Delta\sigma_d(P_{i-1}, Q_{i-1}).$$

Find a plane H_{i-1} satisfying $P_{i-1} \subset H_{i-1}^+$ and $Q_{i-1} \subset H_{i-1}^-$.

If $i > 1$, decrement i by 1 and go to **Step 2A**; otherwise, output $\sigma_d(P_1, Q_1)$, and stop.

Theorem 5.4 *Let P and Q be two convex polyhedra, with n and m vertices, respectively, each given by its hierarchical representation. For any direction d, the algorithm described above computes the directional depth of collision $\sigma_d(P, Q)$ in time $O(\log n \log m)$.*

Proof: We first establish the correctness of the algorithm. Phase 1 finishes when $P_i \cap Q_i \neq \emptyset$, for some i, $1 \leq i \leq k$. One easily verifies that if $P_i \cap Q_i = \emptyset$, then $P_{i-1} \cap Q_{i-1} \neq \emptyset$ if and only if one of the three pairs of polytopes in Step 1B is found to intersect. If none of the pairs intersect, then one also observes that at least one separating plane H tangent to both P'_{i-1} and Q'_{i-1} satisfies the invariant. This proves the correctness of Phase 1.

In Phase 2, we maintain the invariant that P_i and Q_i are in contact, and a plane H_i separating their interiors is known. Suppose that we enter an iteration of this phase with P_i and Q_i. It is easy to see that only the polytopes P'_{i-1} and Q'_{i-1} are of interest for determining whether P_{i-1} and Q_{i-1} intersect. Furthermore, $\Delta \sigma_d(P_{i-1}, Q_{i-1})$ correctly reflects the amount by which Q_{i-1} must be translated to make its interior disjoint from P_{i-1}. We find a plane H_{i-1}, which can be determined easily using the contact point of the two polytopes, and start the next iteration. The invariant is maintained and correctness of Phase 2 has been established.

Finally, we show that each iteration of Phase 1 or Phase 2 can be completed in $O(\log n)$ time. The basic step of computing the polytopes P'_{i-1} and Q'_{i-1} in Steps 1A and 2A takes $O(\log n)$ time, as follows. By the convexity of P_{i-1}, at most one of its vertices lies in H^-. If there is a vertex $v \in V(P_{i-1})$ lying in H^-, then v also forms the first point of contact between P_{i-1} and a plane moving parallel to H_i from infinity. By Lemma 5.2, we can find v, and thus P'_{i-1}, in time $O(\log n)$. The remaining steps take $O(\log n)$ time by Corollary 5.3, since both P'_{i-1} and Q'_{i-1} have bounded complexity. This completes the proof of the theorem. ∎

6 Extensions, Related Problems and Conclusions

The depth of collision is a reasonable measure of negative distance between intersecting objects. In many realistic situations, however, some additional factors might also play a role. Consider, for instance, two intersecting polyhedra P and Q, where P is nonconvex and Q is convex. A translation of Q in direction d by the amount $\sigma_d(P,Q)$ surely makes the interiors of P and Q disjoint; however, Q may still be stuck in one of the "holes" of P. Therefore, one may be interested in moving Q to a position from where it can be translated arbitrarily far from P without intersecting the latter. In terms of the convolution it requires finding a point in the infinite face of the complement of $P \oplus Q^{-1}$. This problem, in fact, turns out to be easier than computing σ_d and σ. We maintain an implicit representation of only the outermost boundary of $R = P \oplus Q^{-1}$. After linear-time preprocessing, a directional depth query can be answered in just $O(\log n)$ time, since it is a ray-shooting query in a simple polygon.

Similarly, practical constraints might require that all translations be along a fixed set of directions. The problems again become simpler: for each direction d, we perform an (implicit) trapezoidal decomposition of R, and preprocess the resulting subdivision for point location. Given a placement of Q^q and direction d, we locate q in the subdivision of R corresponding to d and compute $\sigma_d(P, Q^y)$ by intersecting the ray $q + td$ with the sides of the trapezoid containing q. The procedure takes $O(\log n)$ time.

Finally, the minimum depth of collision between two three-dimensional convex polyhedra P and Q^q can be computed in $O(n^2)$ time by computing their convolution explicitly and then finding the boundary point closest to q.

Several open problems are suggested by our work. Perhaps the most interesting and useful ones are in three dimensions. Given two convex polyhedra P and Q in three dimensions, is it possible to compute $\sigma_d(P, Q)$ in $O(\log n)$ time? Is it possible to compute the minimum depth of collision $\sigma(P, Q)$ more efficiently, say, in $O(n \log^k n)$ time, for some small constant k? Extensions involving nonconvex polyhedra are also worth investigating.

Appendix: Convex Polygons

We present a logarithmic time procedure for computing the directional depth of collision between two convex polygons, based on binary search in the object space. Let P and Q be two intersecting planar convex polygons, with a total of n vertices; each polygon is given as an ordered list of vertices, stored in an array. Given a direction d, represented as a unit vector, we want to compute the depth of collision $\sigma_d(P, Q)$. Without loss of generality, we assume that d points along the positive Y-axis. (This assumption is not crucial but it simplifies the exposition.)

For a point $p \in P$, let $\sigma_d(p, Q)$ denote the minimum distance by which Q must be moved in direction d so that p is in the exterior of Q. Similarly, for any point $q \in Q$, let $\sigma_d(P, q)$ denote the minimum distance by which q must be moved in direction d so that q is in the exterior of P. Observe that the depth of collision between P and Q is

$$\sigma_d(P, Q) = \max_{p \in P} \sigma_d(p, Q) = \max_{q \in Q} \sigma_d(P, q)$$

We begin by locating the leftmost and the rightmost vertices of P and Q, denoted p_l, p_r in P and q_l, q_r in Q. This can be done by a straightforward binary search in $O(\log n)$ time. Let $(q_l, q_{l+1}, \ldots, q_{r-1}, q_r)$ be the lower chain of vertices of the boundary of Q, and let $(p_r, p_{r+1}, \ldots, p_{l-1}, p_l)$ be the upper chain of vertices of the boundary of P. We complete these chains into (unbounded) polygons by adding a point at infinity, namely, the point $(0, +\infty)$ to the Q-chain and the point $(0, -\infty)$ to the P-chain. Let Q_L and P_U denote the resulting (lower) Q-polygon and (upper) P-polygon. The following fact is easy to establish.

Observation 6.1 $\sigma_d(P, Q) = \sigma_d(P_U, Q_L)$.

We compute $\sigma_d(P_U, Q_L)$ by a "coordinated binary search," in which at each iteration we discard roughly half of one of the chains. Each iteration takes $O(1)$ time and the whole procedure runs in $O(\log n)$ time. Let P_i, Q_i denote the chains at the ith iteration, where $P_0 = P_U$ and $Q_0 = Q_L$. Denote by $X(p)$ and $Y(p)$ the X and Y coordinates of a point p.

Lemma 6.2 *In $O(1)$ time, we can replace P_{i-1}, Q_{i-1} by P_i, Q_i such that (1) $\sigma_d(P_i, Q_i) = \sigma_d(P_{i-1}, Q_{i-1})$, and (2) either $|P_i| \leq \left\lfloor \frac{|P_{i-1}|}{2} \right\rfloor + 1$ or $|Q_i| \leq \left\lfloor \frac{|Q_{i-1}|}{2} \right\rfloor + 1$.*

Proof: We start by setting $P_i = P_{i-1}$ and $Q_i = Q_{i-1}$. After a constant amount of work, we discard half the vertices from either P_i or Q_i. Let p_j (resp. q_k) be the vertex of P_i (resp. Q_i) with the median X-coordinate, and assume without loss of generality that $X(p_j) \geq X(q_k)$.

Denote by $r^+(p_j)$ the ray $\overrightarrow{p_j p_{j+1}}$ and by $r^-(p_j)$ the ray $\overrightarrow{p_j p_{j-1}}$. Similarly, denote by $r^+(q_k)$ and $r^-(q_k)$ the rays originating from q_k and directed towards its two neighboring vertices. We consider three cases depending upon the relative positions of these rays and the points p_j and q_k. See Figure 2.

Case 1. *(q_k lies above the ray $r^-(p_j)$.)* See Figure 2A.

In this case, either $r^+(q_k)$ or $r^-(q_k)$ does not intersect $r^-(p_j)$. Discard all vertices of Q_i with X-coordinates bigger (resp. smaller) than $X(q_k)$ in the former (resp. latter) case. This does not alter the depth of collision, since all the discarded vertices lie above P.

Case 2. *(p_j lies below the ray $r^+(q_k)$.)* See Figure 2B.

In this case, either $r^+(p_j)$ or $r^-(p_j)$ does not intersect $r^+(q_k)$. Discard all vertices of P_i with X-coordinates bigger (resp. smaller) than $X(p_j)$ in the former (resp. latter) case. This does not alter the depth of collision, since all the discarded vertices lie below Q.

Case 3. *(Neither Case 1 nor Case 2 holds.)* See Figure 2C.

In this case, p_j and q_k lie in the wedges of each others' rays. Since the apex angle of each wedge is less than π, one of the rays from p_j intersects a ray from q_k.

If $r^+(p_j) \cap r^+(q_k) \neq \emptyset$, then we discard all vertices of P_i whose X-coordinates are larger than $X(p_j)$. This does not alter the depth of collision since an easy geometric argument shows that $\sigma_d(p_j, Q) \geq \sigma_d(p, Q)$, for all the discarded vertices p.

On the other hand, if $r^-(p_j) \cap r^-(q_k) \neq \emptyset$, then we discard all vertices of Q_i whose X-coordinates are smaller than $X(q_k)$. This also does not alter the depth of collision since $\sigma_d(P, q_k) \geq \sigma_d(P, q)$ for all the discarded vertices q.

The test in each of the three cases can be performed in $O(1)$ time. Since p_j and q_k are the median vertices, the size of one of the chains decreases by about half. This completes the proof. ∎

After $O(\log n)$ applications of Lemma 6.2, at least one of the chains, say P_k, is reduced to a single edge. We can find the depth $\sigma_d(P_k, Q_k)$ in $O(\log n)$ time by a binary search among the vertices of Q_k as follows. Let x be the edge that remains in P_k, with endpoints a and b. Let $y \in Q_k$ be the point of contact between x and Q_k when x is translated by $\sigma_d(x, Q_k)$. Then y is either the point of tangency between Q_k and the line determined by x, or it is the intersection of Q_k with the ray through a or b in the direction d. In either case, y can be determined by a standard binary search in $O(\log n)$ time. We summarize our result as follows.

Theorem 6.3 *Given two intersecting planar convex polygons P and Q, with a total of n vertices, we compute $\sigma_d(P, Q)$, for any direction d, in time $O(\log n)$.*

References

[1] P. Agarwal. Ray shooting and other applications of spanning trees with low stabbing number. *Proc. of 5th ACM Symposium on Computational Geometry*, 315-325, 1989.

[2] A. Aggarwal, L. Guibas, J. Saxe, and P. Shor. A linear time algorithm for computing the Voronoi diagram of a convex polygon. *Discrete and Computational Geometry*, 4:591-604, 1989.

[3] T. Asano. An efficient algorithm for finding the visibility polygons for a polygonal region with holes. *Transactions of IECE of Japan*, E-68:557–559, 1985.

[4] J. Bentley and T. A. Ottman. Algorithms for reporting and counting geometric intersections, *IEEE Transactions on Computers*, C-28:643–647, 1979.

[5] C. E. Buckley and L. J. Leifer. A proximity metric for continuum path planning. *Proc. 9th International Joint Conference on Artificial Intelligence*, 1096-1102, 1985.

[6] S. A. Cameron and R. K. Culley. Determining the minimum translation distance between two convex polyhedra. *Proc. IEEE International Conference on Robotics and Automation*, 591-596, 1986.

[7] B. Chazelle. Efficient polygon triangulation. Technical Report, Princeton University, 1990.

[8] D. Dobkin and D. Kirkpatrick. Determining the separation of preprocessed polyhedra—a unified approach. *Proc. of ICALP*, July 1990.

[9] D. Dobkin, J. Hershberger, D. Kirkpatrick and S. Suri. Implicitly Searching Convolutions and Computing Depth of Collision. Manuscript in preparation, 1990.

[10] H. Edelsbrunner, L. Guibas, and J. Stolfi. Optimal point location in a monotone subdivision. *SIAM Journal on Computing*, 15:317–340, 1986.

[11] M. Garey, D. S. Johnson, F. P. Preparata and R. E. Tarjan. Triangulating a simple polygon. *Information Processing Letters*, 7:175–179, 1978.

[12] L. Guibas, M. Overmars, and M. Sharir. Intersecting line segments, ray shooting, and other applications of geometric partitioning techniques. In *Proc. of the First Scandinavian Workshop on Algorithm Theory*, pages 64–73. Springer-Verlag, 1988. Lecture Notes in Computer Science 318.

[13] L. Guibas, L. Ramshaw and J. Stolfi. A kinetic framework for computational geometry. *Proc. 24th Foundations of Computer Science*, 100-111, Nov. 1983.

[14] L. J. Guibas and J. Stolfi. Ruler, compass, and computer: The design and analysis of geometric algorithms. Research Report 37, DEC Systems Research Center, 1989. Also appeared in *Theoretical Foundations of Computer Graphics and CAD*, Springer-Verlag.

180

[15] S. Hertel and K. Mehlhorn. Fast triangulation of simple polygons. *Lecture Notes in Computer Science*, 158, 207-218, 1983.

[16] K. Kedem, R. Livne, J. Pack and M. Sharir. On the union of Jordan regions and collision-free translational motion amidst polygonal obstacles. *Discrete and Computational Geometry*, 1:59–72, 1986.

[17] S. S. Keerthi and K. Sridharan. Efficient algorithms for computing two measures of depth of collision between convex polygons. Technical Report, Department of Computer Science and Automation, IIS, Banglore, India, 1989.

[18] D. Kirkpatrick. Optimal search in planar subdivisions. *SIAM Journal on Computing*, 12:28–35, 1983.

[19] T. A. Ottman, P. Widmeyer and D. Wood. A fast algorithm for Boolean mask operations. Inst. f. Angewandte Mathematik und Formale Beschreibungsverfahren, D-7500 Karlsruhe, Report no. 112, 1982.

[20] S. Suri and J. O'Rourke. Worst-case optimal algorithms for constructing visibility polygons with holes. *Proc. of 2nd ACM Symposium on Computational Geometry*, 14-23, 1986.

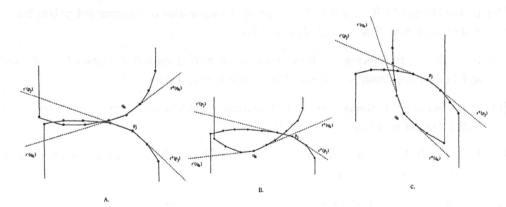

Figure 2. Proof of Lemma 6.2

Characterization for a Family of Infinitely Many Irreducible Equally Spaced Polynomials

Toshiya ITOH

Department of Information Processing
The Graduate School at Nagatsuta
Tokyo Institute of Technology
4259 Nagatsuta, Midori-ku, Yokohama 227, Japan
e-mail: titoh@cc.titech.ac.jp

Abstract:

This paper shows a necessary and sufficient condition for a family of *infinitely* many Equally Spaced Polynomial (**ESP's**) to be irreducible over $GF(2)$ and the *uniqueness* of irreducible ESP's over $GF(2)$, i.e., there exist no distinct irreducible ESP's of the same degree. It is worth noting that these results in this paper *completely* characterize all irreducible ESP's.

1 Introduction

In designing cryptographic systems (see, e.g., [IM], [M1], [Tetal].) or error-correcting codes (see, e.g., [M2], [MS].), one often takes finite field arithmetics especially for $GF(2^m)$ to provide efficient equipments or devices. Basically, finite field arithmetics consist of simple operations such as addition, subtraction, multiplication, and division.

Massey and Omura [MO] proposed a parallel multiplier for $GF(2^m)$ based on normal bases [MS], **Massey-Omura Multiplier (MOM)**, and this multiplier has *structural modularity* and small *circuit depth*. (For the related works, see [P], [W], [Wetal].) Itoh and Tsujii [IT1] and Oorschot and Vanstone [OV] *independently* showed that MOM's for $GF(2^m)$ are available to construct a fast algorithm for computing multiplicative inverses in $GF(2^m)$. Furthermore, Itoh et al. [AIT], [I], [IT3] improved those results to establish fast dividers by using a sequence of subfields of $GF(2^m)$. The idea similar to the above can be found in [P].

Wah and Wang [WW] defined a polynomial over $GF(2)$ of a specified form, **All One Polynomial (AOP)**. In [WW], they also proved a necessary and sufficient condition for AOP's to be irreducible over $GF(2)$, and pointed out that an irreducible AOP provides a small size MOM for $GF(2^m)$. Mullin et al. [Metal] introduced a criterion to measure circuit complexity of MOM's, and showed that MOM's derived from irreducible AOP's satisfy the greatest lower bound for the circuit complexity of MOM's. In [Metal], they also found a sufficient condition to minimize the circuit complexity of MOM's for $GF(2^m)$.

As an extension of an AOP, Itoh and Tsujii [IT2] defined a polynomial over $GF(2)$ of a specified form, **Equally Spaced Polynomial (ESP)**, and showed a necessary and sufficient condition for a family of *finitely* many ESP's to be irreducible over $GF(2)$. In addition, Itoh and Tsujii [IT2] proved the uniqueness of irreducible ESP's in a weak sense, and applied them to design a family of *finitely* many small size parallel multipliers with the structural modularity for a class of $GF(2^m)$. It should be noted that the multiplier based on an irreducible ESP over $GF(2)$ is also applicable to construct a fast algorithm for computing multiplicative inverses for a class of $GF(2^m)$.

This paper further extends the results in [IT2], and gives a complete characterization for a family of *infinitely* many irreducible ESP's over $GF(2)$. The complete characterization for a family of *infinitely* many irreducible ESP's over $GF(2)$ is very simple, thus it can be regarded as a *good* way to generate *infinitely* many irreducible ESP's over $GF(2)$. Furthermore, this paper also shows the uniqueness of irreducible ESP's over $GF(2)$ in a strict sense, i.e., there exist no distinct irreducible ESP's of the same degree. Thus this characterization enables us to construct a family of *infinitely* many small size parallel multipliers with the structural modularity for a class of $GF(2^m)$.

2 Definitions and Known Results

In this section, we give formal definitions of **All One Polynomial (AOP)** and **Equally Spaced Polynomial (ESP)**, respectively, and show a lemma for an AOP to be irreducible over $GF(2)$. Furthermore, we show theorems on a family of *finitely* many irreducible ESP's and the uniqueness of irreducible ESP's in a weak sense.

Our interest is concentrated on irreducible polynomials over $GF(2)$, then in the rest of this paper, we sometimes omit to denote "over $GF(2)$," and simply say as "irreducible AOP's" or "irreducible ESP's."

Definition 2.1 [WW]: *A polynomial $p(x) = x^m + x^{m-1} + \cdots + x + 1$ over $GF(2)$ is called* **All One Polynomial (AOP)** *of degree m.*

Definition 2.2 [IT2]: *A polynomial $g(x) = p(x^s)$ over $GF(2)$ is called s-**Equally Spaced Polynomial** (s-ESP) of degree sn, where $p(x)$ is an AOP of degree n.*

Let $\mathcal{OP}$ denote a set of odd primes and let $\mathcal{Z}_n^*$ denote a set of integers i such that $\gcd(i, n) = 1$. For irreducible AOP's over $GF(2)$, the following lemma is known:

Lemma 2.3 [WW]: *An AOP of degree m is irreducible over $GF(2)$ iff $(m + 1) \in \mathcal{OP}$ and 2 is a generator of $\mathcal{Z}_{m+1}^*$.*

In addition, for irreducible ESP's, the following theorems are known:

Theorem 2.4 (Corollary 1 in [IT2]): *Define $g_i(x) = p(x^{(m+1)^i})$ $(0 \le i \le k-1)$, where $k \ge 2$ and $p(x)$ is an AOP of degree m. Then every $g_i(x)$ $(0 \le i \le k-1)$ is irreducible iff $p(x)$ is irreducible and $2^{m(m+1)^{k-2}} \not\equiv 1 \pmod{(m+1)^k}$.*

Theorem 2.5 (Theorem 3 in [IT2]): *Let $g(x) = p(x^s)$ be an s-ESP of degree sn, where $p(x)$ is an AOP of degree n. Then $g(x)$ is irreducible iff $p(x)$ is irreducible, and for some integer $l \ge 2$, $s = (n + 1)^{l-1}$ and $2^{n(n+1)^{l-2}} \not\equiv 1 \pmod{(n+1)^l}$.*

3 Main Results

In this section, we show a complete characterization for a family of *infinitely* many irreducible ESP's, which is an extension of Theorem 2.4, and the uniqueness of irreducible ESP's in a strict sense, which augments Theorem 2.5.

To do this, we need to show the following lemma:

Lemma 3.1: *Let $p \in \mathcal{OP}$. For any $a \in \mathcal{Z}_p^*$, if $a^{p-1} \not\equiv 1 \pmod{p^2}$, then $a^{(p-1)p^{k-2}} \not\equiv 1 \pmod{p^k}$ for any $k \ge 2$.*

Proof: We prove the lemma by induction. For $k = 2$, the lemma is *trivially* true, because this is the assumption. Assume that for $k = s$, the lemma is true, i.e., $a^{(p-1)p^{s-2}} \not\equiv 1$ (mod p^s). From Euler's theorem, it follows that $a^{(p-1)p^{s-2}} \equiv 1$ (mod p^{s-1}), and this implies that $a^{(p-1)p^{s-2}} = 1 + C_{s-1}p^{p-1}$ for some integer C_{s-1}. Hence $p \nmid C_{s-1}$, because $a^{(p-1)p^{s-2}} \not\equiv 1$ (mod p^s), and thus

$$
\begin{aligned}
a^{(p-1)p^{s-1}} &= (1 + C_{s-1}p^{s-1})^p \\
&= 1 + pC_{s-1}p^{s-1} + C'p^{s+1} \\
&\equiv 1 + p^s C_{s-1} \pmod{p^{s+1}}.
\end{aligned}
$$

Recalling that $p \nmid C_{s-1}$, we can conclude that $a^{(p-1)p^{s-1}} \not\equiv 1$ (mod p^{s+1}), and this implies that the lemma is also true for $k = s + 1$. $\square$

Combining Theorem 2.4 with Lemma 3.1, we can show a complete characterization for a family of *infinitely* many irreducible ESP's.

Theorem 3.2: Let $g_i(x) = p(x^{(m+1)^i})$ for all $i \geq 0$, where $p(x)$ is an AOP of degree m. Then every $g_i(x)$ $(i \geq 0)$ is irreducible iff $p(x)$ is irreducible and $2^m \not\equiv 1$ (mod $(m+1)^2$).

Proof: (*if-part*) Since $p(x)$ is irreducible, $(m+1) \in \mathcal{OP}$. (see Lemma 2.3.) Furthermore, for any $k \geq 2$, $2^{m(m+1)^{k-2}} \not\equiv 1$ (mod $(m+1)^k$), because $2^m \not\equiv 1$ (mod $(m+1)^2$). (see Lemma 3.1.) Combining Theorem 2.4 with this, we can immediately conclude that every $g_i(x)$ $(i \geq 0)$ is irreducible over $GF(2)$.

(*only if-part*) This can be shown from Theorem 2.4 in a straightforward way. $\square$

An irreducible s-ESP of degree sn must satisfy that $(n+1) \in \mathcal{OP}$ and $s = (n+1)^k$ for some positive integer k. (see Theorem 2.5.) Thus we immediately have a theorem on the uniqueness of irreducible ESP's in a strict sense.

Theorem 3.3: Let $g(x)$ and $h(x)$ be an irreducible s-ESP of degree sn and an irreducible s'-ESP of degree $s'n'$, respectively. If $sn = s'n'$, then $g(x) = h(x)$.

Proof: From Theorem 2.5, it follows that

$$
\begin{cases} (n+1) \in \mathcal{OP}; \\ s = (n+1)^{l-1} \ (l \geq 2). \end{cases} \quad \text{and} \quad \begin{cases} (n'+1) \in \mathcal{OP}; \\ s' = (n'+1)^{l'-1} \ (l' \geq 2). \end{cases}
$$

If $sn = s'n'$, then $n(n+1)^{l-1} = n'(n'+1)^{l'-1}$. Here we prove the theorem by contradiction. If $n+1 \neq n'+1$, then $(n+1)^{l-1}|n'$, thus n' must be of the form that $n' = k(n+1)^{l-1}$ for some integer $k \geq 1$. Then

$$n'(n'+1)^{l'-1} = k(n+1)^{l-1}\left\{k(n+1)^{l-1}+1\right\}^{l'-1} > n(n+1)^{l-1},$$

because $l \geq 2$ and $l' \geq 2$. Hence it must be $n+1 = n'+1$, and thus $(n+1) \nmid n'$. This implies that $(n+1)^{l-1}|(n'+1)^{l'-1}$, and thus $l \leq l'$. In a way similar to this, we can also show that $l' \leq l$, and eventually $l = l'$. Hence if $sn = s'n'$, then $n = n'$ and $s = s'$, and this implies that $g(x) = h(x)$. $\square$

4   Applications

In this section, we present two applications of the main results in section 3, one is to design a multiplier for a class of $GF(2^m)$ with structural modularity and small circuit depth, and the other is to design a (fast) divider for a class of $GF(2^m)$ with $O(\log m)$ multiplications in $GF(2^m)$ and $O(\log m)$ permutations over $\{0,1\}^{O(m)}$.

4.1   Multipliers Based on Irreducible ESP's

In this subsection, we show that Theorem 3.2 enables us to design a family of *infinitely many* multipliers for a class of fields $GF(2^m)$ with structural modularity, of size $O(m^2)$ and depth $O(\log m)$. (The weaker result can be found in [IT2].)

Let $p(x)$ be an irreducible AOP of degree m and let $g_i(x) = p(x^{(m+1)^i})$ for all $i \geq 0$. From Theorem 3.2, it follows that if $2^m \not\equiv 1 \pmod{(m+1)^2}$, each $g_i(x)$ $(i \geq 0)$ is irreducible over $GF(2)$. It is shown in [IT2] that for any irreducible ESP of degree m, there exists a parallel type multiplier with structural modularity, of size $O(m^2)$ and depth $O(\log m)$. Thus an irreducible AOP of degree m such that $2^m \not\equiv 1 \pmod{(m+1)^2}$ generates a family of infinitely many parallel type multipliers for $GF(2^{M_i})$ with structural modularity, of size $O(M_i^2)$ and depth $O(\log M_i)$, where $M_i = m(m+1)^i$ $(i \geq 0)$. On details for the construction of the multipliers, see [IT2].

Note that any parallel type MOM for $GF(2^m)$ consists of m modular Boolean circuits of the same structure, and the structure of each modular Boolean circuit is determined by a bilinear form of inputs to the multiplier. Thus the number of terms in the bilinear form $Term(m)$ is one of the good criterion to measure the circuit complexity of the multiplier. In [Metal], Mullin et al. show that for any MOM for $GF(2^m)$, $Term(m)$ is not less than $2m + 1$, and also show a sufficient condition on m for $Term(m)$ to be *exactly* $2m + 1$. Since $Term(m)$ is $O(m^2)$ for almost all m, the parallel type MOM for $GF(2^m)$ has in general size $O(m^3)$ and depth $O(\log m)$. Thus the parallel type multiplier given here can be regarded as a *complementary* candidate for small size fast multipliers.

4.2 Dividers Based on Irreducible ESP's

In this subsection, we show the construction of a fast divider for a class of fields $GF(2^m)$, which is based on an irreducible ESP and has size $O(m^2 \log m)$ and depth $O(\log^2 m)$. The key idea to design such a divider is very similar to that of [IT1].

Let $\{g_i(x)\}_{i \geq 0}$ be a family of infinitely many irreducible ESP's of degree M_i, where $M_i = m(m + 1)^i$ $(i \geq 0)$. For any $a \in GF(2^{M_i})$, let $a(x)$ be

$$a(x) = a_0 + a_1 x + \cdots + a_{M_i-1} x^{M_i-1},$$

where $a_i \in GF(2)$ $(0 \leq i \leq M_i - 1)$. From the Fermat's Theorem, it follows that for any $a \in GF(2^{M_i})$ $(a \neq 0)$, $a^{-1} = a^{2^{M_i}-2}$, thus immediately

$$a^{-1}(x) \equiv \{a(x)\}^{2^{M_i}-2} \pmod{g_i(x)}.$$

Here by padding $(m + 1)^i$ 0's to $a(x)$, we define $A(x)$ to be

$$A(x) = a(x) + \underbrace{0 \cdot x + \cdots + 0 \cdot x}_{(m+1)^i}.$$

It is not difficult to see that the least positive integer e such that $g_i(x)$ divides $(x^e + 1)$ is $(m + 1)^{i+1}$ and $A(x) \equiv a(x) \pmod{x^{(m+1)^{i+1}} + 1}$, thus $a^{-1}(x) \equiv A^*(x) \pmod{g_i(x)}$, where $A^*(x) \equiv \{A(x)\}^{2^{M_i}-2} \pmod{x^{(m+1)^{i+1}} + 1}$.

Let $a^{-1}(x)$ and $A^*(x)$ be

$$a^{-1}(x) = \hat{a}_0 + \hat{a}_1 x + \cdots + \hat{a}_{M_i-1} x^{M_i-1}, \ \hat{a}_i \in GF(2);$$
$$A^*(x) = A_0^* + A_1^* x + \cdots + A_{(m+1)^{i+1}-1}^* x^{(m+1)^{i+1}-1}, \ A_i^* \in GF(2),$$

respectively. Then for s $(0 \le s < (m+1)^i)$ and t $(0 \le t < m)$,

$$\hat{a}_{s+t(m+1)^i} \equiv A^*_{s+t(m+1)^i} + A^*_{s+m(m+1)^i} \pmod{2},$$

because $g_i(x)$ divides $\left\{ x^{(m+1)^{i+1}} + 1 \right\}$ and $g_i(x)$ is $(m+1)^i$-ESP of degree $m(m+1)^i$.

For each j $(0 \le j < (m+1)^{i+1})$, define k_j such that $(x^j)^2 \equiv x^{k_j} \pmod{x^{(m+1)^{i+1}} + 1}$, and define a set S to be $S = \{0, 1, \ldots, (m+1)^{i+1} - 1\}$. Then this constitutes a permutation $\pi : S \to S$ that satisfies $\pi(j) = k_j$ for all $j \in S$. Thus using a technique similar to [IT1], we can derive a fast division algorithm for $GF(2^{M_i})$ that requires $O(\log M_i)$ multiplications in $GF(2^{M_i})$ and $O(\log M_i)$ permutations over S. (see above.)

When we use the multiplier in subsection 4.1, each multiplication in $GF(2^{M_i})$ requires size $O(M_i^2)$ and depth $O(\log M_i)$. In addition, it is immediate to see that the permutation $\pi : S \to S$ requires size $O(M_i)$ and depth $O(1)$ and the transform from $A^*(x)$ to $a^{-1}(x)$, i.e., $a^{-1}(x) \equiv A^*(x) \pmod{g_i(x)}$, requires size $O(M_i)$ and depth $O(1)$.

Hence there exists a family of *infinitely* many (fast) dividers for $GF(2^{M_i})$ $(i \ge 0)$, each of which is based on an irreducible $(m+1)^i$-ESP $g_i(x)$ of degree $m(m+1)^i$ $(i \ge 0)$, of size $O(M_i^2 \log M_i)$ and depth $O(\log^2 M_i)$, where $M_i = m(m+1)^i$ $(i \ge 0)$.

5 Conclusion and Remarks

This paper proved a *stronger* result (see Theorem 3.2.) that completely characterizes a family of *infinitely* many irreducible ESP's. (The result in [IT2] characterized only a family of *finitely* many irreducible ESP's.) This paper also showed the uniqueness of irreducible ESP's in a strict sense (see Theorem 3.3.), i.e., there exist no distinct irreducible ESP's of the same degree. Thus the results in this paper completely characterize all irreducible ESP's. It is worth noting that given an irreducible AOP $p(x)$ of degree m such that $2^m \not\equiv 1 \pmod{(m+1)^2}$, it enables us to generate a family of *infinitely* many irreducible ESP's in the form of $p(x^{(m+1)^i})$ $(i \ge 0)$.

It is known that irreducible *trinomials* over $GF(2)$ sometimes enable us to design multipliers for $GF(2^m)$ of small circuit complexity. Hence Theorem 3.2 can be regarded as one of the *good* ways to generate *infinitely* many irreducible trinomials over $GF(2)$: From the facts that $p(x) = x^2 + x + 1$ is irreducible over $GF(2)$ and $2^2 \not\equiv 1 \pmod{(2+1)^2}$, it follows that $g_i(x) = p(x^{(2+1)^i})$ $(i \geq 0)$ generates a family of *infinitely* many irreducible trinomials over $GF(2)$, i.e.,

$$g_0(x) = p(x^{(2+1)^0}) = x^2 + x + 1;$$
$$g_1(x) = p(x^{(2+1)^1}) = x^6 + x^3 + 1;$$
$$g_2(x) = p(x^{(2+1)^2}) = x^{18} + x^9 + 1;$$
$$g_3(x) = p(x^{(2+1)^3}) = x^{54} + x^{27} + 1;$$
$$g_4(x) = p(x^{(2+1)^4}) = x^{162} + x^{81} + 1;$$
$$g_5(x) = p(x^{(2+1)^5}) = x^{486} + x^{243} + 1;$$
$$g_6(x) = p(x^{(2+1)^6}) = x^{1458} + x^{729} + 1;$$
$$\vdots$$
$$g_i(x) = p(x^{(2+1)^i}) = x^{2 \cdot 3^i} + x^{3^i} + 1,$$
$$\vdots$$

and this enables us to design multipliers (with *structural* modularity) for $GF(2^{M_i})$ of circuit size $O(M_i^2)$, where $M_i = 2 \cdot 3^i$ $(i \geq 0)$.

In [Z], Zierler enumerated all irreducible trinomials over $GF(2)$ of degree ≤ 1000. It is not difficult to see that the main results in this paper, Theorems 3.2 and 3.3, *partially* (but completely) characterize a family of *infinitely* many irreducible trinomials over $GF(2)$ derived from an irreducible AOP $p(x) = x^2 + x + 1$.

Recently, Sugimura and Suetsugu [SS] further extended the results in this paper to the general fields $GF(p)$, where $p \in \mathcal{OP}$. In [SS], they defined an *alternation s-ESP* over $GF(p)$ of degree sn, i.e.,

$$q(x) = x^{sn} - x^{s(n-1)} + x^{s(n-2)} + \cdots + (-1)^i x^{s(n-i)} + \cdots + (-1)^n,$$

and showed a necessary and sufficient condition for a family of *infinitely* many alternation ESP's to be irreducible over $GF(p)$, where $p \in \mathcal{OP}$.

Acknowledgment:

The author would like to thank Mitsunori Ogiwara of Tokyo Institute of Technology for his valuable comments and discussions on this work.

References

[AIT] Asano, Y., Itoh, T., Tsujii, S., "Generalized Fast Algorithm for Computing Multiplicative Inverses in $GF(2^m)$," *Electronics Letters*, Vol.25, No.10, May 1989, pp.664-665.

[I] Itoh, T., "Algorithms for Finite Fields Arithmetics and Their Applications to Public-Key Cryptosystems," *Doctor Thesis*, Tokyo Institute of Technology, August 1988.

[IM] Imai, H. and Matsumoto, T., "Algebraic Method for Constructing Asymmetric Cryptosystems," *Proc. of AAECC*-3, Grenoble France, July 1985.

[IT1] Itoh, T and Tsujii, S., "A Fast Algorithm for Computing Multiplicative Inverses in $GF(2^m)$ Using Normal Bases," *Information and Computation*, Vol.78, No.3, September 1988, pp.171-177.

[IT2] Itoh, T. and Tsujii, S., "Structure of Parallel Multipliers for a Class of Fields $GF(2^m)$," *Information and Computation*, Vol.83, No.1, October 1989, pp.21-40.

[IT3] Itoh, T. and Tsujii, S., "Effective Recursive Algorithm for Computing Multiplicative Inverses in $GF(2^m)$," *Electronics Letters*, Vol.24, No.6, March 1988, pp.334-335.

[M1] McEliece, R.J., "A Public-Key Cryptosystem Based on Algebraic Coding Theory," *DSN Progress Report* 42-44, January/February 1978, pp.114-116.

[M2] McEliece, R.J., "Finite Fields for Computer Scientists and Engineers," *Kluwer Academic Publisher*, 1987.

[MS] MacWilliams, F.J. and Sloane, N.J.A., "The Theory of Error-Correcting Codes," *North-Holland*, New York, 1977.

[MO] Massey, J.L. and Omura, J.K., patent Application of "Computational method and apparatus for finite field arithmetic," submitted in 1981.

[Metal] Mullin, R.C., Onyszchuk, I.M., Vanstone, S.A., and Wilson, R.M., "Optimal Normal Bases in $GF(p^n)$," *Discrete Applied Mathematics*, Vol.22, No.2, 1988/89, pp.149-161.

[OV] Oorschot, P.C. and Vanstone, S.A.,"A Geometric Approach to Root Finding in $GF(q^m)$," *IEEE* Trans. on Inform. Theory, Vol.35, No.2, March 1989, pp.444-453.

[P] Pincin, A., "A New Algorithm for Multiplication in Finite Fields," *IEEE* Trans. on Comput., Vol.38, No.7, July 1989, pp.1045-1049.

[SS] Sugimura T. and Suetsugu, Y., "A Consideration on Existence Condition of Irreducible Cyclotomic Polynomial and Its Property," *Technical Report of the IEICE*, ISEC89-45, March 1990, pp.37-45.

[Tetal] Tsujii, S., Kurosawa, K., Itoh, T., Fujioka, A., and Matsumoto, T., "A Public-Key Cryptosystem Based on the Difficulty of Solving a System of Non-linear Equations," *Electronics Letters*, Vol.23, No.11, May 1987, pp.558-560.

[W] Wang, C.C., "Exponentiation in Finite Fields," *Doctor Thesis*, UCLA, 1985.

[Wetal] Wang, C.C., Truong, T.K., Shao, I.S., Deutsch, L.J., and Reed, I.S., "VLSI Architecture for Computing Multiplications and Inverses in $GF(2^m)$," *IEEE* Trans. on Comput., Vol.34, No.8, August 1985, pp.709-715.

[WW] Wah, P.K.S. and Wang, M.Z., "Realization and Application of Massey-Omura Lock," *Proc. of International Zurich Seminar*, 1989, pp.175-182.

[Z] Zierler, N., "On Primitive Trinomials (mod 2)," *Information and Control*, Vol.12, 1968, pp.541-554.

DISTRIBUTED ALGORITHMS FOR DECIPHERING

Michel COSNARD   Jean-Laurent PHILIPPE
LIP / IMAG - Ecole Normale Supérieure de Lyon
46, allée d'Italie F-69364 LYON CEDEX 07 FRANCE
e-mail : cosnard@frensl61.bitnet

Abstract

Many authors have already presented parallel implementations of the Multiple Polynomial Quadratic Sieve algorithm used to break RSA keys. They only parallelize the sieve step. We present in this paper a theoretical study of the parallelization of all the steps of the MPQS factoring algorithm for a distributed memory multiprocessor. We propose a first solution ensuring no communications during the sieve phase but with a bad load balancing. Then, a new distribution strategy of the polynomials permits us to get a better load balancing. We derive an implementation on the FPS T40 hypercube. We compare various distribution strategies and show how to achieve superlinear speedups.

Keywords

Distributed algorithm, factorization, decipher, quadratic sieve, superlinear speedup.

1 . Introduction

The Quadratic Sieve algorithm [Pom 85] has origins which date back to M. Kraïtchik [Kra 26], but has not been implemented before the middle of 80's. With emerging high power computers, it has been used in the domain of cryptography to break RSA keys. It is a powerful method for factoring large integers up to 100 decimal digits which has been improved by Montgomery and implemented as the Multiple Polynomial Quadratic Sieve by Silverman [Sil 87].

Various authors have already published results about parallel implementations of this algorithm [PST 88], [CaS 88]. These implementations use either a network of workstations [CaS 88], [LeM 89] or very powerful vector processors (Cray-1 and Cray XMP) [Ger 83].

Distributed memory computers appear as an interesting trade-off between these two solutions. In common with the network of workstations, they possess the same ability to increase the number of processors in order to deal with the enormous amount of computations needed by the MPQS, without using low communication networks. In common with the shared memory computers, they could be dedicated to these tasks and high efficiency could be expected.

In [DaH 88], Davis and Holdridge describe an implementation of the MPQS on the NCUBE computer, a distributed memory multiprocessor with 1,024 nodes.

In this paper, we study the algorithmic aspect of a parallel implementation of the MPQS on such distributed memory computers. We compare various data repartitions, computation and communication strategies. As an experimental result of this study, we show that superlinear speedups can be achieved.

In the second paragraph, we rapidly present the MPQS algorithm. Then, we show the problems arising while parallelizing the MPQS algorithm, in a non-constrained memory distributed environment, and then, more reallistically, with a limited memory in a distributed environment. And, we derive an implementation of MPQS for a distributed memory 32-nodes multiprocessor, the FPS T40 hypercube.

2. The Multiple Polynomial Quadratic Sieve Algorithm

The basic quadratic sieve algorithm [DHS 84], [Ger 83], [Pom 82], used to factor N, searches for two integers X and Y such that $X^2 \equiv Y^2 \bmod N$. Then $(X-Y, N)$ is a proper factor of N. The two congruent squares X^2 and Y^2 are constructed from auxiliary congruences of the form $u_i^2 \equiv v_i^2 w_i \bmod N$. If we can find a set I of indices such that $W_I = \prod_{i \in I} w_i$ is a square, we can let $X = \prod_{i \in I} u_i$ and $Y = \left(\prod_{i \in I} v_i\right)\left(\prod_{i \in I} w_i\right)^{\frac{1}{2}}$.

The problem is then to factor W_I. So we need the prime factorization of all the w_i involved in W_I. It is difficult to find enough completely factored w_i such that their product be a square. In order to factor the w_i's, we use a set of small primes called the factor base.

Let us take $u_i = u(x) = ax+b$, $v_i = v(x) = a$ and $w_i = w(x) = a^2x^2+2bx+c$, with $b^2 \equiv N \bmod a^2$, $|b| < a^2/2$, and $b^2 - N = a^2c$. Choosing a close to $\sqrt{\dfrac{\sqrt{2N}}{M}}$ ensures that $|w(x)|$ does not vary too much on [-M, M[. Hence we can use many polynomials of sizes approximately constant with respect to a. The algorithm thus contains three parts: the initialization phase, the sieve phase and the factorization phase.

Let us give the algorithm to factor N :

Algorithm

Initialization :

Compute k (number of elements of the factor base),

B (= 2^m greater than the biggest element of the base),

M (the upper bound of the sieve interval).

Compute the k elements p_i of the base : p_i prime and $\left(\dfrac{N}{p_i}\right) = 1$.

Compute the solutions of $x^2 \equiv N \bmod p_i^\alpha$, for each α and p_i such that $p_i^\alpha \le B$.

Loop :

while not enough factored w(x) *do*

/* Compute a polynomial $w(x) = a^2x^2 + 2bx + c$: */

Compute a: a prime, $a = 4q+3$, a near $\sqrt{\dfrac{\sqrt{2N}}{M}}$, and (N/a) = 1.

Deduce b: $b = N^{\frac{(a(a-1)\,/\,2)\,+\,1}{2}}$, $|b| < a^2/2$. Deduce c: $c = (b^2 - N)/a^2$.

/* Sieve [-M, M[with each p_i of the factor base and each α such that $p_i^\alpha \le B$. */

Initialize the sieve array tab_sieve[-M, M[to 0.

For each p_i^α *do* find the starting point s in [-M, M[*end do*.

Repeat

Add $\lfloor \log_2 p_i \rfloor$ to tab_sieve[s]

$s = s + p_i^\alpha$

until $s \ge M$.

Define a bound V.

For each t in [-M, M[such that tab_sieve[t] $\ge$ V do try to factor w(t).

end do

Perform a gaussian elimination on the matrix of the factored w(x).

Compute the GCD of some dependent lines of the matrix. The GCD is a cofactor of N.

End algorithm

This is the basic version of the MPQS algorithm. Some refinements have been proposed that decrease the execution time. These include the use of a multiplier to bias the factor base towards small primes, the large prime variation to use the nearly but not completely factored w(x), the small prime variation that 'forgets' the low contribution of small primes in the sum of the log's and thus avoids a large number of sieving steps [RLW 88]. But these refinements do not deeply change the algorithmic view of MPQS.

3. Theoretical approach

We study now the theoretical problems when parallelizing all the steps of the MPQS algorithm on a hypothetic multiprocessor where each processor has no limit for memory space and then, on a limited memory multiprocessor.

3.1. The initialization

This step is nearly entirely devoted to the computation of the factor base (k elements p_i and the roots of $x^2 \equiv N \mod p_i^\alpha$). Let $p_k \approx 2k \log(2k)$ be the estimated greatest element of the factor base. Let P be the number of processors (0 to P-1). The interval $[3, p_k]$ is distributed as consecutive sub-intervals among consecutive processors. Each processor i searches for k/P elements in its proper sub-interval $\left[i. \dfrac{2k \log(2k)}{P} + 1, (i+1). \dfrac{2k \log(2k)}{P} \right]$. But as primes and quadratic residues are not equally distributed in small intervals, some of the processors may not find the k/P elements, while other processors may find more than k/P elements. Furthermore, the number of generated elements may be different from k.

The first algorithmic approach consists in computing more elements than needed in order to be sure to get at least k elements. But the computation of an element of the factor base is high power consuming, due to the test of its primality. It is thus not a good solution.

The second algorithmic approach consists in computing all the elements of the factor base that can be found in $[3, p_k]$. As the computation is distributed among the processors, this implies a synchronization in order to compute the total number of generated elements. This communication step can be done in one synchronization and $\log(P)$ communication steps. If we get less than k elements, we determine a new interval for searching for the missing elements. We broadcast this interval (one synchronization and $\log(P)$ communication steps). These steps are repeated until enough elements have been generated.

The third algorithmic approach consists in searching for the elements in $[3, p_k]$, and computing the number of found elements. Then a single processor searches for the missing elements while the other ones begin solving the equations $x^2 \equiv N \mod p_i^\alpha$. No time is lost, but it is difficult to balance the work load.

In the three cases, the equations $x^2 \equiv N \mod p_i^\alpha$ have to be solved for each α and p_i such that $p_i^\alpha \leq B$. The small elements of the factor base have $\alpha \geq 2$. We can prove that all the elements having $\alpha \geq 2$ are on processor 0, provided that N satisfies the following condition:

$$\log \log k - \log k + 2 \log P - 2 \log(\log k - \log P) \leq 0.$$

From now on, let us suppose that all the small elements of the base are on processor 0. Each small element implies α equations to be solved. So if all the processors have k/P elements, processor 0 will have more equations to solve than the other processors. That is why we balance the number of elements of the factor base among the processors such that all the processors have the same number of equations to solve. This balancing implies a new synchronization point and communications. But the computation of the roots can now be achieved with efficiency 1.

3.2. The sieve

The algorithm for the sieve step is a set of four nested loops informally described by the following algorithm:

Algorithm

    while not enough factored w(x) *do*
        for each polynomial w(x) *do*
            for each power α of each element p of the factor base *do*
                for each element x of the interval *do*
                    update value tab_sieve[x]
                end for
            end for
        end for
    end while
end algorithm

In this algorithm, three variables determine the parallelization technique: the generation of new polynomials, the distribution strategy of the factor base and the interval to be sieved, among the processors. One can distribute one of the variables of the three loops, and keep on each processor the two other ones. Hence we get two solutions for parallelizing the sieve: a cooperative computation for factoring each w(x) (distribution of the factor base) or an independent computation of k/P w(x)'s (distribution of the polynomials or of the interval).

3.2.1. Distribution of the polynomials

Each processor has to find k/P factored w(x). It generates its own polynomials, but has in memory the whole factor base and the whole interval. There is no need for communications, since the processors can sieve the whole interval with the whole factor base, search for the w(x) candidates that may be completely factored and factor them without any additional information. When a w(x) factors completely, the parameters (a, b, c, x and the prime factors with their multiplicities) are stored.

Within a limited memory space, it is not always possible to keep both the whole base and the interval in the memory. Therefore, if there is enough memory space to contain the factor base and a subinterval for sieving, it is possible to sieve completely without any communications. But it is not as fast as with an unlimited memory space, because for each subinterval, it is necessary to recompute some already known information. However the processors need no synchronization, except when they finish computing the number of lines of the matrix (i.e. the number of w(x) completely factored on the base), in order to perform the elimination step.

3.2.2. Distribution of the factor base

All the processors have the same polynomials and the whole interval. They sieve with a subset of the factor base. When the sieve with one polynomial is achieved, w(x) may be totally factored on the factor base, but this factorization is distributed among the processors, since the factor base is distributed. To get the final result of the sieve, communication steps are necessary to sum all the results obtained on each processor. So after each sieving phase, we synchronize the processors, communicate and reduce (sum) the results. This implies a waste of time, because each processor works at its own speed.

3.2.3. Distribution of the interval to be sieved

Each processor has the same polynomials, and the whole factor base. This corresponds to applying the first strategy on a subinterval. As all the processors will need the same polynomial at some time, it is a great waste of time that each processor generates this polynomial. So in the initialization phase, all the processors could generate and store all the required polynomials (in a parallel efficient way). This is possible within an unlimited memory space.

Within a limited memory space, it is impossible to store a priori all the polynomials that will be used during the sieve phase. So, these polynomials are to be generated either by each processor (redundancy of high consuming time work) or by a central processor (waste of time when synchronizing processors either one at a time with the central sender on request, or all together if we want a single communication step (broadcast) to occur).

3.2.4. Evaluation

One can see that only the first and third strategies can be really implemented on distributed memory multiprocessors with no limit of memory space, because they do not need any communications during the sieve phase. With a limited memory multiprocessor the third strategy could not be efficiently implemented. Concerning the first strategy, the efficiency depends on the size of both the factor base and the interval. If they fit in memory, no problem. If not, a supplementary work has to be performed to cut the interval into sub-intervals (sequentially sieved) and/or cut the base into subsets, implying much more computation and control.

3.3. Factorization of w(x)

When a w(x) is a candidate to a complete factorization over the factor base, i.e. the value in the interval of sieve at index x is greater than a precomputed value V, two strategies are possible. The first one consists in dividing a w(x) by each element of the factor base. But as these divisions are performed with infinite precision integers, they are high time consuming and most of them are not necessary. The second one consists in testing if x is congruent to the beginning index of the sieving with each element p of the base. If so, p is a prime factor of w(x), otherwise p is not. This strategy avoids almost all the infinite precision divisions, and is therefore much more faster.

3.4. Gaussian elimination

The sieve phase returns the factorizations of the factored w(x). These factorizations are stored in a matrix. Each row corresponds to a completely factored w(x). The columns represent the elements of the factor base. The elements of the matrix are the multiplicities of each element of the base in the factorization of w(x). The matrix is sparse. The elements of the matrix are then reduced to their class in Z/2Z. This matrix is sparser. We perform a Gaussian elimination over this matrix [PaW 84], [CTV 87], [Wie 86]. It can be done efficiently either in a parallel or in a sequential way, depending on the size of the matrix.

3.5. The GCD computation

The Gaussian elimination returns the linear dependencies between some lines of the matrix. It means that the product of the w(x) corresponding to these lines is a square. It is now easy to compute W_I, X and Y. Then gcd(X, Y) is a cofactor of N. This gcd can be computed in a sequential manner, since it does not take much time.

4. Implementation

We implement the multiple polynomial quadratic sieve on a 32-node distributed memory multiprocessor, the FPS T40 hypercube. Each processor is a T414 Transputer and has 1 Mbyte memory space. We use the distribution strategy of the polynomials presented in §3.2.1. We use the multiprecision package provided by J-L Roch in the PaC project (Parallel Algebraic Computing) [Roc 89].

Our aim is to minimize the total execution time. With the distribution of the polynomials, there are no communications during the sieve phase. In [CaS 88], the distribution of the polynomials is such that 2 processors may work with the same polynomials and thus get redundant results. This occurs because they partition the set of polynomials into subsets of polynomials, such that on each processor, the values of a are consecutive.

We propose to partition this set into families of polynomials such that two processors have different families (two processors cannot work with the same polynomials) and such that these families together cover the whole set of required polynomials. Each processor has its own family of polynomials $a^2x^2+2bx+c$. A family is defined by $a = 4Pj + 4i + 3$, i being the identity of the processors. Any two processors have different a's, and gathering these families leads to values of a equivalent to the original form.

4.1. Results

We factor integers N having 2 factors of approximately the same size.

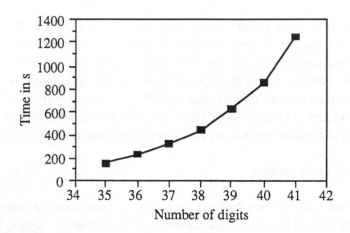

The time needed to factor an integer in the range 35 to 41 decimal digits is shown on the above figure. It increases very rapidly and is coherent with the theoretical execution time $\exp\{(1+o(1)) \sqrt{\ln N \ln \ln N}\}$, to factor N.

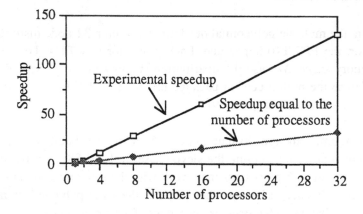

On the above figure, we present the superlinear speedups for the sieve phase of a 41-digit integer. The superlinear speedups come from the growth of memory space when increasing the number of processors. There is less memory management (memory is dynamically allocated and freed), because memory space is larger.

4.2. A new distribution strategy

The time to find k/P rows could be different from one processor to another. This is due to a bad load balancing. For the current algorithm, each processor has to find k/P rows. These k/P rows require a certain number of polynomials. We balance the load through a new task assignment distributing the theoretical number L of polynomials required to get k completely factored w(x). Each processor work with L/P polynomials. Then, we synchronize the processors (it will not lose much time, since the load is balanced), and compute the total number of factored w(x) on all the processors. If it is sufficient, begin the Gaussian elimination, otherwise distribute the task for finding the few missing w(x). The aim of such a technique is that no processor is idle while other ones are working. The algorithm is the following on each processor:

Algorithm

    Initialization phase.
    Compute the approximated number L of polynomials to get about k factored w(x).
    Sieve with the L/P polynomials on each processor, and factor w(x).
    Synchronize the P processors.
    Compute the number nbw of w(x).
    If nbw < k *then*
        Find the (k - nbw) / P missing w(x) on each processor.
    End if.
    Perform Gaussian elimination on the final matrix.
end algorithm.

This new algorithm (new distribution strategy) makes a better use of the power of the processors, since some do not stay idle while other ones work. Furthermore, it decreases the total execution time of the sieve phase. With the previous algorithm, a processor could stay idle for about 50% of the total execution time. With this new algorithm, a processor does not stay idle for more than 5% of the total sieve and factor time. Furthermore, it decreases the total execution time by 25%.

5. Conclusion

The design of the MPQS algorithm makes it well suited to a parallel implementation. Trivial parallelizations have already been implemented. We choose to carefully study all the steps of the algorithm and parallelize each of them. Our aim is to get the best workload for the processors and decrease the total execution time. But it raises problems: it leads to a very bad load balancing in our first algorithm. We then propose a new distribution strategy for the polynomials, that cannot assign the same polynomial to different processors, and that balances the work among the processors. The workload is then better. It also yields to a decrease of the total execution time and a higher efficiency. Algorithmic and mathematical refinements are being studied. They will be implemented in next implementations. Furthermore, this algorithm is currently developed to be implemented on other paralle machines, such as the TNode and the MegaNode distributed memory multiprocessors.

References

[CaS 88]   T. R. CARON, D. SILVERMAN, "Parallel Implementation of the Quadratic Sieve", The Journal of Supercomputing, 1, 1988, pp. 273-290.

[CTV 87]   M. COSNARD, B. TOURANCHEAU, G. VILLARD, "Gaussian Elimination on Message Passing Architectures", Proceedings of ICS 1987, Athens, 1987, Springer Verlag.

[DaH 88]   J. DAVIS, D. HOLDRIDGE, "Factorization of Large Integers on a Massively Parallel Computer", Eurocrypt '88 Abstracts, IACR, 1988, pp. 235-243.

[DHS 84]   J. DAVIS, D. HOLDRIDGE, G. J. SIMMONS, "Status Report on Factoring", Proceedings of EuroCrypt 84, LCNS.

[Ger 83]   J. GERVER, "Factoring Large Numbers with a Quadratic Sieve", Math. Comp., Vol. 41, 1983, pp. 287-294.

[Kra 26]   M. KRAITCHIK, "Théorie des nombres. Tome II", Gauthier-Villars, Paris, 1926.

[LeM 89]   A. K. LENSTRA, M. S. MANASSE, "Factoring by electronic mail", Proceedings Eurocrypt '89, 1989.

[PaW 84]   D. PARKINSON, M. WUNDERLICH, "A Compact Algorithm for Gaussian Elimination over GF(2) Implemented on Highly Parallel Computers", Parallel Computing, 1984, pp. 65-73.

[Pom 82]   C. POMERANCE, "Analyis and Comparison of some Integer Factoring Algorithms", in "Comput. Methods in Numb. Th.", (H.W. Lenstra, J. and R. Tijdeman, eds), Math. Centrum Tracts, n° 154, Part I, Amsterdam, 1982, pp. 65-73.

[Pom 85]   C. POMERANCE, "The Quadratic Sieve Factoring Algorithm", Advances in Cryptology (T. Beth, N. Cot and I. Ingemarrson, eds), Lect. Notes in Comput. Sc., Vol. 209, Springer Verlag, 1985, pp. 169-182.

[PST 88]   C. POMERANCE, J. W. SMITH, R. TULER, "A Pipeline Architecture for Factoring Large Integers with the Quadratic Sieve Factoring Algorithm", SIAM J. Comput., Vol. 17, n° 2, April 1988, pp. 387-403.

[RLW 88] H. J. J. Te RIELE, W. M. LIOEN, D. T. WINTER, "Factoring with the Quadratic Sieve on Large Vector Computers", Report NM-R8805, Centrum voor Wiskunde en Informatica, Amsterdam, 1988.

[Roc 89]   J. L. ROCH, "Calcul Formel et Parallélisme. L'Architecture du Système PAC et son Arithmétique Rationnelle", Thesis, Grenoble, december 1989.

[Sil 87]   R. D. SILVERMAN, "The Multiple Polynomial Quadratic Sieve", Math. of Comp., Vol. 48, n° 177, January 1987, pp. 329-339.

[Wie 86]   D. H. WIEDEMAN, "Solving sparse linear equations over finite fields", IEEE Trans. Inform. Theory, IT-32, 1986, pp. 54-62.

An Efficient Algorithm for Optimal Loop Parallelization (Extended Abstract)

Kazuo Iwano Susan Yeh

IBM Research, Tokyo Research Laboratory*

Abstract

In this paper, we developed an efficient algorithm for obtaining a compact time-optimal schedule for loops without conditional branches. The running time of our algorithm is $O(mn(p_{max}^2+\log(nLD)))$ where n (m) is the number of vertices (edges) in the associated data dependence graph, p_{max} is the denominator of the maximum slope in its irreducible form, L is the maximum number of machine cycles it takes to execute a statement, and D is the maximum number of iterations spanned by a statement. When L and D are polynomially large in n, the running time becomes $O(mn^3)$, which is the best result currently known.

1 Introduction

With the advances in parallel computing, extracting parallelism from programs has become an important topic of both practical and theoretical interest. One of the challenges is for compilers to detect as much parallelism as possible automatically, in order to take advantage of recent powerful approaches to parallelism such as Very Long Instruction Word (VLIW) machines [1,2,3,7,8]. With this motivation, we developed an efficient algorithm for obtaining a compact time-optimal parallel schedule for loops without conditional branches.

We first represent the data dependences among statements in a loop by a directed graph called a *data dependence graph* $G = (V, E)$ with n vertices and m edges. Figure 1 illustrates a sample loop L, and Figure 2 shows the corresponding data dependence graph whose vertices correspond to statements. An edge (u, v) exists if there is a *true dependence* from u to v, that is, statement u writes some memory location and later statement v reads the same memory location. If the graph is disconnected, then we can consider and schedule the connected components independently. Without loss of generality, we can assume that the data dependence graph is connected.

Each edge of the data dependence graph has two integer labels: *length* and *iteration distance*. The *length* of edge (u, v) is the number of machine cycles it takes to execute

---

*5-11 Sanbancho, Chiyoda-ku, Tokyo, Japan 102

statement u. Note that this number is independent of v. We naturally define the *length* of a cycle to be the sum of the lengths of the edges in the cycle. The *iteration distance* of an edge is the number of iterations spanned by the dependence. For example, if statement u writes memory location M at iteration i, and later statement v reads M at iteration j, then the iteration distance of edge (u, v) is $j - i$. Dependences that span one or more iterations are called *loop-carried* dependences.

Cycles in the data dependence graph correspond to essential interlocks between different iterations of a loop. The *slope* of a cycle, defined as the number of machine cycles required to execute all the statements in the cycle divided by the number of iterations spanned by the statements in the cycle, limits the execution rate of the loop. Suppose that statement v is on cycle C of length l and iteration distance d, and let $v(i)$ denote statement v at iteration i. Then statement $v(i + d)$ must be executed at least l machine cycles (or steps) after the step when statement $v(i)$ is executed. We then observe that the maximum slope, the best possible execution rate of statements, can be obtained by using algorithms for the minimum cost-to-time ratio cycle problem [6,10]. In this paper, we denote the maximum slope of a given data dependence graph by l_{max}/p_{max} in its irreducible form.

When we unfold loops in a data dependence graph, we obtain what we call a *dynamic graph*. Figure 3 illustrates the dynamic graph associated with the graph in Figure 2. Dynamic graphs have been well studied by [4,9]. A one-dimensional dynamic graph consists of infinitely many identical basic cells connected in a regular way. Here we can regard each cell (*resp.* each edge connecting cells) as an iteration of a loop (*resp.* a loop-carried dependence edge). We thus make use of a periodic structure of dynamic graphs to obtain a periodic pattern that appears in the optimal parallel schedule. Figure 8 illustrates a periodic pattern of an optimal schedule for the loop in Figure 1.

Finally, the running time of our algorithm for computing a compact time-optimal schedule is $O(mn(p_{max}^2 + \log(nLD)))$, where L (resp. D) is the maximum edge length (*resp.* the maximum iteration distance). When L and D are polynomially large in n, the running time becomes $O(mn^3)$, which is currently the best result.

The organization of this paper is as follows. In Section 2, we discuss previous work on loop parallelization. In Section 3, we formalize our problem and introduce some basic properties. In Section 4, we reduce the problem of finding the maximum slope to the minimum cost-to-time ratio cycle problem. In Section 5, we interpret our problem in the context of the theory of dynamic graphs. In Sections 6 and 7, we present our algorithm for a strongly connected data dependence graph. We illustrate this algorithm with an example in Section 8. In Section 9, we close with some concluding remarks.

2 Previous Work

In the literature, parallelizing loops is sometimes referred as *software pipelining*, although often the underlying models used for computation are different. In the *doacross* method [2], each iteration of the loop is assigned to a virtual processor. Munshi and Simons [8] have shown that the problem of determining the optimal delay on loops without conditional statements in the doacross method is NP-complete. However, as shown later, if we allow parallel execution of multiple statements in the same iteration, then a compact time-optimal schedule can be found in polynomial time.

Earlier work [11,13] on software pipelining loops without conditional statements, allowing parallel execution of statements within the same iteration, has resulted in heuristics that provide "good" but not necessarily optimal (in terms of execution time) schedules. Recently, Aiken and Nicolau [1] proposed a greedy method for examining a loop's execution history and extracting a pattern (or a schedule) of the loop. They showed that the pattern produced by the greedy method provides the shortest execution time for the loop. The greedy method takes polynomial time to detect the emergence of a pattern, but may take exponential time to compute the pattern.

Reiter [12] considered the problem of determining an admissible schedule for parallelizing loops without conditional statements. Reiter provided necessary and sufficient conditions for a schedule to be admissible, and characterized the solution space of the class of periodic admissible schedules achieving the optimal periodic rate. However, he did not discuss the complexity of computing a periodic admissible schedule. This paper improves [12] by addressing the loop parallelization problem in present-day language, and providing an efficient algorithm for computing a compact time-optimal schedule.

Recently, Zaky and Sadayappan [14] devised an algorithm for computing a compact time-optimal schedule for a strongly connected data dependence graph G. Their algorithm first unfolds loops p_{max} times, then obtains the maximum slope by solving the eigenvalue problem in a related path algebra, and finally computes a periodic schedule. The time complexity of the algorithm is $O(n_r m + n_r m_r p_{max} + m p_{max} + n^2 p_{max})$ where n (resp. m) is the number of vertices (resp. edges) in G and m_r (resp. n_r) is the number of edges (resp. the size of a feedback vertex set) in the unfolded graph.

A number of researchers [3,7] have considered the more general and difficult problem of software pipelining loops with conditional statements and proposed heuristics that generated reasonable schedules. The complexity of these heuristics may be high, that is, the running time of the heuristics is not necessarily polynomial time bounded.

3 Problem Statement and Admissibility

A *schedule* S consists of a sequence of *long statements* $\{S_i | i \in Z\}$ executed in order. Each step S_i, called the *i-th step* or *i-th long statement*, consists of statements that are executed in parallel in this step. Note that we assume that the loop has been executed infinitely many times and will also be executed infinitely many times in the future. We also assume the following: (1) A loop has no conditional statements; (2) Resources are plentiful; that is, the system can handle as much parallelism as is available. Let us denote a statement v at iteration j by $v(j)$. When statement $v(j)$ is scheduled at the i-th step in S, we denote this relationship by $S(v(j)) = i$ or $v(j) \in S_i$.

As stated in the Introduction, each edge e in E has two integer labels; that is, the length, denoted by $t(e)$, and the iteration distance, denoted by $d(e)$. We call the pair $(t(e), d(e))$ a *constraint* or *label* of e. A set of labels $A = \{(t(e), d(e)) | e \in E\}$ is called a *constraint set*. We say that a schedule S *satisfies* a constraint $(t(e), d(e))$ for $e = (u, v) \in E$, if for any integer j, statement $v(j+d(e))$ is scheduled at least $t(e)$ steps after the step in which statement $u(j)$ is scheduled; that is, $S(u(j)) + t(e) \leq S(v(j+d(e)))$. If a schedule S satisfies all constraints in a set A, we call S an A-*admissible* schedule. An A-admissible schedule S is called *periodic of period* $(l^*, p^*) \in Z \times Z$, if S satisfies the equation $S(v(j + kp^*)) = S(v(j)) + kl^*$ for each integer k and each v in V. We now have the following theorem:

Theorem 3.1. Let $G = (V, E)$ be a data dependence graph and A a constraint set. For some positive integers l^* and p^*, let us define a new constraint set A_1 as follows: $A_1 = A - \{(t(e), d(e))\} + \{(t(e) + kl^*, d(e) + kp^*)\}$, where k is an arbitrary integer and e is an arbitrary edge in E. If there exists an A_1-admissible periodic schedule S of period (l^*, p^*), then schedule S is also A-admissible. $\square$

4 Minimum Cost-to-Time Ratio Cycle

In this section we show that the maximum slope can be efficiently computed by using minimum cost-to-time ratio cycle algorithms.

Since the slope of a cycle limits the execution rate of the loop as discussed in the Introduction, the maximum slope serves as a lower bound on the execution rate of the loop. In Section 6, we obtain what we call a *time-optimal* schedule; that is, a schedule that achieves an execution rate equal to the maximum slope.

For an edge e, let us regard the length (*resp.* iteration distance) as the *time* (*resp. cost*). We analogously define the *cost* and *time* of a cycle. We can now define the *cost-to-time ratio* of a cycle as the cost of a cycle divided by the time of a cycle. The *minimum cost-to-time ratio cycle* problem is to find a cycle with a minimum cost-to-time ratio. We now observe that the slope of a cycle equals to the inverse of the cost-to-time ratio of a cycle. When the time of each edge is equal to one, the cost-to-time ratio of a cycle is the *mean cost* of a cycle; that is the average edge cost of a cycle. Now the *minimum cycle mean* problem is to find a cycle with a minimum *mean cost*.

Among extensive research done on the minimum cycle mean problem, Karp [6] obtained an $O(mn)$ time algorithm, and Orlin and Ahuja [10] obtained an $O(\sqrt{n}mlog(nD))$ time algorithm. Each algorithm can be extended to solve the minimum cost-to-time ratio cycle problem with an additional $O(\log(nLD))$ factor in time complexity.

5 Dynamic Graphs

A one-dimensional *dynamic graph* G^1 is an infinite graph obtained by repeating what we call a *basic cell* infinitely many times and connecting cells in a regular way. See [4,9] for general discussion on dynamic graphs.

Given a data dependence graph, if we unfold the loop infinitely many times, we can associate the structure obtained with a dynamic graph. Here loop-carried dependence edges correspond to edges that connect different cells. In this sense, we can associate a data dependence graph with a one-dimensional dynamic graph, and can also regard the x-th cell of a dynamic graph as the x-th iteration of an execution.

A dynamic graph is said to be *weakly connected* if there exists an undirected path between any pair of vertices. Orlin [9] showed that each weakly connected component of a dynamic graph is isomorphic to one another. Therefore, if the dynamic graph associated with a given data dependence graph is not weakly connected, we only have to schedule one weakly connected component and reuse the obtained schedule for the other weakly connected components. Thus, we can assume that the dynamic graph induced by a given data dependence graph is weakly connected.

6 Scheduling a Strongly Connected Component

In this section we consider the problem of finding an admissible schedule when a given data dependence graph is strongly connected.

The following theorem is used for converting a data dependence graph into a graph that is suitable for our algorithm in this section.

Theorem 6.1. Let $G = (V, E)$ be a directed graph and $f : E \longrightarrow R$ be an arbitrary one-dimensional edge labeling. Suppose there exists a directed spanning tree T in G rooted at s. Let $dist(v)$ be the length from s to v in T with respect to the label f. Now we define a new labeling f': $f'(e) = f(e) + dist(v) - dist(w)$ for $e = (v, w) \in E$. Then the new labeling f' satisfies the following: (1) for every tree edge $e \in T$, we have $f'(e) = 0$, and (2) for any path P from x to y in G, $f'(P) = f(P) + dist(x) - dist(y)$, where $f(P)$ is the sum of edge labels in P; that is, $f(P) = \sum_{e \in P} f(e)$. In particular, if T is a directed spanning shortest path tree, then the new labeling $f'(e)$ is nonnegative. $\square$

We now consider a strongly connected data dependence graph $G = (V, E)$. Let $A = \{(t(e), d(e)) \mid e \in E\}$ be a given constraint set and s be an arbitrary vertex in G. Note that as a property of a data dependence graph, G does not have negative cycles with respect to either length t or iteration distance d. Therefore, there is a directed spanning shortest path tree T rooted at s with respect to iteration distance d. Let $dist(v)$ be the shortest distance from s to v in G with respect to d, and $length(v)$ be the distance from s to v in T with respect to t. From Theorem 6.1, we define a new length $t_T(e)$ and a new iteration distance $d_T(e)$ for $e = (v, w) \in E$ as follows: $t_T(e) = t(e) + length(v) - length(w)$ and $d_T(e) = d(e) + dist(v) - dist(w)$. Note that $(t_T(e), d_T(e)) = (0, 0)$ for every $e \in T$, and $d_T(e) \geq 0$ for every $e \in E$. Let $A_T = \{(t_T(e), d_T(e)) \mid e \in E\}$ be a new constraint set with respect to a directed spanning shortest path tree T. We now have the following theorem:

Theorem 6.2. Let $G = (V, E)$, $A = \{(t(e), d(e)) \mid e \in E\}$, T, s, $A_T = \{(t_T(e), d_T(e)) \mid e \in E\}$, $dist(v)$, and $length(v)$ be defined as above. If there exists an A_T-admissible schedule S_T, then we can create an A-admissible schedule S from S_T as follows: for each $v \in V$

and each $j \in Z$, we define $S(v(j))$ as follows: $S(v(j)) = S_T(v(j-dist(v))) + length(v)$. □

From the above theorem, we can now devise the following algorithm:

Input: A strongly connected data dependence graph $G = (V, E)$ and a constraint set $A = \{(t(e_i), d(e_i)) | i = 1, 2, \ldots, m\}$.

Output: A periodic pattern $Z_{vertical}$ of an A-admissible schedule; that is, $Z_{vertical} = \{S_{ij} | i \in Z, 0 \le j \le p_{max} - 1\}$, where S_{ij} is the set of statements of the j-th iteration that are scheduled at the i-th step; that is, $S_{ij} = \{v(j) \mid v \in V\} \cap S_i$.

Algorithm SCC:

1. Apply Theorem 6.1. Let s be an arbitrary vertex in V.

   (a) Compute a directed spanning shortest path tree T with respect to d. For each $v \in V$, let the shortest iteration distance from s to v in G be $dist(v)$, and the length from s to v in the spanning tree T be $length(v)$.

   (b) For each edge $e = (v, w) \in E$, compute $d_T(e) = d(e) + dist(v) - dist(w)$ and $t_T(e) = t(e) + length(v) - length(w)$. Let $A_T = \{(t_T(e), d_T(e)) | e \in E\}$ be the obtained constraint set.

2. Compute the maximum slope l_{max}/p_{max}.

3. (a) Compute the *standard constraint set* $A_T^* = \{(t^*(e), d^*(e)) | e \in E\}$ defined as follows: $t^*(e) = t_T(e) - al_{max}$ and $d^*(e) = d_T(e) - ap_{max}$ such that $0 \le d^*(e) \le p_{max} - 1$ and $a \in Z$.

   (b) Compute an A_T^*-admissible schedule S_T^* as described in the next section. (From Theorem 3.1, an A_T^*-admissible periodic schedule S_T^* is also A_T-admissible.)

4. Compute an A-admissible schedule S from an A_T^*-admissible schedule S_T^* by using Theorem 6.2. That is, for each $v \in V$ and each $k \in \{0, 1, \ldots, p_{max} - 1\}$, let $S(v(k)) = S_T^*(v(k - dist(v))) + length(v)$.

7 Computing an A^*-admissible Periodic Schedule

In this section, we compute an A_T^*-admissible schedule S_T^*. For the sake of simplicity, we use A^* and S instead of A_T^* and S_T^*, respectively, in this section. We use the same notation as in Algorithm SCC; that is $G = (V, E)$, l_{max}, p_{max}, and a constraint set $A^* = \{(t^*(e), d^*(e)) | e \in E\}$.

Since our objective is to obtain a periodic schedule of period (l_{max}, p_{max}), we will compute a schedule for statements $v(k)$, where $v \in V$ and $0 \le k \le p_{max} - 1$. Therefore, we create a graph $G^* = (V^*, E^*)$, which consists of p_{max} cells of the dynamic graph induced by the data dependence graph G; that is, $V^* = \{v(j) | v \in V, j = $

$0, 1, \ldots, p_{max} - 1\}$. For each $e = (v, w) \in E$ and $j \in \{0, 1, \ldots, p_{max} - 1\}$, we create a new edge $e(j) \in E^*$ associated with e and j as follows: we first define $d^{**}(e(j))$ and $t^{**}(e(j))$ in such a way that a new edge $e(j)$ is defined by the edge from $v(j)$ to $w(d^{**}(e(j)))$ with length $t^{**}(e(j))$. Note that $w(d^{**}(e(j)))$ is the vertex w in the $d^{**}(e(j))$-th cell in G^*. We define $d^{**}(e(j))$ by $d^{**}(e(j)) \equiv j + d^*(e) \pmod{p_{max}}$, where $j + d^*(e) = d^{**}(e(j)) + a_{e(j)} p_{max}$, $a_{e(j)} \in Z^+ \cup \{0\}$, and then we define $t^{**}(e(j))$ by $t^{**}(e(j)) = t^*(e) - a_{e(j)} l_{max}$. Therefore, a new edge set E^* is defined as follows: $E^* = \{e(j) = (v(j), w(d^{**}(e(j)))) | e \in E, j = 0, 1, \ldots, p_{max} - 1\}$. Intuitively, an edge $e(j)$ represents a set of edges $\{e(j + k p_{max}) | k \in Z\}$ in the dynamic graph associated with a given data dependence graph (See the example in the next section). We now have the following theorem:

Theorem 7.1. G^* does not have a positive cycle with respect to t^{**}. □

From the theorem, we can compute the length $longest(v(j))$ of a longest path from $s(0)$ to $v(j)$ in G^* with respect to t^{**}. We then define a schedule S as follows: for each $j \in Z$, $S(v(j)) = longest(v(j_0)) + k l_{max}$ such that $j = j_0 + k p_{max}$, $k \in Z$, and $0 \le j_0 \le p_{max} - 1$.

Theorem 7.2. The obtained schedule S is A^*-admissible and has a periodic pattern of period (l_{max}, p_{max}). Moreover, S is *time-optimal* and *compact*. Here a schedule is said to be *time-optimal* when every statement is scheduled at the best possible execution rate (that is, l_{max}/p_{max}). A time-optimal schedule S is said to be *compact* if its periodic pattern has the shortest possible length (that is, l_{max}). □

Theorem 7.3. Algorithm SCC runs in $O(mn(p_{max}^2 + \log(nLD)))$ time. □

Note that Theorems 7.1 holds for an arbitrary pair of (l, p) such that $l/p > l_{max}/p_{max}$. Therefore, for any pair (l, p) such that $l/p > l_{max}/p_{max}$, we can create an A-admissible schedule having a period of (l, p). By using this observation, we can naturally extend the results in this section to a (not necessarily strongly connected) data dependence graph. See details in a full version of this paper [5].

8 Example

In this section, we illustrate how Algorithm SCC works for the loop in Figure 1 and its associated data dependence graph G in Figure 2. In Step 1, we compute a directed spanning tree T as indicated by the solid lines in Figure 4. In Step 2, since B, C, and D each have only one incoming edge with label $(0, 0)$, we can contract edges (A, B), (B, C), and (C, D) and obtain a graph $G_{contract}$ in Figure 5. Step 3 creates a graph G^* in Figure 6 from $G_{contract}$. Since the maximum slope in $G_{contract}$ is $7/2$, the graph G^* consists of two cells. Note that the edge $(F(1), E(0))$ with label -6 in G^* is induced by the edge (F, E) with label $(1, 1)$ in $G_{contract}$; the label -6 of edge $(F(1), E(0))$ is obtained by $(1,1) \equiv (-6,-1)$ $(\mod (7,2))$. In G^*, we compute the longest distance from $A(0)$ to $v(k)$, denoted by $longest(v(k))$, for each $v(k) \in V^*$, $k = 0, 1$; this is 0 for $A(0), E(0), F(0)$ and 1 for $A(1), E(1), F(1)$. Thus, we obtain the schedule S^* shown in Figure 7. For example, $E(1)$ is scheduled at time 1 since $longest(E(1)) = 1$. Note that S^* has a period $(7,2)$. Finally, we obtain the desired periodic pattern S as shown in Figure 8. For example, $S(E(1)) = 6$ is obtained by $S(E(0 + dist(E))) = S^*(E(0)) + length(E)$, where $dist(E) = 1$ and $length(E) = 6$.

9 Conclusion

In this paper, we devised an efficient algorithm for obtaining a compact and time-optimal schedule for a loop with no conditional statements. We made use of minimum cost-to-time ratio cycle algorithms, and extracted a periodic pattern of the desired schedule from a dynamic graph efficiently. The running time of our algorithm is $O(nm(p_{max}^2 + \log(nLD))$ where n (m) is the number of vertices (edges) in the associated data dependence graph, p_{max} is the denominator of the maximum slope in its irreducible form, and L (resp. D) is the maximum edge length (resp. the maximum iteration distance). When all iteration distances and lengths are polynomially large in n, our algorithm runs in $O(mn^3)$, which is the best known time bound.

References

[1] A. Aiken and A. Nicolau, Optimal Loop Parallelization, in *Proc. of the SIGPLAN '88 Conference on Programming Language Design and Implementation*, 1988, 308-317.

[2] R. Cytron, Doacross: Beyond vectorization for multiprocessors, in *Proc. of the 1986 Int'l Conf. on Parallel Processing*, Aug. 1986, 836-844.

[3] K. Ebcioglu, A Compilation Technique for Software Pipelining of Loops with Conditional Jumps, in *Proc. 20th Annual Workshop on Microprogramming*, Dec. 1987, 69-79.

[4] K. Iwano and K. Steiglitz, Testing for cycles in infinite graphs with periodic structure, in *Proc. of 19th ACM Symposium on Theory of Computing*, 46-55, 1987.

[5] K. Iwano and S. Yeh, An Efficient Algorithm for Optimal Loop Parallelization, Research Report RT 0043, IBM Tokyo Research Laboratory, March, 1990.

[6] R. M. Karp, A characterization of the minimum cycle mean in a digraph, *Discrete Mathematics*, 23(1978), 309 -311.

[7] M. Lam, Software Pipelining: An Effective Scheduling Technique for VLIW Machines, in *Proc. of the SIGPLAN '88 Conference on Programming Language Design and Implementation*, 1988, 318-328.

[8] A. A. Munshi and B. Simons, Scheduling Sequential Loops on Parallel Processors, in Proc. 1987 Int'l. Conf. on Supercomputing, Lecture Notes in Computer Science, vol. 297, 392-415.

[9] J. B. Orlin, Some problems on dynamic/periodic graphs, in *Progress in Combinatorial Optimization*, ed. W. R. Pulleyblank, 1984, Academic Press, Orlando, Florida.

[0] J. B. Orlin and R. K. Ahuja, New scaling algorithms for assignment and minimum cycle mean problems, Sloan Working Paper 2019-88, Sloan School of Management, M.I.T., 1988.

[1] B. R. Rau and C. D. Glaeser, Some Scheduling Techniques and an Easily Schedulable Horizontal Architecture for High Performance Scientific Computing, in *Proc. 14th Annual Workshop on Micropramming*, Dec. 1981, 183-198.

[2] R. Reiter, Scheduling Parallel Computations, *Journal ACM*, 15 (1968), 590-599.

[3] B. Su, S. Ding, and J. Xie, URPR - An Extension of URCR for Software Pipelining, in *Proc. 19th Annual Workshop on Microprogramming*, Oct. 1986, 94-103.

[4] A. Zaky and P. Sadayappan, Optimal static scheduling of sequential loops on multiprocessors, in *Proc. 18th International Conference on Parallel Processing*, Aug. 1989, III-130-137.

```
for i ← 1 to N do
A:  a[i] ← e[i − 1];
B:  b[i] ← a[i];
C:  c[i] ← b[i − 1];
D:  d[i] ← c[i] * k;
E:  e[i] ← d[i] + f[i];
F:  f[i] ← d[i − 1];
```

Figure 1: A sample loop L. We assume that an assignment takes one machine cycle, while a multiplication in statement D takes three machine cycles.

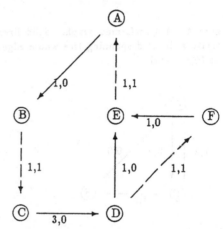

Figure 2: A data dependence graph G corresponding to the loop L. Each edge e has two labels; $length(e)$ and $iteration\ distance(e)$. Dotted lines indicate loop-carried dependences.

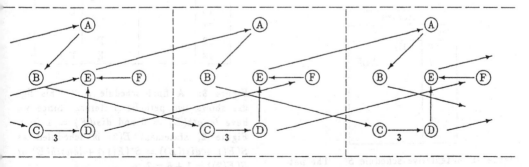

Figure 3: The dynamic graph associated with a graph in Figure 2. Each cell divided by dotted lines corresponds to an iteration of the loop execution. Number on edges indicate edge lengths, while an unlabeled edge has a default length of 1.

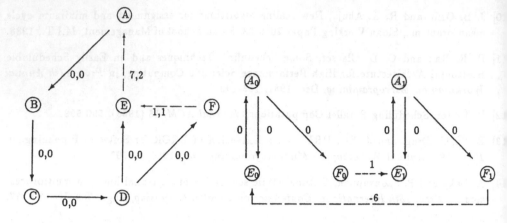

Figure 4: A transformed graph. Solid lines indicate a directed spanning tree whose edges have $(0,0)$ labels.

Figure 6: The graph G^* consists of two cells. The label -6 of edge (F_1, E_0) is obtained by $(1,1) \equiv (-6,-1) \pmod{(7,2)}$. For the sake of simplicity, we use F_i instead of $F(i)$ in the figure.

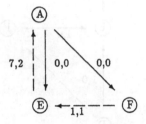

Figure 5: A contracted graph $G_{contract}$. The four vertices A, B, C, D in Figure 4 are contracted to the single vertex A.

	iteration					
time	0	1	2	3	4	5
0	AE	F				
1	B	A				
2		CB				
3			C			
4						
5	D					
6	E	FD				
7			AE	F		
8			B	A		
9				CB		
10					C	
11						
12				D		
13				E	FD	

Figure 8: A final schedule S. This figure shows two periodic patterns. Since we have $length(E) = 6$ and $dist(E) = 1$ from Figure 2, statement $E(2)$ is scheduled as $S(E(1 + dist(E))) = S^*(E(1)) + length(E)$ or $S(E(2)) = 1 + 6 = 7$.

	iteration	
time	0	1
0	AEF	
1		AEF
2		
3		
4		
5		
6		

Figure 7: A tentative schedule S^*. The pattern's size is 7 by 2.

Another view on the SSS* algorithm

Wim Pijls
Arie de Bruin
Erasmus University Rotterdam
P.O.Box 1738, NL-3000 DR Rotterdam
The Netherlands
wimp@cs.eur.nl

Abstract

A new version of the SSS* algorithm for searching game trees is presented. This algorithm is built around two recursive procedures. It finds the minimax value of a game tree by first establishing an upper bound to this value and then successively trying in a top down fashion to tighten this bound until the minimax value has been obtained.

This approach has several advantages, most notably that the algorithm is more perspicuous. Correctness and several other properties of SSS* can now more easily be proven. As an example we prove Pearl's characterization of the nodes visited by SSS* [Pl].

1 Introduction

During the last two decades several algorithms have been developed for computing the minimax value of a game tree. The most famous one is the alpha-beta-algorithm [Kn]. Another well known one is the SSS*-algorithm [St]. In this paper a new version of the latter algorithm will be developed equivalent with SSS* in the sense that the same nodes are examined in the same order. Recursion plays a key role in this version.

The SSS* algorithm originates from Stockmann [St]. The explanation in this paper is rather opaque, the algorithm is presented as a (semi-) parallel search in the "state space" consisting of so called partial (min-) solution trees. Later papers by Kumar and Kanal [KK1, KK2] recognized this algorithm as a special case of Branch and Bound. In [KK2] the observation is made that there is a dual view on SSS*, namely as a (sequential) search over (max-) solution trees. We will expand on this view and give it some formal underpinning. Several authors have studied SSS*, most notably Pearl [Pl] and Ibaraki [Ib1]. These papers give amongst others characterizations of the nodes which are visited by SSS*, and contain proofs of the superiority of SSS* over alpha-beta in this respect. Due to the fact that in these papers the SSS* algorithm is specified essentially as a search these investigations are complicated.

In this paper we present a version of SSS* as a top down recursive algorithm. This leads to a more perspicuous description of the algorithm. We will use this presentation amongst others to present an alternative proof of the fact that SSS* surpasses alpha-beta [Ib1].

After some preliminaries, our version of SSS*, which will be called SSS-2, is introduced in section 3. In section 4 some properties are derived which will be needed for a comparison between SSS-2 and alpha-beta. Some of these results are similar to the ones published in [Pl] and [Ib1]; also a few new results are presented. In section 5 we will make a few remarks on implementation issues, and we will sketch how SSS-2 can be transformed into the usual format. Almost all proofs are omitted by a lack of room. A complete description of all proofs can be found in [PB].

2 Game trees and solution trees

In most literature on games [Ib1], [Ib2], [Kn] and [Pl] e.g. the notions of game tree, max-node, min-node and minimax-value are defined. We adopt the convention that the minimax value of a game tree T, denoted by $f(T)$, is the minimax value of the root of this tree. An important notion in our version of SSS* is that of *solution tree*.

Definition 1 *A (max-) solution tree S in a game tree T is defined as a subtree T with the property that for all non terminal nodes n in S we have:*
-if n is a max node, then all its children are included in S.
-if n is a min node, then exactly one of its children is included in S.

In [KK2] such a tree is called an OR solution tree. The standard explanation of the SSS* algorithm is based on the complementary notion, which we will call a min-solution tree. Such a tree can be defined by interchanging the restrictions for min and max nodes in the above definition. Given a game tree T and a node $n \in T$, we denote the set of all solution trees rooted in n by $\mathcal{M}_T(n)$, or $\mathcal{M}(n)$ if no confusion can occur. $\mathcal{M}_T$ stands for $\mathcal{M}_T(root\ of\ T)$.
We adopt the convention that the minimax function, computed within a solution tree, will be denoted by the letter g; the g-value of the root of a solution tree S is denoted by $g(S)$.

Lemma 1 *Let S be a solution tree. Then for all $x \in S$ we have $g(x) \geq f(x)$.*

Lemma 2 *For each game tree T there exists a solution tree S in $\mathcal{M}_T$ such that $f(T) = g(S)$.*

In the sequel it will be useful to have a kind of genealogical ordering $\gg$ on game trees at our disposal. To be able to establish such an order we assume that in any non-terminal the child nodes have a fixed order, i.e. that we can establish whether one child is older than another one.

Definition 2 *Let n be a node in a game tree T. Let S_1 and S_2 be two different solution trees in $\mathcal{M}_T(n)$. Because $S_1 \neq S_2$ n cannot be a terminal. We define $S_1 \gg S_2$ recursively as follows:*
If n is a max node, then consider the oldest child m of n such that the subtrees $\bar{S}_1$ and $\bar{S}_2$, both rooted in m are different (because $S_1 \neq S_2$, such a subtree must exist). We define $S_1 \gg S_2$ if $\bar{S}_1 \gg \bar{S}_2$.
Suppose n is a min node. Let $\bar{S}_1$ be the subtree of S_1, rooted in the child m_1 of n in S_1, and let $\bar{S}_2$ be the subtree of S_2, rooted in the child m_2 of n in S_2. We define $S_1 \gg S_2$ if either m_1 is older than m_2 or $m_1 = m_2$ and $\bar{S}_1 \gg \bar{S}_2$.

Notice that for every n, we have that $\gg$ is a total ordering on the set $\mathcal{M}_T(n)$.
Related to this ordering is an ordering $\gg$ on nodes which is an extension of the "older than" relation. We say that $m \gg n$ iff there exist two ancestors m' and n' of m and

n respectively, such that m' and n' have the same parent and m' is older than n'. An intuitive explanation of this ordering is as follows. The game tree may be viewed as a genealogy of a royal family or a dynasty. The root is the actual king. The ordering $\gg$ on the nodes in the tree corresponds to their priority in succeeding the king.

3   The new SSS* version

The new version of SSS*, which we call SSS-2, is based on the idea to first establish an upper bound for $f(T)$, the game tree under consideration, and after that to repeatedly transform this upper bound into a tighter one. This is repeated until the upper bound cannot be diminished any more, in which case we have determined the minimax value of T.

We can establish an upper bound on a game tree T by exploiting the following recursive property of such an upper bound. The bottom of the recursion occurs when T consists of a terminal node only. In that case the game value of this node is a good upper bound. If the root of T is a max node, then we can obtain an upper bound for $f(T)$ by establishing an upperbound for each of its children and taking the maximum of these bounds. If the root of T is a min node then we do not need to investigate all its children, an upper bound of any child of the root is also an upper bound for the root itself.

If we turn this description into an algorithm (this will be the procedure "expand" to be defined later on), then it is clear that in order to find an upper bound in this way, we have to construct a solution tree in $\mathcal{M}_T$. We will organize "expand" in such a way that it constructs the oldest solution tree. This will be realized by taking in a min node the oldest child instead of "any child".

Now if we want to tighten the upper bound related to this solution tree, say S, we can realize this by transforming S as follows. First of all, if S consists of only one terminal node, then it is not possible to generate a better upper bound: we have obtained the minimax value. Now if the root of S is a max node, we can obtain a better upper bound only if for all children c of the root with $g(c) = g(root)$ a solution tree in $\mathcal{M}_T(c)$ can be generated with a lower g- value than $g(c)$.

If, on the other hand, the root of S is a min node then there are more possibilities to generate a better upper bound. First of all, one can try to obtain (recursively) a better upper bound for the current child of the root of S. But it is also possible to select another child c' of the root, and to try to establish an upper bound for $f(c')$ by building a new solution tree in $\mathcal{M}_T(c')$ with g- value $< g(S)$. Finding such a new solution tree is in fact a generalization of the expand process mentioned before: the difference is that now we are not satisfied with any new solution tree, we want a solution tree with a g- value better than a given value (i.e. $g(root)$). We will therefore define a procedure $expand(n, v_{in}, v_{out})$ that generates the oldest solution tree in $\mathcal{M}_T(n)$ with g -value $< v_{in}$ (if possible). Notice that by taking $v_{in} = \infty$ the oldest solution tree is obtained. Later in this section we will define the procedure $diminish$ which is based on the description given above on how to obtain from a given solution tree a better one.

The overall idea behind SSS-2 can thus be described as follows. Construct the oldest solution tree S in T. Repeatedly transform this solution tree S into another tree S' such that $g(S') < g(S)$ and $S' \ll S$ in such a way that if this transformation does not succeed, we have obtained the solution tree with the smallest g-value which is, by Lemma 1 and 2 the minimax value of T. The algorithm uses a global variable G which has as a value the current solution tree in $\mathcal{M}_T$.

We now give the main body of SSS-2.

```
begin
  root:= the start position of the game;
  expand(root, ∞, v_out, S);
  G := S;
  repeat
    [ v_in := v_out;
      diminish(root, v_in , v_out );
    ]
  until v_in = v_out ;
end.
```

In order for this program to work correctly the procedures *expand* and *diminish* must have the properties sketched above. We need therefore the following specifications:

Specification of procedure $diminish(n, v_{in}, v_{out})$
input-parameters: n, a node in the solution tree G,
                      v_{in}, a real number,
output-parameter: v_{out}, a real number.

If the call $diminish(n, v_{in}, v_{out})$ is executed in a situation with global solution tree $G = G_0$ such that n is a node in G_0, and $v_{in} = g(G_0(n))$, where $G_0(n)$ is the subtree of G_0 rooted in n, then either this call terminates with $v_{in} = v_{out}$ (no better solution tree found), in which case $v_{in} = v_{out} = f(n)$, or the call terminates with $v_{in} > v_{out}$ (a better solution tree was found), in which case G has a new subtree $G'(n)$, rooted in n, with $g(G'(n)) = v_{out}$, such that $G'(n)$ is the oldest solution tree younger than $G_0(n)$ with g-value $< v_{in}$.

Specifications of the procedure $expand(n, v_{in}, v_{out}, S)$
input-parameters: n, a node in T,
                      v_{in}, a real number
output-parameters: S, a solution tree rooted in n,
                      v_{out}, a real number.

The call $expand(n, v_{in}, v_{out}, S)$ either terminates with $v_{in} \leq v_{out}$ (no suitable tree found), in which case $f(n) \geq v_{in}$, or terminates with $v_{in} > v_{out}$ (there is a suitable tree), in which case S is the oldest solution tree in $\mathcal{M}_T(n)$ with g-value $< v_{in}$, and furthermore $g(S) = v_{out}$.

We now give the code of *diminish* and *expand*.

```
procedure diminish(n, v_in, v_out)
begin
  if terminal(n) then v_out := v_in
  else if type(n)=max then
          [ for c:= firstchild(n) to lastchild(n) do
              [ if g(c) = v_in then diminish(c,v_in, v'_out);
                if v_in = v'_out then exit forloop;
```

```
                    ]
            v_out := max(g-values of all children of n)
        ]
    else if type(n)=min then
        [ c := the single child of n in G;
          diminish (c, v_in, v_out) ;
          if v_in = v_out then
              for b:=nextbrother(c) to lastbrother(c) do
                  [ expand(b, v_in, v'_out, S);
                    if v_in > v'_out then
                        [ detach in G from n the subtree rooted in c
                                and attach S to n;
                          v_out := v'_out;
                          exit forloop;
                        ]
                  ]
        ]
end.

procedure expand(n, v_in, v_out , S);
begin
    if terminal(n) then
        [ if f(n) < v_in then S := the tree consisting only of node n;
          v_out := f(n);
        ]
    else if type(n) = max then
        [ for c := firstchild(n) to lastchild(n)
              [ expand (c, v_in, v'_out, S');
                if v'_out ≥ v_in then
                    [ v_out := v'_out;
                      exit forloop;
                    ]
              ]
          S := the tree composed by attaching
                  all intermediate values of S' to n;
          v_out := max of all intermediate values of v'_out;
        ]
    else if type (n)=min then
        [ v_out := v_in;
          for c := firstchild(n) to lastchild(n)
              [ expand (c, v_in, v'_out, S');
                if v'_out < v_in then
                    [ S := tree with S' attached to n;
                      v_out := v'_out;
                      exit forloop;
                    ]
              ]
        ]
end;
```

Lemma 3 *The procedures diminish and expand meet the specifications stated above.*

Lemma 4 *SSS-2 terminates for every finite tree T.*

Lemma 5 *Consider an execution of SSS-2 on a game tree T. Suppose the main loop performs n iterations, and let G_i be the value of the global solution tree G before the i-th iteration of this loop. Then we have:*
-G_1 is the oldest solution tree in $\mathcal{M}_T$.
-G_{i+1} is the oldest solution tree with g-value $< g(G_i)$ $(i = 1, \ldots, n-1)$.

We call $G_1, \ldots, G_n$ the solution tree sequence related to the execution of SSS-2 on T. Notice that Lemma 5 does not state anything on the final value $\bar{G}$ of G on termination of the algorithm. It is always true $g(G_n) = g(\bar{G})$, but it is not necessarily the case that $G_n = \bar{G}$.

Theorem 1 *For every finite tree T we have $v_{in} = f(T)$ on termination of SSS-2.*

4   Some properties of the new algorithm

In this section we will compare the efficiency of SSS-2 and the alpha-beta algorithm. Before doing so a few useful results will be derived. First a few definitions. For each node n in the game tree the quantities $\beta(n)$, $\alpha_L(n)$ and $\alpha_R(n)$ can be defined. These quantities play an important role in the well known alpha-beta-algorithm. Firstly we need the the notions AMIN, AMAX, AMIN-LC and AMAX-LC. The definitions are quite equal to those in [Pl]; in [Ib1] the same notions are used by a different name.

Definition 3 *For each node n the following quantities are defined:*
$AMAX(n) = \{ x \mid x$ is a max-node and x is a proper ancestor of $n \}$
$AMIN(n) = \{ x \mid x$ is a min-node and x is a proper ancestor of $n \}$
$AMIN\text{-}LC(n) = \{ x \mid x$ is a child of an element in $AMIN(n)$ and $x \gg n \}$
$AMAX\text{-}LC(n) = \{ x \mid x$ is a child of an element in $AMAX(n)$ and $x \gg n \}$
$AMAX\text{-}RC(n) = \{ x \mid x$ is a child of an element in $AMAX(n)$ and $x \ll n \}$

See Figure 1 to illustrate these definitions.

Definition 4 *Suppose n is a node in a game tree. Then we define:*
$\beta(n) = \min\{f(x) | x \in AMIN\text{-}LC(n) \}$
$\alpha_L(n) = \max\{f(x) | x \in AMAX\text{-}LC(n) \}$
$\alpha_R(n) = \max\{f(x) | x \in AMAX\text{-}RC(n) \}$
(we assume $\max(\emptyset) = -\infty$ and $\min(\emptyset) = \infty$).

Further we define the notion of β-ancestor of n. This node is the ancestor of n that is responsible for the value $\beta(n)$, i.e. this ancestor has a child in AMIN-LC with minimal f-value over AMIN-LC.

Definition 5 *Given a node n in a game tree. Suppose m' is a node such that m' in $AMIN\text{-}LC(n)$ and $f(m') = \beta(n)$. Then the β-ancestor of n is defined as the father of m'. In case of ties the node closest to the root is chosen. If $AMIN\text{-}LC(n) = \emptyset$ we take the root of T as the β-ancestor. Notice that for every node n for which $AMIN\text{-}LC(n) \neq \emptyset$ we have that the β-ancestor of n is a min node.*

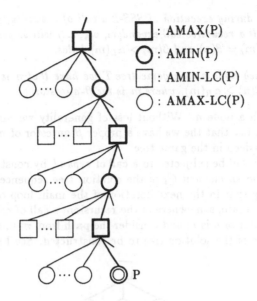

Figure 1: The sets AMAX, AMIN, AMAX-LC, AMIN-LC

We now derive a few results on the execution of SSS-2. Many are straightforward invariants or can be proven using invariants. First, we establish a property of every node n that is subjected to a diminish call during the execution.

Lemma 6 *During execution of SSS-2, before each call diminish$(n, v_{in}, \ldots,)$ we have that $\beta(n) > v_{in} > \alpha_L(n)$ and that $v_{in} \geq \alpha_R(n)$.*

We have a similar result for expand calls.

Lemma 7 *During execution of SSS-2, before each call expand$(n, v_{in}, \ldots,)$ we have that $\beta(n) = v_{in} > \alpha_L(n)$ and that $\beta(n) \geq \alpha_R(m)$, where m is the β-ancestor of n.*

Now the superiority of SSS-2 over alpha-beta follows from these lemma's and the following result by Pearl[Pl]:

Lemma 8 *A node n is examined by the alpha-beta algorithm if and only if $\beta(n) > \alpha_L(n)$.*

Theorem 2 *SSS-2 surpasses alpha-beta in the sense that SSS-2 visits no more nodes than alpha-beta does.*

In [Pl] the superiority of SSS* over alpha-beta is expresssed by comparing the sets of leaves examined during the execution of each algorithm, whereas now the sets of nodes are compared.

The next result that we want to derive is an extension of Lemma's 6 and 7: we want to establish that the property of n given in Lemma 7 is not only a necessary condition for a node to be expanded but a sufficient condition as well. This result gives us a precise characterization of the nodes activated by SSS-2, similar to the result (see Lemma 8) for alpha-beta. We need an auxiliary lemma first.

Lemma 9 *Suppose during execution of SSS-2 a call of expand(n, v_{in}, ...) is performed. Then during this call a recursive call expand(m, v_{in}, ...) will be generated for every child m of n for which $\beta(m) = \beta(n)$ and $\beta(m) > \alpha_L(m)$ holds.*

Theorem 3 *For each node n in a game tree T we have that n is examined by SSS-2 if $\beta(n) > \alpha_L(n)$ and $\beta(n) \geq \alpha_R(m)$, where m is the β-ancestor.*

Proof. Choose such a node n. Without loss of generality we can assume that AMIN-LC(n) is not empty, i.e. that the we have a proper β-ancestor of n. This is the case if n is at least two levels deep in the game tree.

We will prove that n will be subjected to a call of *expand*, by constructing a solution tree $S \in \mathcal{M}_T$ that will be an element G_i in the solution tree sequence (cf. Lemma 5), and furthermore showing that in the next iteration of the main loop of SSS-2 the call *diminish(root,...)* will generate, somewhere in the recursion, a call of *expand(n, ...)* .

Suppose the β-ancestor of n is m and consider the path from m upward to the root. This will be the backbone of the solution tree to be constructed. See Figure 2. Attach to the

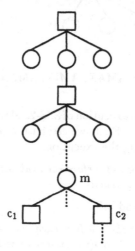

Figure 2: The tree constructed in the proof of Lemma 9

max nodes in this path all its children p (which are therefore members of AMAX-LC(m) and AMAX-RC(m)). We will next describe what the subtrees S_p rooted in these nodes p should be.

For each node p in AMAX-LC(m) we have $f(p) \leq \alpha_L(m) \leq \alpha_L(n) < \beta(n)$. So there exists for each p in AMAX-LC(n) a solution tree S_p rooted in p with the property $g(S_p) \leq \beta(n)$. Attach the oldest solution tree with this property to p.

For each node $p \in$ AMAX-RC(m) we have $f(p) \leq \alpha_R(m) \leq \alpha_R(n) \leq \beta(n)$. Attach to such a p the solution tree S_p defined as the oldest solution tree rooted in p with $g(S_p) \leq \beta(n)$. Because m is the β-ancestor of n, m has a child with f-value $= \beta(n)$. Let c_1 be the oldest child of m with that property. The child of m on the path from m to n is called c_2. Then we have that c_2 is younger than c_1, that the children of m older than c_1 have an f-value $> \beta(n)$, and that the children of m between c_1 and c_2 have f-value $\leq \beta(n)$. Since $f(c_1) = \beta(n)$, there exists a solution tree rooted in c_1 with g-value $= \beta(n)$. Attach the oldest solution tree with this property to m.

This construction has now resulted in a solution tree $S \in \mathcal{M}_T$ with g-value $= \beta(n)$. Moreover for every $S' \gg S$, we have $g(S') > g(S)$. Therefore S is an element of the solution tree sequence discussed after lemma 5.

Now consider the call $diminish(root, \ldots)$ in the main loop of the program which tries to diminish this S. The reader is invited to check that during this call, the procedure expand is activated with input parameters $n = c_2$ and $v_{in} = f(c_1)$.

For all nodes n' on the path from c_2 up to and including n we have the property that $\beta(n') = \beta(n) > \alpha_L(n) \geq \alpha_L(n')$, that n' and n both have m as β-ancestor, and thus that $\beta(n') \geq \alpha_R(m)$. By induction on the length of the path from c_2 to n we can prove, using Lemma 9, that each n' on this path will be expanded. So, n will be expanded. $\square$

As mentioned in Lemma 8 the nodes which are examined by alpha-beta are characterised by one simple condition. We have seen in the previous theorem that the characterisation of the nodes visited by SSS-2, needs one additional condition. In the next theorem we show a particular type of game tree where this additional condition is satisfied automatically for any node that satisfies the alpha-beta condition.

Theorem 4 *Suppose T is a game tree with the property that in each max node n the children $c_1, c_2, \ldots, c_n$ are ordered in such way that $f(c_1) \geq f(c_2) \geq \ldots \geq f(c_n)$. Then the alpha-beta-algorithm and SSS-2 examine the same set of nodes.*

The alpha-beta algorithm (see [Kn]) contains a recursive procedure with three input parameters: a node parameter n and two real parameters called *alpha* and *beta* respectively. The main body of the algorithm calls this procedure with $n = root$ of T (T is a game tree), *alpha* $= -\infty$ and *beta* $= +\infty$. It can be shown that the alpha-beta procedure returns the value $f(T)$ also when the two parameters have values such that *alpha* $< f(T) <$ *beta* holds. The set of examined nodes becomes smaller as the distance between alpha and beta decreases. In the next theorem a smaller alpha-beta window is given that still yields a superset of the node set examined by SSS-2. Before we need a lemma.

Lemma 10 *A node n examined by SSS-2 satisfies the relation $\beta(n) \geq f(T)$.*

Theorem 5 *Given a game tree T such that $f(T) = v_0$. Suppose the procedure alpha-beta is executed with parameters alpha=$v_0 - \varepsilon$ and beta=∞ where ε is any constant > 0. Then the set of nodes examined by the alpha-beta procedure is a superset of the set of nodes that is examined by SSS-2.*

5 Implementation issues

The global variable G should be implemented as a tree where the nodes are labelled by its g-value. In [KK2] it is observed that the list OPEN, used in the original SSS*, represents an OR tree by the set of its leaves. Careful observation reveals that it is possible to use this representation in our version as well.

The procedure *expand* can be adapted to this new representation in a straightforward fashion. The only difference is that the parameter S is now implemented as a list of its terminals. The prodecure *diminish* requires closer inspection. The execution of this procedure can be divided into three stages. During the first stage a terminal is selected in a top down fashion, using the g-value in the internal nodes of G. This terminal is the oldest one with maximal value. In the second phase it is attempted to obtain a better g-value for some ancestors of this terminal. This can be done by a strictly local search: backing up to a father and expanding a younger child. As soon as a better g-value has been found, the execution enters its third phase, which in essence amounts to updating g-values upwards until the root.

If we switch to a new representation the first stage can be reduced to selecting the oldest terminal with the maximum value. During the second stage we walk up the tree performing some expand calls on the way. The third phase can be omitted, since there is no need to update g-values any longer.

Notice that this new description is closer in spirit to the original Stockman version. Notice also that we only deal with a list of terminals ordered with respect to their game value. There is no need for control information as it is done in the Stockman triples, e.g. Solved/Live and g-values.

6 Conclusions

In this paper a new version of SSS* has been presented which is more transparant and "better suited for understanding" whereas the original version is "more convenient for implementation"[Ib1]. This enabled us to prove several properties of the algorithm, e.g its correctness and its surpassing alpha-beta. Our intention is to extend these results towards related algorithms, e.g. the RSEARCH version of SSS*.

References

[Ib1]   T. Ibaraki, *Generalisation of Alpha-Beta and SSS* Search Problems*, Artificial Intelligence 29 (1986) 73-117.

[Ib2]   T. Ibaraki, *Searching Minimax Game Trees under Memory Space Constraint*, to appear.

[KK1]   V. Kumar and L.N. Kanal, *A General Branch and Bound Formulation for Understanding and Synthesizing And/Or Tree Search Procedures*, Artificial Intelligence 21 (1983) 179-198

[KK2]   V. Kumar and L.N. Kanal, *Parallel Branch and Bound Formulations for AND/OR Tree Search*, IEEE Transactions on Pattern Analysis and Machine Intelligence, Vol PAMI-6 no. 6, november 1984.

[Kn]   D.E.Knuth and R.W.Moore, *An Analysis of Alpha-Beta Pruning*, Artificial Intelligence 6 (1975), 293-326.

[PB]   W. Pijls and A. de Bruin, *Another view on the SSS* algorithm.*, Report EUR-CS-90-01, Erasmus University Rotterdam.

[Pl]   I.Roizen and J. Pearl, *A Minimax Algorithm Better than Alpha-Beta? Yes and No.* Artificial Intelligence 21 (1983) 199-220.

[St]   G.C. Stockman, *A Minimax Algorithm Better than Alpha-Beta?*, Artificial Intelligence 12 (1979) 179-196.

Algorithms from Complexity Theory: Polynomial-time Operations for Complex Sets

*Lane A. Hemachandra**
University of Rochester
Department of Computer Science
Rochester, NY 14627, USA

Abstract

Typically, a set is described as complex when its *membership* problem has no efficient solution. Many sets of central interest for data management, design of computer chips, scheduling, fault-tolerant network design, and general optimization—such as the NP-complete sets—are thought to be complex in this sense. Though much effort has been spent accumulating evidence of the infeasibility of *membership testing* for these sets, less effort has been spent determining which operations *can* be efficiently performed on complex sets.

This paper surveys recent advances in the study of the sets for which fundamental operations—data compression and hashing, approximate membership testing, and enumeration—can be performed efficiently. We will see that many sets traditionally viewed as complex nonetheless have efficient algorithms for these operations.

1 Introduction

The difficulty of membership testing is a core issue of computational complexity theory. However, there is abundant evidence that many sets of fundamental importance—such as the NP-complete sets—lack efficient membership testing algorithms. Though work to directly address this impasse—for example, by proving that P = NP—is clearly of central importance, a parallel and related stream of research has quietly been developing that itself sheds great light on the foundational question of complexity theory: What can be efficiently computed? This parallel stream studies not the sets with efficient membership tests; rather, for operations OP other than membership, it seeks to identify the class of sets on which OP can be efficiently performed.

This parallel approach is neither new nor independent of the study of membership testing. The two approaches are symbiotically and synergistically intertwined.

---

*Supported in part by the International Information Science Foundation under grant 90-1-3-228 and the National Science Foundation under grant CCR-8996198 and a Presidential Young Investigator Award.

Both date to the earliest days of computational complexity theory. Within the theme of "complexity of operations other than membership" falls pioneering work on sets with polynomial-time padding algorithms, sets with polynomial-time selector functions, and sets with polynomial-time self-reduction algorithms [BH77,MP79,MY85,Sel79,Sel82, GJY87]; many dozens of papers fall within this research paradigm.

In recent years, the level of research activity along these lines has steadily intensified. The present paper serves as a survey of a few recent advances along these lines. It is by no means an exhaustive survey; rather, we focus on three lines of research. Furthermore, our goal is not to immerse ourselves in details, which can be found in the primary sources. Rather, the next three sections strive to present an overview of the advances that have emerged from recent studies of the sets for which hashing and data compression, approximate membership testing, and enumeration can be efficiently performed.

2 Hashing

Central to the problems of data management, data storage, and data compression is the question of finding an address at which to store a given data item. Hash functions directly address this problem; they are simply functions that map data items to addresses.

Perfect minimal hash functions map items to addresses in such a way that there are no collisions (two items in the set are never mapped to the same address), and there are no "holes" in the storage area (every address has something mapped to it). Static hash functions compute addresses based only on the data item input, and not based on the sequence of items that have already been hashed.

The literature on hash functions is enormous; the area has been, and remains, extremely active. However, one recent paper broadens the scope of this research stream by proposing the study of the class of *infinite* sets that have *efficiently computable perfect minimal static hash functions*. This work of Goldsmith, Hemachandra, and Kunen [GHK90], and earlier work on the ranking functions of Goldberg and Sipser [GS85], will be the focus of this section. Again, following the theme of this survey, we will see that many sets traditionally viewed as complex—including all well-known NP-complete sets—have efficiently computable perfect minimal static hash functions.

Formally, we say that:

Definition 2.1 [GHK90]

1. f is a *polynomial-time perfect minimal hash function* (PMH function) for a set A if:

   (a) f is polynomial-time computable, and

   (b) $f(A) = \Sigma^*$, and

   (c) $(\forall a, b \in A)[f(a) \neq f(b)]$.

2. If A has a PMH function, then we say that A is P-compressible.

3. For a complexity class $\mathcal{C}$, we say that $\mathcal{C}$ is P-compressible if every infinite set in $\mathcal{C}$ is P-compressible.

Note that only infinite sets have can have PMH functions. The definition is the natural analog to the notion of perfect minimal hash functions for finite sets [Knu73,Spr77]—part 1b is a minimality condition and part 1c is a perfection condition.

In considering candidate PMH functions for sets, one particularly simple type of function springs to mind—the ranking function. First studied in a complexity-theoretic setting by Goldberg and Sipser [GS85], the ranking function $r_A(\cdot)$ of an infinite set A is the function $r_A(x) = \|\{y \mid y \in A$ and $y \leq_{lex} x\}\|$. That is, the ranking function finds the "position" within A of a given input string.

Were they often polynomial-time computable for complex sets, ranking functions would make ideal PMH functions. Unfortunately, ranking a set instantly yields a membership test of complexity approximately equivalent to that of the ranking function. Thus, the very best we could hope for would be that all P sets are P-rankable. Unfortunately, even this seems unlikely. In the following theorem, #P [Val79] represents the class of functions that count the accepting paths of nondeterministic polynomial-time Turing machines.

Theorem 2.2

1. [GS85,Huy88,HR90] $P = NP = P^{\#P}$ if and only if all sets in P have polynomial-time ranking functions; indeed, this characterization holds even for all sets in NLOGTIME (in a machine model that makes such a class reasonable [Huy88]).

2. [HR90] If P has small ranking circuits then $P^{\#P} \subseteq \Sigma_2^p$ (and thus the polynomial hierarchy also collapses [KL80]).

3. ([HR90, Theorem 5.2], strengthened via the result of Toda ([Tod90], see [ABG90]) and Cai and Hemachandra [CH90] that even polynomially-enumerative counting is hard) $P = NP = P^{\#P}$ iff the ranking functions of all sets in P are $n^{\mathcal{O}(1)}$-enumerable. ($n^{\mathcal{O}(1)}$-enumerability [CH89] is a form of approximation.)

Thus, even P defies attempts at perfect minimal hashing—exact, approximate, or by circuits—via ranking.

The unfortunate failure of ranking functions as PMH functions is due to the fact that ranking functions must respect lexicographical ordering. PMH functions that are not bound by such constraints have much greater flexibility. [GHK90] proves that every recursive set with an infinite P-printable subset is P-compressible. It follows that:

Theorem 2.3 [GHK90] SAT is P-compressible. Indeed, all standard[1] NP-complete sets are P-compressible.

---

[1] By standard, we mean those sets that are polynomially isomorphic to SAT. Though almost all known NP-complete sets fall within this category [BH77], possible exceptions have been constructed by Joseph and Young [JY85] and Watanabe [Wat90]. See Section 4 for discussion of these potential exceptions.

Thus, P-compressibility (that is, perfect minimal hashing) is a good example of an algorithmically nontrivial operation that can be performed on relatively complex sets. What else can we say about the P-compressible sets? Necessary conditions and sufficient conditions are known, and related oracles exist.

Theorem 2.4 [GHK90]

1. If $E = NE^{NP}$, then NP is P-compressible.[2]

2. If the exponential hierarchy collapses to E, then the polynomial hierarchy is P-compressible.

3. If NP is P-compressible, then all infinite NP sets have infinite sparse[3] NP subsets.

4. There is an oracle A relative to which there are sparse sets in $NP^A - P^A$, yet NP^A is P-compressible.

Let's try to interrelate these result. Part 1 gives a sufficient condition for NP to be P-compressible. However, Hartmanis, Immerman, and Sewelson [HIS85] have shown, via a relativizable proof, that $NP - P$ contains sparse sets if and only if $NE \neq E$. Thus, part 4 says that that part 1's sufficient condition cannot be proven necessary by any proof technique that relativizes. Part 3 gives a necessary condition for the P-compressibility of NP.

In summary, in this section we have seen that many complex sets have efficiently computable perfect minimal hash functions.

3 Approximate Membership Testing

In this paper, our concern is to find operations for which sets of apparently high membership complexity have efficient algorithms. The other two sections—hashing and enumeration—discuss operations very different from membership. In contrast, this section studies the class of sets that have efficient "membership-like" tests—that is, sets where membership can be "approximately" tested.

Two nice notions of approximate membership testing are the P-closeness of Yesha [Yes83] and APT (almost polynomial time) of Meyer and Paterson ([MP79], see also [Ber78,KM81,BS85]).

Definition 3.1

1. [MP79] A set is in APT if it is accepted by a deterministic Turing machine that runs in (some) polynomial time-bound for all but a sparse set of inputs.

2. [Yes83,Sch86] A set is P-close if there is a set $B \in P$ such that they symmetric difference of A and B is sparse.

---

[2] $E = \cup_{c>0}$ DTIME$[2^{cn}]$; NE $= \cup_{c>0}$ NTIME$[2^{cn}]$.

[3] A set A is sparse if, for some polynomial p: for all n, A contains at most $p(n)$ strings of length n.

It follows from the definitions that APT is a subset of P-close. On one hand, P-close sets can be *extremely* complex; for example, any tally set, no matter how complex, is P-close. On the other hand, all P-close sets have small circuits, and if any standard (i.e., paddable [BH77]) NP-complete sets is P-close, then P = NP [Sch86, Theorem 4.9]. A stronger result holds for the smaller class APT: If *any* NP-complete set is in APT, then P = NP [MP79].

We now turn to two recently proposed notions of approximate membership testing: near-testing (NT) and nearly near-testing (NNT).

We've argued that for sets outside of P, it seems natural to ask which properties can be quickly computed, and, conversely, for properties slightly less revealing than membership, it seems natural to characterize the sets for which that property can be quickly computed. The recent work of Goldsmith, Hemachandra, Joseph, and Young [GHJY87], can be viewed from this perspective. Goldsmith, Joseph, and Young defined the class of near-testable (NT) sets.

Definition 3.2 [GJY87] A set L is *near testable* ($L \in$ NT) if, there is a polynomial-time computable function that input x computes which of (a) $(x \in L) \oplus (x_- \in L)$ or (b) NOT$[(x \in L) \oplus (x_- \in L)]$ holds.[4]

NNT will in some sense combines aspects of P and NT. Fix a set L. It is clear that for each $x \neq \epsilon$ exactly two of the following four statements hold:

1. $x \in L$,

2. $x \notin L$,

3. $(x \in L) \oplus (x_- \in L)$,

4. NOT$[(x \in L) \oplus (x_- \in L)]$.

If we could determine in polynomial time which of 3/4 holds, L would be in NT. If we could determine in polynomial time which of 1/2 holds, L would be in P. Clearly, of 1/2 exactly one holds, and of 3/4 exactly one holds. Thus, even though we know that for each x exactly two of 1/2/3/4 hold, if we could compute which two hold in polynomial time then L would be in P.

Nearly near-testable sets (NNT) allow us to capture a more modest amount of information about an given element x. A set is said to be nearly near-testable if for each x we can find, in polynomial time, *one* of 1/2/3/4 that holds. (Clearly, NT $\subseteq$ NNT.)

Definition 3.3 [HH90] A language L is in NNT if there is a polynomial-time computable function f such that for each x either:

• $(f(x) = \text{``}x \in L\text{''})$ and $(x \in L)$, OR

• $(f(x) = \text{``}x \notin L\text{''})$ and $(x \notin L)$, OR

• $(f(x) = \text{``}(x \in L) \oplus (x_- \in L)\text{''})$ and $((x \in L) \oplus (x_- \subset nL))$, OR

---

[4] $\oplus$ represents "exclusive or," and x_- represents the predecessor of x in standard lexicographical order.

- $(f(x) = $ "NOT$[(x \in L) \oplus (x_- \in L)]$") and (NOT$[(x \in L) \oplus (x_- \in L)]$).

Now, in the spirit of this paper, we would hope that some quite complex sets are near-testable or nearly near-testable. Indeed, this turns out to be the case. NT is deeply related to parity polynomial time ($\oplus$P, [PZ83,GP86], see also [BHW89,Tod89,CH]) and NNT related to the optimization complexity class OptP [Kre88,BJY90]. In particular, Theorem 3.6 says that NNT is the same (within the flexibility of $\leq^p_{1-1}$ reductions) as $\oplus$P altered by allowing an OptP function as a second argument to the underlying non-deterministic TM.[5]

Definition 3.4

1. [PZ83,GP86] A set A is in $\oplus$P if there is a nondeterministic polynomial-time Turing machine M such that: for all x, $x \in A$ if and only if $M(x)$ has an odd number of accepting computation paths.

2. [Kre88] An NP *metric Turing Machine*, $\widehat{N}$, is a nondeterministic polynomial-time Turing machine such that every branch writes a binary number and accepts; and for $x \in \Sigma^*$ we write $\mathrm{opt}^{\widehat{N}}(x)$ for the largest value on any branch of $\widehat{N}$ on input x.

3. [Kre88] A function f is in OptP (optimization polynomial time) if there is an NP metric Turing machine $\widehat{N}$ such that $f(x) = \mathrm{opt}^{\widehat{N}}(x)$ for all $x \in \Sigma^*$.

4. [LS86] $count_N(x)$ represents the number of accepting paths of machine N running on input x.

5. [HH90] L is in $\oplus$OptP if and only if there is a nondeterministic polynomial-time Turing machine N, and a polynomial $r(\cdot)$, and a function $f \in$ OptP such that:

$$x \in L \iff count_N(f(x)) \text{ is odd.}$$

Theorem 3.5 [GHJY87]

1. NT $\subseteq \oplus$P.

2. $\oplus$P $\leq^p_m$ NT; indeed, $\oplus$P $\leq^p_{1-1}$ NT. That is, for each language L in $\oplus$P, there is a language L' in NT such that L reduces to L' via a one-to-one reduction computable in deterministic polynomial time.

Theorem 3.6 [HH90]

1. NNT $\subseteq \oplus$OptP.

2. $\oplus$OptP $\leq^p_m$ NNT; indeed, $\oplus$OptP $\leq^p_{1-1}$ NNT.

Thus, approximate membership testing is closely related to the complexity of optimization, as represented by the class OptP, and to the power of parity polynomial time.

---

[5] Where the maximum function is always applied to sets of strings (which are in fact integers encoded in binary), and returns the integer value that the the lexicographically largest string encodes; we follow the convention that $\max(\emptyset) = 0$.

In particular, many sets—for example, $\oplus$SAT—that have traditionally been viewed as complex have, within the flexibility of $\leq^p_{1-1}$ reductions, efficient algorithms to nearly test membership.

4 Polynomially Enumerable Sets

How hard is it to *enumerate* the elements of possibly complex sets? To discuss this, we must first fix a natural notion of "enumerating" the elements of a set. A number of approaches have been proposed in the literature: probabilistic, recursion-theoretic, cryptographic, etc. [Her51,Myh59,You69,Sel78,AAB+90].

We discuss a proposal that studies enumeration in perhaps its most nature sense: is there a computationally feasible, *uniform* way of generating the entire set, one element at a time? In particular, is there a polynomial-time computable function f such that, for some initial element x_0, the set is exactly $\{x_0, f(x_0), f(f(x_0)), \ldots\}$? This complexity-theoretic notion has many analogs in recursive function theory. Loosely, one can think of this as a "polynomially enumerable" set in the same sense that r.e. sets are described as "recursively enumerable."

These notions have been studied by Hemachandra, Hoene, Siefkes, and Young [HHSY]. Their quite lengthy study is intensely algorithmic—it explicitly constructs nontrivial polynomial-time algorithms that operate as generators for complex sets. We survey and interpret the key results from this line of research.

Definition 4.1 [HHSY]

1. A set $L \subseteq \Sigma^*$ is *polynomially enumerable* (we'll abbreviate this as *i-enumerable*) if a polynomial-time computable function $f : \Sigma^* \to \Sigma^*$ and an element $x \in \Sigma^*$ exist such that: $L = \{x, f(x), f(f(x)), \ldots\}$. The function f is called the *enumeration function* for L.

2. A set $L \subseteq \Sigma^*$ is *polynomially bi-enumerable* (*bi-enumerable*) if there are elements $x \in \Sigma^*$ and $y \in \Sigma^*$, and a polynomial-time computable function $f : \Sigma^* \to \Sigma^*$, such that: $L = \{x, f(x), \ldots\}$ and $\overline{L} = \{y, f(y), \ldots\}$.

3. A function f is *invertible* if it is one-one and f^{-1} is polynomial-time computable on $range(f)$.[6] A set $L \subseteq \Sigma^*$ is *invertibly i-enumerable* (*invertibly bi-enumerable*) if it is an i-enumerable set (a bi-enumerable set) with an invertible enumeration function.

4. $A \leq^p_{1,si} B$ if there is a one-one polynomial-time computable function f with $x \in A \iff f(x) \in B$ and $|f(x)| > |x|$ for all x.

5. We will call a set $A \leq^p_{1,si}$-*self-reducible* if $A \leq^p_{1,si} A$.

---

[6]For such a function, $range(f)$ is polynomial-time decidable by checking whether $f(f^{-1}(x)) = x$ holds; thus the two common definitions of f^{-1}—one detecting values not in the range and one failing arbitrarily on such values—are equivalent for the purposes i-enumerability.

The following theorem shows that many complex sets are i-enumerable.

Theorem 4.2 [HHSY]

1. Every nonempty r.e. set that is $\leq^p_{1,si}$-self-reducible is i-enumerable.

2. Every recursive set $\emptyset \neq L \neq \Sigma^*$ that is $\leq^p_{1,si}$-self-reducible is bi-enumerable.

In fact, it follows from this theorem that all honestly paddable NP-complete sets (SAT, Clique, etc.) are bi-enumerable, and, indeed, that even the specific k-creative NP-complete sets proposed by Joseph and Young are bi-enumerable.

Definition 4.3 [JY85] A set L is k-creative if $L \in$ NP and there exists a polynomial-time computable function f such that for all i: $L(M_i) \in \text{NP}^{(k)}$ implies $f(i) \in L \iff f(i) \in L(M_i)$. The function f is called the *productive function* for $\overline{L}$. (All k-creative sets are known to be NP-complete [JY85].)

Corollary 4.4

1. Every set L that is k-creative with an honest one-one productive function for $\overline{L}$ is bi-enumerable.

2. Every set in NP that is honestly paddable is bi-enumerable.

Thus, one may assert that all well-known NP-complete sets are bi-enumerable. Are *all* NP-complete sets bi-enumerable? Watanabe has observed that if f is a one-way function [GS88], there is no obvious way of showing that $f(SAT)$ is P-isomorphic to SAT—yet, $f(SAT)$ will be an NP-complete set [Wat90]. Thus, Watanabe has provided a reasonable scheme for building candidate non-i-enumerable NP-complete sets. A second line of evidence against the i-enumerability of all NP-complete sets is the fact that there are relativized worlds in which some NP-complete sets lack enumeration functions.

Theorem 4.5 There is a recursive oracle A that is reasonable ($\text{P}^A \neq \text{NP}^A$) relative to which one-way functions do not exist and there is NP^A-complete set that is not i-enumerable by any P^A computable enumeration function (and thus not bi-enumerable by any P^A computable enumeration function).

Results similar to those described above hold for invertible i-enumerability.

Theorem 4.6

1. Every nonempty r.e. set that is $\leq^p_{1,si}$-self-reducible via a polynomially invertible function is invertibly i-enumerable.

2. Every recursive set $\emptyset \neq L \neq \Sigma^*$ that is $\leq^p_{1,si}$-self-reducible via a polynomially invertible function is invertibly bi-enumerable.

Thus, the theory of enumerative generation has provided a third and final area in which operations other than membership can be efficiently performed on sets whose membership complexity is thought to be high. In particular, all standard NP-complete sets have polynomial-time generation schemes. In earlier sections, we noted that all standard NP-complete sets have polynomial-time computable perfect minimal static hash functions, and that all $\oplus$P and all $\oplus$OptP sets are efficiently reducible to sets that have, respectively, polynomial-time near-testing and polynomial-time nearly near-testing algorithms.

Acknowledgements

I am very grateful to my friend and colleague Osamu Watanabe for making participation in this conference possible, and to my coauthors on the projects described herein.

References

[AAB+90] M. Abadi, E. Allender, A. Broder, J. Feigenbaum, and L. Hemachandra. On generating solved instances of computational problems. In *Advances in Cryptology – CRYPTO '88*, pages 297–310. Springer-Verlag *Lecture Notes in Computer Science #403*, 1990.

[ABG90] A. Amir, R. Beigel, and W. Gasarch. Some connections between bounded query classes and non-uniform complexity. In *Proceedings 5th Structure in Complexity Theory Conference*. IEEE Computer Society Press, July 1990. To appear.

[Ber78] P. Berman. Relationship between density and deterministic complexity of NP-complete languages. In *Automata, Languages, and Programming (ICALP 1978)*, pages 63–71. Springer-Verlag *Lecture Notes in Computer Science #62*, 1978.

[BH77] L. Berman and J. Hartmanis. On isomorphisms and density of NP and other complete sets. *SIAM Journal on Computing*, 6(2):305–322, 1977.

[BHW89] R. Beigel, L. Hemachandra, and G. Wechsung. On the power of probabilistic polynomial time: $P^{NP[\log]} \subseteq PP$. In *Proceedings 4th Structure in Complexity Theory Conference*, pages 225–227. IEEE Computer Society Press, June 1989.

[BJY90] D. Bruschi, D. Joseph, and P. Young. A structural overview of NP optimization problems. In *2nd International Conference on Optimal Algorithms (1989)*. Springer-Verlag *Lecture Notes in Computer Science*, 1990. To appear.

[BS85] J. Balcázar and U. Schöning. Bi-immune sets for complexity classes. *Mathematical Systems Theory*, 18:1–10, 1985.

[CH] J. Cai and L. Hemachandra. On the power of parity polynomial time. *Mathematical Systems Theory*. To appear.

[CH89] J. Cai and L. Hemachandra. Enumerative counting is hard. *Information and Computation*, 82(1):34–44, July 1989.

[CH90] J. Cai and L. Hemachandra. A note on enumerative counting. Technical Report TR-330, University of Rochester, Department of Computer Science, Rochester, NY, 14627, February 1990.

[GHJY87] J. Goldsmith, L. Hemachandra, D. Joseph, and P. Young. Near-testable sets. Technical Report 87-11-06, University of Washington, Department of Computer Science, Seattle, Washington, November 1987.

[GHK90] J. Goldsmith, L. Hemachandra, and K. Kunen. On the structure and complexity of infinite sets with minimal perfect hash functions. Technical Report TR-339, University of Rochester, Department of Computer Science, Rochester, NY, 1990.

[GJY87] J. Goldsmith, D. Joseph, and P. Young. Self-reducible, P-selective, near-testable, and P-cheatable sets: The effect of internal structure on the complexity of a set. In *Proceedings 2nd Structure in Complexity Theory Conference*, pages 50–59, 1987.

[GP86] L. Goldschlager and I. Parberry. On the construction of parallel computers from various bases of boolean functions. *Theoretical Computer Science*, 43:43–58, 1986.

[GS85] A. Goldberg and M. Sipser. Compression and ranking. In *17th ACM Symposium on Theory of Computing*, pages 440–448, 1985.

[GS88] J. Grollmann and A. Selman. Complexity measures for public-key cryptosystems. *SIAM Journal on Computing*, 17:309–335, 1988.

[Her51] H. Hermes. Zum Begriff der Axiomatisierbarkeit. *Mathematische Nachrichten*, 4:343–347, 1950–1951.

[HH90] L. Hemachandra and A. Hoene. On sets with efficient implicit membership tests. In *Proceedings 5th Structure in Complexity Theory Conference*. IEEE Computer Society Press, July 1990. To appear.

[HHSY] L. Hemachandra, A. Hoene, D. Siefkes, and P. Young. On sets polynomially enumerable by iteration. *Theoretical Computer Science*. To appear.

[HIS85] J. Hartmanis, N. Immerman, and V. Sewelson. Sparse sets in NP-P: EXPTIME versus NEXPTIME. *Information and Control*, 65(2/3):159–181, May/June 1985.

[HR90] L. Hemachandra and S. Rudich. On ranking. *Journal of Computer and System Sciences*, V. 40. To appear, 1990.

[Huy88] D. Huynh. The complexity of ranking. In *Proceedings 3rd Structure in Complexity Theory Conference*, pages 204–212. IEEE Computer Society Press, June 1988.

[JY85] D. Joseph and P. Young. Some remarks on witness functions for nonpolynomial and non-complete sets in NP. *Theoretical Computer Science*, 39:225–237, 1985.

[KL80] R. Karp and R. Lipton. Some connections between nonuniform and uniform complexity classes. In *12th ACM Sym. on Theory of Computing*, pages 302–309, 1980.

[KM81] K. Ko and D. Moore. Completeness, approximation, and density. *SIAM Journal on Computing*, 10(4):787–796, 1981.

[Knu73]   D. Knuth. *The Art of Computer Programming: Sorting and Searching*, volume 3 of *Computer Science and Information*. Addison-Wesley, 1973.

[Kre88]   M. Krentel. The complexity of optimization problems. *Journal of Computer and System Sciences*, 36:490–509, 1988.

[LS86]   T. Long and A. Selman. Relativizing complexity classes with sparse oracles. *Journal of the ACM*, 33(3):618–627, 1986.

[MP79]   A. Meyer and M. Paterson. With what frequency are apparently intractable problems difficult? Technical Report MIT/LCS/TM-126, MIT Laboratory for Computer Science, Cambridge, MA, 1979.

[MY85]   S. Mahaney and P. Young. Reductions among polynomial isomorphism types. *Theoretical Computer Science*, 39:207–224, 1985.

[Myh59]   J. Myhill. Recursive digraphs, splinters, and cylinders. *Mathematische Annalen*, 138, 1959.

[PZ83]   C. Papadimitriou and S. Zachos. Two remarks on the power of counting. In *Proceedings 6th GI Conference on Theoretical Computer Science*, pages 269–276. Springer-Verlag *Lecture Notes in Computer Science #145*, 1983.

[Sch86]   U. Schöning. *Complexity and Structure*. Springer Verlag *Lecture Notes in Computer Science #211*, 1986.

[Sel78]   A. Selman. Polynomial time enumeration reducibility. *SIAM Journal on Computing*, 7:440–457, 1978.

[Sel79]   A. Selman. P-selective sets, tally languages, and the behavior of polynomial time reducibilities on NP. *Mathematical Systems Theory*, 13:55–65, 1979.

[Sel82]   A. Selman. Analogues of semirecursive sets. *Information and Control*, 52:36–51, 1982.

[Spr77]   R. Sprugnoli. Perfect hashing functions: A single probe retrieving method for static sets. *Communications of the ACM*, 20(11):841–850, 1977.

[Tod89]   S. Toda. On the computational power of PP and $\oplus$P. In *Proceedings 30th IEEE Symposium on Foundations of Computer Science*, pages 514–519. IEEE Computer Society Press, October/November 1989.

[Tod90]   S. Toda, February 1990. Personal Communication.

[Val79]   L. Valiant. The complexity of enumeration and reliability problems. *SIAM Journal on Computing*, 8(3):410–421, 1979.

[Wat90]   O. Watanabe. A note on the P-isomorphism conjecture. *Theoretical Computer Science*, 1990. To appear.

[Yes83]   Y. Yesha. On certain polynomial-time truth-table reducibilities of complete sets to sparse sets. *SIAM Journal on Computing*, 12(3):411–425, 1983.

[You69]   P. Young. Toward a theory of enumerations. *Journal of the ACM*, 16:328–348, 1969.

COMPLEXITY CORES AND
HARD PROBLEM INSTANCES

Uwe Schöning
Universität Ulm
West Germany

Abstract

Many intractable problems such as NP-complete problems (provided $\mathbf{P} \neq \mathbf{NP}$) have easy subproblems. In contrast, we investigate the existence and the properties of inherently hard subproblems, called complexity cores. Furthermore, the question is posed whether individual problem instances can be inherently hard (for all algorithms solving the problem), and this question is answered positively.

1 Introduction

Consider a typical **NP**-complete set, like SATISFIABILITY (SAT, for short). It is known that the satisfiability problem can be efficiently solved on the class of Horn formulas (a Horn formula is a propositional formula in conjunctive normal form where each clause consists of at most one positive literal). A similar statement holds for the class of Krom formulas (each clause consists of at most 2 literals). Many applications enforce the use of such "easy" formulas only (consider e.g. logic programming) to escape the **NP**-completeness properties of SAT in general.

We consider the somewhat converse problem here: What are the special cases for SAT—and in general, of any **NP**-complete set—that cause the problem's intractability? We have in mind a restriction (i.e. an infinite subset of Σ^* where Σ is the underlying alphabet) such that on this restricted domain no such "easy" subproblems for the satisfiability problem can be found. (Such a restriction necessarily has to be disjoint from the Horn and Krom formulas.)

In the next section we give the necessary definitional setup to deal with these issues. Such restricted problem domains that are inherently hard, as discussed above, are called "complexity cores". See Figure 1 (the shaded area is a complexity core). The existence and the properties (such as density and maximality) of such complexity cores are then discussed in the further sections.

Complexity cores have to be infinite sets, by definition. Within this definitional setup, finite sets are not relevant. This is due to the fact that any algorithm can be "patched" with a finite amount of additional information to solve a given finite set of instances fast – without increasing the time complexity of the algorithm.

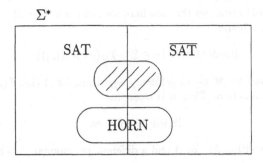

Figure 1: Easy and hard special cases of SAT

But notice that the price to be paid for this "finite patching property" is that the *size* of the algorithm increases by approximately that amount that is necessary to *describe* those finitely many instances. Therefore, to develop a sensible notion of the complexity of individual instances of a problem, it is neccessary to take the size of the algorithms for solving the problem into account. The result is a theory of instance complexity which is on the surface similar to the notion of Kolmogorov complexity, but the difference is that it is not the complexity of a string as such which is measured, but its inherent difficulty *with respect to* a given decision problem.

The notion of a complexity core was introduced by Lynch [13] and further explored, modified, and generalized in [4, 7, 16, 18, 20, 23, 2, 3, 6, 19, 21, 22]. Instance complexity was invented in [10], and further results can be found in [17, 11].

2 Preliminaries

The deterministic multi-tape Turing machine is our model of computation. It is powerful enough to being able to efficiently simulate all other (sequential, one-processor) computation models, and it is simple enough to be used for the definition of time complexity. Let

$$L(M) = \{x \in \Sigma^* \mid M \text{ on input } x \text{ reaches an accepting configuration }\}$$

be the language accepted by M. Let $T_M(x)$ denote the number of computation steps of Turing machine M on input x (until a final (accepting or rejecting) configuration is reached). Define

DTIME$(f(n)) =$
$\{A \subseteq \Sigma^* \mid$ there is a Turing machine M such that $\max\{T_M(x) \mid |x| = n\} = O(f(n))\}$

where $|x|$ denotes the length of string x. Define

$$\mathbf{P} = \bigcup\{\mathbf{DTIME}(p(n)) \mid p \text{ is a polynomial}\}$$

In an analogous way, one can consider non-deterministic Turing machines. This leads to corresponding definitions of the complexity classes **NTIME**$(f(n))$ and **NP**. It is well known that the question **P** =?**NP** is quite famous and plays a central role in Computational Complexity Theory.

We want to study (sets of) instances that are hard for some machine M — with respect to some time bound $t : \mathbb{N} \longrightarrow \mathbb{N}$. Define

$$\text{Hard}(M, t) = \{x \in \Sigma^* \mid T_M(x) > t(|x|)\}.$$

Consider a set $A \notin \mathbf{P}$, and let M be an arbitrary algorithm for A (i.e. $L(M) = A$), and let p be an arbitrary polynomial function. Then it is clear that

$$|\text{Hard}(M, p)| = \infty$$

Consider a different aqlgorithm M' for A and a different polynomial p'. Then it is also clear that

$$|\text{Hard}(M, p) \cap \text{Hard}(M', p')| = \infty$$

Suppose otherwise that this set is finite. Combine M and M' into one single algorithm M''. The complexity of M'' can be polynomially bounded by $\max(p, p')$, and we have

$$\text{Hard}(M'') = \text{Hard}(M, p) \cap \text{Hard}(M', p')$$

which is a finite set. Therefore, $\max\{T_{M''}(x) \mid |x| = n\} = O(q(n))$ for some polynomial q. Thus $A \in \mathbf{P}$ which is a contradiction. We conclude that finite intersections of such Hard-sets of instances for some intractable set A are infinite.

On the other hand, the infinite intersection of all such Hard-sets is empty:

$$\bigcap \{\text{Hard}(M, p) \mid L(M) = A \text{ and } p \text{ a polynomial}\} = \emptyset$$

This is simply true because for any instance x there is a machine M, $L(M) = A$, and a "big enough" polynomial p so that $x \notin \text{Hard}(M, p)$. So, in the next definition it is the "a.e." which makes the big difference.

DEFINITION 1 A (polynomial) *complexity core* for a set A is an infinite set X of instances such that for every machine M, $L(M) = A$, and polynomial p,

$$X \subseteq \text{Hard}(M, p) \text{ a.e.}$$

($X \subseteq Y$ a.e. means that $|X - Y|$ is finite.)

It is not at all clear whether such complexity cores exist and for which sets A they exist. We will see in the next section that the weakest possible assumption about A garantees the existence of complexity cores.

3 The Existence of Complexity Cores

The following theorem is due to Lynch [13]. Our proof follows a much easier argument from [20].

THEOREM 1 Every recursive set $A \notin \mathbf{P}$ has a complexity core.

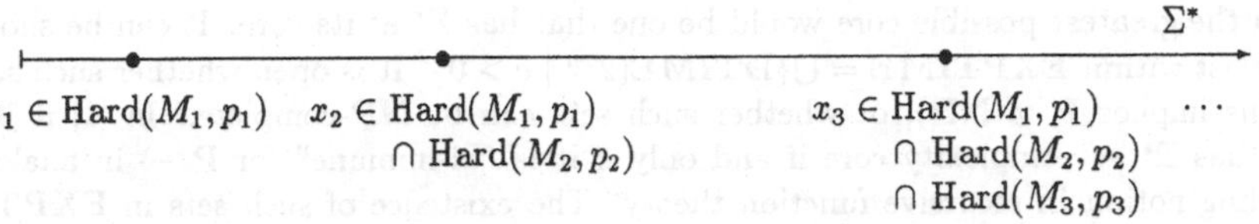

$$\Sigma^*$$

$x_1 \in \text{Hard}(M_1, p_1)$ $x_2 \in \text{Hard}(M_1, p_1)$ $x_3 \in \text{Hard}(M_1, p_1)$ $\cdots$

                 $\cap \text{Hard}(M_2, p_2)$ $\cap \text{Hard}(M_2, p_2)$

                                         $\cap \text{Hard}(M_3, p_3)$

Figure 2: Construction of a Complexity Core

Proof: Let $(M_1, p_1), (M_2, p_2), \ldots$ be a (non-effective) enumeration of all Turing machines M_i which accept A, paired with all polynomials p_i. Construct the complexity core $X = \{x_1, x_2, x_3, \ldots\}$ successively as follows. Let x_1 be the first string in Σ^* (according to lexicographic order) such that $x_1 \in \text{Hard}(M_1, p_1)$. Such string exists since $\text{Hard}(M_1, p_1)$ is infinite. In general, for $i > 1$, let x_i be the first string greater than x_{i-1} such that

$$x_i \in \text{Hard}(M_1, p_1) \cap \ldots \cap \text{Hard}(M_i, p_i).$$

See Figure 2. All these strings x_i exist since all these finite intersections of Hard-sets are infinite (see the discussion above). The so-constructed set X indeed is a complexity core: Let M be any machine with $L(M) = A$ and let p be any polynomial. Then, for some index j, $M = M_j$ and $p = p_j$. By the construction of X, $X \subseteq \text{Hard}(M, p)$ a.e. where the finitely many exceptions are at most $x_1, \ldots, x_{j-1}$. $\square$

In [13], it is additionally shown that such complexity cores can always be constructed to be recursive. Furthermore, in [20] it is shown that the complexity of such cores can be kept just a little above the complexity of A.

COROLLARY 1 IF $\mathbf{P} \neq \mathbf{NP}$ then SAT (as well as any other NP-complete set) has a complexity core.

4   Density of Cores

The construction above is unsatisfying in the sense that there is no assertion about the density of elements in a complexity core. The elements of the above core $x_1, x_2, \ldots$ can be arbitrarily far apart. We will show that complexity cores exist for NP-complete sets which are non-sparse. Let the *density* of a set A be the function on the natural numbers that gives for each n the number of strings in A of length n. A set is called *sparse* if its density can be bounded from above by some polynomial function.

The following theorem relies on a result by Meyer and Paterson [15] (a much stronger result can be found in [14]): If $\mathbf{P} \neq \mathbf{NP}$, then for no NP-complete set A there can be a decision algorithm M ($L(M) = A$) such that for some polynomial p, $\text{Hard}(M, p)$ is sparse. This assertion can be "lifted" into a property of complexity cores as stated below.

THEOREM 2 ([19, 20]) If $\mathbf{P} \neq \mathbf{NP}$ then SAT (as well as any other NP-complete set) has a complexity core that is non-sparse.

A set with the greatest possible core would be one that has Σ^* as its core. It can be shown that such sets exist within $\mathbf{EXPTIME} = \bigcup \{ \mathbf{DTIME}(2^{c \cdot n} \mid c > 0 \}$. It is open whether such sets exist in $\mathbf{NP}$ (this implies $\mathbf{P} \neq \mathbf{NP}$), or whether such sets can be $\mathbf{NP}$-complete. In [1] it is shown that a set has Σ^* as complexity core if and only if it is "bi-immune" for $\mathbf{P}$ — in analogy to a corresponding notion in recursive function theory. The existence of such sets in $\mathbf{EXPTIME}$ – enriched with some additional properties implies the following theorem.

THEOREM 3 ([19, 20]) Every $\mathbf{EXPTIME}$-complete set A has a complexity core whose density is $\Omega(2^{n^\varepsilon})$ (where ε is a constant depending on A).

By modifying the theory of complexity cores (essentially substiting "formal proof system" for "Turing machine", and "proof length" for "time complexity") one can show the following theorem which is of interest for theorem proving.

THEOREM 4 ([23]) Assuming $\mathbf{NP} \neq \text{co-}\mathbf{NP}$, there exists a constant ε and a collection $\mathcal{F}$ of tautologies (of propositional logic) of density at least $2^{\varepsilon \cdot n}$ i.o. such that for every sound proof system S for the tautologies and for every polynomial p, the shortest proof of F in S has length more than $p(|F|)$ for almost every $F \in \mathcal{F}$.

5 Maximal Complexity Cores

Because of the "almost everywhere" in the definition of complexity cores, two cores X_1 and X_2 for some set have to be considered as "the same", or equivalent, if the symmetric difference $X_1 \triangle X_2$ is a finite set. A set $A \notin \mathbf{P}$ does not have a unique complexity core since every infinite subset of a core is a core, too. We consider the question whether, and if so, under which conditions intractable sets have a *maximal* (and greatest) complexity core. More precisely, a set $A \notin \mathbf{P}$ has a maximal core X if X is a complexity core for A and for every other core X' we have $X' \subseteq X$ a.e.

THEOREM 5 The following are equivalent:

1. A has a maximal core.

2. There is a maximal restriction of Σ^* on which the set A is decidable in polynomial time.

Proof: Suppose A has a maximal complexity core X. If A were not decidable in polynomial time on $\Sigma^* - X$ then, by the technique of Theorem 1, a complexity core X' could be constructed being infinitely bigger than X, a contradiction. It is clear that the polynomial time decidable restriction $\Sigma^* - X$ cannot be enlarged into the complexity core X. Therefore it is maximal.

Conversely, let R be a maximal restriction on which A is decidable in polynomial time. If $\Sigma^* - R$ is not a core then there is an infinite extention of R on which A is still polynomial time decidable. Also, the core $\Sigma^* - R$ is not extendible, hence maximal. $\qquad \square$

We see that there is a duality between complexity cores and restrictions on which the set is polynomial time decidable. This is further explored in [21]. We want to know whether sets of interest, like SAT, or **NP**-complete sets in general, do or do not have the property of having a maximal complexity core. One result of [18] states that for "natural" **NP**-complete sets, like SAT, this is not the case.

THEOREM 6 SAT (and any other **NP**-complete set sharing the same "self-reducibility" properties, see [18]) does not have a maximal complexity core.

By Theorem 5, this means that every subset of Σ^* on which SAT can be solved efficiently (like HORN, see Figure 1) can be non-trivially (i.e. by an infinite set) enlarged.

6 Instance Complexity

Complexity cores have to be infinite sets, by definition. Within the framework developed so far, it does not make sense to speak about the hardness of a single instance x for some problem, like SAT. On the other hand, there is a strong intuitive feeling that there should be a notion of individual hardness: for some formulas satisfiability seems to be much easier to decide than for others. For example, a long Horn formula should be considered as an instance for SAT with low "instance complexity". But what are instances with high complexity? Do they exist?

In the world of asymptotic complexity theory (as used for the complexity core notion) there is no place for the complexity of an individual instance. This is because every algorithm for solving a certain problem (like SAT) can be modified to run "fast" on a given instance x (or a finite set of instances): a description of x has to be built into the algorithm. (This is called the "finite patching property": a finite variation of a **P**-set is still a **P**-set, a finite variation of a core is still a core.)

But looking more carefully, one observes that the finite patching process requires that the *size* of the algorithm increases by about the amount that is necessary to "describe" the string x. There is a well developed theory dealing with the descriptional complexity of strings, called Kolmogorov complexity which was developed by Kolmogorov [12] and Chaitin [5]. The definition used here can be summarized as follows.

$$K(x) = \min\{|M| \mid M \text{ is a Turing machine that produces } x, \text{ starting with the empty tape}\}$$

(Actually, one has to be a little more careful here when talking about the *size* of machines, this notion has to be defined relative to some universal machine.)

The above definition can be modified to time-bounded Turing machines, leading to a measure K^t where $t \in \mathbb{N}$ (see [24, 9, 8]).

Kolmogorov complexity is a measure for the complexity of a string x as such, whereas we are looking for a measure for the complexity of a string in connection with solving some decision problem. The same string x might be hard w.r.t. problem A, but easy w.r.t. problem B.

DEFINITION 2 For a recursive set A and $t \in \mathbb{N}$ define

$$IC^t(x : A) = \min\{|M| \mid L(M) = A \text{ and } T_M(x) \leq t\}$$

It is easy to see that (time-bounded) Kolmogorov complexity is always an upper bound to instance complexity:

$$0 \preceq IC^t(x:A) \preceq K^t(x) \preceq |x|$$

Here, $\preceq$ means "less than or equal — up to an additive constant which is uniform for all x (it only depends on A and the underlying programming system and the universal Turing machine, see [17, 11]). We say that a string x is *hard* for some problem A if the instance complexity of x is about the same as its Kolmogorov complexity. For a decision problem A, we say A has hard instances if for any polynomial p there is a polynomial q such that for infinitely many x,

$$IC^{p(|x|)}(x:A) \succeq K^{q(|x|)}(x)$$

Within this framework, sets in **P** have only easy instances:

THEOREM 7 A set A is in **P** if and only if for some polynomial p, $IC^{p(|x|)}(x:A) \preceq 0$.

Proof: The forward direction is clear. Conversely, suppose $IC^{p(|x|)}(x:A) \preceq 0$, i.e. for some constant c and all x, $IC^{p(|x|)}(x:A) \leq c$. This means, among the finitely many machines up to size c which accept A, for each x there is always at least one such machine that is "fast" on x. Combing all these machines into a single one, we obtain a polynomial time algorithm for A. $\square$

On the other hand, it can be shown:

THEOREM 8 If EXPTIME $\neq$ NEXPTIME, then SAT (as well as any **NP**-complete set) has hard instances.

Within EXPTIME very "a.e. complex" sets can be constructed.

THEOREM 9 There exists a set A in EXPTIME such that

$$IC^{e(|x|)}(x:A) \succeq K(x) - 2 \cdot \log_2(K(x))$$

for almost every $x \in \Sigma^*$, where $e(n) = 2^n$.

A consequence of this result is the following corollary.

COROLLARY 2 For every EXPTIME-complete or EXPTIME-hard set A, there is a set of strings $X \subseteq \Sigma^*$ of density at least 2^{n^e} such that for all $x \in X$,

$$IC^{e(|x|)}(x:A) \succeq K(x) - 2 \cdot \log_2(K(x))$$

The results reported in this Section are from [10, 17, 11].

References

[1] J.L. BALCÁZAR AND SCHÖNING, U. Bi-immune sets for complexity classes. *Math. Systems Theory* 18 (1985), 1–10.

[2] R. BOOK AND DU, D.Z. The existence and density of generalized complexity cores. *J. Assoc. Comput. Mach.* 34 (1987), 718–730.

[3] R. BOOK AND DU, D.Z. The structure of generalized complexity cores. *Theor. Comput. Sci.*, to appear.

[4] R. BOOK, DU, D.Z. AND RUSSO, D. On polynomial and generalized complexity cores. *Proc. 3rd Symp. Structure in Complexity Theory*, IEEE, 1988, 236–250.

[5] G.J. CHAITIN. On the simplicity and speed of programs for computing infinite sets of natural numbers. *J. Assoc. Comput. Mach.* 16 (1969), 407–422.

[6] D.Z. DU AND BOOK, R. On inefficient special cases of NP-complete problems. *Theor. Comput. Sci.*, to appear.

[7] S. EVEN, SELMAN, A., AND YACOBI, Y. Hard-core theorems for complexity classes. *J. Assoc. Comput. Mach.* 35 (1985), 205–217.

[8] J. HARTMANIS. Generalized Kolmogorov complexity and the structure of feasible computations. *Proc. 24th Ann. Sympos. Foundations of Computer Science*, IEEE, 1983, 439–445.

[9] K. KO. On the notion of infinite pseudorandom sequences. *Theor. Comput. Sci.* 48 (1986), 9–33.

[10] K. KO, ORPONEN, P., SCHÖNING, U., AND WATANABE, O. What is a hard instance of a computational problem? *Proc. 1st Conf. Structure in Complexity Theory*, IEEE, 1986.

[11] K. KO, ORPONEN, P., SCHÖNING, U., AND WATANABE, O. Instance complexity. Manuscript, 1990, submitted for publication.

[12] A.N. KOLMOGOROV. Three approaches to the quantitative definition of information. *Prob. Info. Transmission* 1 (1965), 1–7.

[13] N. LYNCH. On reducibility to complex or sparse sets *J. Assoc. Comput. Mach.* 22 (1975), 341–345.

[14] S. MAHANEY. Sparse complete sets for NP: solution to a conjecture of Berman and Hartmanis. *J. Comput. Syst. Sci.* 25 (1982), 130–143.

[15] A. MEYER AND PATERSON, M. With what frequency are apparently intractable sets difficult? Techn. Report TM-126, MIT, 1979.

[16] P. ORPONEN. A classification of complexity core lattices. *Theor. Comput. Sci.* 47 (1986), 121–130.

[17] P. ORPONEN. On the instance complexity of NP-hard problems. *Proc. 5th Conf. Structure in Complexity Theory*, IEEE, 1990.

[18] P. ORPONEN, RUSSO, D. AND SCHÖNING, U. Optimal approximations and polynomially levelable sets. *SIAM J. Comput.* 15 (1986), 399–408.

[19] P. ORPONEN AND SCHÖNING, U. The structure of polynomial complexity cores. *Proc. Math. Found. Comput. Sci.* Lecture Notes in Computer Science 176, Springer-Verlag, 1984, 452–458.

[20] P. ORPONEN AND SCHÖNING, U. The density and complexity of polynomial cores for intractable sets. *Inform. and Control* 70 (1986), 54–68.

[21] D. RUSSO AND ORPONEN, P. On P-subset structures. *Math. Syst. Theory* 20 (1987), 129–136.

[22] U. Schöning. *Complexity and Structure*. Lecture Notes in Computer Science 211, Springer-Verlag, 1985.

[23] U. Schöning. Complexity cores and hard-to-prove formulas. *Proc. Computer Science Logic*, Lecture Notes in Computer Science, Springer-Verlag, 1987.

[24] M. Sipser. A complexity theoretic approach to randomness. *Proc. 15th Ann. ACM Symposium on Theory of Computing*, ACM, 1983, 330–335.

Spatial Point Location and its Applications

Xue-Hou TAN   Tomio HIRATA   and Yasuyoshi INAGAKI

Faculty of Engineering, Nagoya University

Chikusa-ku, Nagoya 464, Japan

Abstract

This paper considers the problem of locating a point in a polyhedral sub-division of the space defined by planar polygonal faces. A persistent form of binary-binary search tree is presented so that the point location problem can be solved in $O(\log N)$ query time and $O(N + K)$ space, where N is the total number of edges and K the edge intersections in the image plane. The persistent structure also gives new better solutions for many other geometric problems.

1 Introduction

Locating a point in a planar subdivision is a well-known geometric problem and has received considerably attention (see, e.g., [10, 18]). In contrast, its three-dimensional generalization, called *spatial point location*, has been seldom studied. Suppose the Euclidean space E^3 is partitioned into polyhedra by planar polygonal faces with a total of N edges. Given such a polyhedral subdivision, the spatial point location problem requires to determine the polyhedron containing a given query point.

Chazelle [2] was the first to consider this problem. His Canal Tree technique gives query time $O(\log^2 N)$ and space $O(N)$. However, it is required that the polyhedral subdivision should be *convex* and the vertical dominance on the faces be *acyclic*. Two improvements have been made later. Cole's Similar List method [6] can be used to release the former restriction (it was not actually mentioned but implied in [6]) and yields query time $O(\log N)$ and space $O(N + K)$ where K is the number of edge intersections in the image plane. On the other hand, Preparata and Tamassia [21] release the latter restriction, and achieve query time $O(\log^2 N)$ and space $O(N \log^2 N)$. Their algorithm is based on a combination of the partial persistence-addition technique of Driscoll et al. [9] and the dynamic point-location technique of Preparata and Tamassia [19, 20].

All the solutions above require some restrictions on polyhedral subdivisions. In this paper we release both restrictions, i.e. consider general polyhedral subdivisions. We first reduce the point location problem to storing and searching a sequence of sorted "list-list" structures; each element of the primary list corresponds to a secondary list. A persistent representation of these list-list structures is then presented, which is actually a persistent form of binary-binary search tree. We make use of both partially and fully persistent structures, and achieve a result of $O(log\ N)$ query time and $O(N + K)$ space. Since there exists no restriction on polyhedral subdivisions and the query time is optimal, the data structure developed here can be applied to various geometric problems, not only in E^3 but also in E^2.

For the special subdivision that is formed by n planes, there already exist several solutions. Cole's Similar Lists method [6] gives a result of $O(log\ n)$ query time and $O(n^4/log\ n)$ space, while Chazelle's Canal Tree technique [2] trades space for query time and achieves query time $O(log^2n)$ and space $O(n^3)$. Chazelle and Friedman [4] have recently improved the results to $O(log\ n)$ query time and $O(n^3)$ space by a modification of the random sampling technique of Clarkson [5]. When applied to this special subdivision, our algorithm leads to a result of $O(log\ n)$ query time and $O(n^4)$ space. Although this preliminary result does not attain that of Chazelle and Friedman's, their algorithm can not be applied to general polyhedral subdivisions.

2　Preliminaries

We briefly review the essentials of persistent data structures and the application to planar point location. An ordinary dynamic data structure is called *ephemeral* if an update (insertion or deletion) to the structure destroys the old version, leaving only the new one. A sequence of m updates then produces m versions of the structure, and the ith update generates the ith version. Finding a way to maintain past versions of a data structure is a natural demand. A data structure is called *persistent* if it supports accesses to past versions. The structure is *fully persistent* or *partially persistent* depending upon whether any previously existing version can also be updated or not. The persistent structure embeds all versions of the ephemeral structure, so that access to any version can be effectively simulated.

Driscoll et al. [9] have given general techniques for making an ephemeral linked data structure persistent, provided that each node of the ephemeral structure has bounded in-degree. Specifically, they add the partial persistence to an ephemeral data structure through the method of *node copying*. Each node of an ephemeral structure is expanded to hold k extra pointer slots in addition to the original ones. When a pointer change made to a node v, if there is an empty slot in node v, the new pointer is stored, along with a version stamp indicating when the change occurred; otherwise, a copy $c(v)$ of v is created, which is filled with the newest pointer values of v and thus has k empty slots. Since the copy $c(v)$ requires to store a pointer in the latest parent of v, node copying can ripple backwards the structure. However, amortized over a sequence of pointer changes, there are only $O(1)$ nodes copied per pointer change. If an update operation (insertion or deletion) requires only $O(1)$ pointer changes, the partially persistent structure can then be built with an amortized space cost of $O(1)$ per update. Simulation of an access to the ith version of the structure is simply accomplished by following at each node the appropriate pointer with the maximum version stamp no greater than i.

A main difference between partially persistent structures and fully persistent structures is that the various versions of a partially persistent structure have a natural linear ordering, whereas the versions of a fully persistent structure are only partially ordered. To build a fully persistent data structure, they first impose a total ordering on its versions. The total ordering is represented by a list, called the *version list*. When a new version, say i, is created, it is inserted in the version list immediately after its parent $p(i)$; $p(i)$ is the version that is updated to obtain version i. Thus for any version i, the descendants of i occur consecutively in the version list, starting with i. The version list is represented in Dietz and Sleator's structure [7]. The structure is able to determine, given two versions i and j, whether i precedes or follows j in the version list in $O(1)$ worst-case time, with an $O(1)$ worst-case time bound for insertion. Navigation through the persistent structure is then the same as in the partially persistent case, except that versions are compared with respect to their positions in the version list rather than their numeric values. With a variant of node copying, a fully persistent structure can be obtained with the same bounds as in the partially persistent case. The variant of node copying is called *node splitting*; when a node runs out of slots for new pointers, it is split into two, putting the first half of the pointers in one copy and the remainder in the other. The node splitting process can thus cascade through the structure.

When the persistence-addition techniques are applied to a particular kind of search tree, the *red-black* tree [13], considerable simplifications can be made. In summary, they obtain a way to build both partially and fully persistent search trees with a worst-case logarithmic time per update or access and a worst-case space of $O(1)$ per update.

In the companion paper [22], partially persistent search trees have been applied to give a simple implementation of efficient point location for planar subdivision. Let a vertical line sweep through a planar subdivision and the intersected edges with the sweep line store in an active list. A new version of the active list is created when a vertex of the subdivision is reached. The history of the active list gives all the data for planar point location. The problem is thus reduced to storing and accessing all the versions of the active list in a partially persistent search tree.

In the next section, we will generalize this method to three dimensions. It is easy to see that a much more complex data structure is needed to hold the intersected objects than that in the planar case. We present a persistent form of binary-binary search tree, which is a combination of a partially persistent search tree and a fully persistent search tree.

3 Spatial Point Location

Suppose that the Euclidean space E^3 is subdivided into polyhedra by planar polygonal faces with a total number of N edges. The spatial point location problem aims at data structures and algorithms for determining the polyhedron containing a given query point. For simplicity, we assume that each edge is incident to at most c faces (for some constant c). In fact, this assumption can always be removed by making copies of the edge that is incident to more than c faces (see [16] for details).

We first describe the simple solution due to Dobkin and Lipton [8]. The subdivision edges are projected onto the (x, y) plane. It is assumed that no edge is vertical in the (x, y) plane. These edges are intersected into *intervals*, which define a planar graph G with $O(N + K)$ *regions* (Figure 1), where K is the number of edge intersections in the

$(x,\ y)$ plane. Obviously K is bounded above by $O(N^2)$. For each region, a list of faces whose projections contain that region is created. This face list will be called the region's *face list*. In a region's face list, the faces are sorted into the z order. Associate with each face the polyhedron just below it. Given a query point x, we can perform a planar point location on G to locate the region containing x, and then search the face list of that region for the face immediately above x. Obviously the query time is $O(log\ N)$. If an efficient planar point-location structure, such as Sarnak and Tarjan's [22], is used, the space requirement will be $O((N + K)N)$.

We can reduce the space bound by noticing that the face lists of adjacent regions are similar. Our algorithm is based on the plane-sweep paradigm. A vertical line is swept through the $(x,\ y)$ plane from left to right. The regions crossed by the sweep line are called the *current regions*. For each current region R, we make a note of the two boundary intervals that are intersected with the sweep line. These two intervals are R's *upper* and *lower current intervals*. (The lower current interval is actually redundant, since the upper current interval of the region just below R is exactly R's lower current interval.) A face list is *active* if it is the face list of a current region. During the sweep, the current regions with their face lists are maintained in a data structure for spatial point location. A binary–binary search tree can be used for this purpose: The current regions are vertically stored in a balanced binary search tree, for instance, a red-black tree; each node of this tree denotes a region, whose face list again gets attached to a balanced binary search tree over the z order of the faces. Obviously, our main task is to make this binary-binary search tree persistent.

A distinct data structure is used for the events of the sweep. An event of the sweep occurs when the sweep line meets a vertex of the graph G. The structure is a priority queue; it must be able to answer queries about the next event. The endpoints of all edges are initially sorted into x order. This sorted list is used as the initial event queue. Later a vertex which is a common right end of a current region's intervals is inserted into the priority queue.

When the sweep line is moved left-right through the graph G, the current regions and their face lists change. As a vertex v is passed, the regions left to v are deleted from the set of current regions and their face lists become unactive, the regions right to v are inserted into the set of current regions and the face lists of these newly started regions are created, and the region above or below v changes its lower or upper current interval. Note that the change of the current regions is just the same as that in the planar point location problem. Thus, the first level search tree is implemented as the partially persistent search tree for planar point location developed by Sarnak and Tarjan [22]. The operations on the second level search trees are a sequence of face list creations and cancels. Since region R's face list becomes unactive when region R is deleted from the set of current regions, the remaining task is how to carry out face list creations.

A face list creation needs at first glance up to $O(N\ log\ N)$ time and $O(N)$ space. However, we can reduce the cost to $O(log\ N)$ time and $O(1)$ space. Let v be the vertex being passed, $R_{i+1},\ R_{i+2},\ \cdots,\ R_{i+d}$ the newly started regions at v, ordered from top to bottom, and R_i the region above v (see Figure 2). Observe that R_i's face list has no any change at v and can be used to effectively create the face lists for regions $R_{i+1},\ R_{i+2},\ \cdots,$ R_{i+d}. The following is a detailed description of the face list creations at v:

For $k = i + 1$ **to** $i + d$ **do**

1. R_k's face list is initially copied from R_{k-1}'s face list.
2. R_k's face list is modified with deletions of the faces whose projections do not contain region R_k any longer and insertions of the faces whose projections newly contain region R_k.

od.

Let R_0 denote the leftmost and topmost region in the graph G. All other regions' face lists are then obtained from a sequence of updates on R_0's face list. Figure 3 shows such an example of Figure 1. Numbers on regions show the order of regions' face lists created. Arrows between regions define a binary relation α: $R_i \ \alpha \ R_j$ means that region R_j's face list is obtained by updating region R_i's face list. Observe that relation α gives a partial creating order of regions' face lists. This observation suggests that the sequence of face list creations can be implemented in a fully persistent structure.

From the above description, a face list creation consists of a face list copying, and several (at most c) face insertions and deletions on the copied list. To simplify the discussion, we ignore the issue of face list copying for the moment. Hence a sequence of face list creations, to be exact, a sequence of face insertions and deletions can be implemented in a fully persistent search tree. A face list is thus created in $O(log \ N)$ time and $O(1)$ space, not counting face list copying.

It is now easy to handle face list copying. Let the ith face list in the sequence of creations have *list number i*. Since a face list creation includes several face insertions and deletions, a list number corresponds to several version stamps of the persistent search tree. Note that our interesting is in the last version stamp, which is taken as the real value of that list number. The face list copying operations are then carried out by setting up an one-to-one relationship between the regions and their face lists' numbers. Hence, a face list's number is stored in the nodes of the partially persistent search tree (first level tree) that denote the corresponding region.

Recall that a region in the partially persistent search tree is represented by its upper intervals. Hence a concave region may be represented by more than one interval in the same version of the partially persistent search tree. This shows that a concave region may have several face lists, all of them have the same content.

In summary, the first and second level search trees are implemented as a partially and fully persistent search trees, respectively. The connection between these two different persistent structures is face lists' numbers held by their regions.

By now we have developed a persistent form of binary-binary search tree for spatial point location. From the size of the graph G, we conclude:

Theorem 1 *Given a polyhedral subdivision of E^3 formed by planar polygonal faces, it is possible to preprocess it in $O((N + K)log \ N)$ time and $O(N + K)$ space, so that the point location query can be answered in $O(log \ N)$ time, where N is the total number of edges and K the number of edge intersections in the image plane.*

We end this section with several remarks about the generality of our structure. We note that the faces in the space is allowed to be curved. What is now required is that there exists a total order of faces in each face list and testing whether a point is above or below a face take $O(1)$ time.

Our data structure also supports a generalization of the spatial point location problem in which the queries are of the following form: given a line segment (a, b) parallel to the z axis, report all polyhedra segment (a, b) intersects. It is easy to show that such a query takes $O(log N + t)$ time where t is the number of reported polyhedra.

When applied to the special subdivision induced by a Voronoi diagram of n points in E^3, our algorithm immediately leads to a result of $O(log n)$ query time and $O(n^4)$ space. This can be trivially proved by the fact that the Voronoi diagram has $O(n^2)$ edges in the space and the intersections of these edges in the projection plane are bounded by $O(n^4)$. The best known solutions for this problem so far [2, 21] have query time $O(log^2 n)$, although the optimal space bound $O(n^2)$ can be achieved (see [2]). With the same complexity of $O(log n)$ query time and $O(n^4)$ space, our algorithm can also be applied to the special subdivision that is formed by n planes. Although our preliminary results do not improve upon the best known algorithms developed for these special cases (see also Section 1), it is expected that the space requirement might be reduced.

4 Applications

This section presents applications of the data structure developed in Section 3 to problems that can be formulated in terms of spatial point location, or that relate to such problems; in all cases we are able to improve existing bounds or establish new bounds.

4.1 Next-element searching in three dimensions

In [12], Edelsbrunner et al. introduce the *next-element search problem* in the plane and then apply the solution of it to several problems in computer graphics. Here we generalize the problem to three dimensions.

Let q denote an arbitrary point in the 3-dimensional space and S a set of planar polygonal faces that do not intersect. The next-element of q in S is the one which first intersects with the open z-ray emanating from q in the positive z direction. The next-element search problem in three dimensions requires storing S such that the next-element of a query point q can be determined efficiently.

Clearly, the problem can be reduced to spatial point location (see also [12]). Since our algorithm is based on the image of the original faces, it applies directly to the problem. Using the persistent binary-binary search tree, we achieve:

Theorem 2 *Let S be a set of faces with a total of N edges in the space and K edge intersections in the image plane. There exists a data structure for S that requires $O(N + K)$ space and $O((N + K)log N)$ time for its construction such that the next-element of a query point can be determined in $O(log N)$.*

4.2 Translating a set of faces or polyhedra in three dimensions

Given a set of faces or polyhedra in three dimensions. The *translation problem* requires moving them in a given direction, one at a time, without collisions occurring between them. To solve this problem, we are required to perform a sequence of next-element search operations (see [17]). As an immediate consequence of Theorem 2, we obtain:

Theorem 3 *The translation problem for a set of faces or polyhedra in three dimensions* *can be solved in $O((N + K)\log N)$ time and $O(N + K)$ space, where N is the total number of edges and K the edge intersections in the image plane.*

This result improves Nurmi's result of $O((N + K)\log N)$ time and space [17].

4.3 Hidden surface removal

Given an environment of nonintersecting opaque polyhedra in E^3, we wish to compute the image visible from a given viewpoint. To produce a realistic image of the environment, we first project all the environment's edges onto the image plane. Suppose that the image plane is the (x, y) plane. This establishes a line segment subdivision of the (x, y) plane. Each region of this subdivision is covered by some of the environment's faces. The *hidden surface removal problem* is the task of reporting the frontmost face for each region of the subdivision. Note that the persistent structure developed in Section 3 supports queries not only about the z neighbors of a point but also about the topmost face in the z direction. In order to remove all the subdivision segments whose two incident regions have common topmost faces and all the vertices that are in the interior of an edge or a region, the subdivision of the (x, y) plane must be set up. The planar subdivision can be simply constructed by using Nievergelt and Preparata's algorithm [15]. These yield:

Theorem 4 *The hidden surface removal problem (curved, algebraic faces are allowed) can be solved in $O((N + K)\log N)$ time and $O(N + K)$ space, where N is the total number of the environment's edges and K the number of edge intersections in the image plane.*

Although the same resource bounds have been obtained by Schmitt et al. [23], their algorithm requires various complicated data structures. In [14] McKenna showed that the hidden surface removal problem can be solved in $O(N^2)$ time, which is an improvement to the $O((N + K)\log K)$ bound in cases where $K = \Omega(N^2)$. The drawback of his algorithm is that it requires $O(N^2)$ space.

4.4 Polygonal point enclosure searching

The *polygonal point enclosure search problem* is: given a set of N simple polygons, each with a bounded number of edges, and a query point in the plane, determine all the polygons enclosing the point.

The N polygons consist of $O(N)$ edges. The collection of edges defines a planar graph G. Each region of the graph G is overlapped by some of the original polygons. Our intention is to precompute for each region the list of overlapping polygons. Assume that each polygon has a pre-defined z-coordinate. We can thus built a persistent binary-binary search tree for the graph G. Having located the region containing a given point in the partially persistent search tree, we report all the covering faces of that region in the corresponding version of the fully persistent search tree. This yields:

Theorem 5 *Given a set of N simple polygons in the plane, each with a bounded number of edges, there exists a data structure such that the t polygons that contain a given point can be reported in $O(\log N + t)$ time, using $O(N + K)$ space where K is the number of edge intersections.*

This result improves Edelsbrunner et al.'s result of $O(\log N + t)$ query time and $O(N + N * K)$ space [11].

4.5 Polygon intersection reporting

Let P denote a set of simple polygons in the plane, each with a bounded number of edges. The *polygon intersection problem* requires to report all pairs of polygons which have at least one point in common.

It is obvious that the problem can be reduced to two sub-problems [1]: The line segment intersection problem and the batched polygonal point enclosure search problem. The first subproblem is already solved by Chazelle and Edelsbrunner [3]. Their algorithm runs in $O(N \ log \ N \ + \ K)$ time and $O(N \ + \ K)$ space, where K is the number of edge intersections. The second can be solved by applying Theorem 5. We conclude:

Theorem 6 *All T intersections among N simple polygons can be reported in $O((N \ + \ K) log \ N \ + \ T)$ time and $O(N \ + \ K)$ space, where K is the number of edge intersections.*

A challenging open problem is to determine whether this problem can be solved in optimal time $O(N \ log \ N \ + \ T)$, analogous to the optimal time for computing line segment intersections [3].

5 Conclusions

We have given a simple solution to the spatial point location problem. For a polyhedral subdivision with a total of N edges in the space and K edge intersections in the image plane, our structure supports a point location query with $O(log \ N)$ time, and requires $O(N \ + \ K)$ space and $O((N \ + \ K)log \ N)$ preprocessing time to built. The data structure also gives new better solutions for many geometric problems. Besides these applications, we believe that the data structure is general enough to serve as a stepping-stone for other geometric problems as well.

Acknowledgement

We would like to thank Dr. Takeshi Tokuyama of IBM Tokyo Research Laboratory for his valuable comments and discussions.

References

[1] J.L.Bentley and D.Wood, An optimal worst-case algorithm for reporting intersections of rectangles, *IEEE Trans. on Comput.* **C-29**(1980), 571-577.

[2] B.Chazelle, How to search in history, *Inform. Control* **64**(1985), 77-99.

[3] B.Chazelle and H.Edelsbrunner, An optimal algorithm for intersecting line segments in the plane, in *Proceedings, 29th Annu. IEEE Symp. Found. of Comput. Sci.*(1988), pp. 590-600.

[4] B.Chazelle and J.Friedman, A deterministic view of random sampling and its use in geometry, in *Proceedings, 29th Annu. IEEE Symp. Found. of Comput. Sci.*(1988), pp. 539-548.

[5] K.L.Clarkson, New applications of random sampling in computational geometry, *Discrete Comput. Geometry* **2**(1987), 195-222.

[6] R.Cole, Searching and storing similar lists, *J. Algorithms* **7**(1986), 202-220.

[7] P.Dietz and D.D.Sleator, Two algorithms for maintaining order in a list, in *Proceedings, 19th Annu. ACM Symp. Theory of Computing*(1987), pp. 365-372.

[8] D.Dobkin and R.J.Lipton, Multidimensional search problems, *SIAM J. Comput.* **5**(1976), 181-186.

[9] J.R.Driscoll, N.Sarnak, D.D.Sleator and R.T.Tarjan, Making data structures persistent, *J. Comput. Sys. Sci.* **38**(1989), 86-124.

[10] H.Edelsbrunner, *Algorithms in Combinatorial Geometry*, Springer-Verlag, 1987.

[11] H.Edelsbrunner, H.A.Maurer and D.G.Kirkpatrick, Polygonal intersection searching, *Inform. Process. Lett.* **14**(1982), 74-77.

[12] H.Edelsbrunner, M.H.Overmars and R.Seidel, Some methods of computational geometry applied to computer graphics, *Comput. Vision Graphics Image Process.* **28**(1984), 92-108.

[13] L.J.Guibas and R.Sedgewick, A dichromatic framework for balanced trees, in *Proceedings, 19th Annu. IEEE Symp. Found. of Comput. Sci.*(1978), pp. 8-21.

[14] M.McKenna, Worst case optimal hidden surface removal, *ACM Trans. Graphics* **6**, 1987, 19-28.

[15] J.Nivergelt and F.P.Preparata, Plane sweep algorithms for intersecting geometric figures, *Comm. ACM* **25**(1982), 739-747.

[16] O.Nurmi, A fast line-sweep algorithm for hidden line elimination, *BIT* **25**(1985), 466-472.

[17] O.Nurmi, On translating a set of objects in 2- and 3-dimensional space, *Comput. Vision Graphics Image Process.* **36**(1986), 42-52.

[18] F.P.Preparata and M.I.Shamos, *Computational Geometry*, Springer-Verlag, 1985.

[19] F.P.Preparata and R.Tamassia, Fully dynamic techniques for point location and transitive closure in planar structures, in *Proceedings, 29th Annu. IEEE Symp. Found. of Comput. Sci.*(1988), pp. 558-567.

[20] F.P.Preparata and R.Tamassia, Fully dynamic point location in a monotone subdivision, *SIAM J. Comput.* **18**(1989), 811-830.

[21] F.P.Preparata and R.Tamassia, Efficient spatial point location, in Algorithms and Data Structures (WADS'89), *Lect. Notes in Comput. Sci.* **382**, Springer-Verlag, 1989, pp. 3-11.

[22] N.Sarnak and R.E.Tarjan, Planar point location using persistent search trees, *Comm. ACM* **29**(1986), 669-679.

[23] A.Schmitt, H.Müller and W.Leister, Ray tracing algorithms — theory and practice, Proc. NATO Advanced Study Inst. Theoret. Found. Comput. Graphics and CAD, Springer-Verlag, 1987, pp. 997-1029.

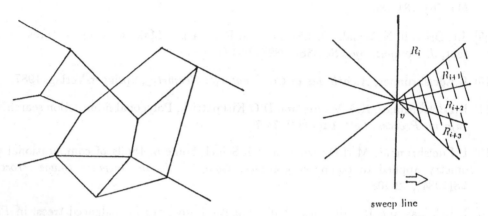

Figure 1. The projection of a spatial subdivision

Figure 2. An event vertex in the projection plane

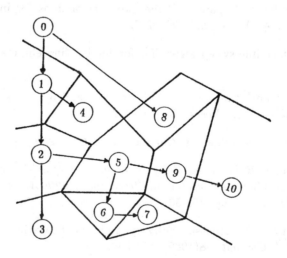

Figure 3. The order of regions' face lists created

Sublinear Merging and Natural Merge Sort

Svante Carlsson*   Christos Levcopoulos*   Ola Petersson*

Abstract

The complexity of merging two sorted sequences into one is linear in the worst case as well as in the average case. There are, however, instances for which a sublinear number of comparisons is sufficient. We consider the problem of measuring and exploiting such instance easiness. The merging algorithm presented, Adaptmerge, is shown to optimally adapt to different kinds of measures of instance easiness. In the sorting problem, the concept of instance easiness has received a lot of attention and is interpreted by a measure of presortedness. We apply Adaptmerge in the already adaptive sorting algorithm Natural Merge Sort. The resulting algorithm optimally adapts to several, known and new, measures of presortedness. We also prove some interesting results concerning the relation between measures of presortedness proposed in the literature.

1 Introduction

It is well known that, in the worst case, $\Omega(m \log((n + m)/m))$ time is necessary for merging two sorted sequences X and Y of length n and m, respectively, $m \leq n$, into one, in a comparison-based model of computation [3]. This lower bound is based on the assumption that X is a randomly chosen subsequence of length n out of a sorted sequence of length $m + n$. In many applications this is not the situation. For example, it might be enough to merge a small portion in the end of X with a small portion in the beginning of Y, that is, there is just a small "overlap" between X and Y, or the sorted output might be obtained by simply splitting X into a small number of parts and inserting parts of Y in between. Such instances of the merging problem are in some sense easier and can be computed faster than the above lower bound indicates. Most worst-case optimal merging algorithms do not take this into account.

In order to be able to take advantage of such "instance easiness", or *preorder*, we must be more precise by what it means. We examine two different approaches for measuring preorder for the merging problem. The first, *Max*, measures the maximum distance that an element is from its correct position. The second, *NS*, tells how many elements in X and Y that receive a new successor in the merged sequence. Further, we provide an algorithm, *Adaptmerge*, which adapts to these measures. Intuitively, a merging algorithm is *adaptive* with respect to a measure of preorder if it merges all instances but performs particularly well on those that have a high degree of preorder (without knowing the value of the measure beforehand). Adaptmerge completes the merge in time the minimum of $O(Max(X,Y))$ and $O(NS(X,Y)\log((n + m)/NS(X,Y)))$, which is optimal with respect to *Max* and *NS*. Here, optimality with respect to a measure means maximum adaptation (in an asymptotic sense). In the worst case Adaptmerge performs $O(m \log((n + m)/m))$ comparisons which matches the aforementioned lower bound. The main idea in Adaptmerge is to use exponential and binary search [7] iteratively to locate the positions in which the two sequences have to be split.

*Algorithm Theory Group, Department of Computer Science, Lund University, Box 118, S-221 00 Lund, Sweden.

It is likewise well known that $\Omega(n \log n)$ time is necessary to sort n elements in both the worst case and the average case, in a comparison-based model of computation. Similarly, this result relies on the assumption that the elements are randomly permuted. Again, in many applications this is not the case, but the sequence to be sorted is already partially ordered (*presorted*). As was the case for merging, most worst-case optimal sorting algorithms do not take existing order within their input into account.

For the sorting problem, we interpret instance easiness by a *measure of presortedness*, which is an integer function on a permutation π of a totally ordered set that reflects how much π differs from the total order. Examples of presortedness measures include the number of runs and inversions. The term presortedness was coined by Mehlhorn [7], who used the number of inversions in the sequence as a measure. Mehlhorn also gave an algorithm, A-Sort, which is adaptive with respect to this measure. Mannila [6] formalized the concept of presortedness and provided an algorithm which is adaptive with respect to several measures. Recently, Altman and Igarashi [1], Levcopoulos and Petersson [4, 5], and Skiena [8] have considered other measures.

Max and *NS* can be used as measures of presortedness as well. *Max* has previously been used in this sense, while *NS* is a new measure which generalizes the measure *Rem*, the minimum number of elements that have to be removed from a sequence in order to leave a sorted sequence, as well as *Exc*, the minimum number of exchanges needed to bring a sequence into sorted order.

Mannila [6] proved that Natural Merge Sort is optimal with respect to the measure *Runs*, that is, the number of consecutive upsequences of maximum length, even if a straightforward merging algorithm is used. In this paper we show how to apply Adaptmerge in Natural Merge Sort to extend its adaptivity. As a result we obtain a sorting algorithm which is optimal with respect to several measures of presortedness, namely *Max*, *NS*, *Rem*, *Exc*, and *Runs*. The algorithm sorts a sequence X of length n in time linear plus the minimum of $O(n \log Max(X))$ and $O(NS(X)(\log Runs(X) + \log(n/NS(X))))$.

2 Measuring Instance Easiness

We start this section by stating some preliminary definitions. Let $X = \langle x_1, \ldots, x_n \rangle$ be a *sequence* of n elements x_i from some totally ordered set. For simplicity it can be assumed that the elements are distinct. For two sequences, $X = \langle x_1, \ldots, x_n \rangle$ and $Y = \langle y_1, \ldots, y_m \rangle$, their *catenation* XY is the sequence $\langle x_1, \ldots, x_n, y_1, \ldots, y_m \rangle$. Further, let $|X|$ denote the *length* of X and $\|S\|$ the *cardinality* of a set S. If $Z = \langle x_{f(1)}, x_{f(2)}, \ldots, x_{f(m)} \rangle$, $m \leq n$, and $f\{1, \ldots, m\} \rightarrow \{1, \ldots, n\}$ is injective and monotonically increasing, then Z is called a *subsequence* of X. In particular, Z is a *consecutive subsequence* of X if there exists an i, $1 \leq i \leq n - m + 1$, such that $Z = \langle x_i, \ldots, x_{i+m-1} \rangle$. For two sequences X and Y, we write $X \leq Y$ if and only if every element in X is no greater than every element in Y. If X and Y are two sorted sequences of length n and m, respectively, we define $rank(x_i, Y) = \max\{\ell \mid 0 \leq \ell \leq m \text{ and } x_i > y_\ell\}$, for $1 \leq i \leq n$, where $y_0 = -\infty$.

2.1 Sorting

Since the concept of a measure of presortedness is well known [6] we will start by defining our measures as such and then, in the next subsection, show how they can be modified to qualify for measuring instance easiness in the merging problem.

For a sequence X of length n, $Max(X)$ is defined to be the maximum distance that an element x_i is from its correct position in the sorted output. More formally, if π is the permutation of $\{1, 2, \ldots, n\}$ such that $x_{\pi(1)} < x_{\pi(2)} < \cdots < x_{\pi(n)}$ then $Max(X) = \max_{1 \leq i \leq n} |i - \pi(i)|$. Some basic properties of Max, which are readily verified, are

Property 1 *For any sequences* $X = \langle x_1, x_2, \ldots, x_n \rangle$ *and* $Y = \langle y_1, y_2, \ldots, y_m \rangle$,

(a) $0 \leq Max(X) \leq n - 1$.

(b) *If* Y *is a subsequence of* X *then* $Max(Y) \leq Max(X)$.

(c) *If* $X \leq Y$ *then* $Max(XY) = \max\{Max(X), Max(Y)\}$.

For a sequence X of length n, let $NS(X)$ be the number of elements x_i that have a different successor in X than in the sorted permutation of X. More formally, $NS(X) = \|\{i \mid 1 \leq i \leq n \text{ and } x_{\pi(i+1)} \neq x_{\pi(i)+1}\}\|$, where x_{n+1} is defined to be an element not in X. The intuition behind NS is that if NS is low, then the sequence to be sorted consists of a small number of consecutive subsequences which are consecutive subsequences in the sorted output as well, and we should be able to take advantage of this to sort faster than $\Theta(n \log n)$.

Property 2 *For any sequences* $X = \langle x_1, x_2, \ldots, x_n \rangle$ *and* $Y = \langle y_1, y_2, \ldots, y_m \rangle$,

(a) $0 \leq NS(X) \leq n$.

(b) *If* Y *is a subsequence of* X *then* $NS(Y) \leq NS(X)$.

(c) *If* $X \leq Y$ *then* $NS(XY) = NS(X) + NS(Y)$.

The concept of an optimal sorting algorithm with respect to a measure of presortedness was introduced by Mannila [6]:

Definition 1 *Let* M *be a measure of presortedness, and* A *a sorting algorithm which uses* $T_A(X)$ *comparisons to sort a sequence* X. *We say that* A *is* M-optimal, *or optimal with respect to* M *if, for some* $c > 0$, *we have, for all* $X = \langle x_1, \ldots, x_n \rangle$,

$$T_A(X) \leq c \cdot \max\{|X|, \log(\|below(X, M)\|)\},$$

where $below(X, M) = \{\pi \mid \pi \text{ is a permutation of } \{1, \ldots, n\} \text{ and } M(\pi(X)) \leq M(X)\}$.

Giving a lower bound for algorithms with respect to a measure of presortedness thus means to bound the cardinality of the *below*-set from below.

2.2 Merging

As mentioned above, we will use both *Max* and *NS* for measuring preorder in the merging problem. However, if X and Y are the sequences to be merged, we let

$$Max(X, Y) = \min\{Max(XY), Max(YX)\} \quad \text{and} \quad NS(X, Y) = \min\{NS(XY), NS(YX)\},$$

since the measure should not take the relative order of the sequences into account. Note that this definition is crucial for the *Max* measure, since there are instances for which $Max(XY) = 0$ while $Max(YX) = n-1$. On the other hand, it is easy to see that $0 \leq |NS(XY) - NS(YX)| \leq 2$.

Readers familiar with measures of presortedness might know that there are certain conditions that a measure has to fulfill in order to qualify as a measure of presortedness [6]. (It is easy to prove that both *Max* and *NS* qualify as measures of presortedness.) One could, of course, formulate similar conditions for measures of preorder for the merging problem. However, we take a looser approach and are satisfied if the measure in some intuitive sense measures instance easiness for the merging problem. We proceed by interpreting the measures, and thereby establish their appropriateness.

In the merging problem, Max tells approximately how many elements that have to be moved. If, for instance, $Max(X, Y) = Max(XY) = k < m$, then the first $n-k$ and the last $m-k$ elements in XY are already in their correct positions and what remains is to merge the $2k$ middle elements. Some basic properties of Max, when applied in the merging problem, are

Property 3 *For any two sorted sequences $X = \langle x_1, \ldots, x_n \rangle$ and $Y = \langle y_1, \ldots, y_m \rangle$, $m \leq n$,*

(a) $Max(X, Y) \geq 0$, with equality if and only if $X \leq Y$ or $X \geq Y$.

(b) $Max(X, Y) \leq n$ if $m > 1$ and $Max(X, Y) \leq \lfloor n/2 \rfloor$ if $m = 1$.

(c) If $x_1 < y_1$ and $x_n < y_m$ then $Max(X, Y) = \max\{n - rank(y_1, X), rank(x_n, Y)\}$.

(d) If $x_1 < y_1$ and $x_n > y_m$ then

$$Max(X, Y) = \min\{\max\{m, n - rank(y_1, X)\}, \max\{m, rank(y_m, X)\}\}.$$

The measure NS has the same motivation in the merging problem as it had in sorting. That is, if X and Y need just be split into a small number of consecutive subsequences for the merging to take place, we should not spend constant time on every single element, but just find the positions where we have to split the sequences. Some basic properties of NS are

Property 4 *For any two sorted sequences $X = \langle x_1, \ldots, x_n \rangle$ and $Y = \langle y_1, \ldots, y_m \rangle$, $m \leq n$,*

(a) $NS(X, Y) \geq 0$, with equality if and only if $X \leq Y$ or $X \geq Y$.

(b) $NS(X, Y) \leq 2m$.

The definition of optimality of an adaptive merging algorithm follows the one for sorting algorithms:

Definition 2 *Let M be a measure of preorder for the merging problem, and B a merging algorithm which uses $T_B(X, Y)$ comparisons to merge two sorted sequences X and Y. We say that B is M-optimal, or optimal with respect to M if, for some $c > 0$, we have, for all $X = \langle x_1, \ldots, x_n \rangle$ and $Y = \langle y_1, \ldots, y_m \rangle$,*

$$T_B(X, Y) \leq c \cdot \log(\|below(X, Y, M)\|),$$

where $below(X, Y, M) = \{\pi \mid \pi$ is a permutation of $\{1, \ldots, m + n\}$ and $\pi(XY) = Z_1 Z_2$, where Z_1 and Z_2 are in ascending order and $M(Z_1, Z_2) \leq M(X, Y)\}$.

Observe that compared to Definition 1 we have removed the permission to spend at least linear time. For sorting algorithms a linear time lower bound is justified, whilst for merging algorithms it is not. Also, note that both our definitions of optimality are just valid in comparison-based models of computation.

3 Maxmerge

To simplify the description and analysis of our main merging algorithm in the next section, we use this section as a warm-up and give a trivial Max-optimal merging algorithm, $Maxmerge$. First, however, let us establish a lower bound for merging algorithms which adapt to Max. By bounding the $below$-set in Definition 2 from below we can prove

Lemma 1 *Any comparison-based Max-optimal merging algorithm needs* $\Omega(Max(X,Y))$ *comparisons to merge two sequences X and Y.*

At first it might seem that even if we have a merging algorithm that performs a sublinear number of comparisons for some inputs, a linear number of assignments is always necessary. This is, however, only true if the resulting sequence is required to be reported in a consecutive order. Otherwise, one can do better. In Maxmerge, as well as in Adaptmerge in the next section, the number of assignments and comparisons are asymptotically equal.

The input to the algorithm consists of two sorted arrays X and Y of length n and m, respectively. The output is an array Z of length at most $2\,Max(X,Y) + 2$. Each entry in Z corresponds to a consecutive subsequence of X or Y. By reporting the elements in these subsequences in the order in which they appear in Z the merged sequence is obtained. (See Figure 1.) Hence, Maxmerge outputs a data structure from which the resulting sequence can be computed in $m + n$ assignments and no comparisons. A similar output format turns out to be crucial when an adaptive merging algorithm is applied in Natural Merge Sort (see Section 5).

$$
\begin{array}{cc}
\begin{array}{c}
\;1\;\;2\;\;3\;\;4\;\;5\;\;6\;\;7\;\;8 \\
X: \boxed{1\,|\,2\,|\,3\,|\,4\,|\,5\,|\,7\,|\,8\,|\,11}
\end{array}
&
\begin{array}{c}
\;1\;\;2\;\;3\;\;4\;\;5\;\;6 \\
Y: \boxed{6\,|\,9\,|\,10\,|\,12\,|\,13\,|\,14}
\end{array}
\end{array}
$$

$$
\begin{array}{ccccccc}
& 1 & 2 & 3 & 4 & 5 & 6 \\
Z: & \boxed{X[1,5]} & \boxed{Y[1,1]} & \boxed{X[6,7]} & \boxed{Y[2,3]} & \boxed{X[8,8]} & \boxed{Y[4,6]}
\end{array}
$$

Figure 1: *Example of input and output of the merging algorithms*

Without loss of generality, we assume that $x_1 < y_1$. Maxmerge starts by an alternating sequential search going both backward and forward in X to find $rank(y_1, X)$, which tells us where the "actual" merge has to start. Then, perform an linear merge of Y and $X[rank(y_1, X) + 1, n]$ until one of the sequences is finished, and fill in Z instead of moving the elements, in the obvious way.

Theorem 2 *Maxmerge is optimal with respect to Max.*

Proof. Omitted due to lack of space. ∎

Adaptmerge

In this section we present the adaptive merging algorithm *Adaptmerge* and prove that it is optimal with respect to *Max* and *NS*. Before that, we provide a lower bound for comparison-based merging algorithms which adapt to the measure *NS*, whose proof is information-theoretic in the proof of Lemma 1.

Lemma 3 *Any comparison-based NS-optimal merging algorithm needs* $\Omega(k\log((m + n)/k))$ *comparisons to merge two sequences X and Y of total length $m + n$, with $NS(X,Y) = k$.*

We note that if $m = 1$, then $k = 1$ or $k = 2$ and Lemma 3 gives the lower bound for searching an ordered sequence. On the other hand, if k is maximized, that is, $k = 2m$ by Property 4b, we obtain the known worst case lower bound on the number of comparisons performed by a merging algorithm, $\Omega(m\log((m + n)/m))$.

The following description of Adaptmerge assumes, without loss of generality, that $x_1 < y_1$. The input/output format is the same as it was for Maxmerge in the previous section. The first version of Adaptmerge we present is only adaptive with respect to NS. It is shown, in the proof of Theorem 4, how a slight modification makes it adapt optimally to Max as well.

Again, consider the example in Figure 1. To compute the array Z we need only find the positions in which X and Y have to be split and what portions of the other sequence should be inserted at these. Equivalently, we compute the elements in X and Y that will receive new successors in the resulting sequence, together with their successors. Now, the intuition behind the algorithm is that if $NS(X,Y)$ is low, there are large consecutive portions in X and Y in which all elements will keep their original successor, and these need thus not be examined entirely. Therefore, we apply exponential and binary search [7] to pass such portions as fast as possible in our search for the next element that will receive a new successor.

Let $x_{i_1}, x_{i_2}, \ldots, x_{i_p}$, $1 \leq i_1 < i_2 < \cdots < i_p \leq n$, and $y_{j_1}, y_{j_2}, \ldots, y_{j_q}$, $1 \leq j_1 < j_2 < \cdots < j_q \leq m$, be the elements in X and Y that will receive new successors. It is clear that $p + q = NS(X,Y)$, $|p - q| \leq 1$, and either $i_p = n$ or $j_q = m$.

Adaptmerge starts by computing $i_1 = rank(y_1, X)$ by an exponential and binary search forward in X. Clearly, $Z[1] = X[1, i_1]$. Second, compute $j_1 = rank(x_{i_1+1}, Y)$ by an exponential and binary search forward in Y, and set $Z[2] = Y[1, j_1]$. Third, compute $i_2 = rank(y_{j_1+1}, X)$ by an exponential and binary search forward in X, starting from $X[i_1 + 1]$, and set $Z[3] = X[i_1 + 1, i_2]$. In this way, we continue to perform exponential and binary searches, alternating between X and Y. We always start the search from where the last element found to receive a new successor is located. When one of the sequences is finished, the next empty entry in Z is set to the remaining portion of the non-empty sequence.

Theorem 4 *Adaptmerge is optimal with respect to Max and NS.*

Proof. Let $NS(X,Y) = k$ and recall that an exponential and binary search finds an element located ℓ positions away in a sorted array in $O(\log \ell)$ time. The time consumed by Adaptmerge is

$$T(n) = O\left(\sum_{\ell=1}^{p} \log(i_\ell - i_{\ell-1}) + \sum_{\ell=1}^{q} \log(j_\ell - j_{\ell-1})\right), \qquad (1)$$

where the first term corresponds to all searches performed in X, the second corresponds to all searches performed in Y, and $i_0 = j_0 = 0$. Here,

$$\sum_{\ell=1}^{p} \log(i_\ell - i_{\ell-1}) = \log\left(\prod_{\ell=1}^{p}(i_\ell - i_{\ell-1})\right) \leq \log\left(\prod_{\ell=1}^{p}\frac{n}{p}\right) \leq \frac{k+2}{2}\log\left(\frac{2n}{k+2}\right) = O\left(k\log\frac{n}{k}\right),$$

because $p \leq k/2 + 1$. Similarly, it follows that the second term in (1) is $O(k\log(m/k))$ because $q \leq k/2 + 1$. Hence,

$$T(n) = O\left(k\log\frac{n}{k} + k\log\frac{m}{k}\right) = O\left(k\log\left(\frac{m+n}{k}\right)\right),$$

which proves the NS-optimality by Lemma 3.

To obtain Max-optimality as well we just have to be a bit careful in the very start of the algorithm. Concurrently with the first forward exponential and binary search we perform an exponential and binary search backward in X for i_1. Note that we do not loose the NS-optimality, because the change made is only affecting the first search, which might now be twice as expensive. ∎

5 Adaptmerge in Natural Merge Sort

Natural Merge Sort [6] is a sorting algorithm which optimally adapts to the measure *Runs*. The number of *runs* in a sequence is the number of consecutive ascending subsequences of maximum length, or more formally, if $|X| = n$,

$$Runs(X) = \|\{i \mid 1 \le i < n \text{ and } x_i > x_{i+1}\}\| + 1.$$

Given a sequence, Natural Merge Sorts starts by finding the runs by a linear time scan. These runs are then repeatedly merged pairwise until there is just one run left which is the sorted sequence. If a straightforward merging algorithm is used, each merge takes linear time in the number of participating elements. Since each element takes part in $\lceil \log Runs(X) \rceil$ merges, the total time consumed by Natural Merge Sort is $O(n + n \log Runs(X))$. Mannila [6] proved that this is optimal with respect to *Runs*.

In this section we show that by applying Adaptmerge in the merges, Natural Merge Sort becomes optimal with respect to *Max* and *NS* as well. Observe that if we want to beat the lower bound for *Runs* on some inputs, while sticking to the idea of repeated pairwise merging in Natural Merge Sort, we are not allowed to spend constant time per participating element in the merges. Hence, Adaptmerge cannot return a sorted sequence stored consecutively in order, since that would cost a linear number of assignments.

The input to Adaptmerge in Natural Merge Sort consists of two doubly linked lists Z_1 and Z_2, together with pointers to the first and last element in each list. Each of Z_1 and Z_2 corresponds to a number of merged runs, located consecutively in the sequence X. For simplicity, let Z_1 correspond to $X[1, p]$ and let Z_2 correspond to $X[p + 1, q]$. The only difference between this input format and the output format of Adaptmerge in Section 4 is the doubly linked list representation instead of an array. Denote by a *block* a subsequence of X that corresponds to one element in Z_1 or Z_2. For example, in Figure 2, where $p = 6$ and $q = 14$, $X[9, 14]$ is a block while $X[3, 6]$ is not. Note that the length of Z_1 is equal to $NS(X[1, 6]) + 1$ and the length of Z_2 is equal to $NS(X[7, 14])$.

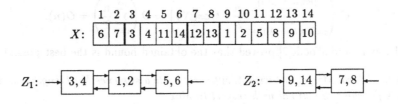

Figure 2: *Example of input of Adaptmerge when applied in Natural Merge Sort. Z_1 corresponds to the sequence obtained if the run in $X[1, 2]$ is merged with the run in $X[3, 6]$.*

Since Adaptmerge is applied repeatedly in Natural Merge Sort the output Z is on the same format as the input. The result of merging $X[1, p]$ and $X[p + 1, q]$ is that some blocks of Z_1 and Z_2 are split into smaller blocks and these appear in their correct positions in the list Z together with the blocks that are not split. For example, the block $X[9, 14]$ in Figure 2 is split into the three blocks $X[9, 10]$, $X[11, 11]$, and $X[12, 14]$, as shown in Figure 3, while the block $X[3, 4]$ remains unchanged.

When merging $X[1, p]$ and $X[p + 1, q]$, we proceed as follows:

1. Find the smallest element x_i in $X[p+1, q]$. This is the first element in the first block of Z_2.

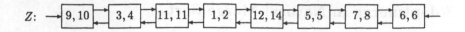

$Z:$ 9, 10 — 3, 4 — 11, 11 — 1, 2 — 12, 14 — 5, 5 — 7, 8 — 6, 6

Figure 3: *Output of Adaptmerge for the input given in Figure 3*

2. Perform a sequential search backward in Z_1 until the block containing the greatest element in $X[1, p]$ which is smaller than x_i is found.

3. Apply Adaptmerge to merge the blocks in $X[1, p]$ which contain elements greater than x_i with $X[p+1, q]$. When a block in Z_1 or Z_2 is finished, just continue with the next one.

In the following two subsections we derive upper bounds on Natural Merge Sort with Adaptmerge in terms of the measures *Max* and *NS*.

5.1 *Max*-optimality

Throughout this subsection it is helpful to think of the algorithm as performing the merges on $\lceil \log Runs(X) \rceil$ levels.

The analysis with respect to *Max* is divided into two parts; the first $\lceil \log Max(X) \rceil$ levels, and the remaining levels. It is clear that the first $\lceil \log Max(X) \rceil$ levels take $O(n \log Max(X))$ time since at most linear time is spent on each level. The remaining levels are shown to take linear time altogether.

If there are no remaining levels, that is, if $Runs(X) \le Max(X)$, we are done. Otherwise, consider a merge of two sequences X_1 and X_2, where $X_1 X_2$ is a consecutive subsequence of X, on the j:th level, $j \ge \lceil \log Max(X) \rceil + 1$. Note that the length of each of these sequences is at least $2^{j-1} \ge Max(X)$. Now, since $X_1 X_2$ is a subsequence of X, $Max(X_1 X_2) \le Max(X)$, by Property 1b. Therefore, only the last $Max(X)$ elements in X_1 and the first $Max(X)$ elements in X_2 are affected by the merge. Hence, merging X_1 and X_2 takes $O(Max(X))$ time. As there are at most $n/2^j$ merges performed on the j:th level, the total time for all remaining levels is given by

$$O\left(\sum_{j=\lceil \log Max(X) \rceil+1}^{\lceil \log Runs(X) \rceil} \frac{n}{2^j} \cdot Max(X)\right) = O\left(\sum_{i=0}^{\infty} \frac{n}{2^i}\right) = O(n).$$

Estivill-Castro and Wood [2] proved that the obtained bound is the best possible, and thus

Theorem 5 *Natural Merge Sort with Adaptmerge sorts a sequence X of length n in time $O(n + n \log Max(X))$, which is optimal with respect to Max.*

5.2 *NS*-optimality

For *NS*, we start our analysis by providing a lower bound, which can be proved by bounding the cardinality of *below*(X, NS) from below.

Lemma 6 *Any comparison-based NS-optimal sorting algorithm needs $\Omega(n + NS(X) \log n)$ comparisons to sort a sequence X of length n.*

To prove that the time consumed by Natural Merge Sort with Adaptmerge matches this lower bound, we again divide the analysis into two parts. This time, however, not by levels, but by different kinds of operations. First, the time spent on operations involving elements in the linked lists, such as traversing pointers or creating new elements, is examined. Second, we study the time spent in searches *in* the blocks.

On the first level there are $Runs(X)$ linked lists containing one block each. The total length of the linked lists is then increasing as we move up the levels and reaches $NS(X)$ or $NS(X)+1$ when the algorithm is finished. On each level no block in a linked list is visited more than twice (once in the search backward in Step 2 and once when Adaptmerge has been invoked in Step 3). Hence, even if the total length of the linked lists is $NS(X)+1$ on the first level, the cost for all linked list operations is bounded by $O(NS(X)\log Runs(X))$.

When analysing the time spent on searches in the blocks, let $NS(X) = k$. What these searches do is splitting a sorted array of length n into $k+1$ blocks. In the worst case, the first search is carried out in the whole array and it splits the array into two blocks. The remaining searches are then divided into two parts; those made in the first block and those made in the second. Denote by k_1 and k_2, $k_1 + k_2 = k - 1$, the number of searches to be performed in the first and second block, respectively. The searches in the first and second block then splits these blocks, and so on recursively. Note that, as the searches are two-way and exponential and binary, the location sought for is found in $2c \log \ell$ time if it is within distance ℓ from the left or right boundary in the block considered. The time consumed by all searches is given by

$$T(n,k) \leq \begin{cases} 2c \log(n/2) & \text{if } k = 1, \\ \max_{1 \leq \ell \leq (n/2)} \{2c \log \ell + T(\ell, k_1) + T(n - \ell, k_2)\} & \text{otherwise.} \end{cases}$$

$T(n,k)$ is asymptotically maximized if the left and right blocks are of approximately the same size, that is, if $\ell \approx n/2$, and if the k positions sought for are evenly distributed in the array, that is $k_1 \approx k_2$. Hence,

$$T(n,k) \leq \begin{cases} 2c \log(n/2) & \text{if } k = 1, \\ 2T(n/2, k/2) + 2c \log(n/2) & \text{otherwise,} \end{cases}$$

which has the solution $T(n,k) = O(k \log(n/k))$.

If we add up the time required for finding the runs, the time spent on linked list operations, and the time spent searching in the blocks, we have

Theorem 7 *Natural Merge Sort with Adaptmerge sorts a sequence X of length n in time $T(n) = O(n + NS(X)\log Runs(X) + NS(X)\log(n/NS(X)))$, which is optimal with respect to NS and Runs. The amount of space needed is $O(NS(X))$ pointers in addition to the input.*

Proof. To see that the upper bound matches the lower bound in Lemma 6, observe that the second term is $O(NS(X)\log n)$ since $Runs(X) \leq n$, and the third term is at most linear. Further, as $NS(X) \leq n$, $T(n) = O(n + n \log Runs(X))$, proving the *Runs*-optimality. Extra space is only required for the linked lists, whose total length is $O(NS(X))$. ∎

6   A Comparison with Other Measures

We study the relation between the new measure NS and other measures of presortedness proposed in the literature.

Together with the number of runs and inversions, the most well known measure of presortedness is Rem, the minimum number of elements that need to be removed from a sequence in order to leave a sorted sequence [6]. More formally,

$$Rem(X) = n - \max\{\ell \mid X \text{ has an ascending subsequence of length } \ell\}.$$

We prove that Rem is in some sense included in the measure NS.

Lemma 8 *For any sequence X, $NS(X) \leq 2\, Rem(X)$.*

Proof. The key observation is that an element contributing to $NS(X)$ either contributes to $Rem(X)$ itself or has a successor which contributes to $Rem(X)$. ∎

The sequence $Y = \langle n/2+1, n/2+2, \ldots, n, 1, 2, \ldots, n/2 \rangle$ shows that there is no constant $c > 0$ such that for any sequence X, $Rem(X) \leq c\, NS(X)$. Hence, NS and Rem are not asymptotically equal.

From Lemma 8, it now follows that any NS-optimal sorting algorithm runs in time

$$O(n + NS(X)\log n) = O(n + NS(X)\log NS(X)) = O(n + Rem(X)\log Rem(X)).$$

Ccombining this with Mannila's result that to sort a sequence X of length n, any comparison-based Rem-optimal sorting algorithm needs $\Omega(n + Rem(X)\log Rem(X))$ comparisons, gives

Lemma 9 *Any NS-optimal sorting algorithm is also Rem-optimal.*

Next, we take the opportunity to show the relation between Rem and a measure of presortedness for which until this paper no algorithm was known to be optimal with respect to, namely $Exc(X) = $ minimum number of exchanges needed to bring X into a sorted sequence. Following the same strategy as above, we can show that $Rem(X) \leq 2\, Exc(X)$ and that any comparison-based Exc-optimal sorting algorithm needs $\Omega(n + Exc(X)\log Exc(X))$ comparisons, which gives

Lemma 10 *Any Rem-optimal sorting algorithm is also Exc-optimal.*

By combining Theorem 5, Theorem 7, Lemma 9, and Lemma 10 we have

Corollary 11 *Natural Merge Sort with Adaptmerge is optimal with respect to the measures Max, NS, Rem, Exc, and $Runs$.*

References

[1] T. Altman and Y. Igarashi. Roughly sorting: sequential and parallel approach. *Journal of Information Processing*, 12(2):154–158, 1989.

[2] V. Estivill-Castro and D. Wood. A new measure of presortedness. *Information and Computation*, 83(1):111–119, 1989.

[3] E. Horowitz and S. Sahni. *Fundamentals of Computer Algorithms*. Computer Science Press, Rockville, Maryland, 1984.

[4] C. Levcopoulos and O. Petersson. Heapsort—adapted for presorted files. In *Proc. 1989 WADS*, pages 499–509. LNCS 382, Springer-Verlag, 1989.

[5] C. Levcopoulos and O. Petersson. Sorting shuffled monotone sequences. In *Proc. 2nd SWAT*. LNCS, Springer-Verlag, 1990. To appear.

[6] H. Mannila. Measures of presortedness and optimal sorting algorithms. *IEEE Transactions on Computers*, C-34(4):318–325, 1985.

[7] K. Mehlhorn. *Data Structures and Algorithms, Vol 1: Sorting and Searching*. 1984.

[8] S.S. Skiena. Encroaching lists as a measure of presortedness. *BIT*, 28(4):775–784, 1988.

Constructing Strongly Convex Approximate Hulls
with Inaccurate Primitives

Leonidas Guibas[1,2]   David Salesin[1]   Jorge Stolfi[2]

[1]*Stanford University*   [2]*DEC Systems Research Center*

1 INTRODUCTION

Finding the convex hull of a set of points is one of the simplest and oldest problems in computational geometry, and yet it is still a surprisingly difficult problem to solve in practice. The difficulty is that the basic geometric tests needed to solve the problem are unreliable or inconclusive when implemented with ordinary floating-point arithmetic. This uncertainty makes it extremely difficult to construct the correct hull, the smallest polygon that is simultaneously convex and that contains every point of the given set.

The main result of this paper is a proof that for every point set there exists a polygon that is convex enough to be found with approximate tests, such as floating-point arithmetic, and that is also very close to containing all the points of the given set. In addition, we present an algorithm for finding such hulls that uses inaccurate primitives and that runs in $O(n^3 \log n)$ time in the worst case, and in $O(n \log n)$ expected time for typical distributions of points. We also give an $O(\log n)$-time algorithm for testing whether or not a point lies inside one of these hulls, using only approximate tests.

The development of robust algorithms using imprecise computations has received a great deal of attention in the last few years [2–13, 15–16]. In particular, Fortune [3] has presented an $O(n \log n)$-time algorithm for computing approximate convex hulls using floating-point arithmetic. One drawback of Fortune's algorithm is that the resulting hulls

are only approximately convex, and therefore do not enjoy many of the nice properties associated with convexity. By contrast, the hulls computed by our algorithm are not only convex, but so convex that many of these desirable properties are preserved in some fashion even when they are tested with floating-point arithmetic.

Recently, Milenkovic and Li [12] have also presented results similar to ours. While their algorithm for constructing approximate hulls has a much better worst-case performance than the one presented here, the existence proof and algorithm in this paper establish tighter bounds on the degree of approximation achievable. Moreover, their algorithm assumes certain properties of floating-point computations, such as monotonicity, whereas the results presented here are based on a very general model of imprecise computations.

In developing our algorithms, we use the *Epsilon Geometry* framework proposed in a previous paper [5]. This framework provides a methodology for developing, proving, and implementing geometric algorithms based on approximate primitives.

Due to limited space, we have been forced to leave out several of our lemmas' proofs in this extended abstract. The proofs that do appear, however, are unusually detailed, partly because we wish to pay special attention to degenerate situations, and partly because we have found that the intuitive arguments common in geometric proofs that assume exact computations can be very misleading when applied to approximate tests [5, 7].

2 PRELIMINARIES

Epsilon Geometry

The Epsilon Geometry framework, introduced in our previous paper [5], defines the notion of an *epsilon predicate* as a means for expressing "approximate tests" in a general setting. An epsilon predicate is defined as follows:

Let $\mathcal{O}$ be a set of geometric objects endowed with some distance metric $\| \ \|$. Let P be a predicate defined on $\mathcal{O}$. Then for any $X \in \mathcal{O}$ and any $\varepsilon \geq 0$, we define $\varepsilon\text{-}P(X)$ as a shorthand for "$P(X')$ is true for *some* $X' \in \mathcal{O}$ such that $\|X, X'\| \leq \varepsilon$." That is, X is at most ε away from satisfying $P(X)$. In the case of an n-ary predicate P, we define $\varepsilon\text{-}P(X_1, \ldots, X_n)$ to mean that $P(X'_1, \ldots, X'_n)$ is true for some $X'_1, \ldots, X'_n$ with $\|X_i, X'_i\| \leq \varepsilon$ for all i.

For $\varepsilon > 0$, we define $(-\varepsilon)\text{-}P(X)$ as a shorthand for $\mathbf{not}\,\varepsilon\text{-}(\mathbf{not}\,P)(X)$. Therefore, $(-\varepsilon)\text{-}P(X)$ means that $P(X')$ is true for *all* $X' \in \mathcal{O}$ such that $\|X, X'\| \leq \varepsilon$. Intuitively, an object X that is $(-\varepsilon)\text{-}P$ is "extremely P," whereas an X that is $\varepsilon\text{-}P$ is only "nearly P." Note that with these definitions, $\varepsilon\text{-}P(X)$ implies $\varepsilon'\text{-}P(X)$ for all $\varepsilon' \geq \varepsilon$, positive or negative.

Implementing Epsilon Predicates

The epsilon predicates defined above are exact mathematical notions that we use in proofs. In programs, a geometric predicate is implemented by a procedure called an *epsilon box*. Typically, an epsilon box will try to evaluate its geometric predicate using floating-point arithmetic. An epsilon box may sometimes fail to decide whether its predicate is true or false.

We consider here three varieties of epsilon boxes. A *T-box* for a predicate $P(X)$ is a procedure $\mathtt{T\_P}(X)$ that computes $P(X)$ and returns either **true**, **false**, or **unknown**. Similarly, an *E-box* for $P(X)$ is a procedure $\mathtt{E\_P}(\varepsilon, X)$ that computes $\varepsilon\text{-}P(X)$ and returns either **true**, **false**, or **unknown**. Thus, the T-box $\mathtt{T\_P}(X)$ is equivalent to the special E-box $\mathtt{E\_P}(0, X)$. Finally, an *I-box* for a predicate $P(X)$ is a procedure $\mathtt{I\_P}(X)$ that returns an estimate of how far X is from satisfying P in the form of an interval $e = (e.lo, e.hi)$ such that $\varepsilon\text{-}P(X)$ is false for $\varepsilon < e.lo$, true for $\varepsilon \geq e.hi$, and unknown for $e.lo \leq \varepsilon < e.hi$.

We use interval arithmetic to combine the outputs of I-boxes. If d and e are two intervals, we define $\max\{d, e\}$ to be the interval $(\max\{d.lo, e.lo\}, \max\{d.hi, e.hi\})$. We define $\min\{d, e\}$ analogously. Occasionally, we will use the notation $e\text{-}P(X)$ with an interval e to mean $\mathbf{not}\,\varepsilon\text{-}P(X)$ for $\varepsilon < e.lo$ and $\varepsilon\text{-}P(X)$ for $\varepsilon \geq e.hi$. Note that the interval e returned by an I-box $\mathtt{I\_P}(X)$ should satisfy $e\text{-}P(X)$.

Accuracy of Epsilon Boxes

In principle, the Epsilon Geometry framework places no limit on the size of the uncertainty interval e returned by a given epsilon box. An epsilon box (of any variety) is considered *correct* as long as it never returns **false** when the predicate is true, or **true** when the predicate is false. For example, the T-box that always returns **unknown** and the I-box that always returns $(-\infty, +\infty)$ are correct implementations of *any* predicate P.

In practice, of course, an epsilon box should keep its uncertainty range reasonably small in order to be useful. We will say that a T-box $\mathtt{T\_P}(X)$ for a predicate $P(X)$ is λ-*accurate* (for $\lambda \geq 0$) if its uncertainty is less than or equal to λ; that is, if for all X the T-box returns **unknown** only when $\lambda\text{-}P(X)$ is true and $(-\lambda)\text{-}P(X)$ is false. Similarly, an E-box $\mathtt{E\_P}(\varepsilon, X)$ is λ-accurate if it returns **unknown** only when $(\varepsilon + \lambda)\text{-}P(X)$ is true and $(\varepsilon - \lambda)\text{-}P(X)$ is false. Finally, an I-box $\mathtt{I\_P}(X)$ is λ-accurate if it always returns an interval e such that $e.hi - e.lo \leq \lambda$.

We will also say informally that an epsilon box is *accurate* if it is λ-accurate for some λ small enough to be useful. Ideally, every epsilon-box implemented with floating-

$\varepsilon\text{-}Coincident(p, q)$	True if p and q lie within ε of a common point, that is, if $\|p, q\| \leq 2\varepsilon$.
$\varepsilon\text{-}Collinear(p, q, r)$	True if p, q, r lie within ε of some common line.
$\varepsilon\text{-}Pos(p, q, r)$	True if p, q, r form a counterclockwise triangle or are ε-collinear.
$\varepsilon\text{-}Neg(p, q, r)$	True if p, q, r form a clockwise triangle or are ε-collinear.
$\varepsilon\text{-}InSegment(z, pq)$	True if z is at most 2ε from the closed segment pq.

Table 1: Some epsilon predicates.

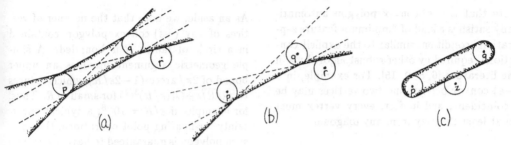

Figure 1: a) $\varepsilon\text{-}Collinear(p, q, r)$. b) $\varepsilon\text{-}Pos(p, q, r)$. c) $\varepsilon\text{-}InSegment(z, pq)$.

point arithmetic should be λ-accurate, with λ equal to a small constant times the machine's floating-point precision. (Our definitions of correctness and accuracy are roughly analogous to Fortune's definitions of "robustness" and "stability" [3].)

Basic Epsilon Predicates

The algorithms we develop in this paper are built on top of the geometric predicates shown in Table 1, which are described in detail in our previous paper [5].

In this paper, we will use the standard Euclidean metric for measuring the size of perturbations. Thus, the predicate $\varepsilon\text{-}Collinear(p, q, r)$ requires that the ε-disk centered at p intersect the ε-butterfly of q and r, defined as the union of all lines connecting a point in the ε-disk of q with a point in the ε-disk of r (Figure 1a). Similarly, the predicate $\varepsilon\text{-}Pos(p, q, r)$ requires that the ε-disk of p intersect the *left ε-basin* of q and r, consisting of the ε-butterfly of q and r and all points to its left (Figure 1b). Finally, the predicate $\varepsilon\text{-}InSegment(z, pq)$ requires that z lie ε-*inside* the segment pq, or equivalently, that the ε-disk of z intersect the ε-*stroke* of

segment pq, defined as the convex hull of the ε-disks of p and q (Figure 1c).

Our algorithms will depend on the following property:

Lemma 1 *Let a, b, and x be three points such that x is inside a rectangle with diagonal ab. Then $\varepsilon\text{-}InSegment(x, ab) \Leftrightarrow \varepsilon\text{-}Collinear(x, a, b)$, for all $\varepsilon \geq 0$.*

3 STRONGLY CONVEX POLYGONS

Definitions

For this paper, we define a *polygon* as a cyclic sequence of vertices $p_0 p_1 \ldots p_{n-1} p_0$. The edges of the polygon are the closed segments $p_i p_{i+1}$, for $0 \leq i < n$ (indices are modulo n). A polygon is *convex* if all its vertices $p_0, \ldots, p_{n-1}$ are distinct, only consecutive edges intersect, and any three distinct consecutive vertices form a strictly positive angle.

According to our definition of epsilon predicates, we say that a polygon is $(-\varepsilon)$-convex, for some $\varepsilon \geq 0$, if it remains convex under any perturbation to its vertices of ε or less. Formally, when measuring convexity, we define the "distance" between two polygons as the maximum distance between a vertex of one polygon and the corresponding vertex of the other. Two polygons with a different number of vertices are defined to be infinitely far apart. We will informally use the term *strongly convex* to mean $(-\varepsilon)$-convex, for some $\varepsilon > 0$ implied by the context.

Note that a $(-\varepsilon)$-convex polygon automatically satisfies a kind of "minimum feature separation" condition, similar to the explicit conditions required by other robust algorithms in the literature [6, 8, 9, 15]. For example, in a $(-\varepsilon)$-convex polygon, no two vertices may be ε-coincident, and in fact, every vertex must lie at least 2ε away from any diagonal.

Properties of Strongly Convex Polygons

We will say that a periodic sequence of real numbers is *unimodal* if it has at most one ascending run and one descending run in one period (allowing for repeated elements). Then the following lemmas characterize strongly convex polygons:

Lemma 2 *A polygon $P = (p_0, ..., p_{n-1})$ is $(-\varepsilon)$-convex for some $\varepsilon > 0$ if and only if all triples $p_i p_j p_k$ with $i < j < k$ are $(-\varepsilon)$-positive.*

Lemma 3 *A polygon $P = (p_0, ..., p_{n-1})$ is $(-\varepsilon)$-convex for some $\varepsilon > 0$ if and only if the x- and y-coordinates of its vertices are unimodal, and all consecutive triples $p_{i-1} p_i p_{i+1}$ are $(-\varepsilon)$-positive.*

Lemma 4 *A polygon $P = (p_0, ..., p_{n-1})$ is $(-\varepsilon)$-convex for some $\varepsilon > 0$ if and only if P is convex and no vertex p_i of P is ε-inside the diagonal $p_{i-1} p_{i+1}$.*

An immediate corollary of Lemma 2 is that any subpolygon of a $(-\varepsilon)$-convex polygon is also $(-\varepsilon)$-convex. Lemma 3 provides an $O(n)$-time algorithm for testing whether a polygon is convex, and if so, for measuring how convex it is. We omit the details for lack of space.

Here is another useful property of strongly convex polygons:

Lemma 5 *Let ε and δ be positive real numbers. For any $(-\varepsilon)$-convex polygon P and any two points y and z, there are at most $2\lceil 2\delta/\varepsilon \rceil$ vertices of P that are δ-inside the segment yz.*

As an aside, we note that the number of vertices of any $(-\varepsilon)$-convex polygon contained in a circle of radius R is bounded. A simple geometric argument provides an upper bound of $2\pi / \arccos(1 - 2\varepsilon/R)$, which reduces to $\pi\sqrt{R/\varepsilon} + O((\varepsilon/R)^{3/2})$ for small ε/R. Thus, for example, if $\varepsilon/R = 10^{-6}$, a typical uncertainty for floating point operations, then any such polygon is guaranteed to have fewer than 3200 vertices.

4 A POINT INCLUSION ALGORITHM

As an application of these lemmas, let's consider the problem of testing whether a given point z lies inside an n-sided convex polygon P. (We consider point z to lie "inside" P if it lies either on the interior or on the boundary of P.)

A classical $O(\log n)$-time algorithm for solving this problem begins by locating the point z in the angle between two consecutive diagonals $p_0 p_k$ and $p_0 p_{k+1}$, using binary search, and then testing z against the line $p_k p_{k+1}$ [14]. Note that this algorithm can be expressed using just the *Pos* orientation test.

Unfortunately, this simple algorithm no longer works if the exact *Pos* test is replaced by an approximate one. Figure 2 illustrates the problem: in both cases, the point z lies approximately within the angle $p_{k+1} p_0 p_k$ and on the positive side of edge $p_k p_{k+1}$, but in one case it lies inside the polygon and in the other case it lies well outside.

The key idea in the classical point location algorithm is that if z is found to lie on the positive side of a diagonal $p_0 p_k$, then the subpolygon $(p_0, p_1, ..., p_k)$ can be eliminated from further consideration. We can still use

---

Algorithm 1: *Given a point z and a $(-\varepsilon)$-convex polygon $P = (p_0, \ldots, p_{n-1})$, for sufficiently large ε, determine whether z is ε-inside P.*

1. Initialize $D \leftarrow P$.

2. While D has 12 or more vertices, do:

    2a. Let the vertices of D be $(d_0, d_1, \ldots, d_{m-1})$. Let $k = \lfloor m/2 \rfloor$.

    2b. Let $t = $ T_Pos(z, d_0, d_k).

    2c. If t is **true** or **unknown**, then delete the vertices $d_3, \ldots, d_{k-3}$ from D.

    2d. If t is **false** or **unknown**, then delete the vertices $d_{k+3}, \ldots, d_{m-3}$ from D.

3. Return T_InConvex(z, D).

---

this key idea—provided that the polygon is sufficiently convex—thanks to the following lemma:

Lemma 6 *Let $P = (p_0, \ldots, p_{n-1})$ be a $(-\varepsilon)$-convex polygon, for some $\varepsilon > 0$. Then the vertices $p_j, \ldots, p_{k-j}$ are more than $(2j - 1)\varepsilon$ away from the left ε-basin of $p_0 p_k$, for any j, k such that $0 < j < k < n$.*

Algorithm 1 describes an accurate $O(\log n)$-time T-box for the *InStronglyConvex* primitive. In addition to the T_Pos box already defined, the algorithm uses the accurate epsilon box T_InConvex(z, D), which tests in $O(n)$ time whether the point z lies inside the convex polygon D, and which is described in our previous paper [5].

Clearly, the algorithm runs in $O(\log n)$ time. Note also that the vertices removed from P comprise at most two chains of consecutive vertices of P, namely $(p_3, \ldots, p_{i-3})$ and $(p_{j+3}, \ldots, p_{n-3})$, for some i, j. Thus, poly-

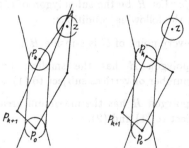

Figure 2: Two indistinguishable cases

gon D can be represented by the original array of vertices, along with the two indices i and j.

The algorithm requires that ε be greater than or equal to the maximum uncertainty λ of the primitive epsilon boxes T_Pos and T_InConvex. Assuming this condition is satisfied, the correctness and accuracy of the algorithm are implied by the following invariants, which hold at the beginning of step 2a:

$$Inside(z, P) \quad \Leftrightarrow \quad Inside(z, D)$$
$$\varepsilon\text{-}Boundary(z, P) \quad \Leftrightarrow \quad \varepsilon\text{-}Boundary(z, D),$$

where $Boundary(z, P)$ means that point z lies on the boundary of P. (Note that the predicate ε-$Boundary(z, P)$ is equivalent to $\bigvee_i \varepsilon$-$Between(z, p_i p_{i+1})$.) These invariants are implied by the following lemma, which can be proved using a straightforward application of Lemma 6:

Lemma 7 *Let $D = (d_0, \ldots, d_{m-1})$ be a $(-\varepsilon)$-convex polygon for some $\varepsilon > 0$, let z be a point such that ε-$Pos(z, d_0, d_k)$ for some $5 \leq k \leq m - 5$, and let D' be the subpolygon obtained by deleting the vertices $d_3, \ldots, d_{k-3}$ from D. Then*

$$Inside(z, D) \quad \Leftrightarrow \quad Inside(z, D')$$
$$\varepsilon\text{-}Boundary(z, D) \quad \Leftrightarrow \quad \varepsilon\text{-}Boundary(z, D').$$

Note that it is possible to design a simpler point location algorithm by first using exact x-comparisons to locate the point in the vertical slab between consecutive vertices of the polygon, and by then using approximate orientation tests only to test the point against the two edges of the polygon that enter the slab. As this example shows, the use of exact comparisons between real numbers often leads to simpler algorithms for solving problems in the plane. One reason for this is that k-dimensional geometric problems can often be solved by reducing them to $(k-1)$-dimensional subproblems, and one-dimensional subproblems can often be solved exactly using exact comparisons. Indeed, Fortune [3] and Milenkovic [9] have chosen to take this approach in their algorithms.

However, we expect exact comparisons to be less useful for solving problems in three or more dimensions. For instance, the naïve three-dimensional generalization of the slab-based algorithm above would not work, because it would require locating the point on a two-dimensional subdivision (the projection of the polyhedron on some plane), which cannot be performed exactly, even assuming exact x- and y-comparisons, and also because there is no three-dimensional analog of Lemma 1, which is used implicitly in the slab-based algorithm to guarantee that the point is ε away from P if and only if it is ε away from the tested edges. In view of these considerations, we have found it worthwhile to develop a point-location algorithm that does not use exact comparisons, in the hope that it may be easier to generalize to higher dimensions.

5 STRONGLY CONVEX HULLS

Given a set of points S, we say that $P \subseteq S$ is a δ-hull for S if every point of S is δ-inside P. This definition agrees with our general notion of epsilon predicates if we define the distance between two polygons to be the maximum distance between a vertex of one polygon and the nearest point on the boundary of the other. Note that this metric is different from the one used for measuring convexity, since it does not require vertices to be matched one-to-one.

The ε-convex δ-hull of a set S is not neces-

sarily unique for $\delta > 0$. However, it is not difficult to prove that any two δ-hulls G and H of S are quite similar in many respects. For one thing, every point of G is δ-inside H, and vice versa. Furthermore, if G and H are also $(-\varepsilon)$-convex, then it follows from Lemma 5 that they have the same number of vertices, to within a constant factor; more precisely, $|G| \leq 2\lceil 2\delta/\varepsilon \rceil |H|$, and vice versa.

Existence of Strongly Convex Hulls

It is by no means obvious that a $(-\varepsilon)$-convex δ-hull always exists for an arbitrary set of points S and arbitrary (positive) ε and δ. In fact, when $\delta < \varepsilon$, there may not exist any such hull. As a counterexample, consider a set of four points arranged on the vertices of a square so that each point is just barely ε-inside the diagonal formed by its neighbors. In order to ensure $(-\varepsilon)$-convexity, only two of the four points can be chosen for the hull, leaving the other two points arbitrarily close to lying 2ε away.

On the other hand, we prove here that a $(-\varepsilon)$-convex δ-hull always exists whenever $\delta \geq (1+\sqrt{2})\varepsilon$. (The existence question is still open for $\varepsilon \leq \delta < (1+\sqrt{2})\varepsilon$.) We begin with a lemma:

Lemma 8 *Let C be a convex polygon with no two vertices $(\sqrt{2}\varepsilon)$-coincident for some $\varepsilon > 0$. Then there is a subpolygon H of C such that H is $(-\varepsilon)$-convex, and every point of C is ε-inside H.*

Proof: Let H be the subpolygon of C satisfying the following conditions:

1) every point of C is ε-inside H;

2) polygon H has the smallest possible number of vertices subject to (1); and

3) polygon H has the maximum area subject to (1) and (2).

(Note that there is at least one subpolygon satisfying (1), namely, C itself.)

Now suppose that H were not $(-\varepsilon)$-convex. Then by Lemma 4 there would be three consecutive vertices a, b, c of H, with vertex b ε-inside segment ac. Assume without loss of generality that ac is horizontal, with b above ac. See Figure 3. Since vertex b is not $(\sqrt{2}\varepsilon)$-coincident with vertices a or c, the angle from ac to ab can be at most 45°. Furthermore, any vertex x of C between a and b must lie ε-inside segment ab, and therefore, by the same argument, the angle between ab and ax is also at most 45°. Arguing similarly for the vertices of C between b and c, we conclude that every vertex of C between a and c must lie between the two perpendiculars to ac at a and c.

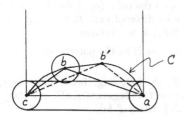

Figure 3: Construction for proof of Lemma 8

By assumption (2), removing b from H would leave some vertex x of C more than 2ε away from the resulting polygon, and therefore from ac. By convexity, point x must come between a and c in the cyclic order of C. Therefore, vertex x must lie higher than b. Assume without loss of generality that x lies to the right of b, that is, between a and b in C. Let b' be the right neighbor of b in C, and let H' be the polygon H with vertex b replaced by b'. We argue that H' is a polygon that satisfies (1) and (2) but has greater area than H, a contradiction.

To see that polygon H' satisfies (1), observe that vertex b' must lie higher than b, and all vertices of C between b and c must lie lower than b, by convexity. Therefore, all vertices of C between b' and c are ε-inside H'. Furthermore, for any vertex x of C between a and b', observe that the shortest segment from x to segment ab must intersect seg-

ment ab', since $axb'b$ is a convex quadrilateral. Thus, point x must lie at least as close to segment ab' as it does to segment ab. Therefore, since by hypothesis x is ε-inside segment ab, it must also lie ε-inside segment ab', and so is ε-inside H'.

Polygon H' also satisfies (2), since it has no more vertices than H. However, polygon H' has greater area than H, since triangle $ab'c$ has the same base as triangle abc but greater altitude. Thus, we have reached a contradiction. We conclude that H is $(-\varepsilon)$-convex. □

Theorem 9 *For any set of points S, there is a subset H of S such that the points of H are the vertices of a $(-\varepsilon)$-convex $((1 + \sqrt{2})\varepsilon)$-hull of S.*

Proof: Let S' be a maximal subset of S with the property that no two points are $(\sqrt{2}\varepsilon)$-coincident. Obviously, any point in S that is not in S' must be at most $2\sqrt{2}\varepsilon$ away from some point of S'.

Let C be the ordinary convex hull of S'. By Lemma 8, there is a subpolygon H of C that is $(-\varepsilon)$-convex and such that every point of C is ε-inside H. By the triangle inequality, every point of S is $((1 + \sqrt{2})\varepsilon)$-inside H. Thus, H is a $(-\varepsilon)$-convex $((1 + \sqrt{2})\varepsilon)$-hull of S. □

Computing Strongly Convex Hulls

Algorithm 2 provides a method for constructing strongly convex approximate hulls, using our approximate geometric primitives. The algorithm takes a set S of n points in the plane and an $\varepsilon \geq 0$, and returns a $(-\varepsilon)$-convex polygon H, and an uncertainty interval d, such that H is a d-hull for S. The algorithm guarantees that $d.hi \leq (5 + \sqrt{2})(\varepsilon + \lambda) + \lambda = O(\varepsilon + \lambda)$, where λ is the maximum uncertainty of the primitive epsilon boxes used.

For simplicity, the description of the algorithm includes exact comparison tests in x and y. However, these exact tests can be replaced by approximate ones through a straightforward modification of the algorithm

Algorithm 2: *Given a set S of n points in the plane and an $\varepsilon \geq 0$, return a $(-\varepsilon)$-convex polygon H, and an uncertainty interval d, such that H is a d-hull for S.*

1. Find the x- and y-extremal points of S. Call them t_0, t_1, t_2, t_3, in counterclockwise order.

2. For each consecutive pair $t_q t_{q+1}$, do:

   2a. Let S_q be the set of all points $s \in S$ such that $\texttt{E\_Pos}(-\varepsilon, t_q, s, t_{q+1}) = \textbf{true}$.

   2b. Make S_q into an xy-monotone chain C_q by sorting the points in x and y and removing all points that are xy-dominated by other points in quadrant q. Let $(c_0, \ldots, c_{m-1})$ be the points of this chain C_q from t_q to t_{q+1}, inclusive.

   2c. Build a graph $G_q = (N, A)$ as follows:

   $$N = \{ (c_i c_j) : i < j \}$$
   $$A = \{ ((c_i c_j), (c_j c_k)) : i < j < k \,\wedge\, \texttt{E\_Pos}(-\varepsilon, c_i, c_j, c_k) = \textbf{true} \}$$

   2d. Compute for each node $(c_i c_j)$ in N a penalty $f(c_i c_j)$, defined as the interval $\max \{ d_{irj} : i < r < j \}$, where $d_{irj} = \texttt{I\_Neg}(c_i, c_r, c_j)$.

   2e. Define the penalty $f(P)$ of a directed path P in this graph to be the maximum of the penalties of its nodes. Let the initial nodes be the pairs $(c_0 c_i)$ for all i, and the final nodes be the pairs $(c_j c_{m-1})$ for all j. Find a directed path P_q from any initial node to any final node for which the penalty $f(P_q).hi$ is minimum.

3. Concatenate the paths P_0, P_1, P_2, P_3 to form a cycle. Let P be the polygon described by that cycle, and let d be the interval $\max_q f(P_q)$.

4. Start with $H = P$. For each extremal point t_q, do the following: Let a and b be the current neighbors of t_q in H, and let d' be the interval $\texttt{I\_InSegment}(t_q, ab)$. If $d'.lo \leq \varepsilon$, then remove t_q from H, and set $d \leftarrow (\max \{d.lo, d'.lo\}, d.hi + d'.hi)$.

5. Output the polygon H and the interval d, asserting that H is a $(-\varepsilon)$-convex d-hull for S.

(involving a cleanup step to ensure that no two point are too close), at the cost of a small increase in the bound on $d.hi$.

The running time of the algorithm is dominated by step 2e, which can be performed in time $O(|A| \log |A|) = O(n^3 \log n)$ time and $O(|N|) = O(n^2)$ space by a standard graph-theoretic algorithm. (The set of arcs A can be generated on the fly and does not need to be stored explicitly.)

Although these bounds appear extravagant, the algorithm can actually be expected to perform quite well in practice because the monotone chains C_q are likely to contain only a small subset of the original points S. A simple probabilistic analysis [1] shows that for many reasonable point distributions (e.g., Gaussian, or uniform in a square), the expected size of the chains C_q is only $O(\log n)$. The running time is then dominated by the construction

of the chains, which can be performed in just $O(n \log n)$ time.

The existence of the paths P_q is guaranteed by the following lemma:

Lemma 10 *Given a set of points S, there exists a convex $(1 + \sqrt{2})(\varepsilon + \lambda)$-hull D for S, with vertices in S, such that:*

1. *the polygon D includes the four extremal points t_0, t_1, t_2, t_3 of S;*

2. *every point of S not inside D is in the bounding box of some edge of D;*

3. *for $0 \leq q \leq 3$, the xy-monotone chain $D_q = (d_0, \ldots, d_k)$ between t_q and t_{q+1} satisfies*

   3a. $\texttt{E\_Pos}(-\varepsilon, t_q, d_i, t_{q+1}) = \textbf{true}$, *and*

   3b. $\texttt{E\_Pos}(-\varepsilon, d_{i-1}, d_i, d_{i+1}) = \textbf{true}$,

   for $0 < i < k$.

Proof: Let $\sigma = (1+\sqrt{2})(\varepsilon + \lambda)$. Let Q be the quadrilateral defined by the four extremal points t_0, t_1, t_2, t_3. For each pair of extremal points $(t_q t_{q+1})$ of S, consider the set S'_q consisting of the points t_q, t_{q+1}, and all points s of S that are more than 2σ away from Q and such that $Pos(t_q, s, t_{q+1})$. Let S''_q be all points of S'_q that are not xy-dominated by other points in quadrant q. By Theorem 9, the set S''_q has a $(-\varepsilon - \lambda)$-convex σ-hull D_q. The polygon D_q must include the extremal points t_q and t_{q+1}, since these points are more than 2σ away from any other point of S''_q.

Now consider the polygon D that is the convex hull of the union of the hulls D_q. Any consecutive triple in any xy-monotone chain of D is a triple of consecutive vertices in some D_q and is therefore $(-\varepsilon - \lambda)$-positive, so the λ-accurate test $\texttt{E\_Pos}(-\varepsilon, d_{i-1}, d_i, d_{i+1})$ is guaranteed to return $\texttt{true}$. Furthermore, every non-extremal vertex of D_q is more than 2σ away from Q, so $\texttt{E\_Pos}(-\varepsilon, t_q, d_i, t_{q+1}) = \texttt{true}$ by Lemma 1.

By the arguments above, polygon D satisfies all three conditions of the lemma. Furthermore, polygon D is also a σ-hull for S since every point of S that is not in some S''_q is either inside D, or at most 2σ away from Q or from some polygon D_q, which are all in D. □

Theorem 11 *Algorithm 2 produces a polygon H that is a $(-\varepsilon)$-convex δ-hull for the given set S, where $\delta = (5 + \sqrt{2})(\varepsilon + \lambda) + \lambda = O(\varepsilon + \lambda)$.*

Proof: First, we argue that the xy-monotone chains D_q of the polygon D, whose existence is proved by Lemma 10, must appear as directed paths in the graphs G_q constructed by our algorithm. Condition 3a of the lemma guarantees that the vertices of D_q will survive in the set S_q constructed in step 2a, and condition 2 guarantees that they will be included in the chain C_q constructed in step 2b. Thus, all edges of D_q will appear as nodes of G_q. Furthermore, condition 3b guarantees that every consecutive triple of vertices of D_q appears as an arc of G_q. Therefore, step 2e will always be able to find some path P_q.

Second, we argue that the polygon P computed in step 3 is a d-hull for S, where d is the interval computed in that step. Observe that any point s of S that is not inside P must lie in the bounding box of an edge $c_i c_j$ of P, because otherwise point s would dominate some vertex of P. Moreover, since P is convex, the distance δ_s between s and P is the same as the distance between s and edge $c_i c_j$. Therefore, by Lemma 1, the penalty of the node $c_i c_j$ computed in step 2d is an interval that contains δ_s. Thus, the interval d computed in step 3 contains the maximum of the distances δ_s for all points s, so P is a d-hull for S.

Third, we argue that the interval d computed in step 3 is bounded above by the quantity $(1+\sqrt{2})(\varepsilon + \lambda) + \lambda$. Since step 2e finds a path with minimal penalty, the path P must have a penalty no greater than that of the chains D_q, which have penalty at most $(1+\sqrt{2})(\varepsilon + \lambda)$ by Lemma 10. On the other hand, the penalties assigned to the paths are measured by a λ-accurate box, so they could be overestimated by at most λ.

Finally, we argue that the polygon H computed in step 4 is a $(-\varepsilon)$-convex $(4(\varepsilon + \lambda))$-hull for the polygon P. For each vertex v of H, denote by $h(v)$ the distance between v and the diagonal connecting its two neighboring vertices. By Lemma 4, the polygon H is $(-\varepsilon)$-convex if and only if $h(v) > 2\varepsilon$ for all v. At the beginning of step 4, the condition $h(v) > 2\varepsilon$ is satisfied for all vertices v except for the extremal points t_q. Moreover, by convexity, deleting a vertex from H cannot decrease the value of $h(v)$ for any other vertex v. Since step 4 deletes all vertices for which $h(v) \leq 2\varepsilon$, the final polygon H is $(-\varepsilon)$-convex. On the other hand, step 4 only deletes a vertex v if $h(v) \leq 2(\varepsilon + \lambda)$. Therefore, the new polygon resulting from each deletion is an $(\varepsilon + \lambda)$-hull for the previous one. Since at most four vertices are deleted, the final polygon H is a $(4(\varepsilon + \lambda))$-hull for the starting polygon P.

We conclude that the returned polygon H is a $(-\varepsilon)$-convex δ-hull, where $\delta \leq 4(\varepsilon + \lambda) + (1 + \sqrt{2})(\varepsilon + \lambda) + \lambda = (5 + \sqrt{2})(\varepsilon + \lambda) + \lambda$. $\square$

Acknowledgements

The Epsilon Geometry framework was inspired by a discussion with Bernard Chazelle, Herbert Edelsbrunner, Michel Gangnet, Ricky Pollack, Franco Preparata, and Micha Sharir. Some of the key ideas used in the convex hull algorithm (in particular, the idea to start with xy-monotone chains of non-dominated points) were suggested by Victor Milenkovic. We would like to thank John Hershberger for his comments on earlier versions of this work. We are grateful to DEC Systems Research Center and the AT&T Foundation for their financial support.

References

[1] O. Barndorff-Nielsen and M. Sobel, "On the distribution of the number of admissible points in a vector sample." Theory of Probability and its Applications, Volume XI, Number 2 (1966), 249–269.

[2] David Dobkin and Deborah Silver, "Recipes for Geometry and Numerical Analysis—Part I: An Emperical Study." Proceedings of the 4th Annual ACM Symposium on Computational Geometry (1988), 93–105.

[3] Steven Fortune, "Stable Maintenance of Point Set Triangulations in Two Dimensions." Proceedings of the 30th Annual Symposium on Foundations of Computer Science (1989), 494–499.

[4] Daniel H. Greene and F. Frances Yao, "Finite-resolution computational geometry." Proceedings of the 27th IEEE Symposium on the Foundations of Computer Science (1986), 143–152.

[5] Leonidas Guibas, David Salesin, and Jorge Stolfi, "Epsilon Geometry: Building Robust Algorithms from Imprecise Computations," Proceedings of the 5th Annual ACM Symposium on Computational Geometry (1989), 208–217.

[6] Christoph M. Hoffman, John E. Hopcroft, and Michael S. Karasick, "Towards implementing robust geometric computations." Proceedings of the 4th Annual ACM Symposium on Computational Geometry (1988), 106–117.

[7] Christoph Hoffman, "The Problems of Accuracy and Robustness in Geometric Computation." Computer , Volume 22 (1989), 31–42.

[8] Victor J. Milenkovic, "Verifiable Implementations of Geometric Algorithms using Finite Precision Arithmetic." Artificial Intelligence, Volume 37 (July 1988), 377–401.

[9] Victor J. Milenkovic, "Verifiable Implementations of Geometric Algorithms using Finite Precision Arithmetic." Ph.D. thesis, Carnegie-Mellon (1988). Available as CMU report CMU-CS-88-168.

[10] Victor J. Milenkovic, "Calculating Approximate Curve Arrangements Using Rounded Arithmetic." Proceedings of the 5th Annual ACM Symposium on Computational Geometry (1989), 197-207.

[11] Victor J. Milenkovic, "Double Precision Geometry: A General Technique for Calculating Line and Segment Intersections using Rounded Arithmetic." Proceedings of the 30th Annual Symposium on Foundations of Computer Science (1989), 500–505.

[12] Victor J. Milenkovic and Zhenyu Li, "Constructing Strongly Convex Hulls Using Exact or Rounded Arithmetic." To appear in Proceedings of the 6th Annual ACM Symposium on Computational Geometry (1990).

[13] Thomas Ottmann, Gerald Thiemt, and Christian Ullrich, "Numerical stability of geometric algorithms." Proceedings of the 3rd Annual ACM Symposium on Computational Geometry (1987), 119–125.

[14] F. Preparata and M. Shamos, Computational Geometry: An Introduction. Springer-Verlag (1985).

[15] M. Segal and C. Séquin, "Consistent Calculations for Solids Modeling." Proceedings of the 1st Annual ACM Symposium on Computational Geometry (1985), 29–38.

[16] K. Sugihara and M. Iri, "Geometric Algorithms in Finite-Precision Arithmetic." Research Memorandum RMI 88-10, University of Tokyo (September 1988).

COMPUTING PUISEUX-SERIES SOLUTIONS TO DETERMINANTAL EQUATIONS VIA COMBINATORIAL RELAXATION
— Extended Abstract —

Kazuo Murota
Department of Mathematical Engineering and Information Physics
University of Tokyo, Tokyo 113, Japan

1. Introduction

Let $A(t, x) = (A_{ij}(t, x))$ be an $n \times n$ matrix with

$$A_{ij}(t, x) = \sum_{s \in \mathbf{Z}} \sum_{r \in \mathbf{Q}} A_{ijrs} t^r x^s, \qquad (1.1)$$

where the summations are assumed to involve a finite number of terms. We are interested in the computational procedure for the Puiseux (=fractional power) series solutions $x = x(t)$ to the equation $\det A(t, x) = 0$. This problem will arise in many different contexts; e.g., sensitivity analysis of eigenvalue x of matrix A which is subject to perturbation t.

If we could explicitly compute the expansion

$$f(t, x) = \det A(t, x) = \sum_s \sum_r f_{rs} t^r x^s$$

to find $r(s) = \min\{r \mid f_{rs} \neq 0\}$ for each s, then we could apply the standard procedure using the Newton diagram. This method plots all the points $(s, r(s))$ on a plane and considers the convex hull of those points; the slope p of the sides of this convex polygon gives the order of the first term $x \sim \gamma t^{-p}$ in the Puiseux-series solutions. The higher-order terms can be determined by repeated application of this procedure.

Example 1.1. For $f(t, x) = -t + (t^2 + t^4)x^2 - t^6 x^4$ the Newton diagram N is shown by • in Fig. 1. We have two possibilities for p, i.e., $p = 1/2, 2$, and obtain four solutions: $x \sim \pm t^{-1/2}, \pm t^{-2}$. □

This work is motivated by the following observations.

1. Even when the matrix A is moderately sized, the explicit enumeration of all the nonzero coefficients f_{rs} would be prohibitive.
2. The exponent p is determined solely by those pairs (s, r) which correspond to the extreme points (or vertices) of the convex hull.
3. There is a close relation between the nonzero terms f_{rs} and the perfect matchings of a bipartite graph associated with A. So long as no accidental numerical cancellation

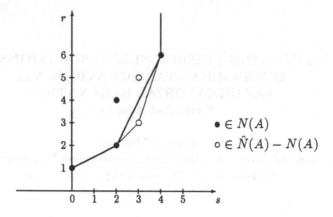

Fig. 1. Newton diagrams $N(A)$ (Example 1.1) and $\hat{N}(A)$ (Example 2.1)

occurs in the determinant expansion, the extreme points (s, r) can be identified by solving a parametric assignment (or weighted bipartite matching) problem (see §2).

This paper establishes a link between the computer algebra and the mathematical programming (combinatorial optimization). We make use of the results in mathematical programming in two different ways. Firstly, the proposed algorithm uses the results from network flow theory in its individual steps; the correctness relies on the duality theorem and the practical efficiency on the fast network-type algorithms. Secondly, the whole algorithm is designed in line with some general methods known in mathematical programming. In particular, we make use of the ideas of "relaxation" typically appearing in integer programming and "artificial variable" in the simplex method for linear programming.

The idea of the proposed algorithm may be described as follows. The exponent in $x \sim \gamma t^{-p}$ is determined from the extreme points of $N(A) = \{(s, r) \mid f_{rs} \neq 0\}$. Instead of $N(A)$, we consider its combinatorial counterpart $\hat{N}(A)$, called the "combinatorial relaxation", based on the relation between the nonzero terms in $\det A$ and the perfect matchings in a bipartite graph associated with A. $\hat{N}(A)$ has the properties that $\hat{N}(A) \supseteq N(A)$ and that $\hat{N}(A) = N(A)$ if no numerical cancellation occurs in the determinant expansion. $\hat{N}(A)$ has the computational advantage that the extreme points can be found by efficient network-type algorithms.

We first solve the easier problem defined by $\hat{N}(A)$, hoping that its solution is also valid for the original problem $N(A)$. If the solution is not good for $N(A)$, we modify A slightly, so that $N(A)$ is kept invariant and at the same time the invalid solution is eliminated from the relaxation. The time complexity of the algorithm is pseudo-polynomial in the worst case and polynomial on the average.

2. Combinatorial Relaxation of Newton Diagram

The structure of A of (1.1) is represented by a bipartite graph $G(A) = G(A; p)$. The vertex set V is the union of row set R and column set C of A. The edge set is $E = \{(ijrs) \mid A_{ijrs} \neq 0\}$, where $\partial^+(ijrs) = i \in R$, $\partial^-(ijrs) = j \in C$. To edge $(ijrs)$ is given a cost $c_{ijrs}(p) = r - ps$ parametrized by p.

Put

$$r(M) = \sum_{(ijrs)\in M} r, \qquad s(M) = \sum_{(ijrs)\in M} s.$$

Suggested by the expansion of determinant, we define

$$\hat{N}(A) = \{(s(M), r(M)) \mid M: \text{perfect matching in } G(A)\}, \tag{2.1}$$

and name it the *combinatorial Newton diagram* or the *combinatorial relaxation* to $N(A)$. $\hat{N}(A)$ is a combinatorial notion in the sense that the numerical values of A_{ijrs} are disregarded. The following statements justify this name.

Proposition 2.1.
(1) $N(A) \subseteq \hat{N}(A)$.
(2) $N(A) = \hat{N}(A)$ if the nonzero coefficients A_{ijrs} are algebraically independent. □

For a line l and a point set N in general, we say l *supports* N if all the points in N lie above (or on) l and $l \cap N \neq \emptyset$; l *tightly supports* (or *t-supports*) N, if, in addition, $|l \cap N| \geq 2$. We call $(s, r) \in \hat{N}(A)$ *genuine* if $(s, r) \in N(A)$, and *spurious* if $(s, r) \notin N(A)$. We say a supporting (resp., t-supporting) line of $\hat{N}(A)$ is *genuine* or *spurious* according as it is a supporting (resp., t-supporting) line of $N(A)$ or not.

The t-supporting lines of $\hat{N}(A)$ can be computed efficiently on the basis of the following relation to the parametric assignment problem on $G(A) = G(A; p)$. Let $c(p)$ denote the minimum cost of a perfect matching in $G(A; p)$. The *breakpoints* are those points at which the slope of $c(p)$ changes. We have the following important statement, a consequence of the well-known point-line duality.

Proposition 2.2.
p is a breakpoint of $G(A; p)$ iff p is the slope of a t-supporting line of $\hat{N}(A)$. □

Example 2.1. Consider a 3×3 matrix

$$A(t, x) = \begin{pmatrix} 1 & 0 & t^2 x^2 \\ 1 + tx & t & t^3 x^3 \\ t^3 x^2 & 1 + t^2 & -1 \end{pmatrix}, \tag{2.2}$$

for which $\det A = -t + (t^2 + t^4)x^2 - t^6 x^4 + \text{spur}(t^3 x^3, t^5 x^3)$, where $\text{spur}(\cdots)$ is the list of terms that are cancelled out. $G(A)$ has 8 perfect matchings, among which two cancelling pairs yield the spurious points $(s, r) = (3, 3)$ and $(3, 5)$. $\hat{N}(A)$ (shown by • and ○ in Fig.1) has three t-supporting lines with slopes $p = 1/2, 1, 3$. Only the first corresponds to the solution $x \sim \pm t^{-1/2}$. □

3. Outline of Algorithm

Here is an intuitive description of the proposed algorithm for determining all possible first-order approximations $x \sim \gamma t^{-p}$ to the solution of $\det A(t, x) = 0$. The algorithm relies on the following observations. (i) The p is determined from the t-supporting lines of $N(A)$. (ii) $N(A)$ is approximated by $\hat{N}(A)$ (cf. Prop.2.1). (iii) T-supporting lines of $\hat{N}(A)$ can be computed by efficient combinatorial algorithms (cf. Prop.2.2). As in Example 2.1, however, not all t-supporting lines of $\hat{N}(A)$ correspond to the solution to $\det A = 0$. To cope with this we take notice of the fact that (row-wise) elimination operations on A can modify $\hat{N}(A)$ without affecting $N(A)$, and show how to modify A by elimination operations so that the particular spurious point in question may be eliminated from $\hat{N}(A)$ (for the modified A). The proposed modification of A by elimination is only local and efficient. Furthermore, this procedure is invoked only rarely, i.e., only when the numerical cancellation results in a spurious extreme point.

We maintain a genuine supporting line l for $\hat{N}(A)$ and a point $P \in l \cap N(A)$. If we have $|l \cap N(A)| \geq 2$ for such l, then l is a t-supporting line for $N(A)$, and hence the slope p of l corresponds to a desired solution. The slope p of l is nondecreasing in the course of the algorithm.

Algorithm (outline)
Step 1 (Initial Point): Find a genuine supporting line l for $\hat{N}(A)$ and a point $P \in l \cap N(A)$ such that $p \approx -\infty$ (i.e., the slope p of l is sufficiently small).
Step 2 (Solution to Relaxation): Rotate l counterclockwise around P so that l t-supports $\hat{N}(A)$; If l is vertical, i.e., parallel to the r-axis, stop; Let P' be the rightmost point (i.e., with the largest s-coordinate) of $l \cap \hat{N}(A)$.
Step 3.1 (Genuine P'): If $P' \in N(A)$ then the slope p of l corresponds to a solution $x \sim \gamma t^{-p}$; Determine γ numerically; $P := P'$; Go to Step 2.
Step 3.2 (Spurious P'): If $P' \notin N(A)$ then modify A by row-wise eliminations so that $P' \notin \hat{N}(A)$, $N(A)$ is not changed and l is a supporting line for $\hat{N}(A)$; Go to Step 2. $\square$

We need to solve the following subproblems.

Subproblem 1 (Initial Point): To find the starting pair (l, P) in Step 1.— Usually we may expect that the *south-west point* $SW(\hat{N}(A)) = (s^*, r^*)$ ($s^* = \min\{s \mid (s, r) \in \hat{N}\}, r^* = \min\{r \mid (s^*, r) \in \hat{N}\}$) is genuine and serves as the initial P; any line l passing through P with the slope $p \approx -\infty$ will do. If, unfortunately, $SW(\hat{N}(A))$ is spurious, we mimic the starting procedure of "artificial variables" in the (two-phase) simplex method for linear programming. Namely, we modify A by introducing a number of "artificial terms". The modified problem has the genuine south-west point and the algorithm can be started. After a number of steps, the artificial terms play no roles and we are solving the original problem.

Subproblem 2 (Solution to Relaxation): To find the t-supporting line for $\hat{N}(A)$ with the next larger slope in Step 2.— This amounts to finding the next larger breakpoint of $G(A; p)$. For the latter problem, a number of efficient algorithms are available, such as the general schemes of [ES], [G] for linear parametric optimization, and the network simplex method [C].

Subproblem 3 (Test for Membership in $N(A)$): To test in Step 3 whether P' of $\hat{N}(A)$ is genuine or spurious.— It is important that only extreme points of $\hat{N}(A)$ are to be tested for membership in $N(A)$. An extreme point of $\hat{N}(A)$ corresponds to the minimum assignment on $G(A; p)$ for some p. Using potentials the membership test is reduced to the test for nonsingularity of a numerical matrix.

Subproblem 4 (Modification of A): To find the adjustment scheme of A in Step 3.2.— It is not trivial to modify A with small amount of computation so that P' is eliminated from $\hat{N}(A)$ while maintaining the condition that l should support $\hat{N}(A)$. An annoying phenomenon is that the elimination operation on A can give rise to new spurious points, which may force the algorithm to run forever in the loop of Step 2 and Step 3.2.

4. Testing for Membership in $N(A)$ — Subproblem 3

Let $D(x) = (D_{ij}(x))$ be an $n \times n$ matrix with $D_{ij}(x) = \sum_{s \in \mathbf{Z}} D_{ijs}\, x^s$. We denote by $\delta(D)$ the maximum degree of a nonzero term in $\det D(x)$, where $\delta(D)$ may possibly be negative. A bipartite graph $G^* = G^*(D)$ is associated with $D(x)$ in a similar manner as $G(A)$ is with $A(t, x)$. The vertex set $V(G^*)$ is the union of row set R and column set C of D, and the edge set is $E(G^*) = \{(ij) \mid D_{ij}(x) \neq 0\}$. To edge (ij) is attached a cost: $c_{ij} = \max\{s \mid D_{ijs} \neq 0\} = \deg_x D_{ij}(x)$. We define $\hat{\delta}(D)$ to be the maximum cost of a perfect matching in $G^*(D)$.

Proposition 4.1.
(1) $\delta(D) \leq \hat{\delta}(D)$.
(2) $\delta(D) = \hat{\delta}(D)$ if the nonzero coefficients D_{ijs} are algebraically independent. □

We say $D(x)$ is *upper tight* (or *u-tight*) if $\delta(D) = \hat{\delta}(D)$. The following procedure tests for u-tightness of $D(x)$ without computing all terms of $\det D(x)$. Let v_i^R ($i \in R$) and v_j^C ($j \in C$) be the potentials associated with a maximum perfect matching. Then $\hat{\delta}(D) = \Delta v$, where $\Delta v = \sum_i v_i^R - \sum_j v_j^C$, and

$$\tilde{c}_{ij} \equiv c_{ij} - v_i^R + v_j^C \leq 0. \tag{4.1}$$

Consider a perfect matching M in $G^*(D)$. M has the maximum cost $\hat{\delta}(D) = \Delta v$ iff (4.1) holds with equality for all $(ij) \in M$. Therefore, a term with strict inequality in (4.1) may be deleted without any influence on the coefficient of $x^{\hat{\delta}(D)}$ in the determinant expansion.

Define $\tilde{D}(x) = (\tilde{D}_{ij}(x))$ and $D^* = (D_{ij}^*)$ by

$$\tilde{D}_{ij}(x) = \begin{cases} D_{ijc_{ij}}\, x^{c_{ij}} & \text{if } \tilde{c}_{ij} = 0 \\ 0 & \text{otherwise,} \end{cases} \qquad D_{ij}^* = \begin{cases} D_{ijc_{ij}} & \text{if } \tilde{c}_{ij} = 0 \\ 0 & \text{otherwise.} \end{cases} \tag{4.2}$$

Proposition 4.2.
$D(x)$ is u-tight iff D^* is nonsingular. □

Let $P' = (s', r')$ be an extreme point (of the convex epigraph) of $\hat{N}(A)$. Then there exists a supporting line l of $\hat{N}(A)$ such that $P' \in l \cap \hat{N}(A)$. We assume that P' is the rightmost point in $l \cap \hat{N}(A)$, as is the case in Step 2.

We can translate these statements into the language of the assignment problem on $G(A; p)$. Since $P' \in \hat{N}(A)$ is an extreme point, there exists a closed interval $[p^{(1)}, p^{(2)}]$ of p and a perfect matching M in $G(A; p)$ such that M has the minimum cost for $p \in [p^{(1)}, p^{(2)}]$ and $(s', r') = (s(M), r(M))$. Let $u_i^R = u_i^R(p)$ and $u_j^C = u_j^C(p)$ be the associated potentials; we have

$$\tilde{c}_{ijrs} \equiv r - ps + u_i^R - u_j^C \geq 0,$$

and the equality holds for $(ijrs) \in M$. Then the line l defined by $r = ps - \Delta u$ supports $\hat{N}(A)$ at P', where $\Delta u = \sum_i u_i^R - \sum_j u_j^C$. Furthermore, P' is the rightmost point of $l \cap \hat{N}(A)$ iff $p < p^{(2)}$.

We extract from $A(t, x)$ those terms which can contribute to the minimum assignment in $G(A)$. Define $D(x) = (D_{ij}(x))$ by

$$D_{ij}(x) = \sum \{A_{ijrs} \, x^s \mid (s, r) \in E_{ij}\}, \tag{4.3}$$

where $E_{ij} = \{(s, r) \mid \tilde{c}_{ijrs} = 0, (ijrs) \in E(G)\}$, and consider the bipartite graph $G^*(D)$ associated with $D(x)$. A perfect matching in $G^*(D)$ corresponds to a perfect matching M' in $G(A)$ such that $(s(M'), r(M')) \in l$, and, conversely, a perfect matching M' in $G(A)$ with $(s(M'), r(M')) = (s', r')$ has a corresponding perfect matching in $G^*(D)$. In particular, the s-coordinate of P' is given by $s' = \hat{\delta}(D) = \Delta v$.

Proposition 4.3.
For the rightmost point P' of $l \cap \hat{N}(A)$, P' is genuine iff D^* is nonsingular. $\qquad \square$

5. Modification of Matrix — Subproblem 4

Suppose we are given a t-supporting line l with slope p of $\hat{N}(A)$ such that $l \cap N(A) \neq \emptyset$ and that the rightmost point $P' \in l \cap \hat{N}(A)$ does not belong to $N(A)$. We are to modify $A(t, x)$ to another matrix $A'(t, x) = (A'_{ij}(t, x))$ such that

(P1) : $N(A') = N(A)$,
(P2) : l supports $\hat{N}(A')$,
(P3) : The rightmost point of $l \cap \hat{N}(A')$ lies to the left of P', i.e., has strictly smaller s-coordinate than that of P'.

The last condition implies $P' \notin \hat{N}(A')$. We require the following additional properties for guaranteed termination and complexity:

(P4) : $\delta^*(A') \leq \delta^*(A)$, where $\delta^*(A) = \max\{\deg_x A_{ij}(t, x) \mid i \in R, \ j \in C\}$,
(P5) : If $A(t, x)$ contains only integer powers of t and x, so does $A'(t, x)$.

We have term $-$ rank $D^* = n$ since $P' \in \hat{N}(A)$ and rank $D^* < n$ since $P' \notin N(A)$ (cf. Prop.4.3). Therefore, there exists a nonzero vector $w = (w_i \mid i \in R)$ such that $w^T D^* = 0$; choose such w with minimal support. Let $i_0 \in R$ be such that $v_{i_0}^R = \min\{v_i^R \mid w_i \neq 0\}$, and define W by

$$W_{ik} = \begin{cases} w_k & \text{if } i = i_0 \\ \delta_{ik} & \text{otherwise.} \end{cases} \tag{5.1}$$

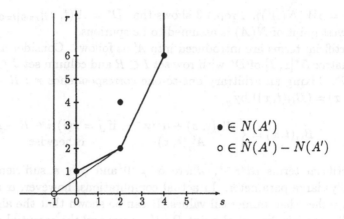

Fig. 2. $\hat{N}(A')$ (Example 5.1)

Using the potentials $\{u_i^R, u_j^C\}$ for $G(A;p)$ and $\{v_i^R, v_j^C\}$ for $G^*(D)$, we define the transformation from A to A' by

$$A'(t,x) = \text{diag}\,(t; -u^R + pv^R) \cdot \text{diag}\,(x; v^R) \cdot W$$
$$\cdot \text{diag}\,(x; -v^R) \cdot \text{diag}\,(t; u^R - pv^R) \cdot A(t,x), \qquad (5.2)$$

where $\text{diag}\,(x; u) = \text{diag}\,(x^{u_1}, x^{u_2}, \ldots, x^{u_n})$, etc. It can be proven that (P1) to (P5) are satisfied.

Example 5.1. Recall Example 2.1. The line l defined by $r = s$ tightly supports $\hat{N}(A)$. We see $P = (2,2) \in l \cap N(A)$ and $P' = (3,3) \notin N(A)$. We may take $w = (1,-1,0)^T$ (having minimal support $\{1,2\}$). We have $i_0 = 1$ if we choose $u_i^R = u_j^C = 0$ $(i,j = 1,2,3)$; $v_1^R = 0, v_2^R = 1, v_3^R = -2, v_1^C = 0, v_2^C = -2, v_3^C = -2$ as the potentials. Then we obtain

$$W = \begin{pmatrix} 1 & -1 & 0 \\ 0 & 1 & 0 \\ 0 & 0 & 1 \end{pmatrix}, \quad A' = \begin{pmatrix} -t^{-1}x^{-1} & -x^{-1} & 0 \\ 1+tx & t & t^3x^3 \\ t^3x^2 & 1+t^2 & -1 \end{pmatrix}$$

with $N(A')$ and $\hat{N}(A')$ shown in Fig. 2. Note that l supports $\hat{N}(A')$, the rightmost point of $l \cap \hat{N}(A')$ is $P = (2,2)$, lying to the left of $P' = (3,3)$, and that $\delta^*(A') = \delta^*(A) = 3$.

6. Starting with Artificial Terms — Subproblem 1

Let M be a minimum assignment in $G(A;p)$ for $p \approx -\infty$, and $u_i^R(p) = r_i^R - ps_i^R$ $(i \in R)$, $u_j^C(p) = r_j^C - ps_j^C$ $(j \in C)$ be the associated potentials. Define $A'(t,x) = (A'_{ij}(t,x))$ by

$$A'_{ij}(t,x) = \sum_s \sum_r A_{ijrs}\, t^{r+r_i^R - r_j^C}\, x^{s+s_i^R - s_j^C}.$$

Note $(0,0) = SW(\hat{N}(A'))$. Prop.4.3 shows that $D^* = A'(t,x)|_{x=0}|_{t=0}$ is singular since the south-west point of $\hat{N}(A)$ is assumed to be spurious.

The artificial terms are introduced into A' as follows. Consider a maximal nonsingular submatrix $D^*[I, J]$ of D^* with row set $I \subset R$ and column set $J \subset C$; $\nu \equiv n - |I| = n - \operatorname{rank} D^*$. Fixing an arbitrary one-to-one correspondence $\pi : R - I \to C - J$, we define $B(t,x) = (B_{ij}(t,x))$ by

$$B_{ij}(t,x) = \begin{cases} A'_{ij}(t,x) + \alpha t^q x^{-1} & \text{if } j = \pi(i), i \in R - I \\ A'_{ij}(t,x) & \text{otherwise} \end{cases} \tag{6.1}$$

using ν artificial terms $\alpha t^q x^{-1}$, where $\alpha \neq 0$ and q is a sufficiently (but pseudo-polynomially) large parameter. In actual computations, however, α and q are treated as symbols rather than numerical values. It can be shown that the algorithm of §3 can start for $B(t,x)$ with the initial point $P = (-\nu, \nu q)$ and the essential portions of $N(A')$ and $\hat{N}(A')$ are kept unchanged.

7. Complete Description of Algorithm

Step 1 [Initial Point]
  (1) : Find a minimum assignment and potentials

$$u_i^R(p) = r_i^R - ps_i^R \ (i \in R), \quad u_j^C(p) = r_j^C - ps_j^C \ (j \in C)$$

  for $G(A;p)$ with $p \approx -\infty$.
  (2) : $A_{ij}(t,x) := \sum_s \sum_r A_{ijrs} \, t^{r+r_i^R - r_j^C} \, x^{s+s_i^R - s_j^C} \ (i \in R, j \in C)$;
  $D^* := A(t,x)|_{x=0}|_{t=0}$;
  Find a maximal nonsingular submatrix $D^*[I, J]$;
  $\nu := n - |I|$. $\qquad\qquad\qquad\qquad\qquad\qquad\qquad$ [$\nu = n - \operatorname{rank} D^*$]
  (3) : If $\nu \neq 0$ then

$$A_{ij}(t,x) := \begin{cases} A_{ij}(t,x) + \alpha t^q x^{-1} & \text{if } j = \pi(i), i \in R - I \\ A_{ij}(t,x) & \text{otherwise} \end{cases}$$

  where $\pi : R - I \to C - J$ is a one-to-one correspondence. $\qquad$ [cf.(6.1)]
  (4) : Find a minimum assignment M for $G(A;p)$ with $p \approx -\infty$;
  $\qquad\qquad\qquad\qquad [P = (s(M), r(M)) = SW(\hat{N}(A)) \in N(A)]$
  $r_{\max} := n \cdot \max\{r \mid A_{ijrs} \neq 0\}$;
  $r_{\min} := n \cdot \min\{r \mid A_{ijrs} \neq 0\}$;
  $p_{\max} := r_{\max} - r_{\min}$. $\qquad\qquad\qquad\qquad\qquad$ [Upper bound on p]
Step 2 [Solution to Relaxation]
  $\qquad\qquad$ [M is a minimum assignment in $G(A;p)$; $P = (s(M), r(M)) \in N(A)$]
  (1) : Solve parametric assignment problem on $G(A)$ to find

$$p := \max\{p' \mid M \text{ is a minimum assignment in } G(A;p')\};$$

  $\qquad\qquad\qquad\qquad\qquad\qquad\qquad\qquad\qquad$ [p is nondecreasing]

  If $p > p_{\max}$ then stop;
  [Line l through P with slope p t-supports $\hat{N}(A)$; $P' :=$ the rightmost point in $l \cap \hat{N}(A)$]

Let u_i^R $(i \in R)$, u_j^C $(j \in C)$ be potentials for $G(A; p)$.

(2) : $D_{ij}(x) := \sum \{A_{ijrs} x^s \mid r - ps + u_i^R - u_j^C = 0\}$ $(i \in R, j \in C)$;

$\qquad\qquad\qquad\qquad\qquad\qquad\qquad\qquad\qquad\qquad\qquad\qquad$ [cf.(4.3)]

Find a maximum assignment M' and potentials v_i^R $(i \in R)$, v_j^C $(j \in C)$ for $G^*(D)$;

$\qquad\qquad\qquad\qquad\qquad\qquad\qquad\qquad\qquad$ [$\Delta v = s$-coordinate of P']

$D_{ij}^* :=$ coefficient of $x^{v_i^R - v_j^C}$ in $D_{ij}(x)$ $(i \in R, j \in C)$. $\qquad$ [cf.(4.2)]

Step 3.1 [Genuine P'] If $\det D^* \neq 0$, do the following.

$\quad(1)$: Find all solutions $x = \gamma$ $(\neq 0)$ to $\det D(x) = 0$ and output $x \sim \gamma t^{-p}$, unless $p \approx -q$. $\qquad\qquad\qquad\qquad\qquad$ [$p \approx -q$ arises from artificial terms]

$\quad(2)$: $M := M'$ (with the understanding that $E(G^*(D)) \subseteq E(G(A)))$.

$\qquad\qquad\qquad\qquad\qquad\qquad\qquad\qquad\qquad\qquad\qquad\qquad$ [$P := P'$]

$\quad(3)$: Go to Step 2.

Step 3.2 [Spurious P'] If $\det D^* = 0$, do the following.

$\quad(1)$: Find w with minimal support such that $w^T D^* = 0$;

$$A_{ij}(t, x) := \begin{cases} \sum_k t^{\rho(i_0, k)} x^{\sigma(i_0, k)} w_k A_{kj}(t, x) & \text{if } i = i_0, j \in C \\ A_{ij}(t, x) & \text{otherwise} \end{cases}$$

where $v_{i_0}^R = \min\{v_i^R \mid w_i \neq 0\}$, $\sigma(i_0, k) = v_{i_0}^R - v_k^R$,
$\rho(i_0, k) = p\sigma(i_0, k) - (u_{i_0}^R - u_k^R)$. $\qquad\qquad$ [cf.(5.2), (5.1)]

$\quad(2)$: Go to Step 2. $\qquad\qquad\qquad\qquad\qquad\qquad\qquad\qquad\qquad\qquad$ $\square$

Now the termination and the complexity of the algorithm are considered. To see the probabilistic behavior let us fix the structure (i.e., the graph $G(A)$) of the input matrix $A(t, x) = A^{(0)}(t, x)$ and regard the numerical values of nonzero coefficients A_{ijrs} as real- (or complex-) valued independent random variables with continuous distributions. Then $\hat{N}(A) = N(A)$ with probability one, since $\hat{N}(A)$ differs from $N(A)$ only because of accidental numerical cancellation. This means that all the exponents p can be determined without any modification of the matrix A in Step 3.2.

Proposition 7.1.
The average time complexity of the algorithm per exponent p is bounded by a polynomial in n, except for the determination of γ. $\qquad\qquad\qquad\qquad\qquad\qquad\qquad$ $\square$

We put $E_0 = |\{(ijrs) \mid A_{ijrs}^{(0)} \neq 0\}|$, $L_0 = \max\{|s|, |r| \mid A_{ijrs}^{(0)} \neq 0\}$, and denote by r_{den} the least common multiple of the the denominators of all r with $A_{ijrs}^{(0)} \neq 0$. Also define $s_{max} = n \cdot \max\{s \mid A_{ijrs}^{(1)} \neq 0\}$, where $A^{(1)}(t, x)$ denotes the matrix A at the end of Step 1.

The following guarantees the finite termination of the algorithm for a general input matrix with fractional powers of t and gives a pseudo-polynomial (i.e., polynomial in n and L_0) bound on the number of steps in the whole algorithm for an input matrix with integer powers of t and x.

Proposition 7.2.
(1) When no artificial terms need be introduced, the points P' produced by the algorithm are all distinct and belong to

$$\mathcal{P} = \{(s,r) \mid s \in \mathbf{Z}, \ r_{\text{den}} r \in \mathbf{Z}, \ 0 \leq s \leq s_{\max}, \ r_{\min} s \leq r \leq p_{\max} s\}.$$

Hence the number of executions of Step 2 is bounded by

$$|\mathcal{P}| \leq (s_{\max} + 1)^2 (p_{\max} - r_{\min} + 1) r_{\text{den}} \leq (2n^2 L_0 + 1)^2 (6n^2 L_0 + 1) r_{\text{den}}.$$

(2) If the given matrix $A = A^{(0)}$ contains only integer powers of t and x, the number of executions of Step 2 is pseudo-polynomially bounded by the input size. (This statement does not preclude the case with artificial terms.) $\qquad\square$

Finally we mention the following proposition for theoretical completeness. It states that the proposed algorithm can be implemented so that its running time has a pseudo-polynomial worst-case bound if the input matrix $A^{(0)}(t, x)$ involes only integer powers of t and x. As stated in Prop.7.2, the number of iterations is pseudo-polynomially bounded. The problem to be considered is that the transformation of the matrix $A(t,x)$ in Step 3.2 may cause an indefinite increase of the number of edges $|E(G(A))|$ in $G(A)$, which is equal to the total number of nonzero terms in $A(t,x)$.

Proposition 7.3.
The proposed algorithm can be implemented to run in time polynomial in n, E_0, L_0 and r_{den}, except for the determination of γ. $\qquad\square$

Part of this work was done while the author stayed at Institut für Ökonometrie und Operations Research, Universität Bonn, supported by Alexander von Humboldt Foundation. The readers are referred to [M] for the complete exposition.

References
[C] W. H. Cunningham: A network simplex method. *Mathematical Programming*, Vol. 11 (1976), 105–116.

[ES] M. J. Eisner and D. G. Severance: Mathematical techniques for efficient record segmentation in large shared database, *Journal of the Association for Computing Machinery*, Vol. 23 (1976), 619–635.

[G] D. Gusfield: Parametric combinatorial computing and a problem of program module distribution, *Journal of the Association for Computing Machinery*, Vol. 30 (1983), 551–563.

[M] K. Murota: Computing Puiseux-series solutions to determinantal equations via combinatorial relaxation, to appear in *SIAM Journal on Computing*. Also: Computing Puiseux-series expansion via combinatorial relaxation, Report No. 89593-OR, Institut für Ökonometrie und Operations Research, Universität Bonn, 1989.

TIGHT LOWER BOUND ON THE SIZE OF PLANAR PERMUTATION NETWORKS

Maria Klawe*
Department of Computer Science, University of British Columbia
Vancouver, BC V6T 1W5, Canada

Tom Leighton**
Department of Mathematics, Massachusetts Institute of Technology
Cambridge, MA 02139, USA

1. INTRODUCTION

We define a **t-permutation network** to be a graph G with t distinguished vertices called terminals, with the property that for any one-to-one pairing $\{(x_i, y_i)\}$ among the terminals there is a set $\{P_i\}$ of vertex-disjoint paths in G with P_i joining x_i to y_i for each i. Permutation networks obviously have many applications in communication networks, but they have also received substantial attention in the context of permutation layouts, a basic tool in the layout of printed circuits and large scale integrated chips (see [CS80], [KKF79], [SH80], [TK82], [AKLLW85], [AKLLW90], [AKS90]).

A **permutation layout** is a permutation network where the graph G is a rectangular (2-dimensional) grid graph. The definition of permutation layout sometimes includes additional assumptions such as that the terminals are partitioned into inputs and outputs with only one-to-one pairings between inputs and outputs considered. Since it is possible to modify our definition and result in a straightforward manner to correspond to these variants we restrict our attention to the case described here.

One of the key questions concerning permutation layouts is how large a rectangle is needed to construct a t-permutation layout, since this influences how densely circuits can be laid out on chips. Examples of rectangular grids with $O(t^3)$ area which contain t-permutation layouts (and simple algorithms for finding the routings of the connecting paths) were given by Cutler and Shiloach in [CS78], who also proved that if all the terminals lie on at most two horizontal lines of the grid, then the rectangle must have area at least $\Omega(t^{2.5})$. Techniques very similar to those given by Cutler and Shiloach are commonly used in circuit layout. In [AKLLW85], Aggarwal et al proved an $\Omega(t^3)$ lower bound on the area of t-permutation layouts, showing that the Cutler-Shiloach techniques are asymptotically optimal. This result raises two obvious questions. Can the area needed be reduced by using multiple layers of grids, or by using some other planar graph instead of rectangular grids? Since area is not an appropriate measure for

* This research partially supported by an NSERC Operating Grant
** This research partially supported by contracts: DARPA N00014-87-K-825 and N00014-89-J-1988, Air Force AFOSR-89-0271, Army DAAL-03-86-K-0171.

planar graphs, in the second question area is replaced by number of vertices as these two measures essentially agree on grids.

The first question is addressed in [AKLLW90] where the $\Omega(t^3)$ lower bound on area is extended to multi-layer grid permutation networks with the restriction that some (arbitrarily small) fixed fraction of the connecting paths do not change layers. The restriction that a fixed fraction of the paths do not change layers is essential, since the standard crosspoint switch is a two layer t-permutation layout with $O(t^2)$ area, in which every routing path changes layers once. In spite of the reduction of area obtainable with the use of layer changes, the practical advantages of avoiding layer changes continue to make planar permutation networks a useful tool in circuit layout. Thus the second question remains a significant issue. The purpose of this paper, is to answer the second question by proving an $\Omega(t^3)$ lower bound on the number of vertices in a planar t-permutation network, showing that the current grid-based techniques are asymptotically optimal.

Like the lower bound for t-permutation grid graphs in [AKLLW85, AKLLW90], our proof uses the permutation property of the graph to simulate a planar embedding of an expanding graph on $\Omega(t)$ vertices and then applies the quadratic lower bound on the crossing number of expanding graphs to get the desired $\Omega(t^3)$ lower bound. However, we also use an additional tool, namely the existence of weight-balanced separators for planar graphs. Combining these two techniques results in a proof which is more general and simpler than the ones for grid graphs given in [AKLLW85] and [AKLLW90].

2. THE LOWER BOUND

Let G be a t-permutation network with n vertices. We first note that we may assume that G has maximum degree 3 since replacing the edges adjacent to each vertex of higher degree with a binary tree connecting the vertex to its neighbours only increases the number of vertices by at most a constant factor, and does not affect the permutation property. In addition we may assume that G is connected since all the terminals must lie in the same connected component of a permutation network, and the connected component will itself be a permutation network. Finally, we may assume that each terminal has degree 1 since if necessary we can hang a new terminal vertex off each original terminal.

We start the section by describing the weight-balanced separator theorem (2.1) and one of its corollaries, culminating with the formulation we will actually apply, the balanced terminal separator lemma (2.2). We then give the version of the lower bound on crossing number (2.3) which we need, and close with the proof of the lower bound on the number of vertices in a planar permutation network (2.4).

The weight-balanced separator is a generalization of the weighted version of the planar separator theorem given in [LT79]. Specifically, the original theorem in [LT79] proves that if every vertex in an n-vertex planar graph has a weight, then there is a set of $O(\sqrt{n})$ edges and vertices whose removal splits the graph into two subsets so that each subset contains at most half the total weight. In the generalization, vertices have several different weights and we want to find a separator which simultaneously splits all the weights in half. The precise statement is as follows.

Theorem 2.1 (Weight-Balanced Separator). Given an n-vertex planar graph where each vertex has a k-vector of weights, the graph can be split into two subsets by removing $O(k\sqrt{n})$ edges and vertices, such that for each component of the weight vector, the total component weight of each subset is at most half the total component weight of the graph.

A weaker form of this theorem was first proved by Leighton in [L82], using a combinatorial result on splitting necklaces of coloured beads. A stronger and very elegant form of the necklace splitting result was proved by Goldberg and West [GW85], though with a rather lengthy and involved proof. Alon and West [AW86] later gave a very simple proof based on the Borsuk-Ulam "ham-sandwich" theorem from topology. The proof of the weight-balanced separator theorem in its full generality can be found in the last two lectures of [LLS89], though in fact the special cases found in [L82], [BL84], and [GW85, p. 104, thm. 4], would suffice for our purposes.

It is well-known, and easy to prove by iteratively applying the original weighted planar separator theorem, that for any p, and any weighted planar graph G of bounded degree, there exist $O(\sqrt{pn})$ edges whose removal splits G into p pieces each having at most $1/p$ of the total weight. Similarly, by using the weight-balanced separator theorem and assigning each vertex a pair of weights, one the vertex's original weight and the second the number of removed edges which are adjacent to the vertex, is not hard to prove the following stronger result. For any p, and any weighted planar graph G of bounded degree, there exist $O(\sqrt{pn})$ edges whose removal splits G into p pieces each having at most $1/p$ of the total weight, and such that each piece is adjacent to $O(1/p)$ of the removed edges. Such decompositions are called fully balanced decompositions and are discussed in detail in the last two lectures of [LLS89] and in [BL84]. Applying the fully balanced decomposition result in the context of planar permutation networks yields the following lemma.

Lemma 2.2 (Balanced Terminal Separator). Given a bounded degree n-vertex planar graph with t terminals each having degree one, there exist $O(\sqrt{nt})$ edges whose removal results in a graph such that each connected component is incident to $O(\sqrt{n/t})$ removed edges and each terminal is its own component.

Proof. First remove the t edges adjacent to terminals. Assign each vertex a weight equal to the number of removed edges adjacent to it. Now taking $p = t$, a fully-balanced decomposition of this weighted graph has the desired properties.∎

Lemma 2.3 (Crossing Pairs).
There exists a constant $c > 0$ such that for each s there is an s-vertex graph H of degree at most 3, such that for each planar embedding of H there are at least cs^2 distinct pairs of edges which cross each other.

Proof. We first note that in any planar embedding of a graph H with the minimum number of edge-crossings, each pair of edges crosses at most once. To see this, suppose we have an embedding and that e and e' are edges which cross each other more than once. Let x and y be consecutive crossings between e and e'. The crossings at x and y can be eliminated by rerouting e' and e so that each follows the other's path between x and y (see Figure 1), and hence the embedding could not have had the minimum number of crossings. Given this observation the lemma follows immediately from the well-known fact that there are expanding graphs of degree 3 and Leighton's quadratic lower bound on the crossing number of expanding graphs [L84]. ∎

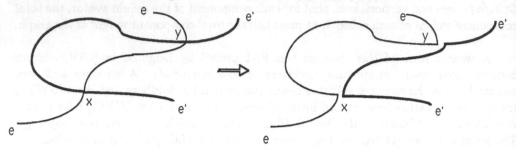

Figure 1

We are now ready to prove the desired lower bound.

Theorem 2.4. If G is a connected n-vertex planar t-permutation network of degree at most 3, then $n = \Omega(t^3)$.

Proof. By the terminal separator lemma, there is a set R of $O(\sqrt{nt})$ edges of G whose removal results in a graph such that each connected component is incident to $O(\sqrt{n/t})$ removed edges and each terminal is its own component. Let $G \backslash R$ be the graph obtained by removing the edges in R from G, and let G^c be the graph of obtained from G by contracting every edge of G which is not in R, and then removing multiple edges. An example is shown in figure 2. It is easy to see that each vertex of G^c corresponds to a connected component of $G \backslash R$, and that G^c is a connected planar graph with $O(\sqrt{n/t})$ maximal degree. We will refer to a vertex of G^c as a **terminal node** if the corresponding connected component of $G \backslash R$ is a terminal. We will assume that we have a fixed embedding of G in the plane.

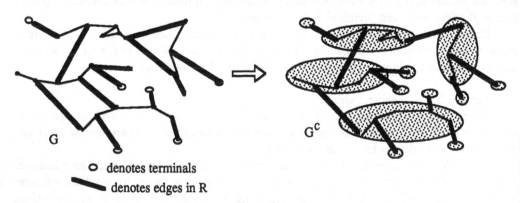

○ denotes terminals

━ denotes edges in R

Figure 2

Let T^c be a subtree of G^c whose leaves are the terminal nodes. Such a tree can be obtained, for example, by taking a spanning tree of G^c and chopping off all branches which contain no terminal nodes. Since each terminal node has degree 1 in G and hence in G^c, it must be a leaf of any spanning tree of G^c and hence the leaves of this tree will be exactly the terminal nodes. Let T be a subtree of G which maps onto T^c, i.e. T is obtained by replacing each edge of T^c with a representative edge in R and replacing each vertex of T^c with a subtree of the component corresponding to that vertex in G^c. An example is shown in figure 3. Note that the leaves of T are the terminals.

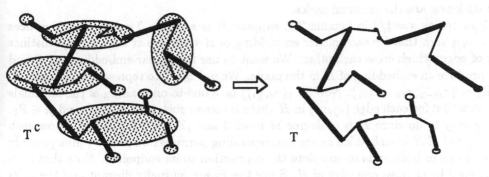

Figure 3

Let Q be a simple curve in the plane connecting the terminals, with Q running alongside the induced embedding of the edges of T in the plane, picking up the terminals as illustrated in Figure 4. Q is assumed to be routed sufficiently closely to T so that it only intersects edges of G when it runs past a vertex of T where it may have to cross an edge in $G \backslash T$ which is adjacent to the vertex. We label the terminals $z_1, \ldots, z_t$ in the order in which they are first visited by Q.

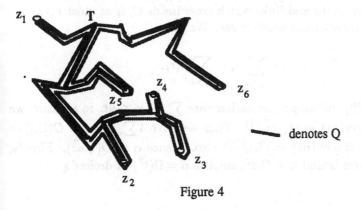

—— denotes Q

Figure 4

We will call each portion of Q joining a pair of consecutive terminals a **link**, and say that a link **runs through a component** of $G \backslash R$ if it runs past some vertex in the component. It will be important to keep in mind that links are not part of any of the graphs but merely simple curves lying in the plane in which the graphs are embedded. Since each link starts and ends at a terminal, and at most two links can run alongside any edge in T, it is easy to see that if y is a vertex of degree d in T^c, then at most $2d$ links can run through the component C_y represented by y. Let $C_1, \ldots, C_m$ be the components of $G \backslash R$, and for each i let n_i be the number of links which run through C_i. We now show that $\sum_{n_i > 4} n_i \le 6t$. Let d_i be the degree of the vertex representing C_i in T^c. We already noted that $n_i \le 2d_i$ and hence it suffices to prove that $\sum_{d_i > 2} n_i \le 3t$. However this is obvious since it is easy to prove that the number of vertices of degree at least 3 in any tree is at most 3 times the number of leaves, and T^c has exactly t leaves since its leaves are the terminal nodes.

Now taking $s = t/3$ in Lemma 2.3, suppose H is a degree 3 graph on $t/3$ vertices $v_1, \ldots, v_{t/3}$ such that for each planar embedding of H there are at least $c(t/3)^2$ distinct pairs of edges which cross each other. We want to use the planar embeddings of G and Q to produce an embedding of H in the plane. We use z_{3h-1} to represent v_h for each h. Let $Z_h = \{z_{3h-2}, z_{3h-1}, z_{3h}\}$. Now let $\{(x_i, y_i)\}$ be a one-to-one pairing of the terminals of G such that for each edge (v_j, v_k) in H there is some i such that $x_i \in Z_j$ and $y_i \in Z_k$. This is easy to do since H is of degree at most 3 and $|Z_h| = 3$ for each h. Now each edge (v_j, v_k) of H is embedded as the corresponding permutation path P_i plus possibly a link at one or both ends to complete the connection to its endpoints. Note that each link is used by at most one edge of H. Since the P_i are mutually disjoint and the links are also mutually disjoint except at possibly the vertices of H, two embedded edges of H can only cross if one of the edges' permutation paths crosses a link used by the other edge. Thus there are $\Omega(t^2)$ distinct pairs of permutation paths and links which cross each other. By the choice of Q, each such crossing can only occur when the link runs past a vertex which is an endpoint of an edge in the permutation path. We will say a permutation path and link cross inside a component of $G \backslash R$ if the vertex is in that component.

For each i let r_i be the number of edges in R incident to the connected component C_i. Since there is at most one permutation path using each edge in R, the number of permutation paths which pass through C_i is at most r_i, and hence the number of distinct pairs of permutation paths and links which cross inside C_i is at most $n_i r_i$. Let α be the total number of distinct pairs which cross. We have

$$\alpha \le \sum_{i=1}^{m} n_i r_i \le 4 \sum_{r_i \le 4} n_i + \sum_{r_i > 4} n_i r_i.$$

Clearly we have $\sum n_i \le 2|R|$, and we proved earlier that $\sum_{r_i > 4} r_i \le 6t$. In addition we have $|R| = O(\sqrt{nt})$ and $\max\{n_i\} = O(\sqrt{n/t})$. Thus we have $4 \sum_{r_i \le 4} n_i = O(|R|) = O(\sqrt{nt})$ and $\sum_{r_i > 4} n_i r_i \le 6t \max\{n_i\} = O(\sqrt{nt})$ also. Hence $\alpha = O(\sqrt{nt})$. Finally, combining this with the lower bound $\alpha = \Omega(t^2)$ implies $n = \Omega(t^3)$ as desired.∎

3. REFERENCES

[A87] N. Alon, Splitting necklaces, Advances in Math. 63(1987) pp.247-253.

[AM85] N. Alon and V. Milman, λ_1, isoperimetric inequalities for graphs, and superconcentrators, J. Comb. Theory B 38(1985), pp. 73 - 88.

[AKS90] A. Aggarwal, M. Klawe and P.Shor, Multi-layer grid embeddings for VLSI, to appear in Algorithmica.

[AKLLW85] A. Aggarwal, M. Klawe, D. Lichtenstein, N. Linial, and A. Wigderson, Multilayer grid embeddings, Proc. IEEE 26th Ann. Symp. Found. Comp. Sci., 1985, pp. 186 - 196.

[AKLLW90] A. Aggarwal, M. Klawe, D. Lichtenstein, N. Linial, and A. Wigderson, A lower bound on the area of permutation layouts,to appear in Algorithmica.

[AW86] N. Alon and D. West, The Borsuk-Ulam theorem and bisection of necklaces, Proc. Amer. Math. Soc., 98(1986) pp. 623-628.

[BL84] S. Bhatt and F. T. Leighton, A Framework for Solving VLSI Graph Layout Problems, JCSS Vol 28 No.2 April 1984, pp. 300 - 343.

[BP77] N. K. Bose and K. A. Prabhu, Thickness of Graphs with Degree Constrained Vertices, IEEE Trans. on Circuits and Systems, Vol. CAS-24, No. 4, April 1977, pp. 184-190.

[CS78] M. Cutler and Y. Shiloach, Permutation Layout, Networks, Vol. 8, 1978, pp. 253-278.

[GW85] C.H.Goldberg and D.B. West, Bisection of circle colorings, SIAM J. Alg. Disc. Math., 6(1985) pp. 93-106.

[KKF79] E. S. Kuh, T. Kashiwabara, and T. Fujisawa, On Optimum Single Row-Routing, IEEE Trans. on Circuits and Systems, Vol. CAS-26, No. 6, June 1979, pp. 361-368.

[Ka80] Y. Kajitani, On Via Hole Minimization of Routings on a Two-Layer Board, Tech. Report, Dept. of Elec. Engineering, Tokyo Institute of Technology, Japan, 1980.

[L81] F. T. Leighton, Layouts for the Shuffle-Exchange Graph and Lower Bound Techniques for VLSI, Ph. D. thesis, Dept. of Math., MIT, 1981.

[L82] F. T. Leighton, A layout strategy for VLSI which is provably good, Proc. ACM Ann. Symp. Theory of Comp., 1982, pp. 85-98.

[L84] F. T. Leighton, New Lower Bound Techniques for VLSI, Math Systems Theory 17(1984), pp. 47-70.

[LLS89] F.T.Leighton, C.E.Leiserson and E.Schwabe, Theory of parallel and VLSI computation, MIT/LCS/RRS 6 Research Seminar Series, Lecture Notes for 18.435 /6.848, March 89.

[LT79] R. J. Lipton and R. E. Tarjan, A Separator Theorem for Planar Graphs, SIAM J. Appl. Math., 36 (1979) 177-189.

[Ri82] R. Rivest, The Placement and Interconnect System, Proc. 19th Design Automation Conference, 1982, pp. 475-481.

[Sh80] I. Shirakawa, Letter to the Editor : Some Comments on Permutation Layout, Networks, Vol. 10, 1980, pp. 179 - 182.

[So74] H.C. So, Some Theoretical Results on the Routing of Multilayer Printed Wiring Boards, 1974 IEEE Int. Symp. on Circuits and Systems, pp. 296-303.

[TKS76] B. S. Ting, E. S. Kuh, and I. Shirakawa, The Multilayer Routing Problem: Algorithms and Necessary and Sufficient Conditions for the SIngle-Row Single Layer Case, IEEE Trans. of Circuits and Systems, Vol. CAS-23, No. 12, 1979, pp. 768-778.

[TK82] S. Tsukiyama and E. S. Kuh, Double-Row Planar routing and Permutation Layout, Networks, Vol. 12, 1982, pp. 287-316.

SIMULTANEOUS SOLUTION OF FAMILIES OF PROBLEMS

by
Refael Hassin

School of Mathematical Sciences
Tel-Aviv University
Tel-Aviv 69978
Israel

Abstract

We are interested in a situation where a large number of problems of some common type must be solved. The problems are defined on some space of "solutions", share the same objective function to be optimized, and differ by their sets of constraints. Thus a solution may be feasible with respect to a set of problems and infeasible with respect to the others. We investigate the possibility of solving the problems simultaneously and presenting the optimal solutions compactly, so that each of these solutions can be easily obtained when needed.

1. Introduction

We are interested in the solution of a set of problems defined on a common solution space. One can imagine a set of candidate *solutions* such as a set of points in R^n. Each solution is associated with a *cost*. There is also a set of *problems* and each of them is characterized by a set of *feasible solutions*. The *value* of a problem is defined to be the minimum cost of a feasible solution to that problem. The solution is then said to be *optimal.*

Solutions with "low" cost will typically be optimal for several problems. For example, a solution with a minimum cost among all candidate solutions will be optimal for all of the problems for which it is feasible. In the following we will investigate the impact of such properties, and show how to exploit them in characterizing and computing optimal solutions to families of problems. We will survey several results which have been developed elsewhere, in particular by Hassin (1988,1989,1990), and extend them. In some cases we will omit the proof of a theorem, and the interested reader is referred to the above mentioned sources.

Situations resembling the structure outlined in the opening paragraph are frequent also in other contexts. A natural example is that where a problem is parametrically solved in order to obtain sensitivity analysis or to solve a more complicated problem. Another is that where the set of constraints changes with time and at each period the problem has to be solved subject to a new set of requirements. To be more specific, let us present an example:

Example 1.1. *Let $G = (N, E)$ be a graph with a vertex set N and an edge set E. Suppose that each edge in E is given a length. Let s and t be given distinct vertices of N. One is often faced with the problem of computing a shortest $s - t$ path in G, as for the purpose of sending a message. Suppose now that the edges of E are not reliable, and some of them may be unserviceable at a given instant. Suppose that whenever a message has to be sent, the decision maker knows the set of serviceable edges. This defines a family of $2^{|E|}$ shortest path problems, one for each possible set of unserviceable edges.*

In Example 1.1, the number of problems is very large, and consequently one may find it best to solve the relevant problem each time without storing past results. A little reflection reveals however that many of these problems share common optimal solutions. For example, let P be the optimal path obtained when all of the edges in E are serviceable. Then clearly P is optimal also for each problem whose set of serviceable edges contains P. This set of problems itself is of a very large size. One can continue this way of thinking by computing the k shortest paths in the graph, for some predetermined k. Given a problem (i.e., a set of serviceable edges), we scan the k best paths, starting from the best one, till we find for the first time one which is feasible for it (i.e., it contains only serviceable edges), and in such a case the best feasible path is optimal for the problem. If k is large enough, we obtain a situation where the set of k best paths "covers" all of the problems, in the sense that it contains a feasible solution for each. Thus, this set contains sufficient information needed to determine an optimal solution for each possible problem. We note however that the size of k may be unnecessarily large since some of the k best paths will not be optimal for any problem, and their proportion in the list may quickly increase with k.

We are concerned therefore with several questions. First, how many distinct optimal solutions may be needed in the worst case to form a cover of the set of possible problems. Second, we are interested in representing this set of solutions compactly, whenever such a representation exists. Finally, we seek an efficient algorithm for computing such a compact representation.

It should be emphasized that it is often the case that the problems in a family are revealed and solved *sequentially*. This is usually the case when the k best solutions must be computed (see, for example, Katoh, Ibaraki, and Mine 1982). Here, in contrast, we deal with the *simultaneous* solution of a set of problems, all of which are known in advance.

2. The Number of Optimal Solutions

Let v be a real valued function on a set X. For simplicity, we assume in the following that X is finite. Let X_i, $i \in S$ be nonempty subsets of X, where S is an index set with finite cardinality. Define the *value* of X_i by

$$V_i = min\{v(x)|x \in X_i\}.$$

In the following the set X will be a finite set of possible solutions to a family of problems, $v(x)$ the value (cost) of x, X_i the set of feasible solutions to Problem i, $i \in S$, and V the value of an optimal solution to Problem i. Let A be an $|S| \times |X|$ binary matrix $A = (a_i(x))$ defining the feasibility relations between X and S; $a_i(x) = 1$ if and only if solution x is feasible for Problem i. We call A a *solution matrix*.

For a given solution matrix, A, let $M(S)$ be the maximum possible number of distinct values in $\{V_i|i \in S\}$, over all functions v. Call a square matrix $B = (b_{ij})$ *triangular* if QBP is triangular in the regular sense, for some permutation matrices Q and P. In other words, we call B triangular if its rows and columns can be permuted so that $b_{ij} = 0$ in the resulting matrix for $j > i$. We call B *proper* if QBP also has $b_{ij} = 1$ for all $i = j$. Let $T(A)$ denote the maximum order of a proper triangular submatrix of A.

Theorem 2.1.

$$M(S) = T(A).$$

Proof:

(i) Let B be a proper triangular submatrix of A of size $m \times m$. Suppose that the columns and rows of B were permuted to make it triangular with a unit main diagonal. Let x_j be the solution corresponding to its j-th column. Let s_i be the problem corresponding to its i-th row. Choose a cost function v such that $v(x_j)$ strictly decreases with j. Then $x_{j+1}, ..., x_m$ are not feasible for s_j while x_j is feasible for this problem. Therefore x_j is optimal for s_j. This shows that $M(S) \geq m$. Choosing B to be of maximum order we obtain that $M(S) \geq T(A)$.

(ii) Suppose that $M(S)$ is realized by a function v and a set $x_1, ..., x_{M(S)}$ of solutions with distinct values $v(x_1) > v(x_2) >, ..., > v(x_{M(S)})$. Let s_i be a problem solved by x_i, $i = 1, ..., M(S)$. Then clearly x_j is not feasible for s_i for all $i < j$ and hence a proper triangular submatrix of A whose order is $M(S)$ exists. It follows that $T(A) \geq M(S)$. $\qquad\qquad\square$

Let us illustrate the possible consequences of Theorem 2.1 by applying it to a generalization of Example 1.1.

Example 2.2. *Let E be a set of elements. Let $S_1, ..., S_r$ be a collection of subsets of E. Let $c(X)$ be a real function defined on the subsets X of E. Let Problem i $i = 1, ..., r$, require to compute $\min\{c(X) | X \cap S_i = \phi\}$.*

To connect Example 2.2 to 1.1, consider the sets S_i as the sets of edges that are not serviceable in Problem i, $c(X)$ is the total length of the edges of X if they constitute an $s - t$ path, and $c(X) = \infty$ otherwise. Clearly it suffices to restrict the search to subsets X such that $|X| \leq |N| - 1$.

Conjecture 2.3. *Suppose that $|S_i| \leq k$ $i = 1, ..., r$, and $c(X) = \infty$ for subsets X such that $|X| > n$. Suppose that $|E| \geq n + k$. Then $T(A) = \binom{n+k}{k}$.*

Note that Conjecture 2.3 claims that $T(A)$ is independent of $|E|$. If correct, it generalizes a theorem of Jaeger and Payan (see, Berge 1973, Ch. 18).

In the following sections we consider upper bounds on $T(A)$ and additional properties of the sets of optimal solutions which lead, in some important cases, to an efficient algorithm for constructing a compact representation of the set of optimal solutions.

3. Solution Bases

Call a set $S' \subseteq S$ *dependent* if there exists $S'' \subseteq S'$ such that $\sum_{i \in S''} a_i(x) = 0$ $(mod 2)$ $\forall x \in X$. Otherwise S' is *independent*. Thus we use the regular definition of linear dependence in the binary field, and S' is independent if and only if the corresponding rows in A are independent. Let $r(A)$, the *rank* of A, be the maximum cardinality of an independent subset of S. With Theorem 2.1 we obtain the following:

Corollary 3.1. $M(S) \leq r(A)$.

Proof: The rows corresponding to a proper triangular submatrix are clearly independent, therefore $T(A) \leq r(A)$. The corollary follows now from Theorem 2.1. $\qquad\square$

Corollary 3.1 is a powerful tool. For example one may use it to obtain the following matroidal result.

Corollary 3.2. *Let A be the cocycle incidence matrix of a matroid $\mathcal{M} = (\mathcal{E}, \mathcal{F})$. Associate with each cocycle C a value $v(C)$. Associate with each element $e \in \mathcal{E}$ a value $V(e) = \min\{v(C) | e \in C\}$. Then the number of distinct element-values is bounded by the rank of $\mathcal{M}$.*

Let $S' \subseteq S$ correspond to a maximal set of independent rows of A, then we call S' a *solution basis*. The cardinality of each solution basis is equal to $r(A)$. Define the *value of a solution basis* S' as $\sum_{i \in S'} V_i$. A *maximum solution basis* is a solution basis with maximum value.

Theorem 3.3. *Let $S' \subseteq S$ be a maximum solution basis. Let $k \in S \backslash S'$, and let $S'' \subseteq S'$ satisfy $a_k(x) = \sum_{i \in S''} a_i(x) \pmod 2 \ \forall x \in X$. Then $V_k = min\{V_i | i \in S''\}$ and there exists $p \in S''$ and $y \in X_p \cap X_k$ such that $v(y) = V_k$.*

Proof: By construction, for every $x \in X_k$ there exists $p \in S''$ such that $x \in X_p$. Therefore $V_k \geq min\{V_i | i \in S''\}$. This inequality cannot be strict since if $V_k > V_p$ for some $p \in S''$ then $(S' \cup \{k\}) \backslash \{p\}$ is a solution basis with value larger than that of S'. Therefore equality holds and there exists $p \in S''$ and $y \in X_k \cap X_p$ as claimed. $\square$

In view of Theorem 3.3, a maximum solution basis contains all the information needed to compute V_k for every $k \in S$. It also gives clues to locate the element of X_k with the value V_k. As a matter of fact, if the values $v(x)$ are distinct for all $x \in X$ then this element is the unique element $y \in X$ satisfying $v(y) = min\{V_i | i \in S''\}$.

When all of the values $V_i \ i \in S$ are given, a maximum solution basis can be computed by the greedy algorithm: Suppose $V_1 \geq V_2 \geq, ..., V_{|S|}$. Insert $i \in S$ to the basis if it does not form a minimal dependent set with any subset $S' \subseteq \{1, ..., i-1\}$.

However, as will be shown in Sections 5 and 6, we need not compute directly all of the values $V_i \ i \in S$ to form a maximum solution basis.

4. Applications to Multiterminal Cut Problems

In this section we apply Corollary 3.1 to a class of problems that are often considered in connection with network flows. Let $N = \{1, ..., n\}$ be a given set. A *cut* of N is a partition (I, J) of N into two disjoint nonempty subsets I and J. For each cut (I, J) let $c(I, J)$ be its *cost* (or *value*), where c is an arbitrary real function defined on the cuts of N. We assume that c is symmetric, i.e., $c(I, J) = c(J, I)$. For a given pair of distinct elements $i, j \in N$ an $i - j$ *cut* is a cut (I, J) such that either $i \in I$ and $j \in J$, or $i \in J$ and $j \in I$. The *minimum $i - j$ cut problem* is to find the $i - j$ cut of minimum cost. An example for such a problem is a scheduling problem on two machines with a constraint that a given pair of jobs must be processed on different machines. A possible objective function is to minimize the makespan. In this case the cost of a cut (I, J) will be $max\{c(I), c(J)\}$ where, for $M \subset N$, $c(M) = \sum_{i \in M} t_i$ and t_i is the processing requirement for job i. A *multiterminal* cut problem requires to solve a set of $i - j$ problems for a set of $i - j$ pairs.

Gomory and Hu (1961) analyzed the multiterminal cut problem for a special additive cost function. Since problems with such costs usually arise with connection to graphs we call them *graphic* problems. Gomory and Hu proved that the $\binom{n}{2}$ graphic cut problems have at most $n - 1$ distinct solutions. They also constructed a data structure a (*cut tree*) that stores these values in a way that allows to determine the solution to any of the problems in linear time. They proved that a cut tree can be constructed from a set of $n - 1$ *noncrossing* cuts, and use this property to compute a cut tree by solving only $n - 1$ cut problems.

The above results were extended by Gusfield and Naor (1988), using a construction of Picard and Queyranne (1980), to a compact representation of *all* of the minimum cuts for each pair of elements. Theorem 4.1 below extends Gomory and Hu's seminal results as well as later extensions by Schnorr (1979) and Cheng and Hu (1988).

In a *k-cut problem*, each partition of N into k disjoint subsets $I_1, ..., I_k$ such that $\cup_{j=1}^{k} I_j = N$ is associated with a cost. In a given problem a set $T = (t_1, ..., t_k) \subseteq N$ is given, and it is required to find a partition of minimum cost such that $t_j \in I_j$ $j = 1, ...k$. The *multiterminal k-cut problem* requires to solve the k-cut problems for or all possible $T \subseteq N$ such that $|T| = k$.

Theorem 4.1. *For the multiterminal k-cut problem* $r(A) = \binom{n-1}{k-1}$.

Proof: We prove the theorem by showing that the submatrix B of A corresponding to the problems $\{Q_T | \ |T| = k, \ 1 \in T\}$ constitutes a solution basis.

1. Independence of the rows of B follows from the following observation:

Let $T = \{t_1, ..., t_{k-1}, 1\}$. Then the k-cut $I_j = \{t_j\}$ $j = 1, ..., k-1$ and $I_k = N \backslash \{t_1, ..., t_{k-1}\}$ is feasible for Q_T but not for any other problem corresponding to a row of B.

2. To prove that B is a basis, we must show that each other row of A not in B forms a dependent set with a subset of rows of B. Let $T = \{t_1, ..., t_k\} \subseteq N$, with $1 \in N \backslash T$ and let $T(j) = T \backslash \{t_j\} \cup \{1\}$. We claim that the sum (mod 2) of the k rows corresponding to $Q_{T(j)}$ $j = 1, ..., k$ is the row of T. This is implied by the following observations:

(i) Consider any feasible k-cut $I_1, ..., I_k$ of Q_T where $t_j \in I_j$ $j = 1, ..., k$. This cut is feasible to $Q_{T(j)}$ if and only if $1 \in I_j$. Therefore, every "1" in the Q_T-row appears in the same column in exactly one $Q_{T(j)}$-row.

(ii) Consider a k-cut $I_1, ..., I_k$ with $t_j \in I_j$ $j \neq i$, and $1 \in I_i$, which is feasible to problem $Q_{T(i)}$ but not for Q_T. Suppose $t_i \in I_l$, then of all problems $Q_{T(j)}$ this solution is feasible only for $Q_{T(i)}$ and $Q_{T(l)}$. Therefore, if the Q_T-row has "0" in a certain column then this column has "1" in either two or none of the $Q_{T(j)}$-rows.   □

Corollary 4.2. *In the multiterminal k-cut problem* $M(S) \leq \binom{n-1}{k-1}$.

In a *k-pair cut problem*, k distinct pairs $\{s_i, t_i\}$ such that $s_i \neq t_i$ $i = 1, ...k$ are given: A k-pair cut is a partition $(I, N \backslash I)$ which is an $s_i - t_i$ cut for all $i = 1, ..., k$. The problem is to obtain the k-pair cut of minimum cost. The *multiterminal k-pair problem*, requires solving all of the m-pair problems with $m \leq k$. Theorem 4.3 bounds $M(S)$ for this problem. For example, it shows that the $\frac{1}{2} \binom{n}{2} [\binom{n}{2} + 1]$ 2-pair problems may have at most $(n-1) + \binom{n-1}{2} = \binom{n}{2}$ distinct solution values.

Theorem 4.3. *For the multiterminal k-pair problem*

$$M(S) \leq \sum_{m=1}^{min(k, n-1)} \binom{n-1}{m}.$$

Define an $i - j$ *zcut* to be a cut (S, T) such that either $i, j \in S$ or $i, j \in T$. A *minimum $i - j$ zcut problem* is to find the $i - j$ xcut of minimum cost. An example for such a problem is a scheduling problem on two machines with a constraint that a given pair of jobs must be processed on the same machine.

Theorem 4.4. *For the multiterminal xcut problem*

$$M(S) \leq n.$$

Remark 4.5. *It can be shown that all of the above bounds are tight.*

Remark 4.6. *Granot and Hassin (1986) extended Gomory and Hu's theorem to graphic cut problems with both node and edge capacities. They present an algorithm for this case, accompanied by a theorem stating that the number of distinct optimal solutions is at most $n - 1$, the same bound as for the problem with edge-capacities only. We note that the theorem is not a special case of either Corollary 4.2 or Theorem 4.3. However, it can also be obtained from Corollary 3.1 by identifying an appropriate solution basis.*

5. Computing Maximum Solution Bases

The algorithm of Gomory and Hu for the graphic multiterminal 2-cut problem relies on a noncrossing property of the optimal cuts (see, Section 7). Gusfield (1990) was the first to observe that this property, though of much interest, is not essential for computing cut trees efficiently. Gusfield used his observation to simplify the algorithm. Cheng and Hu (1988) developed an equally efficient algorithm for solving general (not just graphic) multiterminal 2-cut problems. Their complexity is an improvement over an earlier result by Hassin (1988) for this problem. In this section we further extend this type of results.

Bases of a matrix, and in particular solution bases, define matroids, and consequently the greedy algorithm can be used to compute a maximum solution basis. This fact underlines the algorithm below. To simplify the algorithm we make the following assumption:

Assumption 5.1. *Distinct solutions have distinct costs.*

The algorithm consists of two parts. In the first an initial set of solutions, X', is generated. Each member of this set is optimal for some problem in S. The problems to be solved are chosen sequentially so that none of the previously generated solutions is feasible for the current problem. This guarantees a new optimal solution in each iteration. This initialization algorithm terminates when X' is a *cover* of S in the sense of the following definition:

Definition 5.2. *A subset $X' \subset X$ is said to cover a subset $S' \subset S$ if for every $j \in S'$ X' contains a j-feasible solution.*

The second, and main, part of the algorithm applies a greedy algorithm to compute a maximum solution basis of A. The greedy algorithm finds in each iteration a problem of maximum value whose row in A can be added to the current set of rows and maintain independence of this set.

Algorithm 5.3.

Input: A.

Output: $X' \subset X$ [a cover of S].

5.3.1 Set $X' = \phi$, $S' = S$.

5.3.2 If $S' = \phi$, stop. Else, choose $i \in S'$.

5.3.3 Solve i to obtain an i-optimal solution $\hat{x}$.

5.3.4 Set $X \leftarrow X \cup \{\hat{x}\}$.

Set $S' \leftarrow S' \setminus \{j \in S' | a_j(\hat{x}) = 1\}$.

Go to 5.3.2.

Algorithm 5.4.

Input: A, $X' \subset X$ [a cover of S].

Output: T [a maximum solution basis of A].

5.4.1 Set $T = \phi$.

5.4.2 Set $S' = \{j \in S | T \cup \{j\}$ is independent$\}$.

If $S' = \phi$, stop.

5.4.3 For each $j \in S'$ let $V_j' = \min\{v(x) | x \in X', a_j(x) = 1\}$.

Let $i \in S'$ satisfy $V_i' = \max\{V_j' | j \in S'\}$.

5.4.4 Solve i to obtain an i-optimal solution $\hat{x}$.

Case (i). If $\hat{x} \notin X'$, set $X' \leftarrow X' \cup \{\hat{x}\}$ and go to 5.4.3.

Case (ii). If $\hat{x} \in X'$, set $T \leftarrow T \cup \{i\}$ and go to 5.4.2.

Theorem 5.5. *Algorithm 5.4 outputs a maximum solution basis.*

Proof: The set of solution bases is exactly the set of bases of the binary matroid defined by A. Hence the greedy algorithm computes an optimal solution. To show that Algorithm 5.4 is indeed the greedy algorithm we must prove that in Case (ii) of 5.4.4. $V_i = \max\{V_j | j \in S'\}$. Case (ii) obtains when $\hat{x} \in X'$. Clearly, for all $j \in S'$ $V_j' \geq V_j$, so that $V_i = v(\hat{x}) = V_i' = \max\{V_j' | j \in S'\} \geq \max\{V_j | j \in S'\}$. Since $i \in S'$, equality holds, as required. □

Remark 5.6. *The total number of problems solved in 5.3.3 and 5.4.4 is at most $M(S) + r(A)$, since each problem either has a distinct optimal solution that is used to augment X' (5.3.3 and Case (i) of 5.4.4), or the problem is used to augment the independent set T (Case (ii) of 5.4.4). By Assumption 5.1, the first case can occur at most $M(S)$ times, and by the definition of $r(A)$, the latter can occur at most $r(A)$ times. By Corollary 3.1, the total is at most $2r(A)$.*

6. Representations of Solution Bases

Algorithms 5.3 and 5.4 require $O(|S|r(A))$ operations to execute 5.3.4 and 5.4.3 (note that, after the first time 5.4.3 is executed, V_j' is revised by a single comparison to $v(\hat{x})$ whenever $a_j(\hat{x}) = 1$). Most of the effort may be devoted to solve the $O(r(A))$ single problems in 5.3.3 and 5.4.4, and to construct the sets S' in 5.4.2. Our results are especially useful when the latter effort is small relative to the savings resulting from the reduction in the number of single problems to be solved. For this to be the case one needs to have a good characterization of the matroid defined by the solution matrix so that independence of problems in Step 5.4.2 can be tested efficiently. So far we were able to achieve this goal only in some cases of interest.

To describe such a case we need the following definition.

Definition 6.1. *Let $G = (N, E)$ be a graph with a vertex set N and an (undirected) edge set E. A 2-forest of G is a subgraph of G that either has no cycles or has exactly one cycle and this cycle is of odd cardinality. Let $\mathcal{X}$ be the set of 2-forests of G. Then $\mathcal{M}_2 = (E, \mathcal{X})$ is a matroid (the 2-forest matroid). Its rank (i.e., the cardinality of its bases, the maximal 2-forests) is equal to $|N| - p - \alpha$, where $p + 1$ is the number of components of G and $\alpha = 1$ if G is bipartite, 0 otherwise (see, Conforti and Rao 1987).*

Remark 6.2. *If G is a complete graph with $n \geq 3$ then the rank of $\mathcal{M}_2$ is just $|N|$.*

Let N be a given set. Let E_c and E_x be (not necessarily disjoint) subsets of pairs of elements of N. Consider the multiterminal problem requiring to solve the $i - j$ cut problems for each $i, j \in E_c$ and the $i - j$ xcut problems for each $i, j \in E_x$. We call this problem a *mixed multiterminal problem*. The requirements of the problem can be described by means of a multigraph $\hat{G} = (\hat{N}, \hat{E})$ with a vertex set $\hat{N} = \{1, ..., n\}$ and an edge set $\hat{E} = E_c \cup E_x$. We denote the edges corresponding to E_c as *c-edges* and those corresponding to E_x as *x-edges*. Note that $\hat{G}$ may have pairs of parallel x- and c-edges.

Let A be the solution matrix corresponding to the set of problems defined by $\hat{E}$. As $\hat{G}$ may have parallel edges it should be noted that by a *cycle* of $\hat{G}$ we refer to a sequence of *edges*. We call a cycle *odd* if it has an odd number of x-edges. Otherwise it is *even*. Note that a pair of parallel x- and c-edges defines an odd cycle.

Theorem 6.3. *The set of problems associated with $E \subset \hat{E}$ corresponds to an independent set of rows in A if and only if it is independent in the 2-forest matroid $\mathcal{M}_2$. [Hence, the set of solution bases for the mixed multiterminal problem corresponds to the set of bases of the 2-forest matroid of $\hat{G}$ (i.e., the maximal 2-forests).]*

Let $\hat{G}_x$ be the graph obtained from $\hat{G}$ by condensing all of its c-edges (i.e., considering the end vertices of each c-edge as a single vertex). Note that a pair of parallel c- and x-edges of $\hat{G}$ generates a loop in $\hat{G}_x$ in which case the latter graph is not bipartite.

Corollary 6.4. *Let A be the solution matrix corresponding to a set $\hat{E}$. Let $p + 1$ be the number of components of $\hat{G}$ and $\alpha = 1$ if $\hat{G}_x$ is bipartite, 0 otherwise. Then $r(A) = n - p - \alpha$.*

Remark 6.5. *For the pure 2-cut case clearly $\hat{G}_x$ is bipartite so that $\alpha = 1$. Therefore, a set of problems is independent if and only if the edges of G associated with it do not contain a cycle.*

It is also interesting to note that the least upper bound on the number of distinct solutions in the mixed multiterminal problem which contains all of the $n(n - 1)$ cut and xcut problems is n, as for the pure multiterminal xcut problem.

It follows from Theorem 6.3 that for the mixed multiterminal problem the sets S' of 5.4.2 are easily constructed in polynomial time, without explicitly scanning A. The number of problems to be solved is $O(n)$. This may be a considerable saving in computation time, especially when each problem is NP-hard (for example, in the max-cut case).

7. Graphic Problems

In graphic cut problems the cost function is compactly given by a set of edge-costs σ_{ij} defined on the edge set of a graph. The cost of a cut (I, J) is then $c(I, J) = \sum_{i \in I, j \in J} \sigma_{ij}$.

Graphic cut problems possess a special property. It comes out that this property holds for graphic xcuts as well.

Definition 7.1. *Two cuts (I, J) and (I', J') are said to cross each other if and only if each of the four sets $I \cap I'$, $I \cap J'$, $J \cap I'$, $J \cap J'$ is nonempty.*

Theorem 7.2 (Gomory and Hu 1961). *In a graphic multiterminal cut problem, there is a set of optimal cuts that do not cross.*

Theorem 7.3. *In a graphic multiterminal xcut problem, there is a set of optimal xcuts that do not cross.*

For a set $I \subset N$ we define the $i - I$ xcut problem as the one obtained when the elements of $I \cup \{i\}$ are restricted to be in the same subset of the cut. Let $c_{min} = \min\{c_{ij} | i, j \in N, i \neq j\}$ where c_{ij} is the solution value for the $i - j$ cut problem. Algorithm 7.4 easily follows from Theorem 7.3.

Algorithm 7.4. *(Multiterminal graphic xcut algorithm)*

1. Compute a cut (I, J) with value $c(I, J) = c_{min}$. This is the optimal $i - j$ xcut for all i, j pairs such that $i, j \in I$ or $i, j \in J$.
2. For every $i \in I$ compute an optimal $i - J$ xcut. Let the value of this cut be c_{iJ}. [Execute this step only if $\{i\} \cup J \neq N$, else set $c_{iJ} = \infty$.]
3. For every $j \in J$ compute an optimal $I - j$ xcut. Let the value of this cut be c_{Ij}. [Execute this step only if $\{j\} \cup I \neq N$, else set $c_{Ij} = \infty$.]
4. For every $i \in I$ and $j \in J$ the optimal $i - j$ xcut has value $c_{ij} = \min\{c_{iJ}, c_{Ij}\}$ and an optimal cut is the one producing this value (i.e, the optimal $i - J$ or $I - j$ xcut).

Algorithm 7.4 computes n optimal solutions in addition to the computation of c_{min} in Step 1. The latter can be computed, for example, by solving the multiterminal cut problem on the same network by applying Gomory and Hu's algorithm. If the edge-costs are identical, this subproblem can be solved more efficiently by applying Matula's algorithm (Matula 1987).

8. Concluding Remarks

The triangularity property used in Theorem 2.1 is not computationally useful except for special cases. As a matter of fact, it follows from the work of Yannakakis (1981) and Bartholdi (1982) that even the problem of computing a maximum proper triangular submatrix of a given matrix is NP-hard. In particular, the (greedy) approach where rows are added to a list in an arbitrary order as long as the resulting matrix contains a full rank proper triangular submatrix will not work, as demonstrated by the following example:

$$
\begin{array}{c}
\\
1\\
2\\
3\\
4\\
5\\
6
\end{array}
\begin{array}{cccccc}
\alpha & \beta & \gamma & \delta & \epsilon & \phi \\
\left(\begin{array}{cccccc}
1 & 1 & 0 & 1 & 1 & 1 \\
0 & 1 & 1 & 1 & 1 & 1 \\
1 & 0 & 1 & 1 & 1 & 1 \\
1 & 1 & 1 & 1 & 0 & 0 \\
1 & 1 & 1 & 0 & 1 & 0 \\
1 & 1 & 1 & 0 & 0 & 1
\end{array}\right)
\end{array}
$$

Rows 1 and 2 contain a full rank proper triangular submatrix (with columns β and γ). With any other row added no such submatrix will exist. However, Rows 4, 5, and 6 contain such a submatrix (with Columns δ, ϵ, and ϕ). Therefore, $T(A) = 3$ while the greedy process may stop at a submatrix of order 2.

In view of the above remarks, the question of how many distinct solutions a family of problems may have is only partially answered in this report.

More importantly, we have shown that a maximum solution basis essentially contains all the information needed to solve the problems in the family. However, our algorithm heavily relies on our ability to characterize the underlying matroid and use this characterization for efficient independence tests. In Section 6 we have demonstrated how this can be done for an important case, however we feel that this is just a starting point to a rich terrain for future research. We believe that the investigation of such matroids for additional families of problems will lead not just to efficient algorithms for these problems but also to discoveries of new interesting types of matroids.

A promising direction is doing so for the multiterminal 2-pair cut problem on undirected networks which is associated with the 2-commodity multiterminal flow problem (Hu 1963). Viewing cuts as sets of edges, a problem is given by a pair of (not necessarily distinct) edges, and a feasible solution is a cut containing both of them. As follows from Theorem 4.3 (and the paragraph preceding it), the rank of this matroid is equal to the number of edges in the graph. Another case of interest is the matroid defined by the solution matrix of Example 2.2 and associated with Conjecture 3.2.

It is interesting to note that for graphic cuts, if the cost function is not "degenerate" (i.e., Assumption 5.1 holds), then the bound given by Corollary 3.1 is tight (i.e., $M(S) = r(A) = n - 1$). This property does not hold in general even for 2-cuts or for graphic xcuts. For example consider the graphic xcut problem with $|N| = |E| = 4$, $\sigma_{12} = 5$, $\sigma_{23} = 9$, $\sigma_{34} = 7$, and $\sigma_{14} = 6$.

$\{1\}, \{2, 3, 4\}$ is an optimal $2 - 3$, $2 - 4$ and $3 - 4$ xcut. $\{1, 4\}, \{2, 3\}$ is an optimal $1 - 4$ xcut, and $\{1, 2, 3\}, \{4\}$ is an optimal $1 - 2$ and $1 - 3$ xcut. Thus in this problem $n = 4$ but there are only 3 distinct optimal solutions. Note that if σ_{34} is changed to 8.5 then there will be 4 distinct optimal solutions as $(\{2\}, \{1, 3, 4\})$ will be the $1 - 3$ optimal xcut.

The question of under what conditions Assumption 5.1 guarantees that $M(S) = r(A)$ is open. We conjecture that this is the case at least for the multiterminal 2-pair cut problem.

References

Bartholdi, J.J. (1982), "A Good Submatrix is Hard to Find", *Operations Research Letters* 5, 190-193.

Berge, C. (1973), *Graphs and Hypergraphs*, North Holland Publishing Co., Amsterdam.

Cheng C.K. and T.C. Hu (1988), "Maximum Concurrent Flow and Minimum Ratio Cut", Technical Report Number CS88-141, department of Computer Science and Engineering, University of California, San Diego, December 1988.

Conforti, M. and M.R. Rao (1987), "Some New Matroids on Graphs: Cut Sets and the Max Cut Problem", *Math. of Operations Research* 12, 193-204.

Gomory, R.E., and T.C. Hu (1961), "Multi-Terminal Network Flows", *J. SIAM*, 9, 551-570.

Granot, F. and R. Hassin (1986), "Multi-Terminal Maximum Flows in Node Capacitated Networks",*Discrete Applied Mathematics* 13, 157-163.

Gusfield, D (1990), "Very Simple Methods for All Pairs Network Flow Analysis", *SIAM Journal on Computing* 19, 143-155.

Gusfield, D., and D. Naor (1988), "Extracting Maximal Information on Sets of Minimum Cuts", Computer Science Division, University of California, Davis.

Gusfield, D., and D. Naor (1990), "Efficient Algorithms for Generalized Cut Trees" Proceedings of the First Annual ACM-SIAM Symposium on Discrete Algorithms.

Hassin, R. (1988), "Solution Bases of Multiterminal Cut Problems", *Mathematics of Operations Research* 13, 535-542.

Hassin, R. (1989), "Multiterminal Xcut Problems", presented in the NATO Workshop on *Topological Network Design*, Copenhagen, June 1989.

Hassin, R. (1990), "An Algorithm for Computing Maximum Solution Bases", *Operations Research letters*.

Hu, T.C. (1963), "Multi-Commodity Network Flows", *Operations Research* 11, 344-360.

Katoh, N., T. Ibaraki, and H. Mine (1982), "An Efficient Algorithm for K Shortest Simple Paths", *Networks* 12, 411-427.

Matula, D. (1987), "Determining Edge Connectivity in O(mn)", *Proceedings of the 28th Annual IEEE Symposium on Foundations of Computer Science*, 249-251.

Picard, J.C., and M. Queyranne (1980), "On the Structure of All Minimum Cuts in a Network and Applications", *Mathematical Programming Study* 13, 8-16.

Schnorr, C.P. (1976), "Bottlenecks and Edge Connectivity in Unsymmetrical Networks", *SIAM Journal on Computing* 8, 265-274.

Yannakakis, M. (1981), "Node-deletion Problems on Bipartite Graphs", *SIAM Journal on Computing* 10, 210-237.

Algorithms for Projecting Points to Give the Most Uniform Distribution with Applications to Hashing

Tetsuo ASANO†   Takeshi TOKUYAMA‡

†Osaka Electro-Communication University, Osaka, Japan

‡IBM Research, Tokyo research Laboratory, Tokyo, Japan

Abstract

Given a set S of n points in the plane and an angle θ, a set of $b + 1$ parallel lines $l_0, l_1, \ldots, l_b$ of slope θ (b is fixed) is called a θ-cut of S if $l_0, l_1, \ldots, l_b$ are equally spaced and all the points of S lie between l_0 and l_b. The regions between two consecutive lines are called buckets. In this paper we consider the problem of finding an optimal θ-cut such that points are distributed into b buckets most uniformly, in other words, such that the maximum number of points to be included in one bucket is minimized. Comer and O'Donnell considered the problem under a constraint that l_0 and l_b are two supporting lines of S and presented an algorithm which runs in $O(bn^2 \log bn)$ time and $O(n^2 + bn)$ space. In this paper we present two linear-space algorithm for the problem based on duality transformation. One runs in $O(n^2 + K \log n + bn)$ time, where K is the number of intersections in the transformed plane. It is shown that $K = O(n^2 + bn)$. The other is advantageous if $b < \sqrt{n}$ and runs in $O(b^{0.610} n^{1.695} + bn + K \log n)$ time. Those are improvement of Comer and O'Donnell's algorithm both in time and space complexities. We also show that our algorithm can be extended to the case of buckets in the form of two-dimensional arrays.

Then, we extend our algorithm so as to solve the unconstrained version of the problem (l_0 and l_b may not be supporting lines of S) in polynomial time. First we consider the one-dimensional case and then apply the idea to the two-dimensional case. Especially, the one-dimensional case is important from a standpoint of application to hashing. The problem we consider is as follows: Given a set of n real numbers $x_0, x_1, \ldots, x_{n-1}$ (we assume that they are sorted and $x_0 = 0$ and $x_{n-1} = 1$ without loss of generality). Consider a hash function of the form $h(x) = \lceil \frac{x-L}{w} \rceil \bmod m$, where we assume that $\lceil \frac{0-L}{w} \rceil = 0$ and $\lceil \frac{1-L}{w} \rceil = K - 1$ and K and m are given integers. It is easily seen that the hash function is more flexible than the ordinary form $h(x) = \lceil K \cdot x \rceil \bmod m$ since the latter hash function corresponds to the one in which L and w are fixed to be 0 and $1/K$, respectively.

1   One-Dimensional Case

The unconstrained version of the problem in one-dimensional case is stated as follows.

Fig. 1. Partition of an interval including all the points into a fixed number of subintervals of equal length

For a set S of n real numbers $x_0, x_1, \ldots, x_{n-1}$ and an integer b, a set of $b+1$ real numbers $l_0, l_1, \ldots, l_b$ is called a cut of S if $l_{i+1} - l_i = w, i = 0, 1, \ldots, b-1$ and all the given numbers are between l_0 and l_b, that is, $l_0 \leq x_i < l_b, i = 0, 1, \ldots, n-1$. The problem is to find an optimal cut of S such that those numbers are distributed into b intervals (buckets) most uniformly (see Fig.1).

Since the length of each interval is the same, we can specify a cut by the width w of each interval and the value of $L = l_0$. Here we put a reasonable constraint that the two extreme intervals are not empty. That is, when we denote a set of given numbers included in the i-th interval of a cut (L, w) by $B_i(L, w)$:

$$B_i(L, w) = \{x_j \in S \mid l_i \leq x_j < l_{i+1}\},$$

where $l_i = L + i \cdot w$, the problem is to find an optimal value of (L, w) to minimize

$$\max_{0 \leq i \leq b-1} | B_i(L, w) |.$$

In this paper we assume without loss of generality that $x_0 = 0$ and $x_{n-1} = 1$. Since we also assumed that the two extreme intervals B_0 and B_{b-1} are not empty, we have

$$L \leq 0 < L + w, \quad L + (b-1) \cdot w \leq 1 < L + b \cdot w.$$

Hence, L and w must satisfy

$$L \leq 0, \quad L > -w, \quad L \leq 1 - (b-1)w, \quad \text{and } L > 1 - bw.$$

These relations can be expressed by the four lines as shown in Fig. 2. Since (L, w) must be included in the shaded region of the figure, the region is called a feasible region and if a point represented by (L, w) is contained in the region it is said to be feasible.

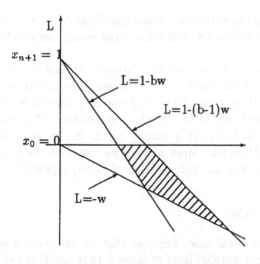

Fig. 2. The feasible region

[Lemma 1] In a cut characterized by a point (L, w) lying between two lines $L = x_j - (k + 1)w$ and $L = x_j - kw$, x_j belongs to the k-th bucket B_k.

Therefore, if we draw b lines $L = x_j - kw, k = 1, 2, \ldots, b$ for each $x_j \in S$, the feasible region is divided into small regions. As is easily seen, there is one-to-one correspondence between those small regions and cuts. Thus, if we enumerate all those regions, we can obtain an optimal cut. Then, how many such small regions may be possible? Although an obvious upper bound on the number of regions is $O(n^2 b^2)$, we can show that it is bounded by $O(n^2)$, which is independent of b.

[Theorem 1] There are at most $O(n^2)$ different cuts for any set of n real numbers.

Since each region resulting from drawing such lines in the feasible region corresponds to a cut, we can find an optimal cut by examining all of them. Note that a topological sweep [EG86] cannot be applied naively, for two successive intersections visited by the topological sweep can correspond to cuts which are different in many places and thus it takes much time to evaluate those cuts. Fortunately, the authors developed an efficient algorithm called topological walk [AGT90] to enumerate all the intersections within a convex polygon. Using the technique, we can find an optimal cut in $O(K + n)$ time and $O(n + b)$ space, where K is the number of intersections within the feasible region.

The next natural question is how many intersections there can be in the feasible region. Unfortunately, we can construct an example with $O(n^2)$ intersections.

The above algorithm can be extended to solve the following problem:

Let S be a set of n real numbers $x_0, x_1, \ldots, x_{n-1}$. We assume without loss of generality that $x_0 = 0 < x_1 < \cdots < x_{n-1} = 1$. For two integers M and m with $M > m$ and a real number $c, 0 \leq c < 1/M$, we can define a hash function

$h(x_i) = \lceil \frac{x_i - L}{w} \rceil \bmod m$,

under the constraints that $\lceil \frac{1-L}{w} \rceil = M - 1$ and $\lceil \frac{0-L}{w} \rceil = 0$. Note that the hash function of the above form is more flexible than that of the usual form

$h(x_i) = \lceil M \cdot x_i \rceil \bmod m$,

since the width of each bucket is fixed to be $1/M$ and L is also fixed to be 0 in the latter hash function.

We evaluate such a hash function by the uniformity of the distribution of hash values. An optimal hash function is defined to be the one minimizing the maximum number of elements having the same hash value.

[Problem] Given a set S as above and specified two integers M and m, find an optimal value of L and w such that the hash function of the above form is optimal.

The problem can be solved in $O(n^2)$ time and $O(n + m)$ space. It should be noted that the problem can be solved independently of M. It is important since M is usually fairly large. m is the hash table size and thus it is usually fixed. So, in general the problem is to find optimal value of (M, L, w). Most naively we can find the optimal value by executing the algorithm for finding an optimal value of (L, w) repeatedly.

2   Two-Dimensional Case

We can extend this result to two-dimensional case. Suppose that we are given a set of points in the plane. Then, a set of $b + 1$ parallel lines of slope θ (b is fixed) is called a θ-cut of S if they are equally spaced and all the points of S lie between two extreme lines. The regions between two consecutive lines are called buckets. Then, the problem is to find an optimal θ-cut such that points are distributed into b buckets most uniformly, in other words, such that the maximum number of points to be included in one bucket is minimized (see Fig. 3).

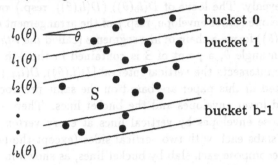

Fig. 3. Partition of the plane into buckets by a set of equally spaced parallel lines. (θ-cut)

We can solve this problem in two different situations. In one situation two extreme lines must be supporting lines of a given point set S and there is no such assumption in the other. Some results are known only for the former case. Comer and O'Donnell [CO82] proposed an efficient algorithm for the former problem that runs in $O(bn^2 \log bn)$ time using $O(n^2 + bn)$ space. In this paper we present two different linear-space algorithms for the constrained version of the problem, based on duality transformation [CGL83]. They first construct an arrangement of n lines dual to given points and partition it into slabs at the vertices of the upper and lower envelopes. These slabs are further decomposed into b trapezoids by $b-1$ equally spaced lines (called *bucket lines*), which are dual to boundaries between adjacent buckets. Then, the optimal angle can be found by enumerating all the intersections between the dual lines and trapezoids. The first algorithm runs in $O(n^2 + K \log n)$ time, where K is the number of intersections reported.

The other algorithm is advantageous if $b < \sqrt{n}$. It performs a simplex range search [CGL83, Ch87, Ed87, EW86, We88, Wi82] in each slab to enumerate all the lines that intersect *bucket lines*, and runs in $O(b^{0.610}n^{1.695} + K \log n)$ time.

2.1   Algorithms for the Constrained Case

First we shall consider the constrained version of the problem, that is, we assume that the two extreme lines l_0 and l_b are two supporting lines of a given point set. Comer and O'Donnell[CO82] used a (circular) plane sweep to solve the constrained version of the problem. Our methods are based on the point-to-line geometric transform [CGL83, Ed87]. A point set is transformed into an arrangement of lines. A line of an angle θ is mapped to a point on a vertical line $x = \tan \theta$. We should sweep on θ in order to solve our problem. The advantage of considering the dual problem is that we can transform a circular plane sweep like radar into an ordinary plane sweep by a vertical sweep-line.

Let S be a set of n points in the plane. For an angle θ, $-\pi/2 < \theta < \pi/2$, let $l_0(\theta)$ and $l_b(\theta)$ be two lines at angle θ supporting S. The region between them is divided by $b-1$ equally spaced lines $l_1(\theta), \ldots, l_{b-1}(\theta)$ (see Fig. 3). Such a set of lines is called a θ-cut of S. The region bounded by $l_i(\theta)$ and $l_{i+1}(\theta)$ is called the i-th bucket, denoted by $B(\theta, i)$.

The cardinality of $B(\theta, i)$ is denoted by $\beta(\theta, i)$, or $\beta(i)$ if we regard θ as a parameter. Throughout the paper the number b is fixed. We denote the geometric transform by D. Thus, $D(l_0(\theta)), \ldots, D(l_b(\theta))$ lie on the the vertical line $x = \tan(\theta)$; moreover, they divide the interval $[D(l_0(\theta)), D(l_b(\theta))]$ evenly. The locus of $D(l_0(\theta))$ ($D(l_b(\theta))$, resp.) over the parameter θ forms the upper envelope (lower envelope, resp.) of the arrangement of lines dual to S. Therefore, that of $D(l_i(\theta))$ forms a chain of line segments (called *the i-th bucket line*) as shown in Fig. 4. For an angle θ, a point of S is contained in the i-th bucket $B(\theta, i)$ if and only if its dual line intersects the vertical interval $[D(l_i(\theta)), D(l_{i+1}(\theta))]$.

The algorithms to be presented in this paper are based on the same preprocessing. We first construct the upper and lower envelopes and the bucket lines. Then, we partition the region bounded by those envelopes by vertical lines at every vertex on the envelopes. This results in $O(n)$ slabs each with two vertical sides (except the leftmost and rightmost ones). Next, we decompose each slab by bucket lines, as shown in Fig. 4. This preprocessing takes $O(n \log n + bn)$ time.

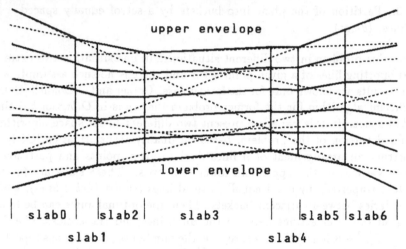

(a) Decomposition into slabs. Dotted lines correspond to given points

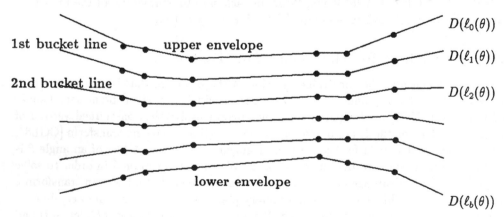

(b) Bucket lines in the dual plane.

Fig. 4. Slab decomposition and bucket lines

Algorithm 1: Naive Plane Sweep

(1) We have n lines and $O(b)$ bucket lines. Enumerate all the intersections by using the intersection reporting algorithm developed by Bentley and Ottman [BO79]. Let K be the total number of intersections.

(2) Sort those intersections in one direction.

(3) Based on the sorted list obtained, perform a plane sweep. At each intersection with the ith bucket line we must decrease $\beta(i-1)$ and increase $\beta(i)$, or increase $\beta(i-1)$ and decrease $\beta(i)$, depending on the slope of the line giving the intersection. We must also compare the distribution obtained with the optimal distribution found thus far.

There are various criteria on the cost of distribution of points into buckets. The time for evaluating the cost, given b buckets, and updating the cost function at each step of the plane sweep, denoted by $T(b)$ and $U(b)$, respectively, depends on the criterion selected. When we want to minimize the maximum bucket size, we can achieve $T(\min(n, b)) = O(b)$ and $U(b) = O(1)$ by using a relatively simple data structure. The space needed for this data structure is $O(b)$ (see Lemma below). If we are interested in minimizing the variance of cardinalities of buckets, a simpler data structure suffices.

[Lemma 2] Given a sequence of integers $(t_1, t_2, \ldots, t_N)$ such that $t_i \in \{-b, -b + 1, \ldots, -1, 1, 2, \ldots, b\}, i = 1, 2, \ldots, N$, for each $t_i, 1 \leq i \leq N$ we can output in constant time in on-line fashion the maximum value among $c(i, 1), c(i, 2), \ldots, c(i, b)$ where

$$c(i, j) = |\{t_k | t_k = +j \text{ and } k \leq i\}| - |\{t_k | t_k = -j \text{ and } k \leq i\}|$$

by using $O(N + b)$ space.

[Theorem 2] Algorithm 1 finds a cut giving the optimal distribution of points in $O((bn + n^2 + K) \log n + K \log K + KU(b) + T(b))$ time and $O(bn + K)$ space.

The time complexity can be slightly improved by using an improved version of Bentley and Ottman's algorithm which is described in the book [PS85]. In the improved version intersections are enumerated in the sorted order without increasing the time and space complexity. Thus, the improved time complexity is $O((bn + n^2 + K) \log n + KU(b) + T(b))$.

2.2 Linear-Space Implementation

In Algorithm 1 we used an intersection reporting algorithm due to Bentley and Ottman [BO79]. If we could enumerate intersections in a smarter way, we could design a better algorithm. In the second algorithm, for each slab we find all the lines intersecting any bucket line within the slab. Once we have such lines, we can dynamically produce all the intersections in increasing order of their abscissa. Such lines are easily found. A line intersects any bucket line within a slab if and only if it belongs to different buckets at the left and right sides of the slab. Since buckets lines are equally spaced, the above condition is tested in constant time.

Algorithm 2: Linear-Space Implementation

[1] The same preprocessing as for Algorithm 1 is performed.

[2] For each slab S_i from left to right do the following:

[2.1]   Partition the slab S_i into b trapezoids with $b-1$ bucket lines.

[2.2]   For each line determine whether it intersects any bucket line.

    If it does, then put the first (leftmost) intersection into a heap.

[2.3]   Perform a plane sweep as follows.
[2.3.1]   Evaluate the distribution of lines into buckets at the left side of
the slab.
[2.3.2]   while(heap is not empty) {
[2.3.2.1]   Extract the leftmost intersection (s_ξ, B_η) from the heap,
where s_ξ is a line and B_η is a bucket line.
[2.3.2.2]   Update the distribution and put the next intersection given by
the line s_ξ into the heap if it lies in S_i.
[2.3.2.3]   If the resulting distribution is better than the current optimum,
then update the current optimum.}

[Theorem 3] Algorithm 2 finds a cut giving the optimal distribution of points in
$O(n^2 + bn + K \log n + KU(b) + T(b))$ time and $O(n + b)$ space.
Proof: We shall concentrate on the time complexity of the algorithm. In the algorithm
we perform a plane sweep slab by slab. For each slab S_i steps 2.1 and 2.2 can be done in
$O(b)$ time and $O(n)$ time, respectively. The plane sweep of step 2.3 can be carried out in
$O(k_i \log n + k_i U(b) + T(b))$ time, where k_i is the number of intersections to be processed
in the slab. The time for evaluating the distribution at the left side can be omitted
except for the leftmost slab, since the distribution at the left side of an intermediate slab
is exactly the same as the one at the right side of the preceding slab. Summing up the
time for each slab, we obtain the time complexity in the theorem.   □

As is easily seen, Algorithm 2 is superior to Algorithm 1 both in time and space
complexities. The leading term in the time complexities of above-described algorithms
may be $K \log n$. Although there seem to be $O(bn^2)$ intersections, we can prove that there
are actually only $O(bn + n^2)$ intersections.

[Theorem 4] Given a set S of n points in the plane and the number b of buckets, there
are at most $O(bn + n^2)$ intersections between the dual lines and bucket lines.

From this theorem, our algorithm requires $O((n^2 + bn) \log bn))$ time and $O(n+b)$ space
for the worst case. Since the algorithm of Comer and O'Donnell requires $O(n^2 b \log bn)$
time and $O(n^2)$ space, the algorithms presented in this paper offer improvements with
respect to both time and space.

2.3   Range-Search Based Algorithm

There are applications where the number of buckets b is much smaller than n, for example, division of an enormous number of keys into a fixed number of blocks in a disk.
Although we have shown the optimal bound of intersections K, we can observe that K
is as small as $O(bn)$ for almost all arrangements. In this section, we give an algorithm
which is efficient for such cases that b and K are small.

The most important operation in the algorithms described thus far is to enumerate all the intersections between bucket lines and infinite lines corresponding to given
points. Considering the problem in its dual plane, it becomes to enumerate all the points
contained in double wedges, since bucket lines are transformed into double wedges and
infinite lines into points. This is a so-called triangular range-searching problem. It is
known that any such range query can be answered in $O(n^{0.695} + k)$ time by using a
Ham-Sandwich tree data structure after $O(n \log n)$-time preprocessing [CGL83, Ch87,

Ed87, EW86, We88, Wi82]. The remaining operations are the same as in Algorithm 2. Thus, at each slab we can enumerate all the lines that intersect any bucket line in $O(bn^{0.695} + k_i)$ time by repeating a triangular range search b times. This leads to an algorithm that runs in $O(bn^{1.695} + K \log n + KU(b) + T(b))$ time and $O(n + b)$ space.

We shall show below that if $b < \sqrt{n}$, then the algorithm can be improved so that it runs in $O(b^{0.610}n^{1.695} + K \log n + KU(b) + T(b))$ time using linear space. The key idea is a combination of the bucketing technique and the erasing subdivision by Ham-Sandwich cuts [Ed87].

Let P be the convex hull of S. For an angle τ, $R(\tau)$ denotes a rectangle, with sides at angle τ, that circumscribes P. Let $R(\lambda)$ be such a rectangle with a minimal area. Then,

[Lemma 3] The area of $R(\lambda)$ is no more than twice of the area of P.

We cut $R(\lambda)$ with two sets of equally spaced $b - 1$ lines at angles of λ and $\lambda + \pi/2$, respectively (see Fig. 5). Then, we have b^2 cells congruent to each other.

[Lemma 4] For any angle θ, a cell is cut with at most two lines of parallel cuts.

For a cell A we store the points of $S \cap A$ using the erasing subdivision by Ham-Sandwich cuts.

[Lemma 5] The number of intersections of the edges of the erasing subdivision in a cell with parallel cuts is $O(k^{0.695})$ for any angle, where k is the number of points in the cell.

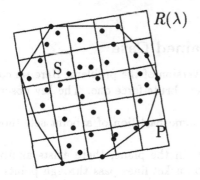

Fig. 5. Minimum enclosing rectangle $R(\lambda)$ and its decomposition into b^2 buckets

We consider an efficient data structure to query the table $B(\theta) = (\beta(\theta, i))_{i=1,...,b}$ for an arbitrary θ. We construct a quad tree of size b^2 on the cells. Each leaf of the tree points to the Ham-Sandwich tree associated with the erasing subdivision of the corresponding cell. Then,

[Proposition] Using the above data structure, we can query the table $B(\theta)$ in $O(n^{0.695} b^{0.610})$ time if $b < \sqrt{n}$.

Proof. The stabbing number of our tree by all lines of a parallel cut is
$$b^2 + \sum_{i=1}^{b^2} ck_i^{0.695} \cdots (1) \qquad (c : \text{a constant})$$
where k_i is the number of points in the i-th cell. The maximum of (1) is attained if the points are distributed evenly. Thus, the stabbing number is not greater than $b^2 + cb^2(\frac{n}{b^2})^{0.695} = b^2 + cb^{0.610}n^{0.695}$. Since $b^2 < b^{0.610}n^{0.695}$, if $b < \sqrt{n}$, we obtain the proposition. □

Using the above data structure we can enumerate all the dual lines that intersect any bucket lines in $O(b^{0.610}n^{0.695} + k_i)$ time for each slab S_i in the dual plane, where k_i is the number of lines reported. This leads to the following theorem:

[Theorem 5] Using the above data structure, an optimal cut is found in $O(b^{0.610}n^{1.695} + K\log n + KU(b) + T(b))$ time using $O(n + b)$ space.

2.4 Distributing Points into Two-Dimensional Buckets

Thus far we have mapped a given set of points into a one-dimensional structure. Here we consider to map them into a two-dimensional structure, that is, two sets of parallel lines of angle θ_1 and θ_2 which partition the quadrangle defined by two set of supporting lines of those angles into b^2 small quadrangles. If the difference between those two angles is fixed, we can find an optimal angles to distribute given points most uniformly into buckets in the same time complexity as Algorithm 1 except the term b^2 being added. The idea is implementation of parallel plane sweep, that is, we sweep the transformed plane using two sweep lines. If the difference between the two angles is fixed, we can find an optimal distribution in the same time complexity as the previous one. What we have to do at each step of the algorithm is to find the next intersection for each sweep line.

2.5 Algorithm for the Unconstrained Case

Thus far we have investigated the constrained version of the problem. Here we consider the unconstrained version. Our algorithm is very brute-force one. The key observation is as follows.

[Definition] If two cuts give the exactly the same partition of a point set, those two cuts are called equivalent.

[Observation] For any cut for a set of points in the plane, there exists another cut which is equivalent with the cut such that two bucket lines pass through points of the given point set.

The algorithm to be presented proceeds as follows. For each pair of points (p_i, p_j) and for each $k, 1 \le k \le b$, we find an optimal cut such that two bucket lines pass through those points and the region between them is divided into k buckets. Note that bucket lines do not depend on the width of the convex hull of a given point set but depend only on those two points. Thus, there are at most $O(bn)$ intersections in the dual plane between bucket lines and lines dual to given points. Roughly speaking, for each pair of points and a value of k, an optimal solution can be found in $O(bn\log n)$ time. Since it is iterated $O(bn^2)$ times, the total time complexity is $O(b^2n^3\log n)$.

3 Concluding Remarks

In this paper we have considered the problem of distributing a set of points most uniformly into buckets and presented two different algorithms for finding the optimal projection of points to give the most uniform distribution. They offer improvements over the existing best algorithm with respect to both time and space complexities. We have considered the projection from a light source lying on the line at infinity. We could deal with some problems on the projection from a light source in the plane. Let us consider the following problem: Given a planar point set, find the projection to give the most uniform distribution of points if the light source moves on a line l and projects planar objects on a screen parallel to l. Since any line can be mapped to the line at infinity by a projective transformation, the problem is reduced to our problem in this paper.

Acknowledgment The authors wish to thank Hiroto Yasuura for giving us a fundamental idea for the unconstrained version of the problem in one dimension. The authors also wish express their thanks Naoki Katoh and Pankaj Agarwal for their discussions.

REFERENCES

[AGT90] Te. Asano, L.J. Guibas and T. Tokuyama: *Walking on an Arrangement Topologically*, manuscript, 1990.

[BO79] J.L. Bentley and T. Ottman: *Algorithms for Reporting and Counting Geometric Intersections*, IEEE Trans. Comput., vol. C-28, pp. 643-647, Sept. 1979.

[CGL83] B. Chazelle, L.J. Guibas, and D.T. Lee: *The Power of Geometric Duality*, Proc. 24th Symposium on Foundations of Computer Science, pp. 217-225, 1983.

[Ch87] B. Chazelle: *Polytope Range Searching and Integral Geometry*, Proc. 28th Symposium on Foundations of Computer Science, pp. 1-10, 1987.

[CO82] D. Comer and M.J. O'Donnell: *Geometric Problems with Applications to Hashing*, SIAM J. Comput., vol. 11, No. 2, pp. 217-26, 1982.

[Ed87] H. Edelsbrunner: *Algorithms in Combinatorial Geometry*, EATCS Monographs in Theoretical Computer Science, vol. 10, Springer-Verlag Berlin Heidelberg, 1987.

[EG86] H. Edelsbrunner and L. J. Guibas: *Topologically Sweeping an Arrangement*, Digital Systems Research Center, Research Report 9, 1986.

[EW86] H. Edelsbrunner and E. Welzl: *Half Planar Range Search in Linear Space and $O(n^{0.695})$ Query Time*, Inform. Process. Lett. vol. 23, pp. 289-193, 1986.

[PS85] F.P. Preparata and M.I. Shamos: *Computational Geometry - An Introduction*, Springer Verlag, New York, 1985.

[We88] E. Welzl: *Partition Trees for Triangle Counting and Other Range Searching Problems*, Proc. 4th Annual Symposium on Computational Geometry, pp. 23-33, Urbana-Champain, IL, June 1988.

[Wi82] D. Willard: *Polygon Retrieval*, SIAM J. Comput., vol. 11, pp. 149-165, 1982.

Topological Sweeping in Three Dimensions

Efthymios G. Anagnostou
Department of Computer Science, University of Toronto

Leonidas J. Guibas
MIT, and DEC Systems Research Center

Vassilios G. Polimenis
Department of Computer Science, University of California, Berkeley

1. Introduction

The sweeping of a collection of geometric objects in both two and three dimensions is one of the earliest and most extensively used techniques in computational geometry. The reason is that sweeping is an extremely versatile method that can serve as a skeleton for many geometric algorithms. In this paper we are interested in sweeping an arrangement of n planes in three-dimensional space. The analogous problem in two dimensions, that is the sweeping of n lines in the plane, has historically attracted considerable attention. There has been a wide variety of problems about lines, or line segments (see [BO], [MS] etc.), all solved by sweeping the plane with a vertical straight line and incrementally updating some kind of intersection structure between the sweeping line and the objects being swept over. Unfortunately this approach does not always yield algorithms optimal in either their time or space requirements, as a line sweep forces us to sort all the pairwise line intersections, even though such sorting may be unnecessary in the final output. When sweeping the full arrangement of n lines in the plane, a straight-line sweep takes $O(n)$ working space and $O(n^2 \log n)$ time. Recently [EG] discovered a novel approach to solve the two-dimensional problem using instead a topological line that is an unbounded simple curve satisfying certain properties milder than straightness. With this approach they were able to achieve optimal $O(n^2)$ time for the sweep, while still keeping the working storage only $O(n)$.

This paper generalizes the [EG] algorithm to solve the sweeping problem for n planes in three dimensions. Our algorithm needs $O(n^3)$ time, which is optimal, and $O(n^2)$ working space. We also give a number of applications of our method to a variety of problems, inspired by the applications listed in [EG].

2. Outline of the Edeslbrunner-Guibas method

By $A(H)$ we will denote a two-dimensional arrangement formed by n lines. The sweeping line is a continuous y-monotone curve with the property that it intersects every line of $A(H)$ in one and only one point. The *cut* is defined as the set of line segments of $A(H)$ intersected by the topological line. Thus there are always n segments in the cut. When two consecutive segments in the cut share a common right end-point, the algorithm can do an *elementary step* at that end-point by advancing the topological line from the left to the right of the point, thus sweeping over it (see Figure 1).

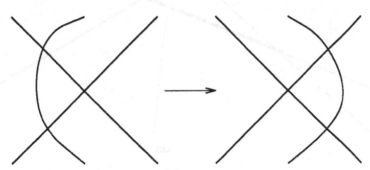

Figure 1: An elementary step

Initially the topological line is a vertical straight line positioned to the left of the leftmost vertex of $A(H)$. As discussed above, the algorithm can do an elementary step iff there exists a vertex with its two left-going edges in the current cut. It has been proved that this is always possible, till all the vertices have been swept.

The major difficulty in implementing the topological sweep is how to discover where in a cut an elementary step can be applied. Here [EG] introduced a novel data structure, namely the *horizon trees*, to help in this task. The *upper horizon tree* of the cut is constructed by starting with the edges of the cut and extending them to the right. When two lines come together at an intersection point, only the one of *higher* slope continues on to the right; the other one stops at that point (see Figure 2). The lower horizon tree is defined symmetrically (i.e., lines of lower slope are the winners).

By considering the n lines in decreasing (increasing) slope the upper (lower) horizon tree can be initialized in $O(n \log n)$ time. The algorithm also maintains the set I of vertices ready to be swept.

3. Preliminaries for the three-dimensional sweep

We now demonstrate how the previous approach can be generalized to a new algorithm for sweeping a three-dimesnional arrangement. Let P be a set of n planes. $A(P)$ denotes the arrangement formed by these planes. We assume that every two planes share a common line and every three of them share a common point.

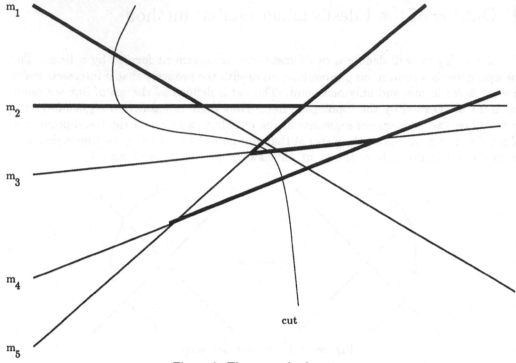

m_1

m_2

m_3

m_4

cut

m_5

Figure 2: The upper horizon tree

Thus there are $\binom{n}{2} = O(n^2)$ such lines and $\binom{n}{3} = O(n^3)$ such points. For the sake of brevity we do not discuss degenerate situations. From now on, when referring to a vertex, we mean one of these $\binom{n}{3}$ points and, when referring to a line, we mean one of these $\binom{n}{2}$ lines.

By analogy with the two-dimensional situation, our algorithm sweeps with a continuous unbounded surface satisfying properties milder than those of a Euclidean plane. This surface intersects every plane of $A(P)$ in a continuous topological line exactly like the one that serves as the sweeping object in the planar algorithm. Consequently the surface shares a common point with every one of the $\binom{n}{2}$ lines of $A(P)$.

Initially the surface is a plane perpendicular to the x-axis positioned to the left of the leftmost vertex of $A(P)$. The algorithm does an elementary step by moving the surface from the left of a vertex to the right of it. A three-dimesnional elementary step consists of three two-dimensional elementary steps, one on each of the three defining planes of this vertex.

We define the current spatial cut to be the set of segments (or rays) of $A(P)$ intersected by the sweeping surface. This space cut is the union of the n planar cuts, as defined by the n topological lines on each plane. The algorithm can do an elementary step iff there exists at least one vertex with *all three* of its left-going edges in the current spatial cut. In extreme cases such an edge may be a left-going infinite ray. At every elementary step the current spatial cut changes: the three left-going edges of the vertex just swept over are replaced by the three right-going ones.

Let us focus our attention to the three defining planes of the vertex where the algorithm is ready to make an elementary step. The three left-going edges of this vertex belong to the current spatial cut. So every two of them belong to the current planar cut of the plane they lie in. Hence one spatial elementary step corresponds to three planar elementary steps on the three defining planes.

We now show that there always exists at least one vertex where an elementary step is possible, till the algorithm sweeps over all vertices of the arrangement.

Lemma 1. *The sweeping surface shares only one common point with every line or, equivalently, every line has only one segment (or ray, in the two extreme cases) in the spatial cut.*

Proof : This property holds for the initial surface. Also, it continues to hold after every elementary step, if it held before (this is obviously true from the definition of the elementary step). □

Lemma 2. *Let P be a vertex of the arrangement. If the surface has not swept P, then it intersects its three defining lines to the left of P. If the surface has passed P, it intersects these lines to the right of P.*

Proof: This property holds for the initial surface, since no vertex has been swept over and all the intersections are left of all the vertices. Suppose that this property holds before an elementary step. We will show that it still holds after it. It certainly holds for the vertices that do not belong at any of the three defining lines of the elementary step because the intersection points (of the surface with the lines) change only for these three defining lines. It also holds for the vertices on the three defining lines that have been passed, because the surface remains to the right of them. For the vertices that were previously to the right of the surface, except for the vertex where the elementary step is currently applied, the assertion also holds because the surface remains to the left of them. For the vertex that has just been swept the assertion holds because the surface moves from its left to its right. □

Now we present the main theorem.

Theorem 1. *The topological surface can always make an elementary step, unless it has swept all the vertices.*

Proof: We will show that (at least) the leftmost vertex among those that have not yet been swept can serve for the next elementary step.

Let us suppose that the leftmost vertex, let it be called S, is not ready to be swept. From Lemma 2 we have that the surface intersects the three defining lines to the left of S. But S is not ready for an elementary step. Hence there must be one defining line which contributes the spatial cut a segment l that is not the left-going edge of the vertex S at this line. But if we consider the right endpoint of the segment l it has not yet been swept (by Lemma 2) and it is located to the left of S, which is a contradiction. □

4. Data Structures and Initialization

We arbitrarily choose a numbering of the n planes $[1, .., n]$.

Let $P[1 : n]$ be the array of plane equations. If the equation of the i-th plane of P is $P_i :\ z = a_i x + b_i y + c_i$ then the i-th entry of $P[]$ is $P[i] = (a_i, b_i, c_i)$. For every plane k we have a set of "local" data structures which we describe below.

The array of plane indices as they define the lines on plane k in increasing slope is denoted by $E_k[1 : n - 1]$. So $E_k[i] = j$ means that the line with the i-th slope on the k-th plane belongs to the j-th plane. Of course all these indices have to be consistent with the previous numbering.

The inverse array of $E_k[]$ is represented by $E_k^{-1}[1 : n]$. $E_k^{-1}[j] = i$ indicates that the intersection of k-th and j-th planes is the i-th line of the k-th plane. $E_k^{-1}[k]$ is arbitrarily set to zero. These inverse arrays can be constructed in $O(n)$ time (for each plane k) from $E_k[]$. We find the ordering of the lines, at any plane by projecting them to the x-y plane. After sorting their slopes we fill in this information in the array $E_k[]$.

The array representing the upper horizon tree is $HTU_k[1 : n - 1]$. The entry $HTU_k[i]$ is a pair (m_i, p_i) of indices indicating the lines that delimit the edge of l_i in the upper horizon tree to the left and to the right, respectively. If this segment is the leftmost on l_i we set $m_i = -1$; if it is the rightmost on l_i we set $p_i = 0$.

The lower horizon tree is represented by $HTL_k[1 : n - 1]$ and is defined similarly.

The array $M_k[1 : n - 1]$ holds the current sequence of indices that form the lines $m_1, ..., m_{n-1}$ of the current planar cut for plane k.

Finally array $N_k[1 : n - 1]$ is a list of pairs of indices indicating the lines delimiting each edge of the above cut. $N_k[i]$ thus encodes the endpoints of the edge on $M_k[i]$. The same convention as that above is used for missing endpoints.

There is also a "global" set of triples I. The triple (i, j, k) is in I iff the point defined by these three planes is ready to be swept (i.e., it is a candidate for an elementary step).

For each plane the upper and lower horizon trees can be initialized in $O(n)$ total time as was showed in [EG] (assuming we know the slope ordering of the lines). After that we can find each N_k in $O(n)$ time since $N_k = HTU_k \cap HTL_k$. M_k initially represents the sequence of lines (on the k-th plane) sorted according to their slopes.

The difficult part of the initialization is the construction of the set I. A point in three dimensional space is ready to be swept if and only if the three left segments belong to the space cut or equivalently every two of them belong to the corresponding plane cut. That means that this point is ready to be swept, in the two-dimensional sense, in each of its three defining planes.

We scan every plane arrangement for planes 1 through n. In $O(n)$ time we can find from N_k the vertices in plane k that are candidates for I. For every such vertex (k, i, j) we look up the other two defining planes, namely i, j. To avoid duplicate entries we do this check only if $k < i$ and $k < j$.

If in the two other planes the same vertex (k, i, j) is candidate for a two-dimensional elementary step, that means that it has to be inserted in the set I. We can make this test in $O(1)$ time by an appropriate use of our data structures.

The initialization of the data structures for each plane costs $O(n)$ plus the time to sort the slopes (i.e. $O(n \log n)$). Hence $O(n^2 \log n)$ time will suffice for all planes. The set I can also be initialized in $O(n^2)$, as was shown above.

5. The main algorithm

At each elementary step the algorithm chooses arbitrarily from I the next vertex to be swept. Let it be $P = (i, j, k)$.

The algorithm does an elementary step at each of the three defining planes i, j, k. That costs $O(1)$ in an amortized sense on each of the planes (as [EG] have shown). After these two-dimensional elementary steps the local data structures have been updated to reflect the new surface. Again the difficult task is to update I. That means to find whatever points become ready to be swept after this elementary step. These points are at most three since they have to be among the right neighbors of P on its three defining lines. This task is very similar to that described at the initialization phase.

We now show how to check if the right neighbor of P, which belong to planes i and j, has to be inserted in I. Let Q be the vertex to the right of P on the line defined by the i-th and j-th planes. Having updated the local data structures for each plane we can now move to plane i and check if Q is ready for a two-dimensional elementary step on plane i. If it is not, we decide that point Q is not ready at this time (but will be later). If it is, we move to plane j and check if the right endpoint (Q) of the edge in the cut of line $E_j^{-1}[i]$ is ready for a two-dimensional sweeping on plane j. If it is, then we insert Q in I.

When we are done with Q we check the other two right neighbors of P. The above tests can all be implemented in $O(1)$ time, since we have three of them and each one requires constant time. This completes the implementation of one spatial elementary step.

6. Complexity analysis of the algorithm

This algorithm takes $O(n^2 \log n)$ for the initialization phase as described previously. Consequently this algorithm takes $O(n^3)$ time in total, since amortized $O(1)$ time is sufficient for each of the $\binom{n}{3}$ vertices (elementary steps). Obviously the algorithm is optimal with respect to time complexity.

As for the space requirements, there are n local data structures each requiring $O(n)$ storage. So the space is $O(n^2)$. The set I cannot ever violate this space complexity because for every line at most one point is ready to be swept at any one time.

7. Applications

1. Longest monotone path

Let P be a set of n planes in general position in E^3. A monotone path p of $A(P)$ is a connected geometric object of alternating edges and vertices of $A(P)$ such that every plane perpendicular to the x-axis intersects p at one exactly point. Therefore p

is unbounded. A vertex v of p is a *turn* if the two incident edges are not collinear. We define the length of p as the number of turns plus one.

Problem : Compute the longest monotone path of $A(P)$.

To compute the longest path we sweep topologically. For each edge e in the current cut we maintain a longest path which extends from e towards the left. The edge e holds the number of turns of this path.

Theorem 2. *The length of the longest monotone path of n planes can be found in $O(n^3)$ time and in $O(n^2)$ storage.*

It is interesting that the topological sweep does not maintain enough information to allow us extract the actual longest path directly because we can not afford to keep around predecessor pointers for all edges and still have a quadratic storage. Otherwise the algorithm would require $O(n^3)$ storage.

2. Stabbing triangles

Let S be a set of closed triangles in E^3 not necessarily disjoint without any degeneracies.

Problem : Find the plane that cuts the maximum number of triangles in S.

In dual space a triangle corresponds to a triple of planes; a plane cuts the triangle iff its dual point lies in some specified double solid angle (out of the eight) which depends on the dual transformation.

Let P be the set of $3n$ planes dual to the $3n$ vertices of the n triangles in S. If the dual points of two planes P_1 and P_2 fall into the same region of the arrangement $A(P)$, then P_1 and P_2 intersect the same triangles and therefore the same number of them.

To solve the above problem we compute for each region of $A(P)$ the number of triangles cut by a plane dual to a point of the region. This piece of information is best computed when the region is first encountered during the topological sweep.

Each manipulation is carried out in constant time.

Theorem 3. *A plane that cuts the maximum number of n given triangles in E^3 can be found in $O(n^3)$ time and in $O(n^2)$ storage.*

In the same way we can solve the following problem:

Problem : Find a plane that cuts no triangles and such that the absolute value of the triangles above the plane minus the number of triangles below is minimum.

To solve this problem we compute for each region the number of triangles in S which lie above the corresponding plane and the number of triangles which are cut by this plane. Again this information can be propagated in constant time from one region to the next during the topological sweep.

Theorem 4. *A plane that avoids all of a given set of n triangles in E^3 and produces a best balance between the number of triangles on either side can be found in $O(n^3)$ time and $O(n^2)$ storage.*

3. Minimum volume tetrahedron

Let S be a set of n points in E^3.

Problem : Determine the four points p, q, r, s such that the tetrahedron $T_{p,q,r,s}$ that they define has minimum volume among all possible ones.

If the four points p, q, r, s define the minimum tetrahedron for each of them (say s) the following holds: Point s is closer to the plane of p, q, r than any other point of S. That means we can move continuously the plane of p, q, r until it hits s without meeting any other point of the arrangement. In the dual arrangement that means that the point defined by the dual planes of p, q, r is in a common region with some face of the plane dual to s. So we propose the following algorithm for the problem:

We make a topological sweeping of the duals of our points and for each region we keep its vertices and its faces. When we leave the region we test every vertex-face pair, compute the the volume of the corresponding tetrahedron and record its volume if it is currently minimum. As shown in [EOS] the total number of vertex-face pairs for the regions is $O(n^3)$. In addition, with a small modification of their proof for Euclidean planes we can prove for our sweeping surface that at every instant of time the total space requirements for all the regions cut by the sweeping surface at this time is $O(n^2)$.

Theorem 5. *The minimum volume tetrahedron defined by n points in E^3 can be determined in $O(n^3)$ time and $O(n^2)$ space.*

References

[BO] J. Bentley, T. Ottmann, "Algorithms for reporting and counting geometric intersections", IEEE Trans. Comput., C 28(1979), pp. 643-647.

[EG] H. Edelsbrunner, L. Guibas, "Topologically sweeping an arrangement", Proc. 18th ACM Symp. on Theory of Computing, 1986, pp. 389-403.

EOS] H. Edelsbrunner, J. O'Rourke, R. Seidel, "Constructing arrangements of lines and hyperplanes with applications", SIAM J. Comput., 15 (1988), pp. 341-363.

[MS] H. Mairson, J. Stolfi, "Reporting and counting intersections between two sets of line segments", Proceedings of the 1987 NATO Conference on Theoretical Foundations of Graphics and Computer Aided Design, Springer-Verlag.

Finding Least-Weight Subsequences
with Fewer Processors

Kwong-fai Chan Tak-wah Lam

Department of Computer Science
University of Hong Kong
Pokfulam, Hong Kong

Abstract

It is known that many problems which can be solved sequentially by dynamic programming, are in the class NC. Indeed most of these problems takes $O(\log^2 n)$ time on parallel models like CREW PRAM, but the number of processors involved is usually a high degree polynomial and the total work (i.e., the processor-time product) is very unfavorable in comparing with the work (i.e., the time) in the sequential case. Recently there has been a lot of progress in speeding up dynamic programming sequentially by restricting the weight functions to satisfy the quadrangle inequalities or the inverse quadrangle inequalities, yet little was heard about any improvement in the parallel complexity. Thus, it is time to see whether such kind of restrictions can lead to any improvement in the processor complexity of these problems. In this paper, we study the least-weight subsequence problem which is a typical problem for one-dimensional dynamic programming. The well known sequential solution to this problem takes $O(n \log n)$ time, while the conventional parallel algorithm uses $O(\log^2 n)$ time on a CREW PRAM with n^3 processors. Our new result is that with the inverse quadrangle inequality, the problem can be solved in $O(\log^2 n \log \log n)$ time on a CREW PRAM with $n/\log \log n$ processors. Notice that the processor-time complexity is close to the sequential time complexity.

1 Introduction

The least-weight subsequence problem was first defined by Hirschberg and Larmore [9] as follows: Given an integer n and a real-valued weight function $w(i,j)$, which can be computed in constant time for all $0 \le i \le j \le n$, find a sequence of integers $\{\alpha_1, \alpha_2, \cdots, \alpha_m\}$ such that $0 = \alpha_1 < \alpha_2 < \cdots < \alpha_m = n$ and $\sum_{1 \le i < m} w(\alpha_i, \alpha_{i+1})$ is minimized. Such sequence is called a least-weight subsequence for the range [0,n].

The weight function is said to be concave, if it satisfies the quadrangle inequality [13], i.e., for all $i_0 \le i_1 < j_0 \le j_1$, $w(i_0,j_0) + w(i_1,j_1) \le w(i_0,j_1) + w(i_1,j_0)$. The weight function is said to be convex, if it satisfies the inverse quadrangle inequality, i.e., for all $i_0 \le i_1 < j_0 \le j_1$, $w(i_0,j_0) + w(i_1,j_1) \ge w(i_0,j_1) + w(i_1,j_0)$.

By imposing the concave property to the weight function, Hirschberg and Larmore [9] improved the sequential time complexity of finding least-weight subsequence from n^2 to $n \log n$.

Later, Galil and Giancarlo [8] further generalized this result to both the concave case and the convex case. Though the quadrangle inequality and its inverse look very similar, a solution for the concave case might not work in the convex case, and *vice versa*. In fact, the best known algorithm for the concave least-weight subsequence problem uses only $O(n)$ time [12], while the convex still requires $O(n \log^* n)$ [10].

Let us switch to the parallel solutions of this problem. Using standard techniques like parallel tree contraction [11, 7], a CREW PRAM with $n^3/\log n$ processors can find the least-weight subsequence in $O(\log^2 n)$ time. (See Section 2.) Recently Atallah, *et. al.* [1] have shown that the multiplication of two matrices satisfying the quadrangle inequality can be done in $O(\log n \log\log n)$ time on a CREW PRAM with $n^2/\log n$ processors. Since the least-weight subsequence problem can be reduced to a series of $\log n$ matrix multiplication, we immediately obtain a parallel algorithm using $\log^2 n \log\log n$ time and $n^2/\log n$ processors for the concave case. In fact, this algorithm can be adapted to the convex case too.

Other than these results, little was previously known on the parallel complexity of the problem when the weight function satisfies either the concave property or the convex property. Perhaps it is due to the sequential nature of the problem, a brute-force approach seems to be the only way to go, that is, a lot of processors are required to perform redundant computation. Actually, as far as we know, parallel solutions to other dynamic programming problems, such as finding the optimal binary search tree, also have this kind of defect. The major contribution in this paper is that the least-weight subsequence problem with any convex weight function can be solved in a processor-efficient way.

The parallel model used in this paper is the CREW PRAM. Details of the model can be found in [6]. In the followings, unless it is stated explicitly, we will assume that the weight function is convex. The organization of the paper is as follows: In section 2, we define the maximal least-weight subsequence, which will be shown to have some nice property to enable us to identify it in parallel easily. Section 3 depicts a straight-forward algorithm implementing the idea in section 2. Finally, in section 4 we further improve our algorithm by making use of a known result in monotone matrix-searching [2, 3].

Before our discussion, let us look at the formal definition on least weight subsequences. We overload the set notation $\{\alpha_1, \alpha_2, \cdots, \alpha_m\}$ to mean a sequence of integers $\alpha_1 < \alpha_2 < \cdots < \alpha_m$.

Definition 1 For any $0 \leq i < j \leq n$, a subsequence in [i,j] is any sequence of integers $S = \{\alpha_1, \alpha_2, \cdots, \alpha_m\}$ with $\alpha_1 = i$ and $\alpha_m = j$.

- Let $\underline{w(S)}$ denote the total weight of the subsequence S, i.e., $\sum_{1 \leq l < m} w(\alpha_l, \alpha_{l+1})$.

- S is said to be a least-weight subsequence (LWS) for $[i,j]$ if $w(S) = \min\{w(S') \mid S'$ is a subsequence in $[i,j]\}$.

- Let $\underline{e(i,j)}$ denote the total weight of any least-weight subsequence for $[i,j]$. Define $e(i,i) = 0$ for all $1 \leq i \leq n$.

2 The Conventional Approach

Figure 1 shows a simple parallel algorithm for the *LWS* problem. It consists of 2 stages, and uses $n^3/\log n$ processors and $O(\log^2 n)$ time. Stage 1 computes the value of $e(i,j)$ for all i,j. Stage 2 constructs a *LWS* for $[0,n]$, which is eventually stored in an array S such that $S[i] = 1$ if and only if i is an element of the *LWS*. Since any *LWS* for $[0,n]$ ends with 0 and n, we initialize $S[0], S[n]$ to 1 and $S[i]$ to 0 for all other i's.

In the algorithm, an array E is used for computing $e(i,j)$'s. For all i, j, $e(i,j)$ is equal to $w(i,j)$ or $e(i,k) + e(k,j)$ for some k strictly between i and j. Initially we initialize each $E[i,j]$ to $w(i,j)$. Then in each iteration the current value of each $E[i,j]$ is compared with all the values $(E[i,k] + E[k,j])$, where k is strictly between i and j. $E[i,j]$ is always updated to the smallest value. If there is a *LWS* of size at most $2^l + 1$ for the range $[i,j]$, then the algorithm guarantees $E[i,j] = e(i,j)$ after the l-th iteration. In other words, each $E[i,j] = e(i,j)$ after the $\lceil \log n \rceil$-th iteration.

Whenever $E(i,j)$ is updated to the value $E(i,k) + E(k,j)$ for some k, we store the value k into the array entry $K[i,j]$, which initially has a value i. Intuitively, the array K keeps the history of how the best value of each $E[i,j]$ is obtained. After stage 1, the value stored in each $E[i,j]$ should become $e(i,j)$. We retrieve the *LWS* for $[0,n]$ from the array K by some sort of backward tracing which takes $O(\log n)$ time with n^2 processors.

Stage 1 — computing $e(i,j)$'s

Repeat $\lceil \log n \rceil$ times

    $\forall i,j$ such that $0 \leq i,j \leq n$ and $j - i \geq 2$ do in parallel

    $\{$ $temp \leftarrow \min_{i<k<j}(E[i,k] + E[k,j])$;

       if $temp < E[i,j]$ then

       $\{$ $E[i,j] \leftarrow temp$;

          $K[i,j] \leftarrow k$, where $temp = E[i,k] + E[k,j]$;

          /* Choose an arbitrary one if more than one such k */ $\}$ $\}$

Stage 2 — backward tracing

$B[0,n] \leftarrow 1$;

Repeat $\lceil \log n \rceil$ times

    $\forall i,j$ such that $0 \leq i < j \leq n$ do in parallel

    if $B[i,j] = 1$ then

    $\{$ $S[K[i,j]] \leftarrow 1$;

       $B[i, K[i,j]] \leftarrow 1$; $B[K[i,j], j] \leftarrow 1$;

       $B[i,j] \leftarrow 0$; /* To avoid concurrent write */ $\}$

Figure 1: The conventional algorithm for finding least weight subsequence

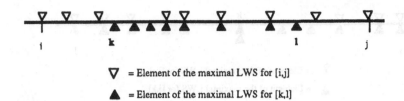

∇ = Element of the maximal LWS for [i,j]

$\blacktriangle$ = Element of the maximal LWS for [k,l]

Figure 2: The relationship between $MS_{i,j}$ and $MS_{k,l}$

3 The Maximal Least-Weight Subsequence

Notice that for any particular interval $[i,j]$, there may be more than one distinct LWS. The convex property of the weight function, however, guarantees the existence of a particular least-weight subsequence S for $[i,j]$ such that any LWS for $[i,j]$ is a subset of S. The details are as follows:

Lemma 2 For any $0 \leq i < j \leq n$, let S_1 and S_2 be any two LWS's for $[i,j]$, then the union of S_1 and S_2 is also a LWS for $[i,j]$.

Due to the length limitation, the proof of Lemma 2 is omitted here, but it can be found in [4].

By applying Lemma 2 repeatedly, we can show that the union of all LWS's for $[i,j]$ is also a LWS, which will be called the maximal LWS for $[i,j]$ and is denoted by $MS_{i,j}$. Indeed, the main result of this paper is that the maximal LWS can be found efficiently with fewer number of processors. Now let us look at some properties of the maximal LWS.

Fact 3 For any $0 \leq i < j \leq n$, let $MS_{i,j} = \{\alpha_1, \alpha_2, \cdots, \alpha_m\}$. Then for any $1 \leq r < s \leq m$, the maximal LWS for $[\alpha_r, \alpha_s]$ is exactly $\{\alpha_r, \alpha_{r+1}, \cdots, \alpha_s\}$.

Moreover, even for those k, l with $i \leq k < l \leq j$, not necessarily chosen from $\{\alpha_1, \cdots, \alpha_m\}$, $MS_{k,l}$ still contains all the elements in the portion of $MS_{i,j}$ within the interval $[k, l]$. Figure 2 depicts this relationship. To be more specific, let us look at the following lemma.

Lemma 4 Given any $0 \leq i < j \leq n$, let x be an element of the maximal LWS for $[i,j]$. Then for all k, l with $i \leq k < l \leq j$ and $k \leq x \leq l$, x is also an element of the maximal LWS for $[k, l]$.

Proof: See the Appendix. $\square$

Perhaps the most interesting property is about the ease of constructing the maximal LWS for a longer interval from two maximal LWS's of consecutive shorter intervals.

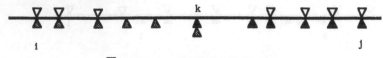

$\triangledown$ = Element of the maximal LWS for [i,j]

$\blacktriangle$ = Element of the maximal LWS for [i,k]

$\blacktriangle$ = Element of the maximal LWS for [k,j]

Figure 3: $MS_{i,j}$ is composed of $MS_{i,k}$ and $MS_{k,j}$

Theorem 5 For any i,k,j with $0 \leq i < k < j \leq n$, let $MS_{i,k} = \{\alpha_1, \alpha_2, \cdots, \alpha_{m_1}\}$ and $MS_{k,j} = \{\beta_1, \beta_2, \cdots, \beta_{m_2}\}$. Then there exist $1 \leq h < m_1$ and $1 \leq l \leq m_2$ such that $MS_{i,j} = \{\alpha_1, \alpha_2, \cdots, \alpha_h, \beta_l, \beta_{l+1}, \cdots, \beta_{m_2}\}$. (See figure 3.)

Proof: Let $MS_{i,j} = \{\gamma_1, \gamma_2, \cdots, \gamma_{m_3}\}$. Let γ_r be the largest element in $MS_{i,j}$ strictly less than k. By Lemma 4, $\gamma_1, \gamma_2, \cdots,$ and γ_r must be elements of $MS_{i,k}$. Given that $MS_{i,k} = \{\alpha_1, \alpha_2, \cdots, \alpha_{m_1}\}$, we let α_h denote the element in $MS_{i,k}$ equal to γ_r.

As γ_{r+1} is the smallest γ in $MS_{i,j}$ greater than or equal to k, by Lemma 4 again, we conclude that $\gamma_{r+1}, \cdots, \gamma_{m_3}$ is in $MS_{k,j}$. Let β_l be the element in $MS_{k,j}$ such that $\beta_l = \gamma_{r+1}$.

By Fact 3, we have $\{\alpha_1, \cdots, \alpha_h\} = MS_{i,\alpha_h} = \{\gamma_1, \cdots, \gamma_r\}$, and $\{\beta_l, \cdots, \beta_{m_2}\} = MS_{\beta_l,j} = \{\gamma_{r+1}, \cdots, \gamma_{m_3}\}$. Therefore, $MS_{i,j}$ is equal to $\{\alpha_1, \alpha_2, \cdots, \alpha_h, \beta_l, \beta_{l+1}, \cdots, \beta_{m_2}\}$. $\square$

Basically, Theorem 5 states that for any $i < k < j$, $MS_{i,j}$ is composed of the first h elements of $MS_{i,k}$ and the last $(m_2 - l + 1)$ elements of $MS_{k,j}$. Now the question is how to locate α_h and β_l. Since $MS_{i,j}$ is the maximal LWS for $[i,j]$, so intuitively, we should find the largest h and the smallest l, which make the total weight of the sequence $\{\alpha_1, \cdots, \alpha_h, \beta_l, \cdots, \beta_{m_2}\}$ be the minimum over all possible choices of h and l. Corollary 6 states formally the conditions for chosing h and l.

Corollary 6 For any i,k,j with $0 \leq i < k < j \leq n$, let $MS_{i,k} = \{\alpha_1, \alpha_2, \cdots, \alpha_{m_1}\}$ and $MS_{k,j} = \{\beta_1, \beta_2, \cdots, \beta_{m_2}\}$. Let h,l be integers with $1 \leq h < m_1$ and $1 \leq l \leq m_2$, satisfying the following two conditions:

1. $w(\{\alpha_1, \alpha_2, \cdots, \alpha_h, \beta_l, \beta_{l+1}, \cdots, \beta_{m_2}\}) =$ $\min_{1 \leq h' < m_1, 1 \leq l' \leq m_2} w(\{\alpha_1, \alpha_2, \cdots, \alpha_{h'}, \beta_{l'}, \beta_{l'+1}, \cdots, \beta_{m_2}\})^{\ddagger}$,

2. For all h', l' with $h \leq h' < m_1$ and $1 \leq l' \leq l$, if $h' \neq h$ or $l' \neq l$, then $w(\{\alpha_1, \alpha_2, \cdots, \alpha_{h'}, \beta_{l'}, \beta_{l'+1}, \cdots, \beta_{m_2}\}) > w(\{\alpha_1, \alpha_2, \cdots, \alpha_h, \beta_l, \beta_{l+1}, \cdots, \beta_{m_2}\})$.

Then $MS_{i,j} = \{\alpha_1, \alpha_2, \cdots, \alpha_h, \beta_l, \beta_{l+1}, \cdots, \beta_{m_2}\}$.

---

‡Note that $MS_{\alpha_1,\alpha_{h'}} = \{\alpha_1, \cdots, \alpha_{h'}\}$ and $MS_{\beta_{l'},\beta_{m_2}} = \{\beta_{l'}, \cdots, \beta_{m_2}\}$. Hence, $w(\{\alpha_1, \cdots, \alpha_{h'}, \beta_{l'}, \cdots, \beta_{m_2}\}) = e(\alpha_1, \alpha_{h'}) + w(\alpha_{h'}, \beta_{l'}) + e(\beta_{l'}, \beta_{m_2})$.

Proof: Let $S = \{\alpha_1, \alpha_2, \ldots, \alpha_h, \beta_l, \beta_{l+1}, \ldots, \beta_{m_2}\}$ where h, l are the integers satisfying the conditions stated above. By Theorem 5, we know that there exist $1 \leq h_0 < m_1$ and $1 \leq l_0 \leq m_2$ such that $MS_{i,j} = \{\alpha_1, \alpha_2, \ldots, \alpha_{h_0}, \beta_{l_0}, \beta_{l_0+1}, \ldots, \beta_{m_2}\}$. Then by the definition of maximal LWS, it is clear that

$$w(MS_{i,j}) = \min_{1 \leq h' < m_1, 1 \leq l' \leq m_2} w(\{\alpha_1, \alpha_2, \ldots, \alpha_{h'}, \beta_{l'}, \beta_{l'+1}, \ldots, \beta_{m_2}\}).$$

Referring to the first condition for chosing h and l, we see that $w(S) = w(MS_{i,j})$. Thus, S is a LWS for $[i,j]$, and S must be a subset of $MS_{i,j}$. In other words, $h_0 \geq h$ and $l_0 \leq l$. Due to the second condition for chosing h and l, if $h_0 \neq h$ or $l_0 \neq l$, we get $w(MS_{i,j}) > w(S)$, which is a contradiction. Therefore, $h_0 = h$ and $l_0 = l$, or equivalently, $MS_{i,j} = \{\alpha_1, \alpha_2, \cdots, \alpha_h, \beta_l, \beta_{l+1}, \cdots, \beta_{m_2}\}$. □

4 Computing the maximal LWS

Corollary 6 enables us to compute the maximal LWS for the range $[0, n]$ in a recursive manner, that is, first compute the maximal LWS's for two shorter ranges $[0, n/2]$ and $[n/2, n]$, and then combine their elements selectively. In the following, algorithm $\mathcal{A}$ is a straight-forward implementation of this idea. It takes $O(\log^2 n)$ time on a CREW PRAM with $n^2/\log n$ processors. Although the improvement of the processor complexity is not significant, it provides a framework to build a better algorithm with linear processor complexity.

For simplicity, we assume n is a power of 2. In algorithm $\mathcal{A}$, there is a recursive procedure *maximal_lws* which on parameters $(i, i + 2^t)$ computes the maximal LWS for $[i, i + 2^t]$, and stores the result in the array $next[i..i + 2^t]$ in such a way that if x belongs to the maximal LWS for $[i, i + 2^t]$, then $next[x]$ points to the next element after x in the maximal LWS for $[i, i + 2^t]$, otherwise $next[x] = \infty$.

Procedure *maximal_lws* consists of 7 steps. Step 1 divides the interval $[i, i + 2^t]$ into two equal halves, and then computes the maximal LWS for each half recursively in parallel. From the two maximal LWS's for $[i, i + 2^{t-1}]$ and for $[i + 2^{t-1}, i + 2^t]$, Step 2 to Step 6 compute the maximal LWS for $[i, i + 2^t]$ according to the conditions stated in Corollary 6. Step 7 updates the array $next$.

Algorithm $\mathcal{A}$:
For all $0 \leq i < n$, initialize $next[i]$ with the value $i + 1$. To find the maximal LWS for $[0, n]$, we call the procedure *maximal_lws*$(0, n)$.

Procedure *maximal_lws*$(i, i + 2^t)$

1. If $t = 0$ then return else $k \leftarrow i + 2^{t-1}$, $j \leftarrow i + 2^t$, and call *maximal_lws*(i, k), *maximal_lws*(k, j) in parallel.

2. For all $i \leq \alpha' < k$, if $next[\alpha'] \neq \infty$, then compute $e(i, \alpha')$, which is the prefix sum of all $w(x, next[x])$, where $i \leq x < \alpha'$ and $next[x] \leq \alpha'$.

3. For all $k \leq \beta' \leq j$, if $next[\beta'] \neq \infty$, then <u>compute $e(k, \beta')$</u>, which is the prefix sum of all $w(x, next[x])$, where $k \leq x < \beta'$ and $next[x] \leq \beta'$.

4. For all $k \leq \beta' \leq j$, if $next[\beta'] \neq \infty$, then <u>compute $e(\beta', j) = e(k, j) - e(k, \beta')$</u>.

5. For all $i \leq \alpha' < k$, if $next[\alpha'] \neq \infty$, <u>compute $mate(\alpha')$</u>, that is the <u>smallest β'</u> in the range $[k, j]$, which <u>minimizes</u> the value $\underline{e(i, \alpha') + w(\alpha', \beta') + e(\beta', j)}$.

6. Compute α_h, that is, the <u>largest α'</u> with $i \leq \alpha' < k$ and $next[\alpha'] \neq \infty$, which <u>minimizes</u> the value $\underline{e(i, \alpha') + w(\alpha', mate(\alpha')) + e(mate(\alpha'), j)}$.

7. $\beta_l, next[\alpha_h] \leftarrow mate(\alpha_h)$.
   For all $\alpha_h < x < \beta_l$, $next[x] \leftarrow \infty$.

To analyze the processor and time bounds, note that Steps 2, 3 and 6 take $O(\log n)$ time with $n / \log n$ processors whereas Steps 4 and 7 take $O(1)$ time with n processors or $O(\log n)$ time with $n / \log n$ processors. Step 5 requires $n^2 / \log n$ processors though the time spent is still $O(\log n)$. As the depth of recursion is $\log n$, the whole algorithm spends time $O(\log^2 n)$ and requires $n^2 / \log n$ processors.

To prove the correctness of algorithm $\mathcal{A}$, we show by induction on t that the procedure $maximal\_lws(i, i + 2^t)$ computes the maximal LWS for $[i, i + 2^t]$ in the array $next$. The basis where $t = 0$ is trivial, since each $next[i]$ is initially set to $i + 1$, and for each i, the maximal LWS for $[i, i + 1]$ is exactly $\{i, i + 1\}$.

Next, we turn to the induction step. From the induction hypothesis, the maximal LWS's for $[i, i + 2^{t-1}]$ and $[i + 2^{t-1}, i + 2^t]$ are stored correctly in the array $next$ after Step 1. Thus, the value of each $e(x, y)$ computed in Step 2 to Step 4 matches with its definition exactly.

In Step 5 and Step 6, α_h and β_l (i.e. $mate(\alpha_h)$) are chosen to minimize the sum $e(i, \alpha') + w(\alpha', \beta') + e(\beta', i + 2^t)$ over all $i \leq \alpha' < i + 2^{t-1}$ and $i + 2^{t-1} \leq \beta' \leq i + 2^t$, so they satisfy the first condition in Corollary 6. Moreover, by the minimization of $mate(\alpha_h)$ in Step 5 together with the maximization of α_h in Step 6, the second condition of Corollary 6 is also satisfied. Then by Corollary 6, we conclude that β_l is the next element after α_h in the maximal LWS for $[i, i + 2^t]$. The subarray $next[i..i + 2^t]$, after the updating at Step 7, stores exactly the maximal LWS for $[i, i + 2^t]$. This completes the induction proof.

5 Reduction to Matrix-Searching

In this section we show how to reduce the number of processors used in algorithm $\mathcal{A}$ to linear. First, we note that all the steps in Algorithm $\mathcal{A}$, except Step 5, can be done in $O(\log n)$ time with $n / \log n$ processors. To improve the processor requirement of Step 5, we reduce the subproblem solved in Step 5 of algorithm $\mathcal{A}$ to a well-known problem, namely the totally monotone matrix-searching problem. Since the latter problem can be solved in $O(\log n \log \log n)$

time with $n/\log\log n$ processors [3], Step 5 can also be solved in $O(\log n \log\log n)$ time with $n/\log\log n$ processors. Therefore, the time complexity of the whole algorithm becomes $O(\log^2 \log\log n)$ while the number of processors used is $n/\log\log n$.

Before we give the reduction, let us review the matrix-searching problem. Let T be an $r \times s$ matrix with real entries. Let $\delta(i)$ be the smallest column index j such that $T(i,j)$ equals the minimum (or maximum) value in the i-th row of T. The matrix T is said to be monotone if for $1 \le i_1 \le i_2 \le r$, $\delta(i_1) \le \delta(i_2)$. T is totally monotone if every 2×2 submatrix of T is monotone. The totally monotone matrix-searching problem is that given a $r \times s$ totally monotone matrix, it has to compute all $\delta(i)$'s.

The subproblem in Step 5 of algorithm $\mathcal{A}$ is to compute $mate(\alpha')$ for each α' with $next[\alpha'] \ne \infty$. Recall that $mate(\alpha')$ is the smallest β' with $next[\beta'] \ne \infty$, which minimizes the sum $e(i,\alpha') + w(\alpha',\beta') + e(\beta',j)$. Consider the subarrays $next[i..k-1]$ and $next[k..j]$. Let $i = \alpha_1 < \alpha_2 < \cdots < \alpha_r < k$ and $k = \beta_1 < \beta_2 < \cdots < \beta_s = j$ be the indices of those entries in the two arrays with value $\ne \infty$. Define a matrix T of size $r \times s$ as follows: for all $1 \le x \le r$ and $1 \le y \le s$, $T(x,y) = e(\alpha_1, \alpha_{r+1-x}) + w(\alpha_{r+1-x}, \beta_y) + e(\beta_y, \beta_s)$. Note that for any $i \le \alpha' < k$ with $next[\alpha'] \ne \infty$, $\alpha' = \alpha_x$ for some $1 \le x \le r$, and $mate(\alpha') = \beta_{\delta(r+1-x)}$. Thus, computing all $mate(\alpha')$'s is equivalent to finding all $\delta(i)$'s of matrix T.

Next we prove the matrix T be totally monotone.

Lemma 7 For any i,k,j with $0 \le i < k < j \le n$, let the maximal LWS for $[i,k]$ be $\{\alpha_1, \alpha_2, \ldots, \alpha_r, k\}$ where $i = \alpha_1 < \alpha_2 < \ldots < \alpha_r < k$, and let the maximal LWS for $[k,j]$ be $\{\beta_1, \beta_2, \ldots, \beta_s\}$ where $k = \beta_1 < \beta_2 < \ldots < \beta_s = j$. Define a matrix T of size $r \times s$ as follows: for any $1 \le x \le r$ and $1 \le y \le s$,

$$T(x,y) = e(\alpha_1, \alpha_{r+1-x}) + w(\alpha_{r+1-x}, \beta_y) + e(\beta_y, \beta_s).$$

Then T is totally monotone.

Proof: To show any 2×2 submatrix of T, say on rows $i_1 < i_2$ and columns $j_1 < j_2$, is monotone, we need to prove that if $T(i_1, j_2) < T(i_1, j_1)$ then $T(i_2, j_2) < T(i_2, j_1)$.

Obviously, $\alpha_{r+1-i_2} < \alpha_{r+1-i_1} < \beta_{j_1} < \beta_{j_2}$, so we can make use of the convex property of $w(i,j)$'s to obtain the following inequality:

$$w(\alpha_{r+1-i_1}, \beta_{j_1}) + w(\alpha_{r+1-i_2}, \beta_{j_2}) \le w(\alpha_{r+1-i_2}, \beta_{j_1}) + w(\alpha_{r+1-i_1}, \beta_{j_2}).$$

If $T(i_1, j_2) < T(i_1, j_1)$, then by the definition of T, we have $e(\alpha_1, \alpha_{r+1-i_1}) + w(\alpha_{r+1-i_1}, \beta_{j_2}) + e(\beta_{j_2}, \beta_s) < e(\alpha_1, \alpha_{r+1-i_1}) + w(\alpha_{r+1-i_1}, \beta_{j_1}) + e(\beta_{j_1}, \beta_s)$. Combining the two inequalities and adding $e(\alpha_1, \alpha_{r+1-i_2})$ to both sides, we obtain

$$e(\alpha_1, \alpha_{r+1-i_2}) + w(\alpha_{r+1-i_2}, \beta_{j_2}) + e(\beta_{j_2}, \beta_s) < e(\alpha_1, \alpha_{r+1-i_2}) + w(\alpha_{r+1-i_2}, \beta_{j_1}) + e(\beta_{j_1}, \beta_s).$$

According to the definition of matrix T, $T(i_2, j_2) = e(\alpha_1, \alpha_{r+1-i_2}) + w(\alpha_{r+1-i_2}, \beta_{j_2}) + e(\beta_{j_2}, \beta_s)$ and $T(i_2, j_1) = e(\alpha_1, \alpha_{r+1-i_2}) + w(\alpha_{r+1-i_2}, \beta_{j_1}) + e(\beta_{j_1}, \beta_s)$. Hence, $T(i_2, j_2) < T(i_2, j_1)$. $\quad\square$

6 Remarks

Though the algorithm shown in this paper can find the maximal LWS for the range $[0..n]$ only, it is possible to adapt the algorithm to compute all maximal LWS's for the ranges $[0..k]$, where $1 \leq k \leq n$. In [4] we show how this can be done without increasing the time and processor complexity of the algorithm.

As we know, the best known sequential algorithms for the convex and concave least-weight subsequence problem use different approaches. Thus, it is not surprising that our parallel algorithm doesn't work when the weight function is concave. Indeed, we think a different kind of technique is required to handle the concave case.

References

[1] M. J. Atallah, S. R. Kosaraju, L. L. Larmore, G. L. Miller, and S-H. Teng, Constructing Trees in Parallel, *Proceedings of the 1989 ACM Symposium on Parallel Algorithms and Architectures*, 1989, 421-431.

[2] A. Aggarwal, M. Klawe, S. Moran, P. Shor, and R. Wilber, Geometric Application of A Matrix-Searching Algorithm, *Algorithmica*, (2) 1987, 195-208.

[3] A. Aggarwal and J. Park, Notes on Searching in Multidimensional Arrays, *Proceedings of the 29th Annual IEEE Symposium on Foundations of Computer Science*, 1988, 497-512.

[4] K.F. Chan and T.W. Lam, Processor-Efficient Parallel Algorithms for Convex Least-Weight Subsequence Problem, Technical Report, Dept. of Computer Science, University of Hong Kong.

[5] D. Eppstein, Z. Galil, R. Giancarlo, G. Italiano, Sparse Dynamic Programming, *Proceedings of the First ACM-SIAM Symposium on Discrete Algorithms*, 1990, 513-522.

[6] S. Fortune and J. Wyllie, Parallelism in Random Access Machines, *Proceedings of the Tenth Annual ACM Symposium on Theory of Computing*, 1978, 114-118.

[7] W. Rytter, Notes on Efficient Parallel Computations for Some Dynamic Programming, *Theoretical Computer Science* 59, 1988, 297-307.

[8] Z. Galil, R. Giancarlo, Speeding up Dynamic Programming with Applications to Molecular Biology, *Theoretical Computer Science* 64, 1989, 107-118.

[9] D.S. Hirschberg and L.L. Larmore, The Least Weight Subsequence Problem, *SIAM Journal on Computing 16*, 1987, 628-638.

[10] M.M. Klawe, Speeding Up Dynamic Programming, Manuscript.

[11] F.L. Miller and J.H. Reif, Parallel Tree Contraction and Its Applications, *Proceedings of the 26th Annal IEEE Symposium on Foundations of Computer Science*, 1985, 496-503.

[12] R. Wilber, The Concave Least Weight Subsequence Problem Revisited, *Journal of Algorithms*, 9(3) September 1988, 418-425.

[13] F.F. Yao, Efficient Dynamic Programming Using Quadrangle Inequalities, *Proceedings of the 12th ACM Symposium on Theory of Computing*, 1980, 429-435.

Appendix

To prove Lemma 4, we first prove the following weaker proposition.

Proposition 8 Given any $0 \leq i < j \leq n$, let x be an element of the maximal LWS for $[i,j]$. Then for all k,l with $i \leq k < x < l \leq j$, $e(k,x) + e(x,l) \leq w(k,l)$.

Proof: Recall that $MS_{i,j}$ denotes the maximal LWS for $[i,j]$. Let $p = \max\{\alpha \in MS_{i,j} \mid \alpha \leq k\}$, $q = \min\{\alpha \in MS_{i,j} \mid \alpha > k\}$, $r = \max\{\alpha \in MS_{i,j} \mid \alpha < l\}$, and $s = \min\{\alpha \in MS_{i,j} \mid \alpha \geq l\}$. Figure 4 shows the relative position of the variables p,q,r and s. Due to the convex

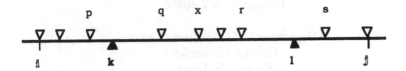

∇ = Element of the maximal LWS for [i,j]

Figure 4: The relative position of the variables p,q,r and s

property of the $w(i,j)$'s, we have the following inequalities.

1. $w(k,q) + w(p,l) \leq w(p,q) + w(k,l)$,

2. $w(r,l) + w(k,s) \leq w(k,l) + w(r,s)$,

3. $w(k,l) + w(p,s) \leq w(p,l) + w(k,s)$.

Furthermore, by the definition of $MS_{i,j}$,

4. $w(p,q) + e(q,x) + e(x,r) + w(r,s) \leq w(p,s)$.

Combining 1 to 4, it is easy to see that $w(k,q) + e(q,x) + e(x,r) + w(r,l) \leq w(k,l)$. Since $e(k,x) \leq w(k,q) + e(q,x)$ and $e(x,l) \leq e(x,r) + w(r,l)$, we can finally conclude that $e(k,x) + e(x,l) \leq w(k,l)$. □

Proof of Lemma 4: Suppose the contrary that for some $i \leq k < l \leq j$, there is an x in $MS_{i,j}$ with $k \leq x \leq l$, but not in $MS_{k,l}$. Obviously, x must be strictly between k and l. Assume $MS_{k,l} = \{\beta_1, \cdots, \beta_{m'}\}$. Let β_r be the largest element in $MS_{k,l}$ smaller than x. Since $\beta_{m'} = l$, r must be smaller than m', and hence β_{r+1} is strictly greater than x. Then by proposition 8, $e(\beta_r, x) + e(x, \beta_{r+1}) \leq w(\beta_r, \beta_{r+1})$.

On the other hand, $\{\beta_1, \cdots, \beta_{m'}\}$ is the maximal LWS for $[k,l]$, so the 2-element sequence $\{\beta_r, \beta_{r+1}\}$ is the maximal LWS and hence the only LWS for the range $[\beta_r, \beta_{r+1}]$. This would mean that $w(\beta_r, \beta_{r+1}) < e(\beta_r, y) + e(y, \beta_{r+1})$ for all $\beta_r < y < \beta_{r+1}$. Thus, contradiction occurs and we have proved Lemma 4. □

Derandomization by Exploiting Redundancy and Mutual Independence

Yijie Han[†]   *Yoshihide Igarashi*[‡]

[†]Department of Computer Science
University of Kentucky
Lexington, KY 40506

[‡]Department of Computer Science
Gunma University
Kiryu, 376 Japan

Abstract

We present schemes for derandomizing parallel algorithms by exploiting redundancy of a shrinking sample space and the mutual independence of random variables. Our design uses n mutually independent random variables built on a sample space with exponential number of points. Our scheme yields an $O(\log n)$ time parallel algorithm for the PROFIT/COST problem using no more than linear number of processors.

1 Introduction

Randomization is a powerful tool in the algorithm design. With the aid of randomization the design of algorithms for many difficult problems becomes manageable. This is particularly so in the design of parallel algorithms. Recent progress in this direction not only results in producing many efficient randomized algorithms, but also provides techniques of derandomization, *i.e.*, to convert a randomized algorithm to a deterministic algorithm through a systematic approach of removing randomness. This paper addresses the techniques of derandomization and presents a fast derandomization scheme which yields a fast parallel algorithm for the COST/PROFIT[KW][L2] problem.

A technique of derandomization due to Spencer[Spencer] is to locate a good sample point by an efficient search of the sample space. For an algorithm using n mutually independent random variables the sample space contains exponential number of sample points[ABI]. If every sample point is tested for "goodness" then the algorithm is derandomized with an explosive increase in the time complexity. In order to obtain a polynomial time algorithm, Spencer's scheme uses efficient search techniques to search the sample space for a good point. In the case when the search is binary the sample space is partitioned into two subspaces, and then the subspace with larger probability of finding a good point is preserved while the other discarded. Such a searching technique enables an algorithm to search an exponential sized sample space in polynomial time. Efficient sequential algorithms have been obtain by using this technique[Rag][Spencer].

Spencer's technique seems to be sequential in nature. A known technique[KW][L1] [ABI] to derandomize parallel algorithms is to reduce the size of the sample space by using limited independence. When the sample space contains polynomial number of points, exhaustive search can be used to locate a good point to obtain a DNC algorithm. This technique has been used successfully by Karp and Wigderson[KW], Luby[L1], Alon et al.[ABI].

In order to obtain processor efficient parallel algorithms through derandomization, Luby[L2] used the idea of binary search[Spencer] on a sample space with pairwise independent random variables. This technique was developed further by Berger and Rompel[BR] and Motwani et al. [MNN], where $\log^c n$-wise independence among random variables was used. Because the sample space contains $2^{\log^{O(1)} n}$ sample points in the design of [BR][MNN], a thoughtfully designed binary search scheme can still guarantee a DNC algorithm.

We show that a derandomization scheme faster than Luby's[L2] can be obtained. One of our observations leading to a faster derandomization scheme is to exploit the redundancy which is a consequence of a shrinking sample space. Initially a minimum sized sample space is chosen for the design of random variables of limited independence. When a binary search technique is used to search the sample space for a good sample point, the sample space is shrunken or reduced. In fact we observe the shrinkage of the sample space when conditional probability is considered. However, when the sample space is shrunken, the original assumption of independence among random variables can no longer hold. That is, dependency among random variables is expected. Such a dependency is a form of redundancy which can be exploited to the advantage of parallel algorithm design. We note that such redundancy has not been exploited before in the previous derandomization schemes[ABI][KW][L1].

Our scheme uses yet another idea of exploiting mutual independence. Previous derandomization schemes for parallel algorithms tried to stay away from mutual independence because a sample space containing n mutually independent random variables has exponential number of sample points[ABI]. We show, on the contrary, that mutual independence can be exploited for the design of fast parallel algorithms. Our design provides a fast algorithm for the PROFIT/COST[KW][L2] problem with time complexity $O(\log n)$ using no more than linear number of processors. It also improves on the time complexity of the $(\Delta + 1)$-vertex coloring algorithm obtained by Luby[L2].

2   Exploiting Redundancy

We consider the scenario of 0/1-valued uniformly distributed pairwise independent random variables in the following setting.

A set of n 0/1-valued uniformly distributed pairwise independent random variables can be designed on a sample space with $O(n)$ points. The following design is given in [ABI][BR][L2]. Let $k = \lceil \log n \rceil$. The sample space is $\Omega = \{0, 1\}^{k+1}$. For each $a = a_0 a_1 ... a_k \in \Omega$, $Pr(a) = 2^{-(k+1)}$. The value of random variables x_i, $0 \le i < n$, on point a is $x_i(a) = (\sum_{j=0}^{k-1} (i_j \cdot a_j) + a_k) \bmod 2$, where i_j is the jth bit of the binary expansion of i.

Typical functions to be searched on have the form $F(x_0, x_1, ..., x_{n-1}) = \sum_{i,j} f_{i,j}(x_i, x_j)$, where $f_{i,j}$ is defined as a function $\{0, 1\}^2 \rightarrow \mathcal{R}$. Function F models the important PROFIT/COST problem studied by Karp and Wigderson[KW] and Luby[L2]. Luby's parallel algorithm[L2] for the PROFIT/COST problem has been used as a subroutine in the derandomization of several algorithms[L2][BRS][PSZ].

The problem can be stated as follows. Given function $F(x_0, x_1, .., x_{n-1}) = \sum_{0 \le i < n, 0 \le j < n} f_{i,j}(x_i, x_j)$, find a point $(x_0, x_1, ..., x_{n-1})$ such that $F(x_0, x_1, ..., x_{n-1}) \ge E[F(x_0, x_1, ..., x_{n-1})]$. Such a point is called a good point. Function F is called the BENEFIT function and functions $f_{i,j}$'s are called the COST/PROFIT functions.

The input of the problem is dense if there are $\Omega(n^2)$ COST/PROFIT functions in F. Otherwise the input is sparse. We denote the number of COST/PROFIT functions in the input by m.

A good point can be found by searching the sample space. Luby's scheme[L2] uses binary search which fixes one bit of a at a time and evaluates the conditional expectations. His algorithm[L2] is shown below.

Algorithm Convert1:
for $l := 0$ **to** k
    begin
        $F_0 := E[F(x_0, x_1, ..., x_{n-1}) \mid a_0 = r_0, ..., a_{l-1} = r_{l-1}, a_l = 0]$;
        $F_1 := E[F(x_0, x_1, ..., x_{n-1}) \mid a_0 = r_0, ..., a_{l-1} = r_{l-1}, a_l = 1]$;
        if $F_0 \ge F_1$ **then** $a_l := 0$ **else** $a_l := 1$;
        /*The value for a_l decided above is denoted by r_l. */
    end
output$(a_0, a_1, ..., a_k)$;

It is guaranteed that the sample point $(a_0, a_1, ..., a_k)$ found is a good point, i.e., the value of F evaluated at $(a_0, a_1, ..., a_k)$ is $\ge E[F(x_0, x_1, ..., x_{n-1})]$.

By linearity of expectation, the conditional expectation evaluated in the above algorithm can be written as $E[F(x_0, x_1, ..., x_{n-1}) \mid a_0 = r_0, ..., a_l = r_l] = \sum_{i,j} E[f_{i,j}(x_i, x_j) \mid a_0 = r_0, ..., a_l = r_l]$. We assume that the input is dense, i.e., $m = \Omega(n^2)$. We will drop this assumption in the next section. We also assume that constant operations(instructions) are required for a single processor to evaluate $E[f_{i,j}(x_i,\ x_j)\ |\ a_0 = r_0,\ ...\ a_l = r_l]$. Algorithm Convert1 uses $O(n^2 \log n)$ operations. The algorithm can be implemented with $n^2/\log n$ processors and $O(\log^2 n)$ time on the EREW PRAM[BH][S] model.

We observe that as the sample space is being partitioned and reduced, the pairwise independence can no longer be maintained among n random variables. Thus dependency among random variables is expected. Such dependency is a form of redundancy which can be exploited.

After bit a_0 is set, random variables x_i and $x_{i\#0}$ become dependent, where $i\#0$ is obtained by complementing the 0-th bit of i. If a_0 is set to 0 then in fact $x_i = x_{i\#0}$. If a_0 is set to 1 then $x_i = 1 - x_{i\#0}$. Therefore we can reduce n random variables to $n/2$ random variables. Since the input is dense, we are able to cut the number of COST/PROFIT functions from m to about $m/4$.

The modified algorithm is shown below.

Algorithm Convert2:
for $l := 0$ to k
    begin
        $F_0 := \sum_{i,j} E[f_{i,j}(x_i,\ x_j)\ |\ a_0 = r_0,\ ...,\ a_{l-1} = r_{l-1},\ a_l = 0]$;
        $F_1 := \sum_{i,j} E[f_{i,j}(x_i,\ x_j)\ |\ a_0 = r_0,\ ...,\ a_{l-1} = r_{l-1},\ a_l = 1]$;
        if $F_0 \geq F_1$ then $a_l := 0$ else $a_l := 1$;
        /* The value for a_l decided above is denoted by r_l. */
        combine($f_{i,j}(x_i,\ x_j)$, $f_{i\#l,j}(x_{i\#l},\ x_j)$,
            $f_{i,j\#l}(x_i,\ x_{j\#l})$ and $f_{i\#l,j\#l}(x_{i\#l},\ x_{j\#l})$; for all i, j);
    end
output($a_0,\ a_1,\ ...,\ a_k$);

Some remarks on algorithm Convert2 is in order. The difference between the scheme used in Convert2 and previous schemes is that previous derandomization schemes use a static set of random variables while the set of random variables used in Convert2 changes dynamically as the derandomization process proceeds. Thus our scheme is a dynamic derandomization scheme while the previous schemes are static derandomization schemes.

The redundancy resulting from the shrinking sample space is being exploited resulting in a saving of $O(\log n)$ operations. Algorithm Convert2 can be implemented with $n^2/\log^2 n$ processors while still running in $O(\log^2 n)$ time, since its computing time is bounded by $(c \log^2 n)(1 + \frac{1}{2^2} + \cdots + \frac{1}{2^{2k}})$ for a constant c.

A bit more parallelism can be extracted from algorithm Convert2 by using idling processors to speed up the later steps of the algorithm. When there are $n^2/2^i$ COST/PROFIT functions left, we can extend a by i bits. Thus the number of iterations will be cut down to $O(\log \log n)$. With some modifications of algorithm Convert2, we are able to obtain the time complexity $O(\log n \log \log n)$ using $n^2/\log n \log \log n$ processors.

We note that Luby's technique[L2] can be used here to achieve the same performance. However, there is a conceptual difference between our technique and Luby's, *i.e.*, ours is a dynamic derandomization technique while Luby's is a static one. Our technique uses the idea of exploiting redundancy. This idea is not present in Luby's technique[L2].

3   Exploiting Mutual Independence

In this section we show how mutual independence can be exploited to the advantage of parallel algorithm design. Our idea is embedded in our design of the random variables which is particularly suited to the parallel searching of a good sample point.

In the previous derandomization schemes[ABI][KW][L1] the main objective of the design of random variables is to obtain a minimum sized sample space. Because small sample space requires less effort to search. Original ideas of the design of small sample space for random variables of limited independence can be found in [Bern][Jo][La]. Luby's result[L2] shows that the design of random variables should facilitate parallel search. Our result presented here carries this idea further so that our design of the random variables facilitates the dynamic derandomization process.

For the problem of finding a good sample point for function $F(x_0, x_1, .., x_{n-1}) = \sum_{i,j} f_{i,j}(x_i, x_j)$, Luby's technique of derandomization yields a DNC algorithm with time complexity $O(\log^2 n)$ using linear number of processors. When the input is dense or the COST/PROFIT functions are properly indexed, Luby's technique yields time complexity $O(\log n \log \log n)$ with linear number of processors. However, indexing the functions properly requires $O(\log^2 n \log \log n)$ time with his algorithm[L2].

Previous solutions[L2][BR][MNN] to the problem uses limited independence in order to obtain a small sample space. A small sample space is crucial to the technique of binary searching if DNC algorithms are demanded. In this section we present a case where mutual independence can be exploited to achieve faster algorithms. Since the sample space has exponential number of sample points, binary search can not be used in our situation to yield a DNC algorithm. What happens in our scheme is that the mutual independence helps us to fix random variables independently, thus resulting in a faster algorithm.

We use n 0/1-valued uniformly distributed mutually independent random variables r_i, $0 \le i < n$. Without loss of generality assuming n is a power of 2. Function F has n variables. We build a tree T which is a complete binary tree with n leaves plus a node which is the parent of the root of the complete binary tree (thus there are n interior nodes in T and the root of T has only one child). The n variables of F are associated with n leaves of T and the n random variables are associated with the interior nodes of T. The n leaves of T are numbered from 0 to $n-1$. Variable x_i is associated with leaf i.

We now randomize the variables x_i, $0 \le i < n$. Let $r_{i_0}, r_{i_1}, ..., r_{i_k}$ be the random variables on the path from leaf i to the root of T, where $k = \log n$. Random variable x_i is defined to be $x_i = (\sum_{j=0}^{k-1} i_j \cdot r_{i_j} + r_{i_k}) \, mod \, 2$. It can be verified that random variables x_i, $0 \le i < n$ are uniformly distributed pairwise independent random variables. Note the difference between our design and previous designs[ABI][L1][L2].

We shall call tree T the random variable tree.

We are to find a sample point $\vec{r} = (r_0, r_1, ..., r_{n-1})$ such that $F|_{\vec{r}} \ge E[F] =$

$\frac{1}{4}\sum_{i,j}(f_{i,j}(0,\,0)+f_{i,j}(0,\,1)+f_{i,j}(1,\,0)+f_{i,j}(1,\,1)).$

Our algorithm fixes random variables r_i (setting their values to 0's and 1's) one level in a step starting from the level next to the leaves (we shall call this level level 0) and going upward on the tree T until level k. Since there are $k+1$ interior levels in T all random variables will be fixed in $k+1$ steps.

Now consider fixing random variables at level 0. Since there are only two random variables x_j, $x_{j\#0}$ which are functions of random variable r_i (node r_i is the parent of the nodes x_j and $x_{j\#0}$) and x_j, $x_{j\#0}$ are not related to other random variables at level 0, and since random variables at level 0 are mutually independent, they can be fixed independently. This apparently saves computing time because random variables can be fixed locally, thus eliminating the need of collecting global status.

Consider in detail the fixing of r_i which is only related to x_j and $x_{j\#0}$. We simply compute $f_0 = f_{j,j\#0}(0,0)+f_{j,j\#0}(1,1)+f_{j\#0,j}(0,0)+f_{j\#0,j}(1,1)$ and $f_1 = f_{j,j\#0}(0,1)+f_{j,j\#0}(1,0)+f_{j\#0,j}(0,1)+f_{j\#0,j}(1,0)$. If $f_0 \geq f_1$ then set r_i to 0 else set r_i to 1. Our scheme will allow all random variables at level 0 be set in parallel in constant time.

Next we apply the idea of exploiting redundancy. This reduces the n random variables x_i, $0 \leq i < n$, to $n/2$ random variables. COST/PROFIT functions $f_{i,j}$ can also be combined, whenever two functions have the same variables they can be combined into one function. It can be checked that the combining can be done in constant time using linear number of processors and $O(n^2)$ space.

We now have a new function which has the same form of F but has only $n/2$ variables. A recursion on our scheme solve the problem in $O(\log n)$ time with linear number of processors.

Theorem 1: A sample point $\vec{r} = (r_0,\, r_1,\, ...,\, r_{n-1})$ satisfying $F|_{\vec{r}} \geq E[F]$ can be found in $O(\log n)$ time using linear number of processors and $O(n^2)$ space. $\Box$

4 Derandomization using Tree Contraction

In this section we outline further improvements on our derandomization algorithm. We show that the derandomization process can be viewed as a special case of tree contraction[MR]. By using the RAKE operation[MR] we are able to cut down the processor and space complexities further.

A close examination of the process of derandomization of our algorithm shows that functions $f_{i,j}$ are combined according to the so-called file-major indexing for the two dimensional array, as shown in Fig. 1. In the file-major indexing the $n \times n$ array A is divided into four subfiles $A_0 = A[0..n/2 - 1, 0..n/2 - 1]$, $A_1 = A[0..n/2 - 1, n/2..n - 1]$, $A_2 = A[n/2..n - 1, 0..n/2 - 1]$, $A_3 = A[n/2..n - 1, n/2..n - 1]$. Any element in A_i proceeds any element in A_j if $i < j$. The indexing of the elements in the same subfile is recursively defined in the same way. The indexing of function $f_{i,j}$ is the number at the i-th row and j-th column of the array. After the bits at level 0 are fixed by our algorithm, functions indexed $4k$, $4k+1$, $4k+2$, $4k+3$, $0 \leq k < n^2/4$ will be combined. After the combination of these functions they will be reindexed. The new index k will be assigned to the function combined from the original functions indexed $4k$, $4k + 1$, $4k + 2$, $4k + 3$. This allows the recursion in our algorithm to proceed.

$$
\begin{array}{cccc}
0 & 1 & 4 & 5 \\
2 & 3 & 6 & 7 \\
8 & 9 & 12 & 13 \\
10 & 11 & 14 & 15
\end{array}
$$

Fig. 1. File-major indexing

Obviously we want the input to be arranged by the file-major indexing. When the input has been arranged by the file-major indexing, we are able to build a tree which reflects the way input functions $f_{i,j}$'s are combined as the derandomization process proceeds. We shall call this tree the derandomization tree. This tree is built as follows.

We use one processor for each function $f_{i,j}$. These COST/PROFIT functions are stored in an array. Let f_{i_1,j_1} be the function stored immediately before $f_{i,j}$ and f_{i_2,j_2} be the function stored immediately after $f_{i,j}$. By looking at (i_1, j_1) and (i_2, j_2) the processor could easily figure out at which step of the derandomization $f_{i,j}$ should be combined with f_{i_1,j_1} or f_{i_2,j_2}. This information allows the tree to be built for the derandomization process. This tree has $\log n + 1$ levels. The functions at the 0-th level (leaves) are those to be combined into a new function which will be associated with the parent of these leaves. The combination happens immediately after the random variables at level 0 in the random variable tree are fixed. In general, functions at level i will be combined immediately after the random variables at level i in the random variable tree are fixed. Note that we use the term level in the derandomization tree to correspond to the level of random variable tree presented in the last section. Thus, a node at level i of the derandomization tree could be at depth $\leq \log n - i$.

Since the derandomization tree has height at most $\log n$, it can be built in $O(\log n)$ time using m processors. If the input is arranged by the file-major indexing, the tree can be built in $O(\log n)$ time using optimal $m / \log n$ processors by a careful processor scheduling.

An example of derandomization tree is illustrated in Fig. 2.

The derandomization process can now be described in terms of the derandomization tree. Combine the functions at level i of the derandomization tree immediately after the random variables at level i of the random variable tree are fixed. The combination can be accomplished by the RAKE[MR] operation which should be interpreted here as raking the leaves at level i instead of raking all leaves. The whole process of the derandomization can now be viewed as a process of tree contraction which uses only the rake operation without using the compress operation[MR].

Since the nodes in the derandomization tree can be sorted by their levels, the derandomization process can be done in $O(\log n)$ time using $m / \log n$ processors.

Theorem 2: A sample point $\vec{r} = (r_0, r_1, ..., r_{n-1})$ satisfying $F|_{\vec{r}} \geq E[F]$ can be found on the CREW PRAM in $O(\log n)$ time using optimal $m / \log n$ processors and $O(m)$ space if the input is arranged by the file-major indexing. $\square$

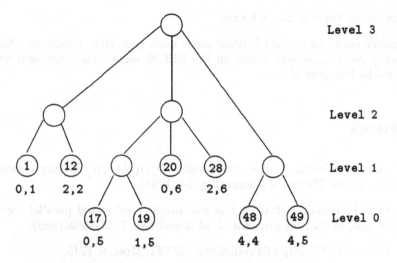

Pairs under leaves are the subscripts of COST/PROFIT functions. The numbers in the circles are the corresponding file-major indexing.

Fig. 2. A derandomization tree

We note that if the input is not arranged by the file-major indexing we could sort the input into the file-major indexing. A parallel integer sorting algorithm suffices here. Unfortunately no deterministic parallel algorithm is known for sorting integers of magnitude $n^{O(1)}$ with time complexity $O(\log n)$ using optimal number of processors. We can, of course, use parallel comparison sorting. Known algorithms[AKS][Co] have time complexity $O(\log n)$ using linear number of processors.

Corollary: A sample point $\vec{r} = (r_0, r_1, ..., r_{n-1})$ satisfying $F|_{\vec{r}} \geq E[F]$ can be found on the CREW PRAM in $O(\log n)$ time using m processors and $O(m)$ space. $\square$

Our algorithm can be used directly to solve the PROFIT/COST problem in $O(\log n)$ time with optimal number of processors and space if the input is arranged by the file-major indexing. If the file-major indexing for the input is not assumed it would in general use linear number of processors. When our algorithm is used in Luby's ($\Delta + 1$)-vertex coloring algorithm[L2] it will improve the time complexity to $O(\log^3 n)$ from $O(\log^3 n \log \log n)$.

5 Conclusions

Our ideas of exploiting redundancy and mutual independence enables us to obtain a faster derandomization scheme. In principle the idea of exploiting redundancy applies to any search schemes which partitions the sample space. In practice how the redundancy could be exploited depends on each case.

Our scheme could be studied further under more restrictive conditions. Such study would usually yield faster algorithms on the CRCW model, although achievable algorithms would be less general.

References

[AKS]. M. Ajtai, J. Komlós, and E. Szemerédi. An $O(N \log N)$ sorting network. Proc. 15th ACM Symp. on Theory of Computing, 1-9(1983).

[ABI]. N. Alon, L. Babai, A. Itai. A fast and simple randomized parallel algorithm for the maximal independent set problem. J. of Algorithms 7, 567-583(1986).

[Bern]. S. Bernstein, "Theory of Probability," GTTI, Moscow 1945.

[BH]. R. A. Borodin and J. E. Hopcroft. Routing, merging and sorting on parallel models of computation. Proc. 14th ACM Symp. on Theory of Computing, 1982, pp. 338-344.

[BR]. B. Berger, J. Rompel. Simulating $(\log^c n)$-wise independence in NC. Proc. 1989 IEEE FOCS, 2-7.

[BRS]. B. Berger, J. Rompel, P. Shor. Efficient NC algorithms for set cover with applications to learning and geometry. Proc. 1989 IEEE FOCS, 54-59.

[Co]. R. Cole. Parallel merge sort. 27th Symp. on Foundations of Comput. Sci., IEEE, 511-516(1986).

[Jo]. A. Joffe. On a set of almost deterministic k-independent random variables. Ann. Probability 2(1974), 161-162.

[KW]. R. Karp, A. Wigderson. A fast parallel algorithm for the maximal independent set problem. JACM 32:4, Oct. 1985, 762-773.

[La]. H. O. Lancaster. Pairwise statistical independence, Ann. Math. Stat. 36(1965), 1313-1317.

[L1]. M. Luby. A simple parallel algorithm for the maximal independent set problem. SIAM J. Comput. 15:4, Nov. 1986, 1036-1053.

[L2]. M. Luby. Removing randomness in parallel computation without a processor penalty. Proc. 1988 IEEE FOCS, 162-173.

[MR]. G. L. Miller and J. H. Reif. Parallel tree contraction and its application. Proc. 26th Symp. on Foundations of Computer Science, IEEE, 478-489(1985).

[MNN]. R. Motwani, J. Naor, M. Naor. The probabilistic method yields deterministic parallel algorithms. Proc. 1989 IEEE FOCS, 8-13.

[PSZ]. G. Pantziou, P. Spirakis, C. Zaroliagis. Fast parallel approximations of the maximum weighted cut problem through Derandomization. FST&TCS 9: 1989, Bangalore, India, LNCS 405, 20-29.

[Rag]. P. Raghavan. Probabilistic construction of deterministic algorithms: approximating packing integer programs. JCSS 37:4, Oct. 1988, 130-143.

[S]. M. Snir. On parallel searching. SIAM J. Comput. 14, 3(Aug. 1985), pp. 688-708.

[Spencer]. J. Spencer. Ten Lectures on the Probabilistic Method. SIAM, Philadephia, 1987.

Planar Separators and the Euclidean Norm

Hillel Gazit[*]
Department of Computer Science
Duke University

Gary L. Miller[†]
School of Computer Science
Carnegie Mellon University &
Dept of Computer Science
University of Southern California

Abstract

In this paper we show that every 2-connected embedded planar graph with faces of sizes $d_1 \ldots d_f$ has a simple cycle separator of size $1.58\sqrt{d_1^2 + \cdots + d_f^2}$ and we give an almost linear time algorithm for finding these separators, $O(n\alpha(n.n))$. We show that the new upper bound expressed as a function of $|G| = \sqrt{d_1^2 + \cdots + d_f^2}$ is no larger, up to a constant factor than previous bounds that where expressed in terms of $\sqrt{d \cdot v}$ where d is the maximum face size and v is the number of vertices and is much smaller for many graphs. The algorithms developed are simpler than earlier algorithms in that they work directly with the planar graph and its dual. They need not construct or work with the face-incidence graph as in [Mil86, GM87, GM].

1 Introduction

Planar graphs have played an important role in both sequential as well as parallel algorithm design. They arise in may areas of computation including: numerical analysis, animation, and VLSI. One of the important properties possessed by planar graphs, but not true for general graphs, is that they have small separators. Historically, a separator, in a graph $G = (V, E)$, is a subset $C \subset V$ such that (1) the remaining vertices can be partitioned into two sets: A and B, (2) $|A|.|B| \le 2/3|V|$, and (3) there are no edges between vertices in A and vertices in B. Lipton and Tarjan were the first to show that paper graphs have $O(\sqrt{v})$ separators, [LT79]. Improvements in the constant have been made by Djidjev and the first author, [Dji81, Gaz86]. There are two important additional properties we require of separators. First, we shall assign a weighting function # to the vertices, edges, and faces of the embedded planar graph and require that the separator will "separate" the weighted graph. Second, we shall require that the separator be a cycle or collection of cycles. Intuitively, this means that we separate the planar graph "drawn" on the plane by cutting along the edges of the graph. Since all the separators we construct in this paper are in fact simple cycles, we will restrict our attention to the simple cycle case. These two restrictions have been addressed by the second author in [Mil86]. We will assume that the reader has some knowledge of this paper. The following definition is from [Mil86].

---

[*]This work was supported in part by grant number N00014-88-K-0623
[†]This work was supported in part by National Science Foundation grant DCR-8713489.

Definition 1.1 *Let G be an embedded planar graph and # an assignment of nonnegative weights to the vertices, edges and faces of G which sums to 1. We say that a simple cycle C of G is a* **weighted-simple-cycle separator** *if both the weight of the interior of C and the weight of the exterior is* $\leq 2/3$.

Let G be an embedded planar graph. The size of a face of G is the number of edges, equivalently the number of faces, on its boundary counting multiplicity. The **Euclid norm of** G equals

$$|G|_2 = \sqrt{d_1^2 + \ldots + d_f^2} \quad \text{where } d_i \text{ is the size of the ith face of } G.$$

We can now state the main theorem of this paper.

Theorem 1.2 *A planar embedded graph G has a weighted-simple-cycle separator of size* $\leq 1.58|G|_2$. *The separator is computable in* $O(n\alpha(n.n))$ *sequential time where* $\alpha(n.n)$ *is the inverse of Ackerman's function.*

We first observe, up to a constant factor that Theorem 1.2 is stronger than the best previous bound of $2\sqrt{d \cdot v}$ where d is the maximum face size and v is the number of vertices, [Mil86].

Lemma 1.3 *The Euclidean norm* $|G|_2 < \sqrt{6d \cdot v}$ *if G has no faces of size* ≤ 2.

Proof: It will suffice to show $(|G|_2)^2 < 6d \cdot v$. We first observe that $\sum_{i=1}^{f} d_i < 6v$. By Euler's formula $e - f < v$. Substituting in the facts that $\sum_{i=1}^{f} d_i = 2e$ and $\sum_{i=1}^{f} d_i \geq 3f$ into the Euler equality from above we get:

$$\sum_{i=1}^{f} d_i = 3\sum_{i=1}^{f} d_i - 2\sum_{i=1}^{f} d_i \leq 6e - 6f < 6v.$$

To finish the proof consider the following equalities:

$$(|G|_2)^2 = \sum_{i=1}^{f} d_i^2 \leq d\sum_{i=1}^{f} d_i < 6d \cdot v.$$

$\square$

2 A New Way to Level a Planar Graph

In our previous papers we performed a BFS numbering in the face-incidence graph. We performed the numbering in this graph so that frontier of the search at any stage will always be a collection of cycles. In the previous papers we assumed that all the faces were basically the same size. Therefore, we added to the new level all the faces adjacent to the frontier at the same time. Here we do not assume that all the faces are the same size. Therefore, we add faces adjacent to the frontier in a more judicious manner. The main new idea in this section is to perform the BFS numbering in G itself, and to introduce artificial edges called level-edges, to be defined later, such that the frontier is again a set of simple cycles. We can not use these edges in the final cycle separator, but they will be important for bookkeeping.

A BFS numbering of G from a subset of vertices $S \subset V$ is an assignment of a number to each vertex of G equal to its distance to the nearest vertex in S.

Let L be BFS numbering of G from a single source s. Let v be some vertex at level $i > 0$. The neighboring vertices of v will be at levels $i - 1$, i or $i + 1$. We introduce two level-edges at v for each consecutive, with respect to the cyclic ordering derived from the embedding of G in the plane, block of level $i + 1$ vertices adjacent to v. If some edge e common to v is also common to another vertex at level i, then mark e level i. Otherwise, let F be the face common to v such that one edge at v is common to a level $i + 1$ vertex, and the other edge is common to one with level number $i - 1$. In this case, a level i edge will be added to the face F and the other attachment of this level-edge is obtained by following the boundary of F, using vertices of BFS number $< i$, around until a vertex of level i is found, see figure 1. Weight on the level-edge (x, y) in face F is the distance between x and y by following the boundary of F in the shorter of the two possible ways.

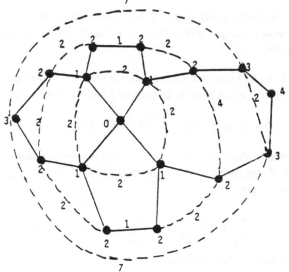

Figure 1: A BFS Number and Its Level-Edges with Weights. The solid edges are the graph edges and the dashed are the level-edges.

In prior separator papers the size of the separator was determined partly in terms of the number of vertices. In this paper it will be solely in terms of the sum of the weights on the level-edges. Let $W(G)$ equal the sum of the weights on the level-edges with respect to some BFS of G. We next show that $W(G)$ is bounded by $(|G|_2)^2$.

Lemma 2.1 *The sum of the level weights* $W(G) \leq \frac{1}{8}(|G|_2)^2$.

Proof: We show the inequality in the lemma holds for each face, and thus for the sum of the faces as a whole. Consider the following combinatorial problem: Let C be a simple cycle of size n drawn in the plane and A a set of noncrossing, vertex disjoint chords of C, i.e., C plus A forms an outer planar graph of degree at most 3. Let the weight of each chord (x, y) in A by equal to the distance between x and y in C. We will show that the sum of the weights of the chords $W(A) \leq n^2/8$. It should be clear that if we prove the last inequality then the lemma follows.

Suppose that $W(A)$ is a maximum over all such sets of chords. We first observe that no interior face of the corresponding outerplanar graph contains more than two chords or two cycle edges. If the chords of A form faces with more chords or cycle edges, we can rearrange the chords of A to strictly increase the value of $W(A)$.

Except for the middle chord, if it exists, each chord has a unique shortest path in C. Thus, we can partition the chords into two sets, two chords belong to the same set if their shortest paths intersect. Thinking of the chords as drawn vertically, we therefore partition the chords in the the left, middle, and right chords. There are several cases depending on whether the left, middle, and right chords have even or odd length and whether or not the middle chord exists. We will only handle the case when the chords all have even length and the middle chord exists. The other cases are similar. Let $2k+2$ be the length of the middle chord. Thus $n = 2(2k+2) = 4(k+1)$. In this case weight of the left chord equals those from the right and therefore we get the following equalities:

$$W(A) = (2k+2) + 2\sum_{i=1}^{k} 2i = (2k+2) + 2(k+1)k = 2(k+1)^2 = n^2/8.$$

$\square$

As in [Mil86] the level edges are formally directed edges or arcs. We decompose the level-edges into a collection of simple cycles called **level-cycles**. Since the edges in the level-cycle need not be edges in G, we will "pull" the level-edges back to paths in G. The **pull-back** of a level-cycle C is defined as follows: Let $e = (x,y)$ be level-edge in C which belongs to a face F. The vertices x and y decompose the boundary of F into two paths P_1 and P_2 from x to y. We replace e by the shorter of P_1 or P_2. If they are equal, we pick the one consisting of smaller BFS numbers, see Figure 2. Observe that the number of edges on the pull-back of level-cycle is at most the weight of the level-cycle. Thus the length of a level-cycle is greater than or equal to its pull-back.

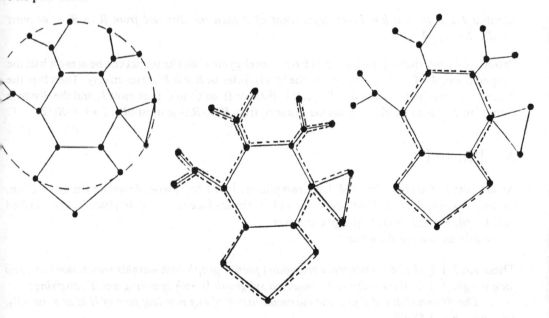

Figure 2: A Level-Cycle, Its Pull-back, and Its Retrack. The first is a graph and a level-cycle, the second is its pull-back, and the third is its retrack-cycle.

In the remainder of the paper we will assume the BFS is performed from the boundary vertices of some face F_0 of G. In this case we get the notion of a retract. The **retract** of a level-cycle C is the subgraph P' contained in the pull-back P of C defined as follows: Let H be all the faces in G which are reachable from F_0 in G^* without using edges that cross edges in P. The boundary, see [Mil86], of H is the retract of C. In the special case when H contains all the faces of G we return some fixed vertex in P such as the vertex with largest level number. If there is more than one we pick the most centered one (if there are two just pick one). In general P' is not a simple cycle but we can decomposed it into simple cycles, as we did to get the level-cycles.

The retrack-cycles form tree structure as did their level-cycles. In section 5 we show how to determine this structure, the length of each cycle, and the weight on the exterior and interior of each cycle. We will assume that we have computed this information.

Let F_0 be the face of G from which we perform the BFS of G and $R_1 \ldots R_k$ be the corresponding set of retract-cycles. The rest of the algorithms will only work with this tree of retrack-cycles. The **retract-weight** of G with respect to F_0 is:

$$|G|_R = |R_1| + \cdots + |R_k| \quad \text{where } |R_i| \text{ is the weighted length of its level-cycle.}$$

Since the retack-cycles form a tree rooted at F_0, we get an ancestor/descendent relation. If R and R' are retrack-cycles where R' is a descendent of R, then the distance from R to R' is the maximum distance from any point in R' to the closest point in R. The distance between retrack-cycles may be quite large even though they are close in the descendent relation. We can bound this distance as follows:

Lemma 2.2 *If R' is a kth direct descendent of R then the distance from R to R' is at most* $k + |R|/2 + |R'|/2.$

Proof: Observe that the distance in G from a level-cycle C and its retract can be at most half the weighted length of C. Let C and C' be the level-cycles of R and R', respectively. Therefore the distance from R' to C' is at most $|R'|/2$, the distance from C' to C is at most k, and the distance from C to R is at most $|R|/2$. Thus the distance from R' to R is at most $|R'|/2 + k + |R|/2.$ □

3 Phase I

As in [Mil86], the algorithm shall have two phases. In the first phase, developed in this section, we find a subgraph H of G with diameter and maximum face size $|G|_2$. In phase 2, as described in [Mil86], we find an $O(|G|_2)$ separator in H.

In this section we show that

Theorem 3.1 *If G is a 2-connected embedded planar graph with weights which sums to 1, no face weight $> 2/3$, there exists a 2-connected subgraph H with spanning tree T satisfying:*

1. The diameter dia of T plus the maximum size h of any non-leaf face of H is at most $|G|_2$, i.e., $dia + h \leq 1.58|G|_2$.

2. The maximum induced weight on any face of H is $\leq 2/3$.

The form of the algorithm follows quite closely that of Section 4 from [Mil86]. We will use the notation and idea from this section. Rather than use the branch cycles, defined in [Mil86], we will use the retract-cycles. As in [Mil86], the trunk $R_1, \ldots, R_t$ is a sequence of retract-cycles where R_1 is the root cycle, R_{i+1} is the direct descendence of R_i, where R_{i+1} has the largest exterior weight amongst their siblings. Let $C = R_r$ be the first cycle in the trunk with interior weight is at least $1/3$. Set $n_1 = |R_1| + \cdots + |R_r|$ and n_2 equal to the sum of the weights of the remaining retract-cycles. By the same arguments used in the proof of Lemma 5 in [Mil86] we get:

Lemma 3.2 *There exists an integer $n_1 \geq 0$ such that some nth ancestor of C, say B, satisfies $n_1 + \frac{3}{4}|B| \leq \sqrt{\frac{3}{2}}\sqrt{n_1}$.*

Proof: The proof is very similar to that of Lemma 5 in [Mil86]. We give the details for completeness.

Suppose the lemma is false and B_i is the ith ancestor of C. In this case $\frac{3}{4}|B_i| > \sqrt{\frac{3}{2}}\sqrt{n_1} - i$ for $0 \leq i \leq \lfloor \sqrt{2n_1} \rfloor$. Now the sum of the weights of the B_i satisfy the following:

$$\sum_{i=0}^{\lfloor \sqrt{\frac{3}{2}}\sqrt{n_1} \rfloor} \lceil \sqrt{\frac{3}{2}}\sqrt{n_1} \rceil - i = \sum_{i=1}^{\lceil \sqrt{\frac{3}{2}}\sqrt{n_1} \rceil} i \geq (\sqrt{\frac{3}{2}}\sqrt{n_1} + 1)(\sqrt{\frac{3}{2}}\sqrt{n_1})/2 > \frac{3}{4}n_1$$

But this contradicts the fact that the sum can be at most $\frac{3}{4}n_1$. □

Let H' be the subgraph consisting of vertices and edges "between" B and C. We next cap the large faces of H' by simultaneously adding on retract-cycles. We stop after n_2 levels where n_2 is given by the following lemma:

Lemma 3.3 *There exists an integer $n_2 \geq 0$ such that after adding all n_2 direct descents of the retract-cycles in H' the maximum retract-cycle size j, which is a leaf in the construction, satisfies $2n_2 + \frac{1}{2}j \leq \sqrt{2}\sqrt{n_2}$.*

Proof: The proof of this lemma is the same as that of Lemma 3.2.

The subgraph H will consist of a subtree of retract-cycles rooted at B plus all vertices and edges in G "between" these cycles. The height of this tree of retract-cycles will be $n_1 + n_2$. Our spanning tree of H will consist of all edges in B but one plus a BFS spanning tree from B into H. By Lemma 2.2 the total distance between the retract-cycles will be at most $n_1 + n_2 + j$. Thus the diameter of the spanning tree will be at most

$$\frac{3}{2}|B| + 2n_1 + 2n_2 + \frac{1}{2}j \leq 2\sqrt{\frac{3}{2}}\sqrt{n_1} + \sqrt{2}\sqrt{n_2} \leq \sqrt{6W(G)} \leq \sqrt{\frac{3}{4}|G|_2}.$$

We next must bound the maximum nonleaf face size of H. The maximum face is either a face from G or a new face in one of the caps. In the latter case, its size can be at most j from Lemma 3.3. In the prior case it will be at most d, the largest size face in G. If we start the BFS from the largest face, then the largest face from G in H can be at most the second largest face of G, which can be at most $|G|_2/\sqrt{2}$. We claim that $j \leq |G|_2/\sqrt{2}$. By Lemma 3.3 we know that $j \leq \sqrt{2n_2} \leq \sqrt{2W(G)} \leq \frac{1}{4}|G|_2$. Thus the diameter of the spanning tree plus the size of the maximum nonleaf face size is at most $\sqrt{\frac{3}{4}}|G|_2 + \sqrt{\frac{1}{2}}|G|_2 \leq (\sqrt{\frac{3}{4}} + \sqrt{\frac{1}{2}})|G|_2 < 1.58$.

The above discussion proves Theorem 3.1.

4 Phase II

In this section we simply observe that we can apply Theorem 5 from [Mil86]. We state the Theorem here for completeness.

Theorem 4.1 *If G is a 2-connected weighted and embedded planar graph with no face weight $> 2/3$ and T is a spanning tree of G then there exists a weight-separator of size at most the diameter of T plus the maximum non-leaf face size.*

Combining Theorems 3.1 and 4.1 we get a proof of Theorem 1.2.

5 Algorithms for Finding the Separator

In this section we discuss implementation details for finding the separator in Theorem 1.2. The sequential algorithm we presented uses Union-Find, and thus is not linear, but may require time $O(nG(n))$ time, see [AHU74]. The main difficulty is finding the tree of retract-cycles and the associated weights. The rest of the construct is straight forward and follows, for the most part, from the algorithms in [Mil86]. First we begin by determining when a face is interior to the ith level and also interior to the ith retract. We start by constructing the geometric dual of G and determining the size of each face. Next, we add the level edge to G in a BFS manner. We determine as we go the weight of each level edge. Observe that when the weight on a level reaches half the face size we use the face size in determining the rest of the edge weights. At this point the face, say F, is interior to the level. We mark the face as captured. To determine if it is interior to the retract, we check if any of the faces adjacent to F in G^* are in the retract. If we find such a face we mark F interior and check if F is adjacent to any faces which are marked as captured but not interior. We mark each such face interior and inspect in a DFS manner their adjacent faces. It follows that we traverse each edge and its dual at most a constant number of times, and determine for each face the level at which its is interior to the retract cycles, its index.

We next compute the tree determined by the retract-cycles. Observe that we cannot even explicitly write out these cycles because the sum of their lengths may be as large as n^2. Thus, we must find this tree without explicitly computing the cycles. By Lemma 3 in [Mil86] each retract-cycle at level i corresponds to the boundary of a connected component in the subgraph G_i^* of G^* induced by the faces of index $> i$. We need the following three pieces of information:

- A rooted tree corresponding to the tree of retract-cycles.

- The length of each retract-cycle.

- The weight exterior to each retract-cycles.

We compute all the information in one pass using one Union-Find. Suppose the face indexes run from 0 to k. We start by doing a Union for each edge between two faces of index k, technically we are working in the dual G^*. Using Finds we determine a representative for each connected component consisting of faces with index k. For a component, we determine its weight and boundary length. Thus, we have computed the three pieces of information for the leaf retact-cycles of index k. Suppose we have computed this information for index $k' + 1$. We compute the information for index k as follows.

1. Perform a Union for each edge between two faces one of index k' and the other of index $\geq k'$.

2. Perform a Find for each face of index k' and each representative from level $k' + 1$, obtaining new representatives for level k'.

3. The weight of each component at level k' will be the sum of the weights of the component from level $k' + 1$ in it, plus the weight contributed to it from its faces of index k' and their boundaries.

4. The new boundary will be the sum of its children's boundary sizes from level $k' + 1$, plus one for each new edge on the boundary of an index k' face, minus one for each old edge.

With this tree, all the remaining algorithms implicit in Phase I can be done in linear time. The algorithms for Phase II are in [Mil86].

6 Applications

6.1 Edge Separators

By considering the dual graph to G, we can construct edge separators.

Definition 6.1 *If* $G = (V.E)$ *is a graph with nonnegative weights on the vertices and edges which sums to one then a subset* $E' \subset E$ *of edges is a* **weighted-edge separator** *if* E' *partitions G into two disjoint subgraphs such that the weight of each is at most* $2/3$.

Theorem 6.2 below improves the previous best upper bound of $\sqrt{k \cdot v}$ obtained by [Mil86, DDSV]. In [Mil86] the weaker bound was proved for 2-connected planar graphs and was extended to all planar graphs in [DDSV]. By arguments similar to those used in Lemma 1.3, we see that this new bound is better up to constant factor.

Theorem 6.2 *A planar graph G with vertices of degree $k_1 \ldots k_v$ has an edge separator of size* $1.58\sqrt{k_1^2 + \cdots + k_v^2}$.

Proof: Suppose G is an embedded planar for which we would like a separator. It is well known that G is 2-connected if and only if its dual G^* is 2-connected, see [Eve79]. Thus, if G is 2-connected, we simply find a simple cycle separator in G^* where the weight on a vertex is zero, the weight on an edge is the weight assigned to its dual, and the weight on a face is the weight assigned to its dual, a vertex in G. It follows that the simple cycle separator of G^* corresponds to an edge separator in G. If G is not 2-connected, we compute the tree T of 2-connected components of G. The vertices of T are either 2-connected components of G or cut points. We weight each vertex of T as in the proof of Theorem 2 in [Mil86]. If T has a separating vertex v, which is also a cut vertex of G, it follows that the edges of v form an edge separator of G of size at most $|G|_2$. If on the other hand, the only separating vertex of T corresponds to a proper 2-connected component C then we proceed as follows: Consider the weight graph C where the weight of each cut vertex x belonging to C includes the weight of the subtree of T which was attached to x. The graph C is 2-connected, no vertex weight greater than $2/3$, and any weighted-edge separator of C is one for G. Thus, we have reduced the general case to the 2-connected case which we handle as above. □

6.2 The Finite Element Graph from Numerical Analysis

Planar separators can be used for direct as well as indirect methods for solving certain linear systems. Possibly the most famous of these methods is nested and parallel nested dissection, [LRT79, PR85].

Definition 6.3 *The Finite Element graph* $\hat{G} = (V. E')$ *of a planar embedded graph* $G = (V. E)$ *has as its edge set*

$$\{(v. w)| \quad v \text{ and } w \text{ share a face of } G.\}$$

By Theorem 1.2 a finite element graph $\hat{G} = (V. E)$ has a separator of size $O(|E|)$. Thus if the nonzero entries of an n by n matrix A form a finite element graph then the linear system $Ax = b$ can be solved in time $O(w(A)^{1.5})$ where $w(A)$ is the number of nonzero entries in A.

7 Open Questions

It is open whether or not the $O(\sqrt{n}\log n)$ time using $\sqrt{n}/\log n$ processors parallel algorithms in [GM] can also be used to find a Euclidean separator. It would be very interesting to know if there is a processor-efficient NC algorithm which finds Euclidean separators, [GM87].

References

[AHU74] A. Aho, J. Hopcroft, and J. Ullman. *The Design and Analysis of Computer Algorithms* Addison-Wesley, 1974.

[DDSV] K. Diks, H. N. Djidjev, O. Sykora, and I. Vrto. Edge separators for planar graphs and their applications. In Borlin, editor, *Proc. of 13th Mathematical Foundation of Computer Science*, pages 280–290, Carlsbad. Springer Verlage. LNCS 324.

[Dji81] H. N. Djidjev. A separator theorem. *Compt. End Acad. Bulg. Sci.*, 34(5):643–645, 1981

[Eve79] S. Even. *Graph Algorithms*. Computer Science Press, Potomac, Maryland, 1979.

[Gaz86] Hillel Gazit. An improved algorithm for separating a planar graph. manuscript, 1986.

[GM] Hillel Gazit and Gary L. Miller. An $O(\sqrt{n}\log n)$ optimal parallel algorithm for a separato for planar graphs. manuscript.

[GM87] Hillel Gazit and Gary L. Miller. A parallel algorithm for finding a separator in plana graphs. In *28th Annual Symposium on Foundations of Computer Science*, pages 238– 248, Los Angeles, October 1987. IEEE.

[LRT79] R. J. Lipton, D. J. Rose, and R. E. Tarjan. Generalized nested dissection. *SIAM J. on Numerical Analysis*, 16:346–358, 1979.

[LT79] R. J. Lipton and R. E. Tarjan. A separator theorem for planar graphs. *SIAM J. of Appl Math.*, 36:177–189, April 1979.

347

Mil86] Gary L. Miller. Finding small simple cycle separators for 2-connected planar graphs. *Journal of Computer and System Sciences*, 32(3):265–279, June 1986. invited publication.

PR85] Victor Pan and John Reif. Efficient parallel solution of linear systems. In *Proceedings of the 17th Annual ACM Symposium on Theory of Computing*, pages 143–152, Providence,RI, May 1985. ACM.

ON THE COMPLEXITY OF ISOMETRIC EMBEDDING IN THE HYPERCUBE

David Avis

School of Computer Science

McGill University

3480 University

Montreal, Quebec

H3A 2A7

A finite metric is h–embeddable if it can be embedded isometrically in the N-cube (hypercube) for some N. It is known that the problem of testing whether a metric is h–embeddable is NP-Complete, even if the distances are restricted to the set $\{2, 4, 6\}$. Here we study the problem where the distances are restricted to the set $\{1, 2, 3\}$ and give a polynomial time algorithm and forbidden submetric characterisation. In fact, we show these metrics are h–embeddable if and only if they are 11–*gonal* and the sum of the distances arround any triangle is even. The so-called truncated metric case, where the distances are chosen from $\{1, 2\}$ was previously settled by Assouad and Deza: the only embeddable metrics arise from the graphs $K_{1,n-1}$, $K_{2,2}$, and $2K_n$ (K_n with all distances 2).

1. Introduction

For an integer n and a finite set $\mathbf{X}=\{x_1, \ldots, x_n\}$, let $(\mathbf{X},d)$ be a metric space. In other words,

$$d(x_i, x_j) = d(x_j, x_i), \quad d(x_i, x_j) = 0, \text{ if and only if } i=j \text{ and}$$

$$d(x_i, x_j) \le d(x_i, x_k) + d(x_k, x_j) \quad \text{for distinct } i,j,k \in \{1,...,n\}.$$

The (Hamming) $N-cube$ Q_N is the set of all binary vectors of length N in R^N. The distance between two vertices of Q_N is the Hamming distance between the two binary vectors. We say that d is h–embeddable if there is a mapping

$$f : X \ \rightarrow \ Q_N$$

that preserves distances. The results in this paper in fact apply to more general *semi–metric spaces* where distinct points may have distance zero. It can easily be verified that such semi-metrics are h–embeddable if and only if the metric formed by contracting points at distance zero to single points is h–embeddable. For simplicity, we deal with metric spaces.

If d is rational, we say that $(\mathbf{X},d)$ is L^1–embeddable if and only if $(\mathbf{X}, kd)$ embeds isometrically into Q_N for some N and some $k>0$. Problems concerning h–embeddable and

ℓ^1–embeddable metrics have a long history, for a survey see Avis and Deza[2]. In this paper we will address some complexity issues related to testing whether a metric is h–embeddable. First we mention a related problem.

We can consider Q_N to be a graph on 2^N nodes by joining two nodes if the corresponding binary vectors differ in exactly one position. The problem of embedding graphs (not necessarily isometrically) into Q_N also has a long history, see for example Garey and Graham[8] for a discussion of forbidden subgraphs. Recently, this problem has again been raised in connection with Hypercube multiprocessors. It is known that the problem is NP-complete for graphs (Krumme et al.[10]), and even for embedding trees (Wagner and Corneil[11]). On the other hand, for the case of isometrically embedding a graph, the problem turns out to have a good characterisation and polynomial time algorithm (Djokovic[7]). When the graph structure is missing, the problem turns out to be NP-complete. This was shown by Chvátal[3] in the setting of intersection patterns, which we now review.

An *intersection pattern* is specified by an integer matrix $C=(c_{ij})$ of dimension $n \times n$. C is an intersection pattern if there exist finite sets A_i, $i=1, ..., n$ such that

$$|A_i \cap A_j| = c_{ij} \quad 1 \le i < j \le n.$$

The sequence $A_i, i=1,...,n$ is called a *realization*. Define $N = |\bigcup_{i=1}^{n} A_i|$ to be the size of the realization. The relationship between intersection patterns and hypercube embedding, pointed out by Deza[5], is given by the following.

Proposition 1.1

a) If $C=(c_{ij})$ is an $n \times n$ intersection pattern with given N, then d is a metric on $n+1$ points embeddable in a Q_N, where

$$d(x_i, x_j) = c_{ii} + c_{jj} - 2c_{ij}.$$

b) If d is a metric on $n+1$ points that is embeddable in Q_N, then $C=(c_{ij})$ is an $n \times n$ intersection pattern, with a realization of size N, where

$$c_{ij} = \frac{1}{2}(d(x_i, x_{n+1}) + d(x_i, x_{n+1}) - d(x_i, x_j)).\square$$

Results about the complexity of hypercube embedding follow from the following result on intersection patterns.

P1. Intersection Pattern.

Instance: Integer $n \times n$ matrix C.

Question: Is C an intersection pattern?

Complexity: NP-complete, even if each entry $c_{ij} \leq 3$. Solvable in polynomial time if each entry $c_{ij} \leq 2$, see Chvátal[3].

This result can be reformulated as a graph theory problem. Interpret C to be the adjacency matrix of a multigraph with n vertices and c_{ij} edges between vertices i and j. If C is an intersection pattern, then the edges of the multigraph can be partitioned into cliques. The sets A_i realizing the intersection pattern correspond to a list of the cliques containing the vertex i. The size N of the realization corresponds to the number of cliques in the partition. By the above result, the clique partition problem is still hard even if each vertex is contained in at most 3 cliques. If each vertex is to be contained in at most 2 cliques, then the problem reduces to line graph recognition, which is solvable in polynomial time. Using the equivalence between intersection patterns and hypercube embedding given in Proposition 1.1 and contracting points at zero distance, we obtain the following.

P2. Hypercube Embedding

Instance: Integer metric d on n points.

Question: Is d embeddable in a Hamming N-cube?

Complexity: NP-complete, even if each $d(x_i, x_j) \in \{2, 4, 6\}$. Solvable in polynomial time if each $d(x_i, x_j) \in \{2, 4\}$ and some point is at distance 2 from all points. Also solvable in polynomial time for metrics arising from bipartite graphs, Djokovic[7].

Using the graph theoretic interpretation given above, other complexity results for edge partitions of graphs can be restated in terms of hypercube embedding or intersection patterns. For example, Hoyler[9] showed that it is NP-complete to determine if the edges of a graph can be partitioned into complete subgraphs of fixed size t, for all $t \geq 3$. In terms of intersection patterns, this implies that it is NP-complete to decide if a matrix is an intersection pattern realizable by subsets $A_i \subseteq \{1, ..., N\}$ such that each element is contained in exactly t subsets.

In this paper, we extend the cases for which P2 is solvable in polynomial time. We first consider the case where the distances are chosen from $\{1, 2\}$. This is the so-called truncated graph metric: for any graph (connected or not) on n vertices the distance between adjacent vertices is one and between non-adjacent vertices it is two. We give a simple characterisation of h–embeddable metrics of this form. Our main result in an extension to metrics where the distances are chosen from $\{1, 2, 3\}$. This shows the vital importance of scaling, given the above mentioned NP-completeness for metrics with distances from $\{2, 4, 6\}$. In conclusion we note that the complexity of determing if a metric is L^1–embeddable is unknown.

2. Preliminaries

Although no necessary and sufficient conditions are known for testing whether a metric is h–embeddable there are two useful necessary conditions, first discovered by Deza[4]: the even condition and the hypermetric condition.

Even Condition. If a metric d is h–embeddable then the sum of the distances arround any triangle must be even. □

Hypermetric Condition If a metric d is h–embeddable then

$$\sum_{1 \le i < j \le n} b_i b_j d(x_i, x_j) \le 0$$

for all integers $b_1, ..., b_n$ satisfying

$$\sum_{i=1}^{n} b_i = 1. \quad □$$

A metric d is k–*gonal* if it satisfies the hypermetric condition for all integers b_i such that

$$\sum_{i=1}^{n} |b_i| \le k.$$

As an illustration, Figure 1 shows that $K_{2,3}$ does not satisfy the hypermetric condition, and in fact is not 5-gonal. The labels on the vertices are the integers b_i. Observe that all metrics are 3-gonal.

The above conditions are often useful for showing that a metric is *not* h–embeddable. In order to show that a metric is h–embeddable it suffices to give the embedding. For this it is convenient to assign sets, rather than binary vectors, to points in the metric space. Suppose a metric d is embeddable in Q_N. We represent this by choosing a set of N *atoms* $\{a, b, c, \cdots \}$, where a corresponds to the first coordinate of the embedding, b corresponds to the second coordinate of the embedding, etc. Each point in the metric space has a corresponding binary vector of length N defining its vertex in Q_N. We replace this vector by the subset of atoms formed by choosing the coordinates of the vector containing a one, see Figure 2. Now if vertices x_i and x_j receive as labels, sets S_i and S_j, we have $d(x_i, x_j) = |S_i \Delta S_j|$, where Δ represents symmetric difference.

An embedding, or labelling as we will call it, is certainly not unique. Clearly we can permute the set of atoms. Less trivially, let $S_1, \cdots, S_n$ be the label sets assigned to the n points, and let T be any subset of atoms. Then we get another valid set of labels

$S_i \, \triangle \, T$, $i=1,..,n$. Call a metric *rigid* if all labellings are equivalent up to permutations of the labels and symmetric difference. In the next section we use the fact[6] that the metric induced by 5 points with all interpoint distances two, denoted $2K_5$, is rigid. The reader may discover that $2K_4$ is not rigid.

In order to illustrate the above concepts, let us consider the case where all distances are chosen from the set $\{1, 2\}$. This problem was previously settled by Assouad and Deza[1] Consider a metric d on n points with this property. Let $G=(V,E)$ be the graph defined on the n points, where two points are adjacent if and only if they have distance one. It is immediate from the even condition that G is a complete bipartite graph (possibly with no edges, ie $K_{0,n}$). Let the vertices in the two parts be denoted S and T. We may assume that S is non-empty and $|S| \geq |T|$. Observe that all inter-point distances in S or T are two. If $|T| = 1$ then the graph is $K_{1,n-1}$ which is h–embeddable: assign a unique atom to each point in S and the empty set to the point in T. If T is empty, the metric has all distances 2, which we denote $2K_n$. This is easily seen to be h–embeddable: assign a unique atom to each vertex. I $|S| = |T| = 2$ then we have $K_{2,2}$ which is h–embeddable: assign labels $\{a\}$ and $\{b\}$ to S and labels $\{a,b\}$ and $\varnothing$ to T. Finally if $|S| > |T| \geq 2$ then we have as a subgraph $K_{2,3}$ which does not satisfy the hypermetric condition so is not h–embeddable. We can summarize the discussion in the following theorem.

Theorem 2.1[1]. A metric with distances one or two is h–embeddable if and only if it is on of $K_{1,n-1}$, $2K_n$, or $K_{2,2}$. $\square$

3. Metrics with distances $\{1, 2, 3\}$

Throughout this section we restrict ourselves to metrics d on n points in which all distances are from the set $\{1, 2, 3\}$, and to avoid some special cases we assume that $n \geq 9$. We will characterize which metrics are h–embeddable in terms of forbidden submetrics, and give an efficient embedding algorithm. We procede as in the discussion before Theorem 2.1 by defining a graph $G = (V, E)$, where V is the set of n points in a metric d and two points are adjacent if and only if they have distance one or three. We will also refer frequently to the subgraph $H = (V, \overline{E})$ of G consisting of just the edges corresponding to distance one in the metric. It is again immediate from the even condition that G is again a complete bipartite graph. Let the parts be $S = \{s_1, s_2, ..., s_{|S|}\}$ and $T = \{t_1, t_2, ..., t_{|T|}\}$. As before we may assume that S is non-empty and $|S| \geq |T|$. Since $n \geq 9$ we have $|S| \geq 5$. Observe that all inter-point distances in S or T are two. As was remarked in the introduction, the submetric induced by S is rigid and so has a unique labelling (up to permutation and symmetric difference). We will use the atoms $\{a_1, ..., a_{|S|}\}$ assigning s_i the label $\{a_i\}$.

In the rest of this section, we concentrate on the subgraph H based on a metric d that is 1–embeddable. Therefore adjacency should be interpreted with respect to H, and the notation $N(v)$ for $v \in V$ refers to the subset of vertices adjacent to v in H. We call a vertex $v \in T$ a *universal* vertex if $N(v) = S$. The following lemmas describe the structure of H.

Lemma 3.1 There is at most one universal vertex. $\square$

If there is a universal vertex, it will receive the empty set as a label, which is seen to be consistent with the labelling of S. Let T_i denote the vertices of T which each have exactly i neighbours in H.

Lemma 3.2 For $i=3,4,...,|S|-1$, $T_i = \emptyset$. At most one of T_0 and T_2 is non-empty. $\square$

Lemma 3.3 If $u,v \in T_2$ are distinct, then $|N(u) \cap N(v)| = 1$. $\square$

We define a label for each $u \in T_2$ as follows: if $N(u) = \{s_i, s_j\}$ then label u with the set $\{a_i, a_j\}$. It follows from the Lemma 3.3 that the labels in T_2 are mutually consistent and it can be checked that they are consistent with the other vertices labelled so far.

Lemma 3.4 If $|T_1| \geq 2$ then either

a) all $v \in T_1$ are adjacent to some common vertex $s_i \in S$, or

b) all $v \in T_1$ have distinct neighbours in S. $\square$

Depending on which case of Lemma 3.4 applies, we apply labels to T_1. In case (a) we give labels $\{a_i, b_0\}$ $\{a_i, b_1\}$ ... $\{a_i, b_{|T_1|-1}\}$. In case (b) we give labels $\{a_{i_1}, b_0\}$ $\{a_{i_2}, b_0\}$ ... $\{a_{i_{|T_1|}}, b_0\}$, where we assume the vertices in T_1 are adjacent to $s_{i_1}, ..., s_{i_{|T_1|}}$. In case $|T_1| = 1$, both labellings are identical. It is quite easy to check that the labelling of T_1 is consistent within T_1 and with S and the possible universal vertex. Its consistency with T_2 follows from the next lemma.

Lemma 3.5 If T_1 and T_2 are non-empty then for all $v \in T_1$ and all $w \in T_2$, $N(v) \cap N(w) \neq \emptyset$. If case (b) of Lemma 3.4 applies, we must have $|T_1| = 2$ and $|T_2| = 1$. $\square$

Note that the labelling given above is valid for the metrics satisfying the second statement of Lemma 3.5. We also observe that case(a) of Lemma 3.4 does not cause any difficulty and the given labelling gives a valid embedding, as illustrated in Figure 4(i).

Finally we label vertices in T_0 with the labels $\{b_0, b_1\}$, $\{b_0, b_2\}$,..., $\{b_0, b_{|T_0|}\}$. It is quite easy to check that the labelling of T_0 is consistent within T_0 and with S and the possible universal vertex. Since both T_0 and T_2 cannot be simultaneously non-empty (Lemma 3.2) no inconsistency can arise between these two subsets. Its consistency with T_1 follows from the next lemma.

Lemma 3.6 If T_0 and T_1 are non-empty and case (a) of Lemma 3.4 applies, then $|T_1| = 2$ and $|T_0| = 1$. $\square$

Note that the labelling given above is valid for the metrics satisfying the statement of Lemma 3.6. We also observe that case(b) of Lemma 3.4 does not cause any difficulty and the given labelling gives a valid embedding, as illustrated in Figure 4(ii). Together, these lemmas can be used to prove the following theorem.

Theorem 3.1 Let d be a metric on $n \geq 9$ points with distances chosen from $\{1, 2, 3\}$, and which satisfies the even condition. Then the following are equivalent:

(i) d is h–embeddable

(ii) d is L^1–embeddable

(iii) d is hypermetric

(iv) d is 11-gonal

(v) d does not contain any of the submetrics shown in Figure 3.

In $O(n^2)$ time a hypercube embedding for d can be found if one exists, or else one of the forbidden submetrics can be exhibited. $\square$

References

1. P. Assouad and M. Deza, "Espaces Metriques Plongeables dans un Hypercube: Aspects Combinatoires," *Annals of Discrete Math.*, vol. 8, pp. 197-210, 1980.

2. D. Avis and M. Deza, "L^1-Embedability, Complexity and Multicommodity Flows," *Research Memorandum RMI 88-11*, Dept. of Math. Engineering and Instrumentation Physics, University of Tokyo, September 1988.

3. V. Chvatal, "Recognizing Intersection Patterns," *Annals of Discrete Maths.*, vol. 8, 1980.

4. M. Deza(Tylkin), "On Hamming Geometry of Unit Cubes," *Doklady Akad. Nauk. SSR.*, vol. 134, pp. 1037-1040, 1960. English translation in Soviet Physics Dokl. 5 (1961) 940-943.

5. M. Deza, "Matrices de formes quadratiques non negatives pour des arguments binaires," *C.R. Acad. Sc. Paris*, vol. 277, pp. 873-875, 1973.

6. M. Deza and N.M. Singhi, "On Rigid Pentagons in Hypercubes," *Graphs and Combinatorics*, vol. 4, pp. 31-42, 1988.

7. D.Z Djokovic, "Distance Preserving Subgraphs of Hypercubes," *J. Comb. Theory B*, vol. 14, pp. 263-267, 1973.

8. M.R. Garey and R.L. Graham, "On Cubical Graphs," *J. Comb.Theory B*, vol. 18, pp. 84-95, 1975.

9. I. Hoyler, "The NP-Completeness of Some Edge-Partition Problems," *SIAM J. Computing*, vol. 10, pp. 713-717, 1981.

10. D. W. Krumme, K. N. Venkataraman, and G. Cybenko, "Hypercube Embedding is NP-Complete," in *Hypercube Multiprocessors 1986*, pp. 148-157, SIAM, 1986.

11. A. Wagner and D.G. Corneil, "Embedding Trees in a Hypercube is NP-complete," in *TR 197/87*, University of Toronto, 1987.

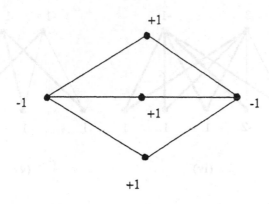

Figure 1: K$_{2,3}$

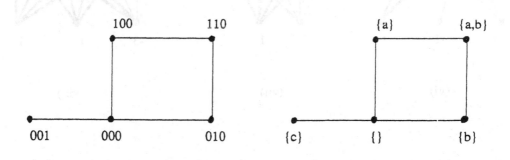

Figure 2: (a) Hypercube Embedding   (b) Equivalent Labelling

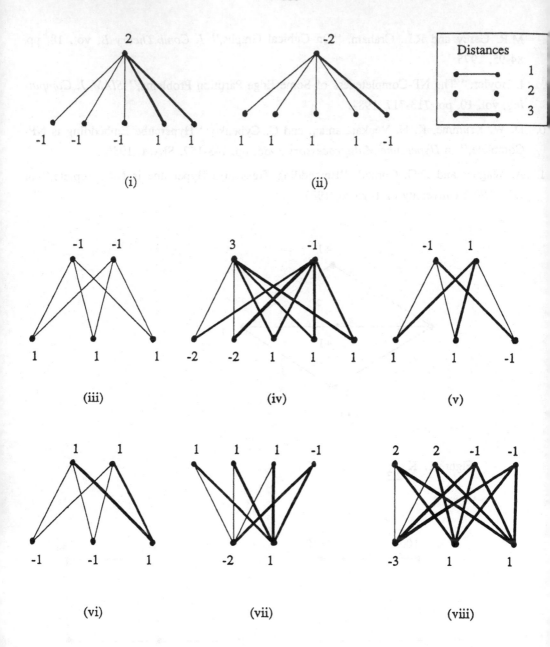

Figure 3: Forbidden Submetrics

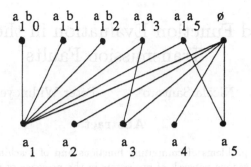

(i) Case (a) of Lemma 3.4

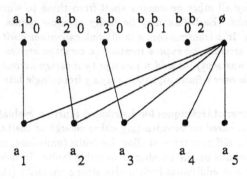

(ii) Case (b) of Lemma 3.4

Figure 4: Valid Labellings

Distributed Function Evaluation in the Presence of Transmission Faults

Nicola Santoro*   Peter Widmayer†

Abstract

We consider the problems of computing functions and of reaching an agreement in a distributed synchronous network of processors in the presence of dynamic transmission faults. We characterize the maximum number of transmission faults per clock cycle that can be tolerated for the computation of arbitrary or specific functions, with several types of faults. The n processors communicate by sending messages through dedicated communication links. Each processor has a one-way link to each other processor. In each clock cycle, each processor may send one message. The message is received in the same clock cycle by all other processors apart from those to which it travels on faulty communication links. Each link may be faulty at some points in time, and operate correctly at others. In a transmission, a faulty link can either *omit* a message (a message is sent, but none arrives), *corrupt* a message (a message arrives that is different from the message that was sent), or *add* a message (a message arrives, but none was sent). Messages are words over a finite alphabet, varying from single bits to strings of arbitrary length.

We propose a number of techniques for distributed function evaluation in the presence of transmission faults, based on broadcasting either enough of the function's arguments or the result value. For different types of allowable faults (omissions, corruptions, additions), we derive upper bounds on the number of tolerable faults. In most cases, these bounds are tight: already one additional fault makes strong majority (the weakest meaningful form of agreement) unachievable. We show that if out of $n(n-1)$ messages received by n processors per clock cycle,

- at most $n-2$ are omissions, an arbitrary function can be computed in a constant number of cycles (in contrast, with at least $n-1$ omissions strong majority is impossible);
- at most $n-2$ are omissions or corruptions, an arbitrary function can be computed;
- at most $n(n-1)$, i.e. all, messages are corruptions, an arbitrary function can be computed;
- at most $\lfloor n/2 \rfloor - 1$ are arbitrary faults, or are corruptions and processors always transmit, strong majority can be reached in a constant number of cycles (in contrast, with at least $\lfloor n/2 \rfloor$ corruptions, where processors always transmit, strong majority is impossible);
- at most $\lceil n/4 \rceil - 1$ are arbitrary faults, or are corruptions and processors always transmit, unanimity can be reached in a constant number of cycles.

For specific functions, we show how the number of cycles needed for the computation can be reduced significantly, as compared to the evaluation of an arbitrary function. Altogether, we draw quite an extensive map of possible and impossible computations in the presence of transmission faults.

---

*Carleton University, Ottawa, Canada, and Università di Bari, Italy
†Freiburg University, Freiburg, West Germany

1 Introduction

The problems of evaluating a function or reaching an agreement among the processors of a faulty distributed system are fundamental in distributed computing. They have been studied under a variety of assumptions in many forms: transaction commit (see e.g. [4, 5, 6, 8, 11, 13, 16, 17]), clock synchronization (see e.g. [7, 10, 12, 15]), distributed firing squad (see e.g. [2, 3]), reliable broadcast (see e.g. [1, 20]); see [9] for a survey. In most of these investigations, transmission faults are modelled in terms of processor failures: the loss or corruption of a message is inscribed to a faulty behavior of either the sending or the receiving processor. Unfortunately, this assumption may lead to undesirable conclusions; for instance, the entire system will be considered faulty even if only one message from each processor may be lost. Similar undesirable situations arise in models that include link failures in addition to processor failures, see e.g. [14, 18]. Since our focus is on transmission faults rather than processor failures, we model the former explicitly. This allows us to avoid undesirable characterizations and to concentrate on the inherent complexity of computing in the presence of such faults. Moreover, the transmission fault model can easily represent also link and processor failures; for instance, loss of all messages sent and received by a processor can be used to describe a crash failure of that processor. That is, the traditional processor failures model can be seen as a special, static case of the transmission faults model, where all the transmission faults involve only messages sent and received by a fixed (but a-priori unknown) set of processors.

Consider n processors distributed over a network, where each processor is directly connected to each other processor. The processors operate in synchrony: within each clock cycle (of a global clock), each processor may perform some local computation, send a message, receive a message of each other processor, and perform some additional local computation, in this order. For each processor, the sent message is broadcast through $n-1$ communication links to all $n-1$ other processors. Each processor has a unique identity, e.g. one of the numbers 1 to n, and stores some local values and some program specifying its computation; the program is the same for all processors. Initially, each processor stores some input value. The objective of a computation is either to compute a non-constant function on all inputs, or to reach an agreement among the processors on a value, depending on the inputs. Throughout this paper, by the term function we always mean (Turing-)computable functions.

Let $\sum_I$ be the non-empty alphabet of possible input values, and let $\sum_O$ be the non-empty alphabet of possible computed values (output). We consider functions $f : \sum_I^n \longrightarrow \sum_O$, where the result of a function computation must be known to each processor after a finite number of clock cycles; such a function is said to be *computable* in our setting. For some problems it is not necessary that a function is computed over the inputs, but instead agreement of a sufficiently large number of processors on some value suffices. We say that the processors reach a *p-agreement*, $1 \leq p \leq n$, if at least p processors declare the same value as their output. In order to be significant, the agreement must depend on the input values; we request that if all input values are the same, the agreement must be on that value. Let us call n-agreement *unanimity*, and $(\lceil n/2 \rceil + 1)$-agreement *strong majority*. Strong majority is the weakest form of agreement that is meaningful; agreement of fewer processors can be reached e.g. for a binary input alphabet if each processor declares its input value to be its output. Clearly, agreement is a weaker computational accomplishment than function computation.

We consider function computation and agreement in the case of a limited number of faults per clock cycle, on dynamically changing communication links. Since each processor is connected to each other processor, we altogether have $n(n-1)$ unidirectional communication links. Let Φ denote the maximum number of faulty links in each clock cycle. A fault has one out of three possible types:

- an omission (O): a message is sent, but none is received;

- a corruption (C): a message is sent, and an arbitrary different message is received;

- an addition (A): no message is sent, but an arbitrary message is received.

Links are faulty in a dynamically changing fashion: each link may be faulty at arbitrary clock cycles (and intact at all other clock cycles), with arbitrary types of faults, as long as in each cycle, at most Φ links are faulty. To better understand the effects of faults, we consider situations in which only subsets of the types of faults are allowed. For any subset $F \subseteq \{O, C, A\}$ of fault types, let an F-system denote a distributed system as defined above, where only faults of type in F may occur. The messages sent through the system relate to input values and (partial) computed values. As an alternative to sending a message, a processor in a clock cycle may decide not to send anything at all; we denote this *empty message* by δ. Let Σ_S be the non-empty alphabet of messages that can be sent; unless stated otherwise, we assume $\Sigma_S \subseteq \{\delta\} \cup \Sigma_I^\dagger \cup \Sigma_O$. Similarly, let Σ_R be the non-empty alphabet of messages that can be received; unless stated otherwise, we assume $\Sigma_R = \Sigma_S$. A transmission relation $t \subseteq \Sigma_S \times \Sigma_R$ relates the sent messages to the received messages. A pair $(m, m') \in t$ denotes the possibility of m' being received after m has been sent. We are interested in the number of faults of the different types that can be tolerated in this setting.

We will consider two basic strategies of computing a function. In the first strategy, each processor is made to know enough input values to evaluate the desired function; in general, here each processor needs to know all input values. It then computes the function value locally. Since transmission faults may occur, we need to make sure that for each processor enough information on input values is available to allow for a correct function evaluation. In the second strategy, the function value is computed at some processor and made known to all others. Here we make use of the fact that for not too many transmission faults, some processor will for sure know of all the input values; then, the function value needs to be communicated to all processors. For both strategies, it is essential that we can propagate a value to all processors, i.e., *broadcast* a value. Let us look at faults in more detail to see how broadcasting works.

2   Broadcasting in $\{O\}$-systems

In an $\{O\}$-system, messages can only be lost, but not corrupted or added. That is, $t = \{(m, m) | m \in \Sigma_S\} \cup \{(m, \delta) | m \in \Sigma_S\}$. Hence, the only problem is the propagation of a message through the system. Not even simple functions can be computed for $\Phi \geq n - 1$, since one processor may be unable to ever propagate its input value to the others, and this value may be the deciding one (for instance, the only *false* for $\Sigma_I = \{true, false\}$ and the Boolean *and*, see [19]). This bound is tight, since $n - 2$ random faults can be tolerated, as the following lemma shows.

Lemma 2.1: In an $\{O\}$-system with $\Phi \leq n - 2$, all inputs can be broadcast in at most 5 clock cycles.

Proof: Let $\sum_S = \sum_I^+$, i.e., one or more inputs are sent in one message by each processor in each cycle. Initially, each processor knows its own input value. Let each processor send all values it knows in each clock cycle. We assume that the receiver can interpret which part of the message represents the input of which processor (e.g., by using a global order on the processors, and composing a message of all inputs in this order, with special symbols for unknown values). Let us trace the propagation of the input of a particular processor, say p_0. In 1 clock cycle, at least two processors know the value, since at most $(n-2)$ of the $(n-1)$ links leaving p_0 have been faulty. In general, if at some stage k processors send a value, $1 \leq k \leq n$, $k \cdot (n-1)$ corresponding messages are sent, out of which at most Φ may be faulty. In the case that is worst with respect to fast propagation of the sent value, all $k(k-1)$ messages that the k processors exchange among themselves are correct, and each processor (except perhaps one) outside the group of the knowing k processors receiving a correct message from some of these k processors also receives a correct message from all others. Then $\lceil (k(n-1) - \Phi - k(k-1))/k \rceil$ new processors will learn the sent value. Together with the k processors already knowing the value, the number of processors increases from k to at least $n - \lfloor \Phi/k \rfloor$ in one clock cycle. Therefore, from 2 processors in 1 clock cycle we get to at least $\lceil (n+2)/2 \rceil$ in 2 cycles, to at least $n - \lfloor (n-2)/\lceil (n+2)/2 \rceil \rfloor \geq n-2$ in 3 cycles, at least $n-1$ in 4 cycles, and n in 5 cycles. Since this holds for the input of each processor, the claim is proved. □

We immediately get

Theorem 2.2: In an $\{O\}$-system with $\Phi \leq n-2$, strong majority can be reached in 2 clock cycles.

Proof: Let the processors agree on some binary property of their input, e.g. whether the length of the input string is even. Each processor determines the corresponding binary value for its input and propagates it to at least $\lceil (n+2)/2 \rceil$ processors altogether in 2 clock cycles (see Lemma 2.1). □

Theorem 2.3: In an $\{O\}$-system with $\Phi \leq n-2$, any function can be computed in at most 5 clock cycles.

Proof: Broadcast all inputs to all processors in at most 5 cycles, according to Lemma 2.1. Compute the function at each processor. □

Note that this result holds for dynamically changing faults, occurring at arbitrary links in each cycle. In contrast, in [19] it has been shown that for $\Phi \geq n-1$, not even strong majority can be reached at all, even if faults are restricted to always occur at the links from only one processor per cycle. Hence, the bound of Theorem 2.3 on the number of permissible faults is tight.

3 Function computation in $\{O, C\}$-systems

The situation becomes more complex, as messages can be lost and corrupted, but not added. That is, the transmission relation $t = \{(m, m') | m \in \sum_S - \{\delta\}, m' \in \sum_R\} \cup \{(\delta, \delta)\}$. The upper bound of $n-2$ on the number of tolerable dynamic transmission faults carries over from $\{O\}$-systems; we will show that indeed all functions can be computed in this case. To investigate the situation more closely, let us look at the number of clock cycles sufficient for various settings. Since the corruption of arbitrary strings over $\sum_I$ into arbitrary strings is hard to treat directly, let us solve our problems by bitwise transmission; more precisely, let $\sum_I = \{0, 1\}^+$ and $\sum_S = \{0, 1, \delta\} = \sum_R$. Here we have $t = \{0, 1\} \times \{0, 1, \delta\} \cup \{(\delta, \delta)\}$; hence, upon receiving 0 or 1 we can conclude that a 0 or 1 must have been sent, but we cannot distinguish between the two. We decide to deliberately restrict $\sum_S$ to $\{0, \delta\}$.

For some functions, it is enough to broadcast some specific input value; for instance, for the boolean *and* of all (boolean) inputs, it is sufficient to broadcast input value *false*. In this way, message 0 being sent represents a specific input value, fixed by the program specifying the computation of each processor.

Lemma 3.1: In an $\{O, C\}$-system with $\Phi \leq n - 2$, any specific fixed value can be broadcast in 5 cycles.

Proof: To broadcast value v, each processor holding a v sends a 0 in the first cycle; all others are silent, i.e., they send δ. Each processor receiving a 0 or 1 knows that the sent value represents v. In the following cycles, each processor knowing of a value v propagates a 0. In at most 5 cycles (see Lemma 2.1), all processors know of v, if there is a v in the system. □

In the general case, i.e., if all input values may need to be broadcast, a 0 being sent may represent a 0 or a 1, alternating with each clock cycle. If clock cycles are numbered with consecutive cardinals, message 0 may e.g. represent a 0 on even numbered and a 1 on odd numbered clock cycles; we call this mode of sending 0's and 1's *alternation*, and we call such a pair of two consecutive cycles an *alternating cycle*.

3.1 Input bits

Consider the special case of $\sum_I = \{0, 1\}$, and the computation of specific functions and of functions in general. Let 0 represent the Boolean value *false*, and let 1 represent *true*.

Theorem 3.1.1: In an $\{O, C\}$-system with $\Phi \leq n - 2$, where input symbols are bits, the Boolean *and* and the Boolean *or* of all inputs can be computed in 5 cycles.

Proof: We do not use alternation. For *and*, broadcast 0 in at most 5 cycles (see Lemma 3.1). Then, each processor returns a 0 as the result of the computation, if it knows of a 0 in the system; otherwise, it returns a 1. Similarly, the Boolean *or* is implemented as the broadcast of input symbol 1. □

But also seemingly more complex functions can still be computed quickly.

Theorem 3.1.2: In an $\{O, C\}$-system with $\Phi \leq n - 2$, where input symbols are bits, for a given, fixed k with $1 \leq k \leq n$ it can be determined in 5 clock cycles whether there are at least (alternatively: at most) k 1's (alternatively: 0's) in the input.

Proof: To see whether at least k 1's occur in the input, each processor holding a 1 sends it in the first clock cycle. At least two processors will have received all the 1's that occur in the input of other processors, in addition to its own input value; we say such a processor *knows* all the 1's. However, no processor can detect at this stage whether it knows all the 1's. In the second cycle, each processor knowing of at least k 1's starts broadcasting a 1 through the network. That is, the fixed value 1 is broadcast as outlined in the proof of Lemma 3.1. Since at least two processors start this broadcast in the second cycle, in at most 5 cycles (see Lemma 2.1) each processor knows the result: iff a 1 has reached the processor, the answer is affirmative. Similarly, the other variants of the problem are solved (note that with a binary alphabet, at least k 1's is equivalent to at most $n - k$ 0's). □

The computation of a function in general is more difficult, since either the vector of all inputs or the result of the computation, a string of bits, must be sent to all processors.

Theorem 3.1.3: In an $\{O, C\}$-system with $\Phi \leq n-2$, where input symbols are bits, any function on the multiset of inputs (as opposed to the input vector) can be computed in $1 + 4\lceil\log n\rceil$ clock cycles. (In this paper, log denotes the base 2 logarithm.)

Proof: It is sufficient for each processor to know the number of zeroes and ones in the input. This can be determined by a binary search over the integers from 1 to n for the precise number of 1's in the input. To this end, each processor holding an input 1 sends it in the first cycle. Then, the four cycles to decide whether there are at least k 1's in the input are repeated exactly $\lceil\log n\rceil$ times, for values of k defined by binary search. At the end, each processor knows the number of 1's and 0's in the input, and can compute the function. □

In the situation in which a function on a multiset of the inputs can be computed, as described above, all processors know the number of 1's in the input, say k_1. All processors that know of k_1 1's in the first cycle can detect at this stage that they also know the input vector; there are at least two of these processors. These processors can now compute an arbitrary function f on the inputs I, and then broadcast the result $f(I)$ (a string) bit by bit. Here only processors having computed $f(I)$ will initiate the broadcast of bits; all other processors will only join in the broadcast. Various possibilities exist for broadcasting a bitstring:

1. Broadcast bits by alternation. Since alternation requires two cycles (one alternating cycle) per bit to be transmitted, and we have at least two initiating processors, in 4 alternating cycles a one-bit message will have been broadcast to every processor; if a processor has not received a message in these 4 alternating cycles, δ has been sent. We can use the freedom in the encoding of one message bit in two binary symbols to cope with the problem of terminating the bitstring, by encoding the bits as follows. In an alternating cycle,

00	stands for	0, and the message continues;
$\delta 0$	stands for	1, and the message continues;
0δ	stands for	0, and the message ends;
$\delta\delta$	stands for	1, and the message ends.

   Since we have at least 2 initiators, in at most $4\lceil\log f(I)\rceil$ alternating clock cycles, i.e., in $8\lceil\log f(I)\rceil$ cycles, the result reaches all processors.

2. Instead of sacrificing an alternating bit for each bit of the entire message, we might as well first send the length of the message (with alternation, as described in 1.), and then send the message by bitwise broadcast. In this case, each bit of the message can be encoded in one binary symbol of $\{0, \delta\}$ to be broadcast. The length of $f(I)$ being $\lceil\log f(I)\rceil$, we need to broadcast $\lceil\log\lceil\log f(I)\rceil\rceil$ bits by alternation, plus an additional $\lceil\log f(I)\rceil$ bits without alternation, in $8\lceil\log\lceil\log f(I)\rceil\rceil + 4\lceil\log f(I)\rceil$ cycles altogether.

3. The process of first sending the length of a message by alternation, and then sending the message itself without alternation may be applied repeatedly, in a fashion minimizing the number of cycles needed, if desired.

Theorem 3.1.4: In an $\{O, C\}$-system with $\Phi \leq n - 2$, where input symbols are bits, any function can be computed in $1 + 4\lceil\log n\rceil + 8\lceil\log f(I)\rceil$ clock cycles (or, alternatively, in $1 + 4\lceil\log n\rceil + 8\lceil\log\lceil\log f(I)\rceil\rceil + 4\lceil\log f(I)\rceil$ cycles).

Proof: From the above discussion. □

3.2 Input strings

In the more general case of input strings, i.e., for $\sum_I = \{0,1\}^+$, we still stick to transmitting messages bitwise, due to possible corruption.

Theorem 3.2.1: In an $\{O,C\}$-system with $\Phi \leq n-2$, any function can be computed in $5((n+1)\lceil \log x_{max} \rceil + \lceil \log \lceil \log x_{max} \rceil \rceil + \ldots)$ cycles, where x_{max} is the length of a longest input string.

Proof: To compute a function, we collect all input strings at all processors, and evaluate the function at each processor. If not known already, we first guess the length x_{max} of a longest input string by means of unbounded search for x_{max}. For each probe with value x, each processor holding an input of length at least x initiates the broadcast of a message. In 5 cycles, the outcome of the probe is known to each processor. Since unbounded search terminates after $\lceil \log x_{max} \rceil + \lceil \log \lceil \log x_{max} \rceil \rceil + \ldots$ probes, $5(\lceil \log x_{max} \rceil + \lceil \log \lceil \log x_{max} \rceil \rceil + \ldots)$ cycles suffice. Then, all inputs are adjusted to maximum length (e.g., by adding leading zeroes if they represent binary numbers). Finally, each processor in turn, as determined e.g. by a global order on processor identities, transmits its input string by initiating the propagation bit by bit. Again, transmitting a message and not transmitting a message are used to transmit the symbols 0 and 1 of $\sum_I$. This consumes $5 \cdot \lceil \log x_{max} \rceil$ cycles for each of the n inputs, adding up to the claimed total. □

Specific functions, again, can be computed faster.

Theorem 3.2.2: In an $\{O,C\}$-system with $\Phi \leq n-2$, the maximum (minimum) value mv of all inputs can be determined in $5(\lceil \log mv \rceil + \lceil \log \lceil \log mv \rceil \rceil + \ldots)$ cycles.

Proof: We use unbounded search for mv, as we did for x_{max} in the proof of Theorem 3.2.1. □

Theorem 3.2.3: In an $\{O,C\}$-system with $\Phi \leq n-2$, the rank r of a given value v, i.e. the number of inputs x_i with $x_i \leq v$, can be computed in $1 + 4 \cdot \lceil \log n \rceil$ cycles.

Proof: In the first cycle, each processor whose input is at most v sends; all others are silent. In the second cycle, each processor computes the number n_i of messages it has received, and increments it if its own input value is at most v. The largest n_i equal r and are known to at least two processors. Since $1 \leq n_i \leq n$, r can be determined by binary search in $4 \cdot \lceil \log n \rceil$ cycles, starting at cycle 2 (each probe in binary search is similar to the probes of unbounded search, cf. the proof of Theorem 3.2.1). □

By using similar techniques, we can select the k-th largest input value, compute the sum of the input values (interpreted as binary numbers), and compute various statistical functions on the inputs, like mean, variance, standard deviation, momentum, within similar time bounds, i.e., faster than arbitrary functions.

4 Other types of faults

To see how the allowable types of faults influence the computing capabilities, let us sketch a few systems' properties. The following theorem shows that for systems with arbitrary corruption, but no loss or spontaneous occurrence of messages, any function can be computed even if all communication links are faulty at all times.

Theorem 4.1: In a $\{C\}$-system with $\Phi \leq n(n-1)$, any function can be computed in $x_{max} + \lceil \log x_{max} \rceil + \lceil \log \lceil \log x_{max} \rceil \rceil + \ldots$ cycles, where x_{max} is the length of a longest input string.

Proof: Bits can be transmitted without ambiguity, by e.g. representing a 0 by the transmission of a 0 or 1, and by representing a 1 by δ. To compute a function, all inputs are collected in all places, by first guessing x_{max} with unbounded search in $\lceil \log x_{max} \rceil + \lceil \log \lceil \log x_{max} \rceil \rceil + \ldots$ cycles, and then transmitting simultaneously for all processors the input strings bit by bit, filled with leading or trailing zeroes or ones, as appropiate. □

Theorem 4.2: In a $\{C\}$-system with $\Phi \leq n(n-1)$, any function of the form $f(I) = |\{x \in I : P(x) \text{ holds}\}|$ for some (Turing-)computable predicate P, can be computed in 1 cycle.

Proof: Each processor computes a bit, depending on whether P holds for its input. Since bit messages are unambiguous, the appropriate bits can be collected by each processor in one cycle. □

Similarly, all links may be faulty at all times, if only spontaneous messages may occur, but no corruption or loss of messages.

Theorem 4.3: In an $\{A\}$-system with $\Phi \leq n(n-1)$, any function can be computed in 1 cycle.

Proof: Each processor sends its input string without ambiguity. After receiving all inputs, each processor computes the desired function. □

Instead of voluntarily always transmitting non-empty messages to avoid the occurrence of additions, the F-system we study may be restricted to always transmit; in that case, we call it an F^+-system. Compared with an F-system, an F^+-system loses the power of sending δ; e.g., the arguments presented in Section 3 for $\{O, C\}$-systems do not apply to $\{O, C\}^+$-systems. From Theorem 4.3 we conclude that function computation in $\{A\}^+$-systems is the same as in $\{A\}$-systems. We say that the *fault tolerance* of two types of systems *is the same*, if they can tolerate the same number of dynamic transmission faults to compute an arbitrary function.

Theorem 4.4: The fault tolerance of $\{C, A\}$-systems is the same as that of $\{O, C\}$-systems.

Proof: For $\sum_S = \{0, 1, \delta\} = \sum_R$, in $\{C, A\}$-systems we have $t = \{0, 1, \delta\} \times \{0, 1\} \cup \{(\delta, \delta)\}$. This is isomorphic to the transmission function of $\{O, C\}$-systems in that δ and $\{0, 1\}$ exchange their roles. □

Theorem 4.5: The fault tolerance of $\{O, A\}$-systems is the same as that of $\{O\}$-systems.

Proof: We decide to always transmit and therefore get an $\{O, A\}^+$-system, which is as tolerant as an $\{O\}^+$-system, which in turn is as tolerant as an $\{O\}$-system, since broadcast of a value can be accomplished by always transmitting (see Lemma 2.1). □

By similar arguments, we can show that the fault tolerance of $\{O, C, A\}$-systems, $\{O, C, A\}^+$-systems, $\{O, C\}^+$-systems and $\{C\}^+$-systems is the same; an upper bound on the number of tolerable faults, in this case, is $\lfloor n/2 \rfloor - 1$, since with $\lfloor n/2 \rfloor$ faults, not even strong majority can be achieved (see [19]). The situation for these types of systems differs somewhat from the previous ones, since from a received value, nothing can be inferred about the sent value. Hence, we argue only on the number of messages that have to be correct; a sufficient number of correct messages for each processor will allow a correct decision.

Theorem 4.6: In a $\{C\}^+$-system with $\Phi \leq \lfloor n/2 \rfloor - 1$, strong majority can be achieved in one cycle.

Proof: Each processor computes the value of a binary property of its input (see Theorem 2.2), and stores it as a special input bit. A strong majority decides on the boolean *or* of the input bits in the following way. In the first cycle, each processor sends its input bit. In one cycle, let a processor decide on 1, iff it knows of an input bit value 1 (by receiving it or storing it as its own value). To see the correctness of this procedure, distinguish two cases. First, there is a 1 in the input bits. Then at most $\lfloor n/2 \rfloor - 1$ of the n processors may not know of this fact, as a result of message corruption. This leaves $\lceil n/2 \rceil + 1$ processors to take the correct decision and agree on 1. Second, there is no 1 in the input bits. Similarly, at most $\lfloor n/2 \rfloor - 1$ processors can be made to decide on 1, which leaves $\lceil n/2 \rceil + 1$ processors to take the correct decision and agree on 0. □

To achieve unanimity, we take two cycles and transmit the following bits, based on the locally computed input bits as in Theorem 4.6. In the first cycle, each processor sends its input bit. In the second cycle, each processor knowing of a 1 sends a 1; all others send 0. In the second cycle, a processor decides on 1 iff it knows of at least $n - 2(\lceil n/4 \rceil - 1)$ 1's.

Theorem 4.7: In an $\{C\}^+$-system with $\Phi \le \lceil n/4 \rceil - 1$, unanimity can be achieved in two cycles.

Proof: For the algorithm defined above and Φ faulty transmissions per cycle, distinguish two cases. First, there is a 1 in the input bits. After the first transmission, at most Φ processors do not know of a 1. After the second transmission, each processor must know of at least $n - 2\Phi$ 1's, because at most Φ messages from the at least $n - \Phi$ processors knowing of a 1 in the second cycle can have been corrupted. Second, there is no 1 in the input bits. Then after the first transmission, at most Φ processors may know of a 1, due to corruption. After the second transmission, a processor may know of at most 2Φ 1's, namely Φ 1's from the processors knowing of a 1 after the first transmission, and Φ 1's from other processors, due to corruption in the second transmission. To distinguish between the two cases it is sufficient to have $n - 2\Phi > 2\Phi$, yielding $\Phi < \lceil n/4 \rceil$. □

Note that there is an unresolved gap between the number $\lceil n/4 \rceil - 1$ of faults that we can tolerate for unanimity (Theorem 4.7) and the number $\lfloor n/2 \rfloor$ of faults that can be shown to entail impossibility of strong majority (see [19]).

Acknowledgement

The authors would like to thank John Dumovich, Andreas Hutflesz, Shy Kutten, Jan van Leeuwen, Jan Pachl, Gabriele Reich and Shmuel Zaks for helpful discussions on these results, and an unknown referee for helpful comments. This work has been supported in part by the Natural Sciences and Engineering Research Council (Canada).

References

[1] K. Birman and T. Joseph, Reliable communication in the presence of failures. ACM Trans. Comp. Syst. 5, 1 (Feb. 1987).

[2] J. Burns and N. A. Lynch, The Byzantine firing squad problem. In Adv. Comp. Res., Vol 4: Parallel and Distributed Computing, JAI Press Inc. Greenwich, Connecticut.

[3] B. Coan, D. Dolev, C. Dwork and L. Stockmeyer, The distributed firing squad problem. In Proc. 17th ACM Symp. on Theory of Computing, Providence, May 1984, 335-345.

[4] D. Dolev, The Byzantine Generals strike again. J. Algorithms 3, 1 (1982), 14-30.

[5] D. Dolev, C. Dwork and L. Stockmeyer, On the minimal synchronism needed for distributed consensus. J. ACM 34, 1(Jan. 1987), 77-97.

[6] D. Dolev, M. L. Fisher, R. Fowler, N. A. Lynch and H. R. Strong, Efficient Byzantine agreement without authentication. Inf. Control 52,3 (1982), 256-274.

[7] D. Dolev, J. Y. Halpern and H. R. Strong, On the possibility and impossibility of achieving clock synchronization. In Proc. 16th ACM Symp. on Theory of Computing, Washington, May 1984, pp. 504-510.

[8] D. Dolev, H. R. Strong, Authenticated algorithms for Byzantine agreement. SIAM J. Computing 12, 4 (Nov. 1983), 656-666.

[9] M. J. Fisher, The consensus problem in unreliable distributed systems (a brief survey). Dept. Comp. Sci. Tech. Rep. 273, Yale University, June 1983.

[10] J. Y. Halpern, B. Simons, H. R. Strong and D. Dolev, Fault tolerant clock synchronization. In Proc. 3rd ACM Symp. on Principles of Distributed Computing, Vancouver, Aug. 1984, 89-102.

[11] L. Lamport, The weak Byzantine Generals problem. J. ACM 30, (July 1983), 668-676.

[12] L. Lamport and P. M. Melliar-Smith, Synchronizing clocks in presence of faults. J. ACM 32, 1 (Jan. 1985), 52-78.

[13] L. Lamport, R. Shostak and M. Pease, The Byzantine Generals problem. ACM Trans. Prog. Lang. Syst. 4, 3 (July 1982), 382-401.

[14] F. Ling, T. Kameda, Byzantine agreement under network failures. Tech. Rep. LCCR 87-18, Simon Fraser University, 1987.

[15] J. Lundelius and N. A. Lynch, A new fault-tolerant algorithm for clock synchronization. Inf. Control 62, 2 (1984), 190-204.

[16] M. Pease, R. Shostak and L. Lamport, Reaching agreement in presence of faults. J. ACM 27, 2 (April 1980), 228-234.

[17] K. J. Perry, A framework for agreement. In Proc. 2nd Int. Workshop on Distributed Algorithms, Amsterdam, July 1987, 57-75.

[18] K. J. Perry and S. Toueg, Distributed agreement in the presence of processor and communication faults. IEEE Trans. Software Engineering SE-12, 3 (March 1986)

[19] N. Santoro, P. Widmayer, Time is not a healer. In Proc. 6th Ann. Symposium Theor. Aspects of Computer Science, Paderborn, February 1989, LNCS 349, 304 - 313.

[20] T. K. Srikanth and S. Toueg, Simulating authenticated broadcasts to derive simple fault-tolerant algorithms. Distributed Computing 2, (1987), 80-94.

OPTIMAL LINEAR BROADCAST

(Extended Abstract)

Sara Bitan and Shmuel Zaks

Department of Computer Science
Technion, Haifa, Israel

ABSTRACT

As the communication lines and switching hardware in distributed networks become much faster, a new trend of algorithms is needed to utilize it. Traditional broadcast algorithm send one packet along each communication line at a time, and propagate it by replicating and sending it on all the outgoing lines. This method does not use the high bandwidth or the switching hardware, and it overloads the processor. Our routing algorithms send several packets simultaneously on one communication line, and each packet is sent along a linear route. In this paper we assume that the switching hardware has limited strength. We present algorithms that compute an optimal broadcast routing, using a bounded number of linear routes. We prove that a greedy algorithm solves the problem, and present an improved algorithm for the same problem.

1. INTRODUCTION

An *asynchronous distributed network* consists of a set of processors connected by communication lines. *Broadcasting* in such a network is the action of propagating information from a *source* processor to all the others. Traditional broadcast algorithms in packet-switching networks usually propagate information using a spanning tree in the following way [S]: the source node sends the information to all its neighbors in the tree, which in turn send it to all their neighbors, except the one who sent the packet, and so on. In this algorithm every internal processor has to replicate the packet and send it over its tree links. In a high bandwidth network such as PARIS [CiG], this algorithm may be inefficient, since it routes only one message along each communication line, thus not utilizing the high bandwidth.

In high bandwidth networks very fast (fiber-optics) communication lines are used, so the traditional assumption that neglects processing time in each node does not hold. Therefor, a processor should not waist processing time on messages that are not destined to a it. The designers of PARIS solved this problem by building a processing element that contains a Network Control Unit (NCU), which is a general purpose processor,and a Switching Subsystem (SS), a special hardware that executes the switching functions [CiG]. Using a special routing technique developed in PARIS, Automatic Network Routing (ANR), the SS can forward a packet without involving the NCU, and send a copy of the packet to the NCU if needed, thus off-loading NCU. An example of an algorithm that uses the special properties of the SS can be found in [CS] and [CGK].

The SS does not support replication of a packet, and simultaneous send of the same packet to several nodes, that are required by the traditional broadcast algorithm ([S]). The SS can only switch a packet from one link to another, and/or send it to the NCU. It supports what is called *linear routing* [ChG], in which each packet is sent along a linear path. Linear broadcast routing means sending a packet from a one node to all the nodes in the network, using only linear routes. In such routing most of the lines carry more than one message simultaneously. Since the communication lines are fast and the switching is done in hardware we assume that the sending a packet more than once through the same line is cheaper than accessing the general purpose processor. We also say that a packet can go through a route between any two NCUs in the network in one time unit (see [CGK]), so the broadcast process terminates in one time unit.

Although the switching is done in hardware there is still a bound on the number of packets that can be sent simultaneously through one SS. This bound depends on the number of buffers in the SS. We develop optimal linear broadcast routing algorithms in a given network for a given broadcast node, under the above limitation. Our algorithms assume the existence of a spanning tree in the network, and the linear broadcast routing covers the tree by paths, each corresponding to a packet. The problem of linear broadcast routing was suggested in [ChG], who showed it is NP-complete for general graphs, and solvable in polynomial time for trees. They present an algorithm that solves the tree linear broadcast problem in $O(N^2)$, where N is the number of nodes in the network. They, however, ignore the node simultaneous send limitations.

In this paper we present algorithms for finding an optimal broadcast in case where the processing unit simultaneous transmission capability is limited, i.e. it can transmit at most a bounded number, k, of packets simultaneously. First an algorithm, A1, whose time complexity is $O(k \cdot N)$, is presented. It is presented mainly because it demonstrates the nature of linear broadcast routing, and therefor is easy to proof. It is then improved to a $O(N)$ algorithm (A2). Both algorithms solve the tree linear broadcast problem, presented in [CHG].

The rest of this paper is organized as follows : in Section 2 we present definitions and notations. In Section 3 we present algorithm A1. Proof of correctness and analysis are given in Section 4. Algorithm A2 is given in Section 5. Applications, including a distributed version of our algorithms, are discussed in Section 6. Most of the proofs are omitted in this Extended Abstract, and can be found in [BZ].

2. PRELIMINARIES

2.1. Basic definitions

A communication network is viewed as an undirected graph $G=(V,E)$, where the vertices in V represent the processors in the network, and the edges in E represent the (bidirectional) communication lines between the processors.

In this paper all the graphs are trees. A *linear broadcast problem* (T,w,r) is specified by a tree $T=(V,E)$, a *weight* function $w:E \to N$ where $N=\{1,2,3,...\}$ and a *broadcast vertex* $r \in V$, from which the broadcast starts. As explained in the Introduction, our goal is to cover the tree by (not necessarily disjoint) paths, all starting at r, and each corresponding to one packet.

Given a linear broadcast problem (T,w,r) we use the following definitions and notations. *degree* (v) denotes the degree of a vertex $v \in V$. The (simple) path connecting the vertices u and v is denoted by $p(u,v)$. We identify the path with the sequence of vertices and edges; namely, the path $v_0 \overset{e_1}{—} v_1 \overset{e_2}{—} ... \overset{e_n}{—} v_n$ is denoted by $p(v_0,v_n)=<v_0,e_1,v_1,e_2, \cdots ,e_n,v_n>$. The set of vertices $\{v_0,v_1, \cdots ,v_n\}$ that appear in a path p, will be denoted $V(p)$. *father* (v) is the vertex predecessor of v in the path $p(r,v)$. *sons* (v) is the set of nodes u such that $v=father(u)$. Given two paths p and q , we say that $p \leq q$ if p is a prefix of q.

For a node v, *subtree* (v) is the subtree spanned by the set of nodes u such that $p(r,v) \leq p(r,u)$. The *weight* $w(p)$ of a path p is the sum of the weights of its edges; namely, for a path $p(v_0,v_n)$ as above, $w(p(v_0,v_n))=\sum_{i=1}^{n} w(e_i)$. The *depth* of a vertex v is $w(p(r,v))$.

Given two paths $p=<v_0,e_1,...,e_n,v_n>$ and $q=<v_n,e_{n+1},...,e_m,v_m>$, we define their *concatenation* as $p \cdot q=<v_0,e_1, \cdots ,e_n,v_n, e_{n+1},...,e_m,v_m>$.

Definition : A *trail* is a concatenation of $l \geq 1$ simple paths, and it will be denoted by $t(v_0,v_1,...,v_l)= p(v_0,v_1) \cdot p(v_1,v_2) \cdot ... \cdot p(v_{l-1},v_l)$. The *weight* $w(t)$ of a trail $t=t(v_0,v_1,v_2...,v_{l-1},v_l)$ is the total weight of the paths that define it; namely, $w(t)=\sum_{i=1}^{l} w(p(v_{i-1},v_i))$.

Definition : Given a trail $t=t(v_0,v_n)= <v_0,e_1, \cdots ,e_n,v_n>$ and an edge $e \in t$, then we say that t *goes down* e, if for some i, $0 \leq i<n$, $e=(v_i,v_{i+1})$, and $depth(v_i)<depth(v_{i+1})$. We say that t *goes up* e, if $depth(v_i)>depth(v_{i+1})$. t *goes up a vertex* v if t goes up the edge $(v,father(v))$. Note that some edges/vertices can appear more than once in t, and thus t can go up and down those edges/vertices.

Definition : A *broadcast set* for a broadcast problem (T,w,r) is a set of trails $\psi=\{t_1,t_2,\ldots,t_m\}$ such that each trail t_i starts at r, and every node $v \in V$ appears in at least one of the trails. The *weight* $w(\psi)$ of a broadcast set $\psi=\{t_1,t_2,\ldots,t_m\}$ is the sum of the weights of the trails that define it; namely, $w(\psi)=\sum_{i=1}^{m} w(t_i)$. An *optimal* broadcast set is a minimum weight broadcast set. A broadcast set that contains a single trail that traverses each edge $e \in E$ exactly twice, is *a Depth First Search set*, shortly denoted as *a DFS set*.

Example 1 : Consider the tree $T_0=(V,E)$ with weights as shown in Figure 1. The path connecting the vertices r and a is $p(r,a)=<r,e_1,i,e_2,a>$, and its weight is $w(p(r,a))=w(e_1)+w(e_2)=12$. $t=t(r,a,i,b)=p(r,a) \cdot p(a,i) \cdot p(i,b)$ is a trail connecting r and b, with weight $w(t)=w(p(r,a))+w(p(a,i))+w(p(i,b))=25$. t goes down e_3,e_2 and k, it goes up a and e_2.

$\psi_D=\{t(r,a,b,c,d,e,f,g,h,r)\}$ is a DFS set for T_0. $w(\psi_D)=2 \cdot \sum_{i=1}^{13} w(e_i)=92$. $\psi_0=\{t(r,a),t(r,c,b),t(r,g,f,e,d),t(r,h)\}$ is an optimal broadcast set, and $w(\psi_0)=66$.

2.2. *Problem formulation*

We study the problem of finding an optimal broadcast set for a given broadcast problem. First we establish the notion of colored edges, vertices, and trees. Clearly, in any broadcast set for a linear broadcast routing problem (T,w,r), at least one trail goes down every edge $e \in E$. In addition the following holds for every optimal broadcast set.

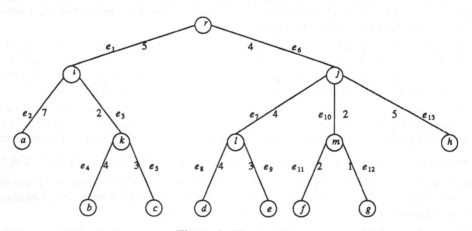

Figure 1: The tree T_0

Lemma 1 [ChG] : For every optimal broadcast set for a broadcast problem (T,w,r), and for any edge e in T, either there exists a one trail going up e and down e, or no trail is going up e.

According to Lemma 1, the edges in E can be partitioned into two disjoint subsets with respect to an optimal broadcast set ψ. We say that an edge e or a vertex v is *red* w.r.t. to ψ if one trail is going up through it, and *blue* w.r.t. to ψ otherwise. Clearly, the color of a vertex v, is the same as the color of the edge $(v, father(v))$, and thus we associate with each broadcast set ψ the set of the blue vertices in ψ, $B(\psi)$, and the set $blue\_l(\psi)$ of its blue leaves. Now we turn to the definition of a colored tree and we show the relation between colored trees and broadcast sets.

Definition : Let $T=(V,E)$ be a tree, a coloring $T_c=(T,B)$ of T is partition of V to two disjoint subsets, where B is the set of blue vertices, and $V-B$ is the set of red vertices.

Definition : Let ψ be a broadcast set for a broadcast problem (T,w,r), and $T_c=(T,B)$ a coloring of T. We say that ψ is *induced by* T if $B=B(\psi)$. We denote the set of broadcast sets induced by a colored tree T_c, $BS(T_c)$.

Given a broadcast set ψ it is straight forward to find the colored tree T_c that induces it.

Observation [ChG]: For any colored tree $T_c=(T,B)$, $BS(T_c)\neq\varnothing$ if and only for every $v\in V$, if v is red, then all the vertices (and edges) in $subtree(v)$ are red, in other words, for all $v\in V$ if v is blue, then $father(v)$ is also blue.

Corollary [ChG]: Let ψ be an optimal broadcast set for a broadcast problem (T,w,r), then for T_c the colored tree that induces ψ the following holds :

(1) The broadcast vertex r is always blue.

(2) All the paths from the root r to the leaves in T_c, have a non-empty blue prefix, and a (possibly empty) red suffix.

(3) Every trail ends in a blue leaf. In other words, every blue vertex has at least one blue son.

A *legal* colored tree for a broadcast problem (T,w,r) is one satisfying properties (1), (2) and (3) in the Corollary. All the colored trees in the sequel, except $(T,\varnothing)$ $((T,\{r\}))$ (that induces a DFS set) are legal.

With each colored tree, T_c we associate the set of its blue leaves $blue\_l(T_c)$ in a similar way to the broadcast set case.

Lemma 2 : Let $T_c=(T,B)$ and $T_c'=(T,B')$ be colored trees for a broadcast problem (T,w,r), then $blue\_l(T_c)=blue\_l(T_c')$ if and only if $B=B'$.

It follows from the above Lemma, that a colored tree can be specified either by its blue vertices set, or by its blue leaves set.

Definition : Let T_c and $T_{c'}$ be two colored trees for a broadcast problem (T,w,r). Two broadcast sets $\psi\in BS(T_c)$ and $\psi'\in BS(T_{c'})$ are *equivalent* (denoted $\psi\sim\psi'$) if $T_c=T_{c'}$; namely, $\psi\sim\psi'$ if they are induced by the same colored tree.

Lemma 3 : Let ψ and ψ' be two equivalent broadcast sets for a broadcast problem (T,w,r) then $w(\psi)=w(\psi')$.

Corollary : Let $T_c=(T,B)$ be a colored tree, and $\psi\in BS(T_c)$ then :

(1) If an edge e is red in T_c, then e appears twice in ψ.

(2) If an edge $e=(father(v),v)$ is blue, then e appears x times in ψ, where x is the number of blue leaves in $subtree(v)$.

From Lemma 3 it follows that the weight of a broadcast set is determined only by the colored tree that induces it, hence we define the weight of a colored tree, and an optimal colored tree.

Definition : Let T_c be a colored tree, then $w(T_c)=w(\psi)$ where $\psi \in BS(T_c)$. An *optimal colored tree* is a minimal weight colored tree.

From the above discussion it follows that we can reduce the problem of finding an optimal weight broadcast set to that of finding an optimal colored tree T_c. From Corollary [ChG] (3) it follows that if T_c is a colored tree that contains k blue leaves, there exist a broadcast set $\psi \in BS(T_c)$ that contains exactly k trails. We use a recursive routine *create_broadcast_set* (see [BZ]), that gets a colored tree (T,B) with k blue leaves, and returns a broadcast set $\psi \in BS((T,B))$, that contains k trails.

Lemma 4 : The subroutine *create_broadcast_set*, with parameters $(r, \psi, blue\_leaves)$, returns $\psi \in BS(T_c=(T,blue\_leaves))$, where T_c is a colored tree for (T,w,r). If T_c contains k blue leaves, then ψ contains k trails.

Lemma 5 : Let $T_k=(T,B)$ be a colored tree, where $T=(V,E)$ and $N=|V|$. Then the complexity of *create_broadcast_set* on T_k is $O(N)$.

Example 2 : Consider again T_0 (Figure 1), and the broadcast set ψ_0 defined in Example 1. The edge e_1 and the vertex k are blue, while e_5 and c are red with respect to ψ_0. The set of blue vertices in ψ_0 is $B(\psi_0)=\{r,i,a,k,b,j,l,d,h\}$; thus the colored tree $T_{0_c}=(T,\{r,i,a,k,b,j,l,d,h\})$ induces ψ_0. $blue\_l(\psi_0)= \{a,b,d,h\}$. $\psi'_0= \{p(r,a), t(r,c,b), t(r,e,d), t(r,g,f,h)\}$ is another broadcast set for the same broadcast problem, and $blue\_l(\psi')=blue\_l(\psi_0)$ and $B(\psi'_0)=B(\psi_0)$. Thus ψ_0, $\psi'_0 \in BS(T_{0_c})$, $\psi_0 \sim \psi'_0$, and $w(\psi)=w(\psi')=w(T_{0_c})$.

We present two algorithms (A1,A2) to solve the following problem :

TLBS (*Tree Linear Broadcast with Simultaneous transmission limitations*): Given a broadcast problem (T,w,r), and $1 \leq k \leq |V|$, find an optimal broadcast set that contains at most k trails.

Note, that if $k=|V|$ we get the TLB (*Tree Linear Broadcast with no limitations*) problem, studied in [ChG].

3. ALGORITHM A1

Let (T,w,r) be a broadcast problem, and $1 \leq k \leq |V|$. The following algorithm computes an optimal broadcast set ψ_k, using at most k trails. The algorithm works in two stages. In the first stage (steps (1) - (3)) it builds an optimal colored tree for (T,w,r), T_k, that contains at most k blue leaves. In the second stage the algorithm builds a broadcast set $\psi_k \in BS(T_k)$, i.e. optimal by the definition of an optimal colored tree.

The algorithm uses the following data structures :

L :　　　　　a constant set containing all the leaves in $T(V,E)$.

$blue\_leaves$:　an array of size k, containing the leaves that have been colored blue. $blue\_leaves(i)$ is the leaf that has been colored blue in stage i of the algorithm.

B :　　　　　a set containing the blue vertices.

db :　　　　　an array containing $db(l)$.

l :　　　　　a variable containing in stage i the leaf that changes from red to blue.

b :　　　　　a variable containing $db(l)$ in stage i.

Δ :　　　　a variable containing the decrease in the weight of T_k in stage i of the algorithm.

ψ_k :　　　　the broadcast set.

The algorithm employs the following procedures :

update_db (b,l) :   this procedure receives two parameters, an internal vertex b, and a leaf l. It updates the *db* of the red leaves in *subtree* (b) due to the change of $p(b,l)$ from red to blue.

create_broadcast_set (v,ψ,B) :
     this procedure constructs a broadcast set $\psi \in BS(T,B)$, see above.

Algorithm A1

(1)	if $|V|=1$, then STOP.	*{ T contains only one vertex }*
(2)	Initializations :	
	$L \leftarrow \{v \in V \mid degree(v)=1\}$;	*{ L contains all the leaves }*
	$i \leftarrow 0$;	
	$B \leftarrow \{r\}$;	*{ B contains the blue vertices }*
		{ Initially all the vertices except r are red }
	$(\forall l \in L)\, db(l) \leftarrow r$;	*{ The deepest blue vertex in each path is r }*
(3)	**repeat**	*{ The coloring stage }*
(3.1)	$i \leftarrow i+1$;	

$$\Delta \leftarrow \max_{x \in L-B} (w(p(db(x),x))-w(p(r,db(x))));$$

Let $x=l$ be the vertex for which the above maximum is obtained, and $b=db(l)$.

(3.2)	if $\Delta \leq 0$, **break**;	*{T_k is optimal for any $k \geq i-1$}*
		{ It contains $i-1$ blue leaves }
(3.3)	*blue_leaves* $(i) \leftarrow l$;	*{ Add l to the blue leaves array }*
	$B \leftarrow B \cup (V(p(b,l))-b)$	*{ Color all the vertices on $p(b,l)$ in blue }*
	update_db (b,l) ;	*{ Update *db* for all the leaves in *subtree*(b)}*
	until $i=k$;	
(4)	*create_broadcast_set* $(r, \psi_k, blue\_leaves)$.	

update_db (b,l)

    Let $p(b,l) = <v_0=b, e_1, v_1, ..., v_{t-1}, e_t, v_t=l>$;
      for $j \leftarrow 1...t$ let $\{u_1,...,u_k\}=sons(v_j)-p(b,l)$
        for $m \leftarrow 1...k$
          $\forall l' \in (subtree(u_m) \cap L)$ do
            $db(l') \leftarrow v_j$;
    $db(l)=l$;

In the discussion of algorithm A1, we will use x^i, where x is any variable used by algorithm A1, to denote the value of variable x after stage i.

Example 3 : Consider the run of the Algorithm A1 with inputs T_0 (Figure 1), and $k=3$. Initially $(T,B^0)=(T,\{r\})$. In the first step of the algorithm, since a is a deepest leaf in T_0, $\Delta^1=12$, and $l^1=a$, hence the path $p(r,a)$ turns to blue, $B^1=\{r,i,a\}$, and $db(b)^1=db(c)^1=i$. In the second step $\Delta^2=12$, and d turns to blue (Note that d is also a deepest leaf in T_0, and that the order of coloring d and a could be changed). Now the set of blue vertices is $B^2=\{r,i,a,j,l,d\}$, so $db(b)^2=db(c)^2=i$, $db(f)^2=db(g)^2=db(h)^2=j$, and $db(e)^2=l$. $\Delta^3= w(p(i,b)) -w(p(r,i))= w(p(j,h))-w(p(r,j))=1$ so l^3 can be either b or h. The coloring stage terminates at step (3), and B^3 can be either $\{r,i,a,j,l,d,k,b\}$, or $\{r,i,a,j,l,d,h\}$. ψ_3 can be either $\{t(r,a),t(r,c,b),t(r,h,g,f,e,d)\}$, or $\{t(r,c,b,a),t(r,e,d),t(r,g,f,h)\}$, with respect to the selection of the third blue leaf. Running the same algorithm for $k=4$ will result in

$\psi_4=\{t(r,a),t(r,c,b),t(r,e,d),\ t(r,g,f.h)\}$. For $k>4$ the algorithm terminates at step (3.2) and $\psi_k=\psi_4$.

4. *PROOF OF CORRECTNESS AND ANALYSIS*

The next lemma shows that in each step of the algorithm the weight of the colored tree (T,B^i) decreases in the amount calculated in Δ in stage i of the algorithm.

Lemma 6: Let $T_k^i=(T,B^i)$ then

$$w(T_k^{i-1})-w(T_k^i)=\Delta^i=depth(l^i)-2\cdot depth(b^i) \qquad (*)$$

Lemma 7: $\Delta^i\geq\Delta^{i+1}$ for every $i\geq1$.

The coloring stage clearly terminates, since the number of blue leaves increases with each iteration. In the trivial case the algorithm terminates in step (1), otherwise the coloring terminates either in step (3.2) of the algorithm, or in step (3.3). The first case is discussed in Theorem 1, and the second in Theorem 2.

Theorem 1 : Let $T_k=(T,B^k)$ be the colored tree built by algorithm A1 , given a broadcast problem (T,w,r), and an integer $1\leq k\leq|V|$, in a run where the coloring terminates in step (3.3), then T_k is optimal.

Sketch of Proof : The proof is by induction. We start with an optimal colored tree T_{opt} for the broadcast problem (T,w,r), with $k'\leq k$ leaves. We build inductively a sequence of equal weight colored trees $\{T_{opt}^i\}$ each with exactly k' blue leaves. $T_{opt}^0=T_{opt}$ and T_{opt}^{i+1} is built from T_{opt}^i in the following way: if $l^{i+1}=blue\_leaves(i+1)\in blue\_l(T_{opt}^i)$ then $T_{opt}^{i+1}=T_{opt}^i$. Otherwise T_{opt}^{i+1} is obtained from T_{opt}^i by adding l^{i+1} to the blue leaves set, and removing from it a leaf $l'\in blue\_l(T_{opt}^i)-\{l^1,l^2,\cdots,l^{i-1}\}$. We show that $k'=k$, hence, by Lemma 2, $T_{opt}^k=T_k$. From the fact that $\forall i\ 0\leq i\leq k,\ w(T_{opt}^i)=w(T_{opt}))$, it follows that T_k is optimal. $\square$

We now discuss the correctness of algorithm A1 in case in terminates in step (3.2)

Theorem 2 : If algorithm A1 on inputs (T,w,r) and k terminates in step (3.2) after $k'<k$ stages, then the colored tree $T_{k'}$ it builds is optimal.

Theorem 3 : The broadcast set ψ_k algorithm A1 outputs for inputs (T,w,r) , a broadcast problem and $k\leq|V|$, is an optimal broadcast set, containing at most k trails. Moreover, the colored tree $T_k^i=(T,B^i)$ created by A1 after i steps for (T,w,r), is an optimal colored tree for (T,w,r) using i trails.

We now discuss the properties of algorithm A1.

Theorem 4 : The weight of the optimal colored tree for a broadcast problem (T,w,r), using k trails is $w(T_k)=2\cdot w(E)-\sum_{i=1}^k\Delta^i$.

Theorem 5 : The series $\{\Delta^i\}$ created by algorithm A1 for a broadcast problem (T,w,r) and k, is independent of the decisions made during the run.

In the next theorem we show that any optimal colored tree for a broadcast problem (T,w,r) is a possible output of algorithm A1. For the proof of the Theorem we need the following algorithm.

Algorithm $A1'$: receives as input a colored tree T_k for a broadcast problem (T,w,r), with k blue leaves. Algorithm $A1'$ is ordering the leaves in $blue\_l(T_k)$ in the order A1 would have colored them. $A1'$ is forced to color all the blue leaves in T_k, and only them. The algorithm outputs $blue\_leaves_{T_k}$, an array containing all the blue leaves in T_k in the above order, and an array Δ_{T_k}, that contains the appropriate profits. (see [BZ] for more details on $A1'$).

For example let T_k' for T_0 and r (see Figure 1) be a colored tree such that $blue\_l(T_k')=\{c,d,e,f\}$, then $A1'$ for T_k' outputs $blue\_leaves_{T_k'}= <d,c,f,e>$, and $\Delta_{T_k'}=<12,10,0,-5>$.

Theorem 6 : Let $T_{opt}=(T,blue\_leaves_{opt})$ be an optimal colored tree for (T,w,r) with k trails. There exists a run of algorithm A1 on inputs (T,w,r) and k that produces a colored tree T_k such that $T_k=T_{opt}$.

Sketch of Proof : The proof is very similar to the proof of Theorem 1. Let $\Delta_{T_{opt}}$ and $blue\_leaves_{T_{opt}}$ be outputs of algorithm A1' in a certain run with input T_{opt}. And let Δ^i and $blue\_leaves(i)$ be the value of Δ in stage i of A1, and the i'th leaf colored blue by A1 respectively. We prove that for each $1 \le i \le k$, there exist a run of algorithm A1 such that:

(1) $\Delta^i = \Delta_{T_{opt}}(i)$.

(2) $blue\_leaves(i) = blue\_leaves_{T_{opt}}(i)$.

The Theorem follows directly from the above two claims. □

Since A1 can output any optimal broadcast set using k trails, in particularly it can output the same broadcast sets algorithm TreeRouting [ChG] outputs.

Lemma 8: Algorithm with input (T,w,r) and k halts after at most $\min(k,|V|)$ stages.

Theorem 7 : The complexity of algorithm A1 with inputs (T,w,r) and k is $O(k \cdot N)$, where $N=|V|$.

5. ALGORITHM A2

The next algorithm we present is using the same technique developed in A1. This algorithm is simpler and more efficient , its complexity is $O(N)$ which is better than $O(k \cdot N)$ for all k's. As we said in the introduction we presented A1 mainly for the sake of proof, and the proof of A2 will be done by reduction to A1.

A2 gets the same inputs as A1, i.e. a broadcast problem (T,w,r) and k a bound on the number of trails to be used, like A1 it creates an optimal colored tree, and then uses *create_broadcast_set* to construct an optimal broadcast set. The algorithm makes three passes on the tree. The first pass is a *DFS* pass, during this pass it computes the depth of all the vertices in the tree, and *max_leaf(v)*, the *id* of the deepest leaf in *subtree(v)*.

To understand better the second pass of algorithm A2 we state here three observations on A1's behavior:

(1) the profit from coloring a certain leaf l in blue depends only at its depth and the depth of the deepest blue vertex in $p(r,l)$, $db(l)$. In algorithm A2 we save the depth of the current deepest internal blue vertex in a variable *deepest_blue*. So the profit from coloring a leaf l in blue is $depth(l)-2 \cdot deepest\_blue$ (compare with (*)).

(2) if you color a path $p(b,l)$ in blue the only vertices that are affected are those in *subtree(b)*. Due to this fact all the vertices in *subtree(b)* can be colored independently of other vertices.

(3) since A1 is greedy, and the leaf with the maximal profit is chosen in each step, the first leaf to be colored in *subtree(v)*, for each internal vertex v, is the deepest leaf in *subtree(v)*, i.e. *max_leaf(v)*. The same leaf, is the only one that might cause v's coloring in blue.

The second pass computes for each leaf l the profit from coloring it in blue, according to the above three rules. The traversal starts in r, and it traverses only vertices that might be colored in blue, i.e. vertices that the profit from coloring them in blue is positive. Traversing a certain path means coloring its vertices and edges in blue.

According to observation (3), when the algorithm reaches a vertex v for the first time it colors the path $p(v, max\_leaf(v))$. The depth of the deepest internal blue vertex in the current path, i.e. *deepest_blue*, is carried along the path. Initially, *deepest_blue*$=0$, and the path $p(r, max\_leaf(r))$ is the first to be colored in blue. When a leaf l is reached, it is added to the list of potential blue leaves, *blue_leaves*, and the profit, whose value (according to observation (2)) is $depth(l) - 2 \cdot deepest\_blue$, to *profits*. After the coloring of $p(b,l)$ is done the traversal moves to *father*(l). When the traversal reaches a vertex $v \in p(b,l)$ for the second time, it is already blue (the first traversal did the coloring). Also for all the vertices u in *subtree*(v) except those in $p(b,l)$, v is the deepest internal blue vertex in $p(r,u)$. So the value of *deepest_blue* is set to $depth(v)$. Now for each of v's sons the algorithm checks if the profit from coloring $max\_leaf(u)$ is greater than zero; namely if $(depth(max\_leaf(u)) - 2 \cdot deepest\_blue) > 0$. If the condition is true, then $p(v, max\_leaf(u))$ is colored in blue in the same way and *subtree*(u) is traversed recursively. Otherwise, *subtree*(u) will not be traversed. The traversal of *subtree*(v) terminates after all is sons are checked and traversed (if necessary). The traversal of the tree terminates when r searched all of its sons.

Before the third *DFS* pass begins, the k most profitable leaves among those in *blue_leaves* and *profits* are selected, and only those leaves are left in *blue_leaves* and *profits*. The third pass creates a broadcast set ψ_k such that $blue\_l(\psi_k) = blue\_leaves$, it calls *create_broadcast_set* routine.

Example 4 : The second pass of algorithm A2 on T_0 (Figure 1) starts at r. Since the deepest leaf in *subtree*(r) is a, and $depth(a) - 2 \cdot depth(r) = 12 > 0$, the traversal continues to a. From a it backtracks up to i. The deepest leaf in *subtree*(i) that has not yet been traversed is b. Since $depth(b) - 2 \cdot depth(i) = 1$, $p(i,b)$ is traversed, and from b the *DFS* continues to k. Since the profit for changing the path to the next deepest leaf c is $(depth(c) - 2 \cdot depth(k)) = -4$ it backtracks back to i, that already searched its two sons, so it returns back to r. r has one son j that has not been traversed yet, and $depth(max\_leaf(j)) - depth(r) > 0$, thus the traversal continues to j, and to $d = max\_son(j)$. It sets $profit(d) = 12$. From d it backtracks to i, and then to j (since $depth(e) - deepest\_blue < 0$). From j it goes down to h, and sets $profit(h) = 1$, and finally it backtracks to r. The results of the second *DFS* path will be: $blue\_leaves = \{a, b, d, h\}$ and $profits = \{12, 1, 12, 1\}$. The resulting ψ_k will be equal to ψ_0 (Section 2.1).

We now turn to the proof of correctness.

Lemma 10 : Let *blue_leaves* be the ordered list of blue leaves produced by A2 for a broadcast problem (T, w, r). Then there exist a run of algorithm A1 with $k = |V|$ such that for every $v \in V$ the first leaf l that was colored blue in *subtree*(v) in A2's run (i.e. joined *blue_leaves*), if such a leaf exist, is the first from *subtree*(v) that joined *blue_leaves* in A1's run.

Lemma 11 : For each leaf l in $T = (V, E)$, let $profit(l)$ be the amount assigned by algorithm A2 during the second *DFS* pass, then there exist a run of algorithm A1 with $k = |V|$, such that for each l, if $profit(l) > 0$ then $l = l^t$ for some $t \leq k$ in A1's run and $profit(l) = \Delta^t$, (t is the stage of the A1 in which l turned to blue), and l was not colored blue in A1's run otherwise.

Theorem 7 : The colored tree $(T, blue\_leaves)$ algorithm A2 outputs for inputs (T, w, r) and k is optimal.

Theorem 8 : The complexity of algorithm A2 is $O(N)$ where $N = |V|$.

It is clear that algorithms A1 and A2 solve the *TLB* problem studied in [ChG] for $k = |V|$. This result does not contradict the lower bound $\Omega(N^2)$ that was presented in [ChG], since their algorithm operates on edges, while ours works with paths, with the right data structure implementation the complexity of $O(N)$ is achieved.

6. APPLICATIONS

The linear broadcast routing algorithms introduced in this paper can be used in high bandwidth network, with switching hardware. In this algorithms the only work done by the general purpose processors (besides the broadcasting node) is receiving the packet when the broadcast is done, all the switching is done by the switching hardware. This algorithm enables transmission of a packet several times along certain communication line thus utilizing the high bandwidth of the communication lines. If we assume that a transmission of a packet between any two general purpose processors in the network takes one time unit, the broadcast terminates during one time unit.

The versions we introduced are centralized algorithms. They can be ran by the node initiating the broadcast. The output of this algorithms is a set of trails. Each trail should be used as a header that contains the switching information (the ANR header in PARIS [CiG]). During algorithm A2 we assume that we can use the vertex id instead of the whole path. This can be done if we use a tree in which the ids of all the vertices in $subtree (v)$ are less than or equal to v. In this case the links can be marked by the ids of v's neighbors, and the switching can be done by a simple comparator.

Algorithm A2 can be very easily modified to a distributed algorithm. In this case there is no need that the broadcasting node will know the topology of all the network, because it can be collected during the first DFS pass. In the distributed case, the passes correspond to messages traversing the network. Instead of the selection, each vertex can send to its father the list of the best k leaves, and the father can use a merge to decide which k leaves it should send up. During the third pass the trails has to be send, each trail contains at most $O(N)$ edges. The distributed algorithm message complexity is $O(k{\cdot}N)$, and the maximum message length is $O(N{\cdot}\log d)$ where d is the maximal vertex degree. The advantages of the distributed version is that no vertex should know the network topology, otherwise the centralized algorithm is more efficient.

REFERENCES

[AHU]   A.V. Aho, J.E. Hopcroft and J.D. Ullman, *The Design and Analysis of Computer Algorithms*, **Addison-Wesley**. June 1976,pp. 97-99.

[BZ]   S. Bitan and S. Zaks, *Optimal Linear Broadcast*, **Technion - Israel Institute of Technology, Department of Computer Science, Technical Report # 623,** May 1990.

[ChG]   C.T. Chou and I.S.Gopal, *Linear Broadcast Routing,* **Journal of Algorithms,** Vol 10(4), 1989, pp.490-517.

[CiG]   I. Cidon and I.S. Gopal, *PARIS: An Approach to Private Integrated Networks,* **Journal of Analog and Digital Cabled Systems,** June 1988.

[CGK]   I. Cidon, I.S. Gopal and S. Kutten, *New Models and Algorithms for Future networks,* **Proceedings of the 7'th Annual ACM Symposium on Principles of Distributed Computing,** Toronto, CANADA, August 1988, pp. 75-89.

[CS]   R. Cohen and A. Segall, *A Distributed Query Protocol For High-Speed Networks,* **Proceedings of the 9'th International Conference on Computers and Communication,** Tel-Aviv, ISRAEL, October 1987, pp. 299-302.

[S]   A. Segall, *Distributed Networks Protocols,* **IEEE Trans. on Information Theory,** IT-29(1), January 1983, pp. 23-35.

GRAPH AUGMENTATION PROBLEMS FOR A SPECIFIED SET OF VERTICES

Toshimasa Watanabe Yasuhiko Higashi and Akira Nakamura
Faculty of Engineering, Hiroshima University
Saijo-cho, Higashi-Hiroshima, 724 Japan

1. Introduction

The paper discusses the *k-edge-connectivity* or *k-vertex-connectivity* or *strong-connectivity augmentation problem* for a *specified set of vertices*: "Given a complete undirected or directed graph $G=(V,E)$, a spanning subgraph $G_0=(V,E')$, a cost function $c:E \rightarrow Z^+$(nonnegative integers) and a specified subset $S \subseteq V$, find a set $A \subseteq E-E'$ of the minimum total cost $c(A)$ such that, for any given pair of vertices in S, the graph $G_0=(V,E' \cup A)$ has at least k *edge-disjoint* or at least k *internally-disjoint* paths between them or at least one directed cycle containing them, where $c(e)=0$ $(\forall e \in E')$ and adding multiple edges to a graph is prohibited." They are abbreviated as **k-ECA-SV, k-VCA-SV** and **SCA-SV**, respectively. If we set S=V then these problems are usual augmentation problems discussed in [1,2,4-8].

In this paper we consider the most fundamental problems 2-ECA-SV and 2-VCA-SV, that is, the case with k=2. The NP-completeness of these problems with G_0 restricted to a tree and S=V have been shown in [2]. Three $O(|V|^2)$ approximation algorithms STC, BRC and BIC for SCA-SV, 2-ECA-SV and 2-VCA-SV, all with S=V, were proposed in [2], respectively. The idea of STC is as follows: First choose a vertex r and finds a minimum cost spanning reverse arborescence T_{in} with the sink r, with respect to the initial cost c. Modify c into c' by setting $c'<u,v>=0$ for $\forall <u,v> \in E(T_{in})$ and $c'<w,v>=\infty$ for any $<w,r>$ entering into r. Again finds a minimum cost spanning arborescence T_{out} with r as the root, with respect to c'. Clearly $E''=(E(T_{in}) \cup E(T_{out}))-E'$ is a solution to SCA-SV with S=V. Time complexity of STC is $O(|V|^2)$, and c(E'') is no more than twice the optimal. The idea of combining T_{in} and T_{out} is used in both BRC and BIC, since a strongly connected graph will be 2-edge-connected if directedness is neglected, where the handling of cutvertices are incorporated in BIC.

We propose an $O(|V|^2)$ approximation algorithm BRA-SV for 2-ECA-SV, and an $O(|V^3|)$ one BIA-SV for 2-VCA-SV. It is shown that the worst approximation of BRA-SV (BIA-SV, respectively) is no more than twice (less than four times) the optimal if G_0 is connected. Although these algorithms run on G_0 which is disconnected, the theoretical estimate of their worst approximations in this case is left for future research. It should be noted that if $c(e)=c(e')$ for any e, $e' \in E$ then BRA-SV (BIA-SV, respectively) can be modified into an $O(|V|+|E'|)$ algorithm proposed in [1] ([1,4]) for 2-ECA-SV (2-VCA-SV) with S=V, where an optimum solution is obtained in each case.

For SCA-SV we show that it is NP-complete even if every edge e of E has $c(e)=1$. An approximation algorithms STA-SV and its variations based on either shortest paths or minimum arborescens, as well as their experimental results, are presented.

Let 2-ECA[G_0=H, S=W] denote the 2-ECA-SV with G_0 and S set to a graph H and a subset W$\subseteq$V, respectively. A similar notation will be used for 2-VCA-SV.

2. Preliminaries

Technical terms not specified here can be identified in [3]. A *graph* G=(V,E) consists of a set V of vertices and a set E of edges. V and E are also written as V(G) and E(G), respectively. Definitions are given mostly for the undirected case; those for the directed case is similar. Two paths P and P' are *edge-disjoint* (*internally disjoint*, respectively) if E(P)∩E(P)=∅ (they have no vertices except their endvertices in common). The *edge-connectivity* ec(G) (the *vertex-connectivity* vc(G), respectively) of G is the minimum number of edges (vertices) whose deletion result in a graph with more components than G or in a single vertex, where a *component* is a maximal connected subgraph. G is *k-edge-connected* (*k-vertex-connected*, respectively) if ec(G)≥k (vc(G)≥k). It is known that ec(G)≥k (vc(G)≥k, respectively) if and only if, for any two vertices u,v of G, G has at least k *edge-disjoint* (k *internally-disjoint*) (u,v)-paths, (where a (u,v)-path is a path connecting u and v). Let G be a directed graph. A vertex r is called a *root* of (a *sink*, respectively) of G if there in a <r,v>-path, a directed path from r to v, (a <v,r>-path) for any vertex v of G. A *spanning arborescence* with the root r (A *spanning reverse arborescence* with the sink r) is a directed acyclic spanning subgraph of G with r, as the root (as the sink), having no incoming (outgoing) edges, and all other vertices having exactly one incoming (outgoing) edge.

3. An $O(|V|^2)$ approximation algorithm BRA-SV

We propose an $O(|V|^2)$ approximation algorithm BRA-SV for 2-ECA-SV and show that the worst approximation is no greater than twice the optimal if G_0 is connected.

We provide an example of transforming graphs to help the reader understand our idea. Let G_0 be an initial graph with edges written in solid lines in Fig. 1.1. We first construct a complete graph $G_s=(V_s,E_s)$ and a tree $T=(V_s,E_s')$ (the graph with edges written in bold solid or fine solid lines in Fig. 1.2) from G and G_0, respectively, by shrinking each bridge connected component of G_0 into a different vertex, and then define a cost function $c_1:E_s \to Z^+$ by

$c_1(u,v)=\min\{\{\infty\}\cup\{c(u',v')|(u',v')\in E-E', u'\in S_1, v'\in S_2\}\}$,

where bridge connected components S_1 and S_2 are shrunk into u and v, respectively. The edge(u',v') so used is kept by a backpointer $b_1(u,v)=(u',v')$, where if $c_1(u,v)=\infty$ then $b_1(u,v)=(u,v)$. Let $V_a \subseteq V_s$ be the set of vertices that correspond to bridge connected components including vertices of S. Each edge on a (u,v)-path of T with $u,v\in V_a$ is called a *path-edge*, and each connected component consisting of edges that are not path-edges is called an *external-path-component* of T. Let $G_c=(V_c,E_c)$ and $T_c=(V_c,E_c')$ (T_c is written in bold solid lines in Fig. 1.2) be the graphs determined from G_s and T_s by shrinking each external-path-component to a distinct vertex, respectively. We call T_c a *path-tree*. A path-tree is a minimal subtree including all bridges that exist in T_s between any two specified vertices of V_a, and every leaf of T_c is in V_a. We define a cost function $c_2:E_c \to Z^+$ by

$c_2(u,v)=\min\{c_1(u',v')|u'\in S',v'\in S''\}$,

where external-path-components S' and S" are shrunk into u and v, respectively. The edge(u',v') so used is kept by a backpointer

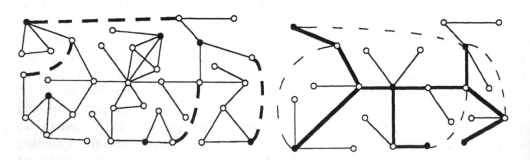

Fig. 1.1. $G_0=(V,E')$, $S \subseteq V$(denoted by ●) and A (broken bold lines).

Fig. 1.2. $T_s=(V_s,E_s')$, T_c (bold lines), $V_a \subseteq V_s$ (black circles) and a solution to 2-ECA[$G_0=T_s$, $S=V_a$],(broken fine lines).

$b_2(u,v)=(u',v')$. G_c, T_c and c_2 can be constructed in $O(|V|^2)$ time. The $O(|V|^2)$ approximation algorithm BRC proposed in [2] for 2-ECA-SV with $S=V$ can be applied to 2-ECA[$G0=T_c$, $S=V_c$] with the cost function c_2. Let A_c be an optimal solution and A_c' be an approximate one given by this BRC, both to 2-ECA[$G0=T_c$, $S=V_c$], respectively. Then [2] shows that $c_2(A_c') \leq 2c_2(A_c)$. Neither A_c or A_c' has multiple edges, and transformation by means of b_2 or b_1 does not produce multiple edges: none of the four sets $A_s=b_2(A_c)$, $A=b_1(A_s)$, $A_s'=b_2(A_c')$, $A'=b_1(A_s')$ cotains multiple edges. A_s and A_s' (A and A', respectively) are an optimal solution and an approximate one to each problem. It is easy to see that $c_2(A_c)=c_1(A_s)=c(A)$ and $c_2(A_c')=c_1(A_s')=c(A')$, showing that $c(A') \leq 2c(A)$. Hence we have the following theorem.

Theorem 1. The algorithm BRA-SV finds a approximate solution for 2-ECA-SV in $O(|V^2|)$ time and the worst approximation is no more than twice the optimal if $G0$ is connected.

4. An $O(|V|^3)$ approximation algorithm BIA-SV

We propose an $O(|V|^3)$ approximation algorithm BIA-SV for 2-VCA-SV and show that the worst approximation is less than four times the optimal if $G0$ is connected.

The problem is reduced to a 2-VCA[$G0=T_c$, $S=V_c$] to which the $O(|V|^2)$ approximation algorithm BIC proposed in [2] can be applied, where T_c is a tree with the vertex set V_c defined later. An approximate

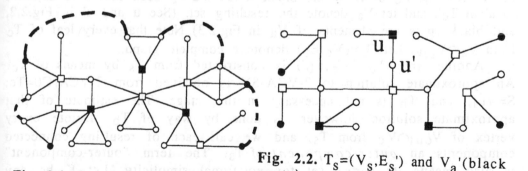

Fig. 2.1. $G0=(V,E)$, $S \subseteq V$(black circles and squares) and a solution A (bold broken lines), where open circles and open squares denote block-vertices and cutvertices, respectively.

Fig. 2.2. $T_s=(V_s,E_s')$ and V_a'(black spots).

solution to the original 2-VCA-SV is obtained from a solution given by this BIC. Since the reduction seems hard to understand at the first reading, we provide a schematic explanation to outline our idea.

Suppose that a connected initial graph G_0 of Fig. 2.1 with edges written in solid lines are given. We construct the *block-cutvertex-tree* $T_S=(V_S, E_S')$ from G_0 (Fig.2.2), where $V_S=V_{cut}\cup V_{blk}$, V_{cut} is the set of all cutvertices of G_0 and V_{blk} is a set of block-vertices of G_0. E_S' is a set of edges that connect a block-vertex u to a cutvertex v if and only if v is in the block B_u corresponding to u. Let $G_S=(V_S,E_S)$ be the complete graph with $E_S=\{(u,v)| u,v\in V_S, u\neq v\}$, and define a cost function $c_1:E_c\rightarrow Z^+\cup\{\infty\}$ as follows, where u and v correspond to blocks B_u and B_v, respectively:

$$c_1(u,v)=\begin{cases} c(u,v) & \text{if } u,v\in V_{cut}, \\ \min\{\{\infty\}\cup\{c(u',v) \mid u'\in B_u \text{ and } u'\notin V_{cut}\}\} & \text{if } u\in V_{blk} \text{ and } v\in V_{cut}, \\ \min\{\{\infty\}\cup\{c(u',v') \mid u'\in B_u, v'\in B_v \text{ and } u',v'\notin V_{cut}\}\} & \text{if } u,v\in V_{blk}, \end{cases}$$

A block-vertex and a cutvertex are denoted by a circle and a square in Figures 2.1–2.4, respectively. A black spot denotes a specified vertex of S. Let

$$V_a'= (V_{cut}\cap S)\cup\{u\in V_{blk} \mid (B_u-V_{cut})\cap S\neq\emptyset\}.$$

We first define a tree $T_c=(V_c,E_c')$, called the *path-tree*, from T_S as follows. Let T_c' be a maximal subtree determined by the edges on (u,v)-paths of T_S with $u,v\in V_a'$. Let $L=V_{cut}\cap\{\text{leaves of } T_c'\}$ and delete all vertices of L from T_S. Let T_c be the resulting tree (Fig.2.3). Replace any cut-vertex $u\in V_c$ of L by the block-vertex $u'\in V_c$ such that u is adjacent to u' in T_c', and let V_a denote the resulting set. (See u and u' in Fig.2.2, and black spots are elements of V_a in Fig.2.3) Note that every leaf of T_c is in $V_a\cap V_{blk}$. Let $G_c=(V_c,E_c)$ denote a complete graph.

Another tree $T_b=(V_b,E_b')$ is constructed from T_S by means of T_c. An approximate solution to 2-VCA-SV is obtained from 2-VCA[G0=Tc, S=Vc], and T_b is not necessary in this stage. The estimate of that approximate solution, however, is done by way of T_b. Delete every vertex of $V_{cut}\cap V_c$ from T_S. and we call each of resulting connected components an *outer-component* of T_S. The term "outer-component" is also means its vertex set for notational simplicity. Let K be any outer-component (a vertex set) of T_S with $|K|\geq 2$. Then T_S has either a block-vertex v_K such that $\{v\}=K\cap V(T_c)$ or a bridge (u_K,v_K) connecting a block-vertex $u_K\in K$ (that is, $u_K\notin V(T_c)$) and a out-vertex $v_K\in V(T_c)$. In the latter case K is called an *attachment* of v_K. We define the *guasi-path-tree* T_b by shrinking each outer-component K with $|K|\geq 2$ into v_K if v_K is a block-vertex or into u_K if v_K is a cut-

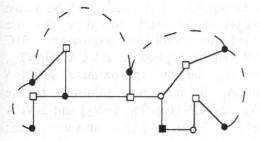

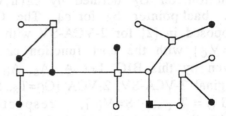

Fig. 2.3. $T_c=(V_c,E_c')$, V_a(black spots) and a solution to 2-VCA[$G_0=T_c$, $S=V_c$] (broken fine lines).

Fig. 2.4. $T_b=(V_b,E_b')$.

vertex.(Fig.2.4). If an attachment K of some cut-vertex is shrunk into a block-vertex u then u is called an *outer-point*. Let K_u denote the set of vertices shrunk into outer-point u. Note that every leaf of T_b is a block-vertex. Let $G_b=(V_b,E_b)$ be a complete graph on V_b, and define a cost function $c_2:E_b\rightarrow Z^+\cup\{\infty\}$ (a backpointer $b_2:E_b\rightarrow E_s$: is defined similarly):

$$c_2(u,v)=\begin{cases} c_1(u,v) & \text{if } u,v\in V(T_S), \\ \min\{c(u',v) \mid u'\in K_u\} & \text{if } u \text{ is an outer-point and } v\in V(T_S), \\ \min\{c(u',v') \mid u'\in K_u,v'\in K_v\} & \text{if } u, v \text{ are outer-points.} \end{cases}$$

We define another cost function $c_3:E_b\rightarrow Z^+\cup\{\infty\}$ and a back-pointer $b_3:E_b\rightarrow$ {paths of G_b represented by sequences of vertices} as follows. First set $c_3\leftarrow c_2$ and $b_3(u,v)\leftarrow u'v'$(a sequence of vertices) for each $(u,v)\in E_b$, where $b_2(u,v)=(u',v')$. Then, for each outer-point $v_f\in V_b-V_c$, excute the following procedures (a),(b) and (c) in this order, where $v\in V_c$ denotes the cut-vertex adjacent to v_f.

(a) For every pair $u',v'\in V_b$ such that T_b has a (u',v')-path containing v, set

$$c_3(u',v')\leftarrow c_3(u',v_f)+c_3(v_f,v') \text{ and } b_3(u',v')\leftarrow b_3(u',v_f)b_3(v_f,v')$$

$$\text{if } c_3(u',v_f)+c_3(v_f,v')<c_3(u',v'),$$

where $b_3(u',v_f)b_3(v_f,v')$ denotes the concatenation of the two sequences with the first vertex of $b_3(v_f,v')$ deleted.

(b) For each edge $(v_f,w)\in E_b$ with $w\in V_c$ and $w\neq v$, set

$$c_3(v,w)\leftarrow c_3(v_f,w) \text{ and } b_3(v,w)\leftarrow b_3(v_f,w) \text{ if } c_3(v_f,w)<c_3(v,w).$$

(c) For each $(v_f,w_f)\in E_b$ with $w_f\in V_b-V_c$ and $w_f\neq v_f$, set

$$c_3(v,w)\leftarrow c_3(v_f,w_f) \text{ and } b_3(v,w)\leftarrow b_3(v_f,w_f) \text{ if } c_3(v_f,w_f)<c_3(v,w),$$

where $w \in V_c$ is the cut-vertex adjacet to w_f. Let $c_4: E_c \to Z^+ \cup \{\infty\}$ be a cost function on G_c defined by $c_4(u,v) = c_3(u,v)$ for each $(u,v) \in Ec$. We use the backpointer b_3 for c_4. The $O(|V|^2)$ approximation algorithm BIC proposed in [2] for 2-VCA-SV with $S=V$ can be applied to 2-VCA [$G_0=T_c$, $S=V_c$] with the cost function c_4. Let A_c' be an approximate solution given by this BIC. Let A, A_s, A_b and A_c be minimum solutions to the original 2-VCA-SV, 2-VCA [$G_0=T_s$, $S=V_a$], 2-VCA [$G_0=T_b$, $S=V_c$] and 2-VCA [$G_0 = T_c$, $S=V_c$], respectively. Then [2] shows that $c_4(A_c')=c_3(A_c') \leq 2c_3(A_c)=2c_4(A_c)$. Let A' be an approximate solution to the original 2-VCA-SV determined from A_c' by using backpointers b_3, b_2 and b_1 in this order, where multiplicity is deleted if any. Then $c(A') \leq c_3(A_c')=c_4(A_c')$. It is easy to see that $c(A)=c_1(A_s)=c_2(A_b)$. Now we consider A_b and A_c. The crucial point of the estimate is to show that a solution $A_c'' \subseteq E_c-E_c'$ to 2-VCA[$G_0=T_c$, $S=V_c$] with $c_4(A_c'')=c_3(A_c'')$ $<2c_2(A_b)$ can be determined from A_b. Before proving this statement we show that if this is true then the desired estimate follows: since $c_4(A_c) \leq c_4(A_c'')$, $c_4(A_c') \leq 2c_4(A_c)$ (by [2]) and $c(A') \leq c_4(A_c')$, we have

$$c(A') \leq c_4(A_c') \leq 2c_4(A_c) \leq 2c_4(A_c'') < 4c_2(A_b)=4c_1(A_s)=4c(A).$$

We provide Fig. 2.5 to explain our idea in proving the crucial point mentioned above. Suppose that A_b consists of edges denoted by halftone lines, and let H_b denote the subgraph induced by A_b, where we can assume that H_b is a tree if A_b is nonempty. The bold solid lines shows the tree T_c, and black spots on H_b corresponds to leaves of T_c. Let R_{uv} denote the path connecting u and v in H_b. The point is that we can find a path P (shown by dotted fine lines) as follows: P consists of

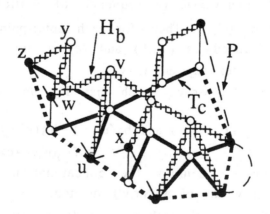

Fig. 2.5. A schematic explanation in constructing A_c'' to 2-VCA[$G_0=T_c,S=V_c$] from A_b to 2-VCA[$G_0=T_b$, $S=V_c$].

edges not in H_b and containing only black spots on H_b, each appearing only once. Furthermore for each edge e of H_b, P has at most two edges (u,v) and (u',v') such that both R_{uv} and $R_{u'v'}$ pass through e. For the edge $(u,v) \in E(H_b)$ of Fig.2.5, P has two edges (u,w) and (u,x) for which R_{uw} and R_{ux} exist. On the other hand H_b has at least one edge, say (y,z), for which P has only one edge, (z,w), for which R_{zw} exists. Let $A_c"$ be the set of edges denoted by dotted bold lines in the figure. Note that there is one-to-one correspondence between $E(P)$ and $A_c"$. Clearly $A_c"$ is a solution to 2-VCA$[G_0=T_c, S=V_c]$. The difinition of c_3 means that $c_3(a,b) \leq c_3(E(R_{ab}))$ for each $(a,b) \in E(P)$. Hence

$$c_3(E(P)) = \Sigma_{(u,v) \in E(P)} c_3(u,v) \leq \Sigma_{(u,v) \in E(P)} c_3(E(R_{uv}))$$

$$< 2\Sigma_{(u',v') \in A_b} c_3(u',v') = 2c_3(A_b) \leq 2c_2(A_b).$$

Since the definition of c_3 also implies that $c_3(A_c") \leq c_3(E(P))$, we have $c_3(A_c") < 2c_2(A_b)$. The $O(|V|^3)$ time complexity comes from the definition of c_3. Thus we obtain the next theorem.

Theorem 2. The algorithm BIA-SV finds an approximate solution for 2-VCA-SV in $O(|V|^3)$ time and the worst approximation is less than four times the optimal if G_0 is connected.

5. Strong-connectivity augmentation problem

The problem is formulated as a decision problem as follows:

(SCA-SV with unity weights)

Instance: A directed complete graph $G=(V,E)$, a spanning subgraph $G_0=(V,E')$, a specified subset $S \subseteq V$, and an integer k.

Question: Is there a set $A \subseteq E-E'$ with $|A| \leq k$ such that, for any pair of vertices in S, G_0+A has at least one directed cycle containing them ?

The reduction is from the *set covering* problem, a well-known NP-complete problem, defined as follows:

(Set Cover)

Instance: A collection C of sets, all subsets of the universal set U, and an integer k.

Question: Is there a subcollection $C' \subseteq C$ with $|C'| \leq k$ such that C' covers U, that is, the union of all sets in C' is equal to U ?

We can prove the following theorem.

Theorem 3. SCA-SV with unity weights is NP-complete.

We propose an $O(|V|^3)$ approximation algorithm STA-SV and its variations. Only one of them, STA-SV, is explained: it has two stages. It first constructs a spanning arborescence T_{out} rooted at a vertex $r \in S$. This is done by finding a shortest $<r,v>$-path for each vertex $v \in V$. In the second stage set $c<u,v>=0$ for any edge $<u,v>$ of T_{out}, and STA-SV again constructs another reverse arborescence T_{in} rooted at r by finding <v,r>-shortest path for each v∈ V with respect to the current cost c. We obtain an approximate solution $A=E(T_{out})\cup E(T_{in})-E(G_0)$.

STA-SV

Input: G=(V,E) a complete directed graph with a cost function c:E→Z^+(,where if <u,v>∈ E then c<u,v>=0), G_0=(V,E') a subgraph of G and S⊆V a specified set of vertices.

Output: an approximation solution A⊆E-E'.

1. For each pair of vertices u,v∈ V, find a shortest path P<u,v> and the distance   d<u,v> from u to v.
   {the first stage}
2. Choose any vertex r in S as a candidate for the root. Set S:={r}, S':=S-{r}, and   A:=∅.
3. **If** S':=∅ **then go to** step 6.
4. Find u_1∈ S,v_1∈ S' with d<u_1,v_1>=min{{d<u,v> | u∈ S,v∈ S'}∪{d<w,r> | w∈ S'}}.
   If d<u_1,v_1>=d<w,r> is greater than d<r,w> **then** set r:=u_1.
5. Set S':=S'-{v_1}, A:=A∪E(P<u_1,v_1>), S:=S∪V(P<u_1,v_1>) and **go to** step 3.
   {the second stage}
6. Set c<u,v>:=0 for each edge <u,v> in A.
7. For each pair of vertices u,v∈ V, find a shortest path P<u,v> and the distance   d<u,v> from u to v.
8. Set S:={r}, S:=S-{vertices which are not leaves of T_{out}}, S':=S-{r}, A':=A.
9. If Sv'=∅, then set A:=A'-E', and Stop.
10. Find u_1∈ S',v_1∈ S such that d<u_1,v_1>=min{d<u,v> | u∈ S',v∈ S}.
11. Set S':=S'-{u_1}, A':=A'∪E(P<u_1,v_1>), S:=S∪V(P<u_1,v_1>), and **go to** step 9.

5.4. Experimental results. There is an example of the problem whose approximation solution given by the algorithm STA-SV may close to |S| times the optimal. (The theoretical analysis of worst case approximations is, however, left for future reserch.) Some of experimental results are shown in Tables 1 and 2. AL1 denotes SCA-SV, AL2 does an algorithm in which AL1 is repeated |S| times by taking each vertex of S as a starting candidate for the root, and AL3, AL4 are other approximation ones not explained here. OPT is an optimal cost. Each upper integer is an approximate total cost and the lower denotes the CPU time in second. All algorithms are implemented on NEC EWS-4800(CPU:68020, 16MHz) by means of C language.

References

[1] K. P. Eswaran and R. E. Tarjan, Augmentation problems, SIAM J.

Table 1. Experimental results for G with |V|=16. Optimum solutions are known.

DATA	AL 1	AL 2	AL 3	AL 4	OPT	DATA	AL 1	AL 2	AL 3	AL 4	OPT
1 0 1	1 0	1 0	1 0	2 3	1 0	1 0 6	2 2	2 2	2 3	2 5	2 2
	0.183	0.533	0.166	0.433			0.183	1.049	0.316	0.416	
1 0 2	1 3	1 1	1 1	2 3	1 1	1 0 7	2 4	2 4	2 4	2 4	2 4
	0.166	0.533	0.166	0.466			0.183	1.066	0.316	0.416	
1 0 3	1 7	1 5	1 5	2 0	1 5	1 0 8	1 5	1 4	1 4	1 8	1 4
	0.183	0.549	0.166	0.449			0.183	1.066	0.299	0.483	
1 0 4	1 6	1 6	1 6	1 8	1 6	1 0 9	2 6	2 5	2 6	2 5	2 5
	0.183	0.533	0.149	0.516			0.183	1.066	0.316	0.433	
1 0 5	1 1	8	8	1 1	8	1 1 0	1 9	1 9	1 9	2 3	1 9
	0.183	0.549	0.166	0.433			0.183	1.049	0.299	0.499	

Table 2. Experimental results for G with |V|=50. G and costs are randomly generated. Optimum solutions are unknown.

data	AL 1	AL 2	AL 3	AL 4	AL 1	AL 2	AL 3	AL 4	AL 1	AL 2	AL 3	AL 4
	|S| = 2 0				|S| = 3 0				|S| = 4 0		
5 0 1	4 4	3 8	3 6	5 1	5 3	4 8	4 8	5 9	6 4	6 1	6 3	6 1
	4.583	49.981	7.016	10.599	4.583	77.396	14.316	10.416	4.85	112.56	26.82	9.48
5 0 2	2 0	1 6	1 9	2 4	2 2	1 8	2 3	2 4	2 5	2 2	2 4	2 4
	4.416	49.464	6.433	18.115	4.566	76.346	12.749	18.549	4.73	107.88	23.15	18.62
5 0 3	2 4	2 4	2 4	3 2	3 0	2 9	2 9	3 4	3 6	3 4	3 7	3 4
	4.433	49.298	7.183	13.099	4.583	78.046	13.732	13.249	4.85	110.93	25.08	12.68
5 0 4	2 3	2 3	2 3	3 0	4 9	4 7	4 7	4 5	4 7	4 5	4 2	4 5
	4.416	48.814	6.899	12.382	4.599	77.863	15.016	11.816	4.82	111.40	26.27	11.77
5 0 5	2 1	2 0	2 0	2 3	2 5	2 2	2 2	2 3	3 2	2 8	2 8	2 5
	4.433	48.998	6.966	9.816	4.616	77.863	14.649	11.532	4.78	110.40	27.22	11.73
5 0 6	2 3	2 1	2 3	3 2	3 5	2 8	3 0	3 2	3 9	3 3	3 7	3 0
	4.383	48.214	6.383	16.482	4.566	76.080	13.199	16.432	5.03	114.41	23.60	15.37

Comput. , 5(1976), 653-655.
[2] G. N. Frederickson and Ja'Ja', Approximation algorithms for several graph augmentation problems, SIAM J. Comput. , 10(1981), 270-283.
[3] F. Harary, "Graph Theory", Addison-Wesley, Reading, MA (1969)
[4] A. Rosenthal and A. Goldner, Smallest augmentations to biconnect a Graph, SIAM J. Comput. , 6(1977), 55-66.
[5] T. Watanabe and A. Nakamura, Edge-connectivity augmentation problems, J. Comput. and System Sci. , 35(1987-08), 96-144.
[6] T. Watanabe and A. Nakamura, 3-connectivity augmentation problems, Proceedings of 1988 IEEE Int. Sympo. on Circuits and Systems(1988-6), 1847-1850.
[7] T. Watanabe, T. Narita and A. Nakamura, 3-Edge-connectivity augmentation problems, 1989 IEEE Int. Sympo. on Circuits and Systems (1989-05), 335-338.
[8] T. Watanabe and A. Nakamura, A Smallest augmentation to 3-connect a graph, Discrete Applied Mathematics, to appear.

A Heuristic Algorithm for
the k-Center Problem with Vertex Weight[1]

Qingzhou Wang and Kam Hoi Cheng
Computer Science Department, University of Houston
Houston, Texas USA 77204-3475

Abstract

In this paper, the NP-hard k-center problem with vertex weight is investigated. The problem is to choose k vertices as service centers so that the maximum weighted service delivery distance to any vertex is minimized. The notions of a neighbor vertex set and a d-restricted digraph based on weighted distance are introduced. The proposed algorithm uses a greedy strategy to choose the vertex with maximum vertex weight as the next service vertex. We have proved that the proposed algorithm generates results that are guaranteed to be no greater than twice the optimal solution values. This is the best possible polynomial time heuristic unless P=NP.

1. Introduction

In many practical situations of providing services to a large population, determining locations to place a given number of service stations is an important design consideration. The goal often is to minimize the delivery time of the service. An example of such placement problem involves the construction of fire stations in a given region. If there are n towns in the region, and available resources only allow us to construct k fire stations, a solution to this placement problem should identify k towns to build fire stations so that fire engines can be dispatched to any town in the shortest possible time. Similar problems of placing ambulance dispatching centers and other emergency repair centers all have the same solution criterion of providing service in the shortest possible time. A common name for this problem is the *k-center problem*. If we assume that the time required to deliver service from point u to point v is proportional to the distance between them, we can use service delivery distance instead of delivery time in our discussion.

---
[1]This research is based in part upon work supported by the Texas Advanced Research Program under grant 1028-ARP.

A generalized version of the problem, the *k-center problem with vertex weight*, is considered in this paper. A *usage* weight is associated with each location to represent its likelihood of requesting emergency service. This weight could be the local population or the known accident rate. Service centers should be assigned close to those locations that are likely to require service. Formally, let $G = (V, E)$ be a weighted graph where a usage weight $\omega(v)$ is associated with each vertex v, $v \in V$. For any two vertices u and v in V, the distance $\delta(u, v)$ is defined to be the length of the shortest path linking u and v, i.e., if the shortest path linking u and v is $e_1, e_2, \ldots, e_l$, then $\delta(u, v) = \sum_{i=1}^{l} |e_i|$ where $|e_i|$ is the length of edge e_i. The *weighted* distance $\delta_\omega(u, v)$ from u to v is defined to be

$$\delta_\omega(u, v) = \delta(u, v) \cdot \omega(v). \tag{1.1}$$

Note that $\delta_\omega(u, v)$ is generally not the same as $\delta_\omega(v, u)$. This weighted distance can easily be extended to a subset $S \subseteq V$. The weighted distance from S to v is defined as

$$\delta_\omega(S, v) = \min\{\delta_\omega(s, v) : s \in S\}. \tag{1.2}$$

The k-center problem with vertex weight is to choose the vertex subset S, $S \subseteq V$ and $|S| = k$, such that the function

$$f(S) = \max\{\delta_\omega(S, v) : v \in V\} \tag{1.3}$$

achieves its minimum value, $r_\omega(G, k)$. The value $r_\omega(G, k)$ is called the *weighted radius* of k centers in G, i.e.,

$$r_\omega(G, k) = \min_{|S|=k} f(S). \tag{1.4}$$

The k-center problem with vertex weight is NP-hard [GARE79]. Since polynomial time optimal solutions are unlikely, we are interested in polynomial time approximation algorithms. An approximation algorithm is called a ξ-approximation algorithm if for all problem instances, it can guarantee that

$$\frac{S_H}{S_O} \leq \xi \tag{1.5}$$

where S_O and S_H are the optimal and the approximate solutions respectively.

The k-center problem, its variations and generalizations have been studied in several literatures. Formulations on many of its variations can be found in [HAND79]. Algorithms for the optimal solutions of various k-center problems have been developed by Kariv and Hakimi [KARI79]. Several 2-approximation algorithms have been proposed for the simple k-center problem [GONZ85], [HOCH86] and [FEDE88]. 3-approximation algorithms have also been developed in [HOCH86] and [FEDE88] for the *weighted k-center problem*. Note that our k-center problem with vertex weight is different from the weighted k-center problem where a *cost* weight is associated with the construction cost of a service center at each vertex. The weighted k-center problem is to choose a vertex set as service centers which minimizes the service delivery time/distance under the constraint that the total construction cost is no more than k.

Throughout the paper, let $|V| = n$. We assume that both edge length and vertex weight are positive and the graph to be connected. So

$$0 \le \delta(u,v) < \infty \quad \text{and} \quad 0 \le \delta_\omega(u,v) < \infty, \quad \forall u, v \in V. \tag{1.6}$$

The rest of the paper is organized as follows. In §2, the notions of a d-restricted digraph (directed graph) and a neighbor vertex set are introduced. These are the basis of our heuristic algorithm for the k-center problem with vertex weight. Our heuristic algorithm is presented and shown to be of polynomial time complexity in §3. §4 shows that our proposed algorithm is a 2-approximation algorithm which is the best possible heuristic for the problem unless P=NP.

2. Preliminaries

For a weighted graph $G = (V, E)$ of n vertices, it is easy to see that there are at most n different vertex weights, $n(n-1)/2$ different edge lengths and $n(n-1)$ different values of weighted distances. Let m be the number of different weighted distances, then $m \le n(n-1)$. Let these different weighted distances be arranged in increasing order of values, i.e.,

$$d_0 < d_1 < \ldots < d_i < \ldots < d_{m-1}. \tag{2.1}$$

For an optimal selection of k-centers on the graph G, its weighted radius $r_\omega(G, k)$ must be one of the m weighted distances in Eq. (2.1), i.e.,

$$\exists d_j \text{ such that } d_j = r_\omega(G, k). \tag{2.2}$$

Let d be a non-negative number. For a weighted graph G, the *d-restricted digraph* $G(d) = (V, E(d))$ has the same vertex set V as in G. The edge set $E(d)$ is defined as follows:

$$E(d) = \{\vec{e}_{(u,v)} : \delta_\omega(u,v) \le d\} \tag{2.3}$$

where $\vec{e}_{(u,v)}$ is the directed edge in $G(d)$ from vertex u to vertex v. From the definition of

$G(d)$, it is easy to see that

$$E(d_{i+1}) - E(d_i) = \{\vec{e}_{(u,v)} : \delta_\omega(u,v) = d_{i+1}\}. \tag{2.4}$$

Hence for any d, $d \in [d_i, d_{i+1})$, we have $E(d) = E(d_i)$ and thus

$$G(d) = \begin{cases} G(0), & d \in [0, d_0); \\ G(d_i), & d \in [d_i, d_{i+1}), \quad i = 0, \ldots, m-2; \\ G(d_{m-1}), & d \in [d_{m-1}, \infty). \end{cases} \tag{2.5}$$

In other words, there are $m + 1 \leq n(n-1) + 1$ different d-restricted digraphs that can be derived from any graph G with n vertices.

Next, we define the notion of a neighbor vertex set. Since it will only be used on d-restricted digraphs, we define it on a vertex of a d-restricted digraph. Let $u \in V$ be an arbitrary vertex in a d-restricted digraph $G(d) = (V, E(d))$. The set of all u's neighbor vertices, $N(u)$, is defined as

$$N(u) = \{u\} \cup \{v : \exists \vec{e}_{(u,v)} \in E(d)\} \cup \{v : \exists x \in V, \; \vec{e}_{(x,u)}, \vec{e}_{(x,v)} \in E(d)\}. \tag{2.6}$$

The justification of this definition should become clear in §4 when we prove that our heuristic algorithm is a 2-approximation algorithm.

3. A Polynomial Time Heuristic

We first present a greedy algorithm to select a set of vertices each of which has the maximum vertex weight at the time of its selection. This greedy algorithm is formally presented in Algorithm 1. The input to the greedy algorithm can be any digraph with vertex weight, however it is always a d-restricted digraph for our purposes. The chosen set of vertices, $I_g(G(d))$, is called the greedy set and for any pair of vertices in $I_g(G(d))$, they can not be neighbor vertices in the digraph $G(d)$. Algorithm 1 as well as the construction of a d-restricted digraph are easily seen to take polynomial time.

Our proposed heuristic algorithm applies Algorithm 1 to a sequence of d-restricted digraphs of the graph G. It is formally presented in Algorithm 2 and is similar to the algorithm proposed in [HOCH86] for the simple k-center problem. However, the use of a greedy algorithm, the newly defined neighbor vertex set and the d-restricted digraph make the resulting algorithm more powerful. Since the **while** loop in Algorithm 2 executes $m \leq n(n-1)$ times and each iteration takes polynomial time, Algorithm 2 is easily seen to take polynomial time.

> input *d-restricted digraph* $G(d) = (V, E(d))$
> construct neighbor vertex set $N(u), \forall u \in V$
> $I_g(G(d)) \leftarrow \emptyset$
> **repeat**
>   select $u \in V$ with $\omega(u) = \max\{\omega(v) : v \in V\}$
> $I_g(G(d)) \leftarrow I_g(G(d)) \cup \{u\}$
> $V \leftarrow V - N(u)$
> **until** $V = \emptyset$
> **return** $(I_g(G(d)))$

Algorithm 1. A Greedy Algorithm

> construct the sequence d_i's, $i = 0, 1, \ldots, m - 1$, as in Eq.(2.1)
> $DONE \leftarrow$ **false**; $i \leftarrow 0$
> **while** $i < m$ **and** *not DONE* **do**
>   construct the *d*-restricted digraph $G(d_i)$
>   select a greedy set $I_g(G(d_i))$ using Algorithm 1
>   **if** $|I_g(G(d_i))| \leq k$ **then**
> $S \leftarrow I_g(G(d_i))$; $DONE \leftarrow$ **true**; $\tilde{d} \leftarrow d_i$
>   **endif**
> $i \leftarrow i + 1$
> **endwhile**
> **if** $|S| < k$ **then** add $k - |S|$ vertices to S

Algorithm 2. The Heuristic Algorithm with Error Ratio 2

4. Analysis of the Heuristic Algorithm

First we prove in Theorem 1 that when the **while** loop in Algorithm 2 terminates, $\tilde{d} \leq r_\omega(G, k)$.

Theorem 1. When the **while** loop in Algorithm 2 terminates, $\tilde{d} \leq r_\omega(G, k)$.

Proof: If we can show that $|I_g(G(d_j))| \leq k$ where $d_j = r_\omega(G, k)$, then the **while** loop must terminate after the j_0^{th} iteration with $j_0 \leq j$. Hence $\tilde{d} = d_{j_0} \leq d_j = r_w(G, k)$. Furthermore, the corresponding greedy set $I_g(G(\tilde{d}))$ has no more than k vertices.

To show that $|I_g(G(d_j))| \leq k$, it is sufficient to show that for any two vertices $u, v \in I_g(G(d_j))$, they can not be served by the same service center under an optimal k-center arrangement. This is because there are only k service centers, so $|I_g(G(d_j))| \leq k$ if no two vertices are served by the same service center.

The proof is by contradiction. Suppose the statement that vertices u and v must be served by different centers under the optimal arrangement is not true, i.e., vertices u and v are served by the same center. Without loss of generality, assume $w(u) \geq w(v)$ and so vertex u is selected before vertex v in Algorithm 1. There are three possible cases.

Case 1. Vertex u is a service center under the optimal k-center arrangement, and v is served by it. We must have

$$\delta_w(u, v) \leq r_w(G, k) = d_j. \tag{4.1}$$

This implies the existence of edge $\vec{e}_{(u,v)}$ in $G(d_j)$, hence $v \in N(u)$ and vertex v will not be selected into the greedy set $I_g(G(d_j))$; a contradiction.

Case 2. Vertex u is served by the center at vertex v under the optimal arrangement. We have

$$\delta_w(v, u) \leq r_w(G, k) = d_j. \tag{4.2}$$

However,

$$\delta_w(u, v) = \delta(u, v) \cdot w(v) \leq \delta(v, u) \cdot w(u) = \delta_w(v, u), \tag{4.3}$$

hence

$$\delta_w(u, v) \leq d_j, \tag{4.4}$$

which implies that $v \in N(u)$; a contradiction.

Case 3. Both vertices u and v are served by the same service center x under the optimal arrangement. We have

$$\begin{aligned} \delta_w(x, u) &\leq r_w(G, k) = d_j, \\ \delta_w(x, v) &\leq r_w(G, k) = d_j. \end{aligned} \tag{4.5}$$

These imply the existence of edges $\vec{e}_{(x,u)}$ and $\vec{e}_{(x,v)}$ in $G(d_j)$. By the definition of $N(u)$ in Eq. (2.6), $v \in N(u)$; a contradiction.

Hence for any two vertices $u, v \in I_g(G(d_j))$, they are served by different service centers under an optimal k-center arrangement. $\square$

Now suppose the **while** loop in Algorithm 2 terminates after the j_0^{th} iteration. From the proof of Theorem 1, we know that $j_0 \leq j$. Let k_0 be the number of vertices in the greedy set and let them be represented by $u_i, 0 \leq i \leq k_0 - 1$. We observe that during the execution of the greedy algorithm, the vertex set V is replaced repeatedly by $V - N(u_i)$. Mathematically, this process can be represented by the following recurrence relations,

$$V_{i+1} = V_i - N(u_i), \quad \text{with } \omega(u_i) = \max_{v \in V_i} \omega(v), \quad i = 0, \ldots, k_0 - 1. \tag{4.6}$$

Note that it is a monotonically decreasing sequence of sets, i.e.,

$$V = V_0 \supset V_1 \supset \ldots \supset V_{k_0-1} \supset V_{k_0} = \emptyset; \tag{4.7}$$

and the set being removed from V_i is actually $N_i(u_i)$ where

$$N_i(u_i) = N(u_i) \cap V_i, \quad i = 0, \ldots, k_0 - 1. \tag{4.8}$$

It is clear that both $N_i(u_i) \subseteq V_i$ and $N_i(u_i) \subseteq N(u_i)$. It is also important to note that the construction of $N(u)$ for each u in Algorithm 1 is done before the **repeat** loop because of the following scenario. The set $N_i(u_i)$ may contain neighbors of u_i if all vertices in the original V are considered, but they are no longer neighbors of u_i if only vertices in V_i are considered. This is the result of the way we define the neighbor vertex set in Eq. (2.6). The next two theorems establish that our heuristic algorithm is a 2-approximation algorithm.

Theorem 2. Let u_i be the service center for all the vertices in $N_i(u_i)$, $i = 0, \ldots, k_0 - 1$. For any $v \in N_i(u_i)$, the weighted distance from u_i to v will not exceed $2\tilde{d}$, i.e., $\delta_\omega(u_i, v) \leq 2\tilde{d}$.

Proof: Since $N_i(u_i) \subseteq V_i$ and $\omega(u_i) = \max_{v \in V_i} \omega(v)$, we must have

$$\omega(u_i) \geq \omega(v), \quad \forall v \in N_i(u_i). \tag{4.9}$$

From the neighbor set definition in Eq. (2.6), v $(\neq u_i)$ is included in $N_i(u_i) \subseteq N(u_i)$ for one of the following two reasons:

$$v \in N_i(u_i), \ v \neq u_i \begin{cases} \text{reason 1:} & \exists \vec{e}_{(u_i,v)} \in E(\tilde{d}), \\ \text{reason 2:} & \exists x \in V, \text{ such that } \vec{e}_{(x,u_i)}, \vec{e}_{(x,v)} \in E(\tilde{d}) \end{cases} \tag{4.10}$$

where $E(\tilde{d})$ is the edge set of $G(\tilde{d}) = (V, E(\tilde{d}))$. If vertex v is included for the first reason, we have $\delta_\omega(u_i, v) \leq \tilde{d} \leq 2\tilde{d}$. Now, suppose vertex v is included for the second reason. The existence of $\vec{e}_{(x,u_i)}$ and $\vec{e}_{(x,v)}$ in $G(\tilde{d})$ implies

$$\begin{aligned} \delta_\omega(x, u_i) &= \delta(x, u_i) \cdot \omega(u_i) \leq \tilde{d}, \\ \delta_\omega(x, v) &= \delta(x, v) \cdot \omega(v) \leq \tilde{d}. \end{aligned} \tag{4.11}$$

We know that $\omega(u_i) \geq \omega(v)$ from Eq. (4.9), and $\delta(u_i, v) \leq \delta(x, u_i) + \delta(x, v)$ from the shortest path definition of δ in §1. These two facts combined with Eq. (4.11) imply

$$\delta_\omega(u_i, v) = \delta(u_i, v) \cdot \omega(v) \leq \delta(x, u_i) \cdot \omega(u_i) + \delta(x, v) \cdot \omega(v) \leq 2\tilde{d}. \tag{4.12}$$

Therefore, we can conclude that for any $v \in N_i(u_i)$, $\delta_w(u_i, v) \leq 2\tilde{d}$. $\square$

Theorem 3. Let the set S produced in Algorithm 2 be the chosen set of k-centers. Let $r_{w,H}(G, k)$ be the maximum weighted distance from S to any vertex in V, then

$$\frac{r_{w,H}(G, k)}{r_w(G, k)} \leq 2. \tag{4.13}$$

In other words, Algorithm 2 is a 2-approximation algorithm.

Proof: In Theorem 2, we let vertex u_i serve all the vertices in $N_i(u_i)$, and the weighted service distance does not exceed $2\tilde{d}$. Since $I_g(G(\tilde{d})) \subseteq S$, then for any $v \in V$, we must have $\delta_w(S, v) \leq \delta_w(I_g(G(\tilde{d})), v)$. Hence,

$$\begin{aligned}
r_{w,H}(G, k) &= \max\{\delta_w(S, v) : v \in V\} \\
&\leq \max\{\delta_w(I_g(G(\tilde{d})), v) : v \in V\} \tag{4.14} \\
&\leq 2\tilde{d} \leq 2d_j = 2r_w(G, k).
\end{aligned}$$

Therefore, Eq. (4.13) holds. $\square$

Hochbaum and Shmoys [HOCH86] have shown that unless P=NP, a 2-approximation heuristic is the best possible polynomial time algorithm for the simple k-center problem. When all vertex weights are equal, our general problem reduces to the simple k-center problem. It is obvious that our 2-approximation heuristic is the best possible polynomial time algorithm for the k-center problem with vertex weight.

5. Conclusion

We proposed in this paper a polynomial time heuristic algorithm to solve the k-center problem with vertex weight. The notion of a neighbor vertex set is defined and the notion of a d-restricted digraph based on weighted distance is introduced. The proposed algorithm uses a greedy strategy to choose a set of service centers each of which has the maximum vertex weight at the time of its selection. It is a 2-approximation algorithm which is the best possible polynomial time algorithm unless P=NP.

6. References

[FEDE88] T. Feder and D. H. Greene, Optimal Algorithms for Approximate Clustering, *ACM Symposium on Theory of Computing*, 1988, pp. 434–444.

[GARE79] M. R. Garey and D. S. Johnson, Computers and Intractability: A Guide to the Theory of NP-Completeness, *Freeman, San Francisco*, 1979.

[GONZ85] T. F. González, Clustering to Minimize the Maximum Intercluster Distance, *Theoretical Computer Science*, Vol. 38, 1985, pp. 293–306.

[HAND79] G. Y. Handler and P. B. Mirchandani, Location on Networks: Theory and Algorithms, *MIT Press, Cambridge, MA*, 1979.

[HOCH86] D. S. Hochbaum and D. B. Shmoys, A Unified Approach to Approximation Algorithms for Bottleneck Problems, *Journal of ACM*, Vol. 33, 1986, pp. 533–550.

[KARI79] O. Kariv and S. L. Hakimi, An Algorithmic Approach to Network Location Problems. Part I: The p-Centers, *SIAM Journal of Appl. Math.*, Vol. 37, 1979, pp. 513–538.

Parallel Convexity Algorithms for Digitized Images on a Linear Array of Processors

Hussein M. Alnuweiri

Department of Computer Engineering, Box 1729
King Fahd University of Petroleum & Minerals
Dhahran 31261, Saudi Arabia

V. K. Prasanna Kumar

Department of EE-Systems, SAL-344
University of Southern California
Los Angeles, CA 90089-0781, U.S.A.

Abstract

Efficient implementation of global computations on a linear array of processors is complicated due to the small communication bandwidth and the large communication diameter of the array. This paper presents efficient parallel techniques for partitioning, movement, and reduction of data on linear arrays. Also, efficient data structures are used to enable fast sequential access of query points within each processor. This combination of serial and parallel techniques is used to derive an optimal parallel algorithm for computing the convex hull of each connected region in an $n \times n$ image. The algorithm takes $O(n^2/p)$ time on a linear array with p processors, where $1 \leq p \leq n/\log n$. This result is processor-time optimal since an optimal sequential algorithm takes $O(n^2)$ to solve the problem. Thus, a linear array with $n/\log n$ processors can solve the above problem in $O(n \log n)$ time. In comparison, a two dimensional mesh-connected array of processors can solve this problem in $O(n)$ time using n^2 processors. The processor-time product for the mesh is $O(n^3)$, which is not optimal.

1 Introduction

The convex hull is a fundamental geometric structure which is used as a tool for solving a wide class of problems in computational geometry [15]. Convex hull computations are of central importance in many applied areas including computer graphics, image analysis, computer vision, pattern classification, mathematical programming, statistics, and operations research. Because of its significance and interesting structure, the convex hull problem has generated extensive algorithmic research culminating in a number of elegant sequential algorithms [15], and more recently, in efficient parallel algorithms for the problem [1,3,4]. Examples and a survey of applications of the convex hull can be found in [14,15].

This paper presents a new optimal parallel algorithm for computing the convex hull of each connected set of pixels in a digitized image on a linear array of processors. This problem has numerous applications in image analysis and middle-level vision. For example, by computing the convex hull of each connected image region, a smallest enclosing rectangle and the diameter of each region can be computed efficiently. Also, the linear separability of any two regions can be tested efficiently. Other applications include shape approximation, detecting concavities of regions, computing farthest pairs (c.f. [13,14,15]). The proposed algorithm takes $O(n^2/p)$ time on a linear array with p processors, where $1 \leq p \leq n/\log n$. Each processor has a local memory of $O(n^2/p)$ words to store (n^2/p) pixels of the input image. The algorithm is optimal in the sense that its *processor-time product* is $O(n^2)$ which equals (to within a constant factor) the optimal sequential complexity of the problem.

Linear arrays of processors with nearest neighbor connections have several features that make them attractive for VLSI implementation. These features include simplicity of interconnection, modularity, amenability to simpler fault tolerance techniques, and low I/O bandwidth requirements as compared to other array processors with more complex interconnection networks. Parallel architectures based on linear arrays of processors have been proposed or constructed for a wide range of applications [5,7,8,9,10,12]. Most of the applications considered on linear arrays have simple and regular data communications patterns which can be mapped directly onto the linear array structure. However, several factors can complicate the implementation of applications with global data movement requirements on linear arrays. The main sources of inefficiency are the small communication bandwidth and the large communication diameter. To overcome such shortcomings, we propose efficient algorithmic techniques for data reduction and movement on a linear array. The purpose is to let the total computation time be dominated only by the time of the local computations performed by each processor. Furthermore, powerful sequential techniques are used by each processor when performing computations on the data in its local memory. In particular, efficient data structures are utilized to provide for fast access of query points within each processor. These serial and parallel techniques are incorporated within a divide-and-conquer procedure to compute the convex hull of each connected region in the image.

1.1 Problem definition

The convex hull of a set S of planar points, denoted $CH(S)$, is defined as the minimum *convex* set containing S. A point $p \epsilon S$ is said to be an *extreme point* of S (and $CH(S)$), if $p \notin CH(S - \{p\})$. The extreme points of the hull will be enumerated in counterclockwise order starting from the top leftmost point. The edges of the hull are the line segments connecting pairs of consecutive extreme points according to the enumeration order.

For applications in image analysis and computer vision, the input is usually a digitized image. Digitization leads to a more complicated definition of convexity (c.f. [11]). The image points (pixels) can be viewed either as an $n \times n$ array of squares with the image value constant within each such square, or as the points in the plane with integer coordinates (i, j), where $0 \leq i, j \leq n - 1$. The latter view will be adopted in this paper since it leads to a simpler presentation. However, the proposed techniques apply equally well to the former view. Thus, computing the convex hull of a set of pixels will be done first by determining the extreme points of the set according to the above definition, then enumerating them and constructing the edges connecting pairs of consecutive extreme points.

1.2 Model of Computation

The model we use consists of a linear array of p processors, $1 \leq p \leq n$, indexed 0 through $p-1$, and each processor P_i is connected by bidirectional links to its immediate left and right neighbors if they exist (see figure 1). The bandwidth of each link is $\log n$ bits. Each processor also has a local memory of $O(n^2/p)$ words, each word consisting of $\Theta(\log n)$ bits. Also, each processor can access one word from its memory or communicate one word of data to one of its neighbors in $O(1)$ time. Finally, a processor can perform a basic logic or arithmetic operation on a fixed number of words in $O(1)$ time.

Central to achieving processor-time optimality for the given problem, is the efficient partitioning of the image pixels among the processors. The partitioning must be done with the purpose of reducing the communication distance among processors as well as reducing the amount of traffic on a single link of the array during subcomputations. The following image mapping scheme will be used in this paper.

Shuffled row major distribution: In this distribution, the image is partitioned into a $\sqrt{p} \times \sqrt{p}$ array of squares, each of size $n/\sqrt{p} \times n/\sqrt{p}$. The square in the ith row and jth column is denoted by $S(i, j)$. Let $q = \log \sqrt{p} - 1$, and let $(i_q i_{q-1} ... i_0)$ and $(j_q j_{q-1} ... j_0)$ be the binary representations of the numbers i and j, respectively. In the shuffled row-major assignment, $S(i, j)$ is assigned to processor P_k, where the binary representation of k is $(i_q j_q i_{q-1} j_{q-1} ... i_0 j_0)$. An example of this assignment is shown in figure 1.

2 Divide-and-Conquer Techniques for Linear Arrays

Divide-and-conquer is the underlying technique for solving the main problem defined in section 1.1. The basic approach is to recursively partition the image into four disjoint subsquares, solve the problem within each subsquare, and then merge the results from the four subsquares. For p processors, the recursive algorithm can be completed in $\log \sqrt{p}$ iterations. In iteration i, $1 \leq i \leq \log \sqrt{p}$, the image is partitioned into $p/4^i$ disjoint squares each of size $(2^i n/\sqrt{p}) \times (2^i n/\sqrt{p})$. Introducing some notation which will be used extensively in later discussion is now in order.

Let $S(i) = 4^i$, where $0 \leq i \leq \log \sqrt{p}$. An i-partition of the linear array is a partition of the processors into $p/S(i)$ processor-blocks, each consisting of $S(i)$ processors. In iteration i, the jth such block is denoted by $\text{BLOCK}^i(j)$, $0 \leq j \leq \frac{p}{S(i)} - 1$, and it consists

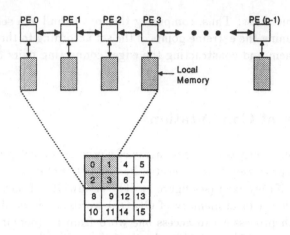

Figure 1: Linear array model and image mapping

of processors $P_{jS(i)}, P_{jS(i)+1}, \cdots, P_{(j+1)S(i)-1}$. Similarly, an *i-partition* on an $n \times n$ image I is a partition of I into $p/S(i)$ squares each of size $2^i \frac{n}{\sqrt{p}} \times 2^i \frac{n}{\sqrt{p}}$ pixels. The squares of an *i*-partition are indexed according to the shuffled row-major indexing scheme mentioned earlier. We denote the square with index j by $Q^i(j)$. The squares of an *i*-partition are mapped onto the processor blocks so that $Q^i(j)$ is stored in $\text{BLOCK}^i(j)$. In many cases it will be more convenient to define $\text{BLOCK}^i(j) \equiv \text{BLOCK}(Q^i(j))$, i.e. $\text{BLOCK}^i(j)$ is just the processor block containing $Q^i(j)$. In general, we will let $\text{BLOCK}(Q)$ denote the processor block containing the image square Q. Note that each square Q^0 of the image is mapped onto the local memory of one processor according to the shuffled row-major indexing scheme. The pixels on the first and last row and the first and last column of an image square Q are called the *boundary* of Q, and are denoted by ΔQ. Also, each square Q^i can be decomposed into four equal quadrants Q^i_0, Q^i_1, Q^i_2, and Q^i_3, provided that $|Q^i| \geq 4$. Each processor in $\text{BLOCK}(Q)$ containing a portion of the boundary of Q is called a *boundary processor*. The set of all boundary processors in $\text{BLOCK}(Q)$ is denoted by $\text{BOUND}(Q)$. Also, the processor with the smallest index in $\text{BLOCK}(Q)$ is called the *leader processor* of that processor block, and denoted by $\text{LEADER}(\text{BLOCK}(Q))$.

2.1 Data Movement

As the computation proceeds in certain applications, a subset of processors may need to read the data stored in another subset of processors. For such applications we define a special data movement operation. Let S be a set of adjacent processors in the linear array. Also, let $T \subseteq S$, and let M be a set of data elements distributed among the processors in T. Define $\text{COPY}(M, T \to S)$ as the operation whose application results in copying the elements of M into each processor in S. Let $|S|$, $|T|$, and $|M|$, denote the number of elements in S, T, and M, respectively, then we have the following result [5]:

Theorem 1 *Given S, T, and M as described above, the operation $COPY(M, T \rightarrow S)$ can be performed in $\Theta(|S| + |M|)$ time.*

In certain cases it may be desired to copy the data M into a subset S^* of processors in S. This still can be done in $\Theta(|S| + |M|)$ time. This operation will be denoted by $COPY(M, T \longrightarrow S^*)$ to distinguish it from the previous COPY operation.

2.2 Application of the COPY Operation

As mentioned earlier, the divide-and-conquer technique consists of $\log \sqrt{p}$ iterations. In iteration i, $0 \leq i \leq \log \sqrt{p} - 1$, the processors of each $BLOCK^i(j)$ solve the problem on image square $Q^i(j)$ and then each four adjacent processor-blocks merge their results. When each square is of size $(n/\sqrt{p}) \times (n/\sqrt{p})$, then it is local to one processor which can solve the problem sequentially. In the merge procedure only information pertaining to the boundary pixels of each $Q^i(j)$ needs to be considered. The data movement operations used involve moving the boundary information to and from the leader processor in each processor-block.

Consider an i−partition of the image. Essential to our techniques are two versions of the COPY operations, one for moving the pixels on the boundary of the four quadrants of a square $Q^i(j)$ into $LEADER(BLOCK^i(j))$, and one for moving these boundary pixels from $LEADER(BLOCK^i(j))$ back into their original processors in $BLOCK^i(j)$. In the following we describe the details of these operations and consider their time complexity.

- **COPY-I** : This is a $COPY(M, T \rightarrow S^*)$ operation, $T = \bigcup_{\ell=0}^{3} BOUND(Q_\ell^i(j)))$ is the union of the sets of boundary processors of the four quadrants of $Q^i(j)$, $M = \bigcup_{\ell=0}^{3} \Delta Q_\ell^i(j)$ is the set of pixels on the boundaries of the four quadrants of $Q^i(j)$, and $S^* = LEADER(BLOCK^i(j))$. Note that $|S| = 4^i$. Also, since the number of pixels on the boundary of each quadrant of $Q^i(j)$ is $4(2^{i-1} \frac{n}{\sqrt{p}} - 1)$, then $|M| = 4(2^{i+1} \frac{n}{\sqrt{p}} - 4)$; and finally, $|T| = 4(2^{i+1} - 4)$. From theorem 1, the total time taken by this operation is $\Theta(|S| + |M|) = \Theta(2^i \frac{n}{\sqrt{p}})$.

- **COPY-II** : This is a $COPY(M, T \rightarrow S^*)$ operation, $T = LEADER(BLOCK^i(j))$, $S^* = \bigcup_{\ell=0}^{3} BOUND(Q_\ell^i(j))$, i.e. the set of the boundary processors of the four quadrants of $Q^i(j)$, and M is the set of pixels now stored in $LEADER(BLOCK^i(j))$ which were originally on the boundary of the four quadrants of $Q^i(j)$. This operation also takes $\Theta(|S| + |M|) = \Theta(2^i \frac{n}{\sqrt{p}})$ time. Note that each processor needs only to copy in the pixels belonging to the image square it contains.

3 Convexity Algorithms on Linear Arrays

This section presents the main algorithm for computing the convex hull of each figure (connected region) in an $n \times n$ image. We assume that the pixels of each figure are labeled distinctly from all other figures. It has been shown in [5], that an $n \times n$ image can be labeled in $O(n^2/p)$ time on a linear array with p processors, for $1 \leq p \leq n$. Thus, labeling the image can be done within the time and processor bound required for computing convexity. Also, it is assumed that each pixel (i, j) is represented by a record p_{ij} which contains other relevant information about the pixel such as its label, information about its neighbors, flags to indicate whether the pixel is an extreme point or a boundary point, etc.

The algorithm consists of $\log \sqrt{p}$ iterations. At the beginning of iteration i, $1 \leq i \leq \log \sqrt{p}$, the convex hulls of all figures within each quadrant of Q^i have been computed. Note that there are at most 4 disjoint convex hulls having the same label in Q^i, one in each quadrant. In iteration i, we compute the convex hulls of all figures within Q^i by merging their respective hulls in the four quadrants of Q^i. To do this, we only need to consider convex hulls of figures with one or more pixels on the boundary of a quadrant (such figures will be called *boundary figures*). Let $F(Q_\ell^i)$ denote the set of labels of boundary figures in quadrant Q_ℓ^i, $\ell = 0, 1, 2, 3$. Also, let $CH_\ell(l_r)$ denote the convex hull of the figure labeled l_r within quadrant Q_ℓ^i. Note that each processor can have at most $\Theta(n/\sqrt{p})$ such figures.

The merging is done in two major steps. In the first step, we merge each pair of hulls having the same label in Q_0^i and Q_1^i and, similarly, for each pair of hulls in Q_2^i and Q_3^i. In the second step, we merge each pair of hulls having the same label in the upper rectangle (formed by $Q_0^i \cup Q_1^i$) and the lower rectangle (formed by $Q_2^i \cup Q_3^i$) of Q^i. In the following, we present the procedure for merging the hulls Q_0^i and Q_1^i.

Initial Computations:
Using a sequential algorithm, each processor P_j constructs the the convex hull of each figure within $Q^0(j)$. Also, each processor enumerates the extreme points of each hull in counter-clockwise order starting from the top leftmost point. For the merge procedure, only convex hulls of boundary figures need to be considered. Figures completely contained within the local image $Q^0(j)$ of a processor P_j need no further processing.

Before the merge, each processor moves the extreme points of each hull $CH(l_r)$, such that $l_r \epsilon F(Q^0)$, into contiguous memory locations. Recall that $F(Q^0) = \{l_1, l_2, \cdots, l_{k(0)}\}$ is the set of labels of the boundary figures of Q^0. Also, each processor P_j constructs a *local list* $L^0(j)$ with $k(0)$ entries $< l_r, M_r >$, $1 \leq r \leq k(0)$, where $l_r \epsilon F(Q^0(j))$ and M_r is the memory address of the first extreme point of $CH(l_r)$. The entires in this lest are sorted by their l_r values. All these operations can be performed in $O(n^2/p)$ time using well known sequential techniques [2,15].

Recursive Merge Procedure:
The algorithm proceeds by recursively merging each pair of hulls having the same label and in two adjacent quadrants. To merge two disjoint hulls A and B with K points each, the two common *lines of support* of A and B must be computed. This can be done using $O(\log K)$ steps of binary search [15]. The merge is done by considering the upper two halves and the lower two halves of the hulls separately. We will only consider merging the upper two halves, since merging the lower two can be done in a similar manner.

Before we present the details of the merge procedure, we present an efficient operation for fast access of extreme points. Later, we show that following some preprocessing such operation can be performed by any processor in $O(1)$ time. This operation is defined as follows:

Definition 1 [*Operation GET*] : *Let $p_r(u)$ be an extreme point labeled l_r and has enumeration order u. When performed by a processor P_j, the operation $GET(R_r(u), Y)$ stores the point $p_r(u)$ in the field $R_r(u)$ and sets a flag Y to 1 (true). If $p_r(u)$ is not in the local memory of P_j then Y is set to 0 (false), and $R_r(u)$ remains null.*

Before using the above operations, a special data structure is set up within each processor. Using this data structure, a processor can perform operation GET in $O(1)$ time. Consider two quadrants Q_0^i and Q_1^i of an image square Q^i. First, we construct a *common list CL* of labels common to both quadrants. This can be done sequentially by sending the boundary pixels of the two quadrants to the leader processor of BLOCK(Q_0^i). Since CL has $\Theta(2^i \frac{n}{\sqrt{p}})$ items, the above data movement takes $\Theta(2^i \frac{n}{\sqrt{p}})$ time using the COPY-I operation of section 2.2. Using the processors in BLOCK(Q_0^i) and BLOCK(Q_1^i), the list CL can be sorted in $O(2^i \frac{n}{\sqrt{p}})$ time [7]. The sorted list is then broadcast to all processors in BLOCK(Q_0^i) and BLOCK(Q_1^i) in $\Theta(2^i \frac{n}{\sqrt{p}})$ time using a COPY-II operation. Each processor P_j receiving CL, associates with each label l_r in CL a memory address M_r pointing to the first pixel of $CH(l_r)$ in its local memory. If such a label is not in P_j, then M_r is set to *null*. The copy of the list in CL which is in processor P_j, together with its associated pointers, will be denoted by $CL(j)$.

Details of the Merge Procedure

As was mentioned earlier, the recusive merge technqiue consists of $(\log \sqrt{p})$ stages. In the following, the implementation details of stage i, $1 \leq i \leq \log \sqrt{p}$, are presented. Consider merging the two quadrants Q_0 and Q_1 of Q^i. Let CL be the common list of labels between the two quadrants. With each label $l_r \epsilon CL$, associate the record $R(l_r) = [\triangle_0^u, \triangle_1^v]$, where $\triangle_0^u$ and $\triangle_1^v$ are the triangles with vertices numbered $(u-1, u, u+1)$ and $(v-1, v, v+1)$ on $CH_0(l_r)$ and $CH_1(l_r)$, respectively. Due to convexity, a sufficient condition for the line $\overline{uv}$ to be tangent to the two hulls at points u and v, is that it must intersect the triangles $\triangle_0^u$ and $\triangle_1^v$ only at points u and v (see figure 2). It is clear that once the two triangles are known, the above condition can be checked in $O(1)$ time. To simplify our presentation, we first consider merging a pair of convex hulls with the same label, then we show how the method is applied to merge all pairs of hulls (with the same label) concurrently.

Stage i of the merge procedure consists of $\log 2^i \frac{n}{\sqrt{p}}$ iterations of binary search. Each iteration can be performed in two phases as follows:

- **Phase 1:** Consider merging the two hulls $CH_0(l_r)$ and $CH_1(l_r)$ in Q_0 and Q_1, respectively. The computation starts by the leader processor sending the record $R(l_r) = [\triangle_0^m, \triangle_1^w]$, where m and w are the median extreme points on $CH_0(l_r)$ and

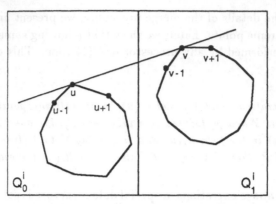

Figure 2: Merging two convex hulls

$CH_1(l_r)$, respectively. This record moves through the processors of BLOCK($Q_0 \cup Q_1$) and then back to the leader processor of (BLOCK(Q_0)). Each processor P_j in BLOCK($Q_0 \cup Q_1$) receiving this record, checks if it has one or more points of the triangles $\triangle_0^m$ and $\triangle_1^w$; if it does it puts it in the corresponding field of $R(l_r)$. This can be done by using the operation GET($R(l_r), Y$). The record is then moved to the next processor.

- **Phase 2:** Upon receiving the record $R(l_r)$ again, the leader processor decides whether the line $\overline{mw}$ is a common line of support to $CH_0(l_r)$ and $CH_1(l_r)$. If it is, then we are done. Otherwise, the information given by the record $R(l_r)$ determines which half of $CH_0(l_r)$ or $CH_1(l_r)$ is to be discarded from further computation. It can be shown [15], that the points to be discarded and the edges connecting them form a connected subchain C of the convex hull boundary. Such a chain can be specified by a pair (b, e), where b and e are the enumeration numbers of the first and last points of C, respectively. The leader processor determines the pair (b, e) for $CH_0(l_r)$ and $CH_1(l_r)$, and sends these pairs to all the processors in BLOCK(Q_0) and BLOCK(Q_1), respectively. Each processor having points on C, can eliminate these points in $O(1)$ time. This can be done by using the common list CL, and by letting each processor maintain two pointers to the active list of extreme points in its local memory. In the next iteration, the median extreme point will be computed with respect to the active list of extreme points.

Time complexity

Note that throughout each phase, only one processor in BLOCK(Q_0^i) and BLOCK(Q_1^i) is active at a time. Also, each processor is active for a fixed number of time units. The remaining processors can be utilized to merge the other hulls concurrently in a pipelined fashion. Each GET operation in Phase 1 can be performed in $O(1)$ time by using the list CL. The leader processor generates one record for each label in CL, and transmitts the records in the order by which their labels appear in CL. For each record $R(l_r) = [l_r, p(h)]$ received by a processor P_j the address of the query point $p(h)$ is given by $M(l_r) + h - 1$, where $M(l_r)$ can be obtained directly from the list $CL(j)$. The records received by each P_j are sorted by their label. Therefore, records received by P_j in consecutive time periods have consecutive entries in the list $CL(j)$. Thus, the set of memory accesses can be obtained by reading the entries in CL one by one in a sequential fashion. The first entry in the list can be located in $O(1)$ time, and each following entry can be also located in $O(1)$ time. Using the memory address in each entry, a GET operation can be completed in $O(1)$ time.

The above operations are performed concurrently by all processors within the same processor-block. The records generated by the leader processor are rotated among the processor in a pipelined fashion until they reach the leader again. After an initial delay of $O(4^i)$ time in each phase, the leader starts receiving one record every $O(1)$ time units. The leader does not start Phase 2 until all records are received. Since the number of these records is $O(2^i \frac{n}{\sqrt{p}})$, the time to perform each phase is $O(4^i + 2^i \frac{n}{\sqrt{p}}) = O(2^i \frac{n}{\sqrt{p}})$. Stage i, of the merge procedure consists of performing the above two phases $\log(2^i n^2/p)$ times. Thus, the total time taken by stage i, is $O(2^i \frac{n}{\sqrt{p}} \log(2^i \frac{n}{\sqrt{p}}))$. The total time taken by the recursive-merge procedure is

$$\sum_{i=1}^{\log \sqrt{p}} O(2^i \frac{n}{\sqrt{p}} \log(2^i \frac{n}{\sqrt{p}})) = O(n \log n).$$

Since the time taken by the initial step is $O(n^2/p)$, the total time taken by the algorithm is $O(\frac{n^2}{p} + n \log n)$. Clearly, this is optimal for $1 \leq p \leq \frac{n}{\log n}$. The following theorem summarizes our results.

Theorem 2 *Given an $n \times n$ black and white image in which all figures have been labeled, the convex hull of each figure can be constructed for all figures simultaneously, in $O(n^2/p)$ time on a fixed size linear array with p processors, for $1 \leq p \leq \frac{n}{\log n}$.*

Remarks

Note that for any $p > \frac{n}{\log n}$, the time taken by the above algorithm is $O((p+n) \log n)$. Using more than n processors degrades the time performance of the algorithm due to the large communication distance. Thus, the minumum possible time by our algorithm is $O(n \log n)$, and this time is achieved with $p = n/\log n$ processors. It is interesting to investigate whether better than $O(n \log n)$ time can be achieved for the above problem by any linear array algorithm.

In a divide-and-conquer algorithm the image is partitioned into approximately two equal parts, where one part is stored in the processors of one half of the linear array and the other part is stored in the processors of the second half. In any such partition of the image, $\Omega(n)$ figures may have convex hulls in both parts of the partition. It is well known that $O(\log n)$ operations are needed to merge two convex hulls each having $O(n)$ vertices [15]. Each such operation requires the inspection of a fixed number of vertices from each convex hull. Each vertex is specified by a $O(\log n)$-bit number indicating its coordinates. Thus, merging two convex hulls, one from each partition, requires the transfer of $\Omega(\log^2 n)$ bits across the single link connecting the two halves of the array.

Since this may be done for $\Omega(n)$ pairs of hulls, the total number of bits to be transferred across this link is $O(n \log^2 n)$. Also, since only $O(\log n)$ bits may be moved across this link in one unit of time, this data transfer operation requires $\Omega(n \log n)$ time. Thus, any linear array algorithm for the above problem, based on merging disjoint convex hulls, requires $\Omega(n \log n)$ time.

References

[1] A. Aggarwal, B. Chazelle, L. Guibas, C. O'Dunlaing, and C. Yap, "Parallel Computational Geometry", Proc. 1985 Symp. on Foundations of Computer Science.

[2] A. Aho, J. Hopcroft, J. Ullman, *Design and Analysis of Computer Algorithms*, Addison Wesley, 1974.

[3] S. Akl, "Parallel Algorithms for Convex Hulls", Dept. Computer Science, Queens University, Ontario, Canada, 1983.

[4] M. J. Atallah and M. T. Goodrich, "Efficient Parallel Solutions to Some Geometric Problems", *J. of Parallel and Distributed Computing*, 3, pp. 492-507, 1986.

[5] H. M. Alnuweiri and V. K. Prasanna Kumar, "Optimal Geometric Algorithms on Fixed-Size Linear Arrays and Scan-Line Arrays", Proc. Ann. Conf. on Computer Vision and Pattern Recognition, 1988.

[6] H. M. Alnuweiri and V. K. Prasanna Kumar, "An Efficient VLSI Architecture with Applications to Geometric Problems",*Parallel Computing*, 12, 1989, pp. 71-83.

[7] G. Baudet and D. Stevenson, "Optimal Sorting Algorithms for Parallel Computers", *IEEE Trans.on Computers*, C-27, January 1978.

[8] K. A. Doshi and P. J. Varman, "Optimal Graph Algorithms on a Fixed-Size Linear Array", *IEEE Trans.on Computers*, C-36, April 1987.

[9] A. I. Fisher, "Scan Line Array Processors for Image Computations", Int'l Symp. on Computer Architecture, 1986.

[10] E. B. Hinkle, J. L. C. Sanz, A. K. Jain, and D. Petkovic, "P^3E New Life for Projection-Based Image Processing", *J. of Parallel and Distributed Computing*, 4, 1987, pp. 45-87.

[11] C. E. Kim, "On The Cellular Convexity of Complexes", *IEEE Trans. on Pattern Analysis and Machine Intelligence*, 1981, pp. 617-625.

[12] H. T. Kung and J. A. Webb, "Mapping Image Processing Operations onto a Linear Systolic Machine", *Distributed Computing*, 1, 1986.

[13] R. Miller and Q. F. Stout, "Geometric Algorithms for Digitized Pictures on a Mesh-Connected Computer", *IEEE Trans. on Pattern Analysis and Machine Intelligence*, March 1985.

[14] F.P. Preparata and D.T. Lee, "Computational Geometry-A Survey", *IEEE Trans.on Computers*, 33, pp. 1072-1100, 1984.

[15] F. Preparata and M. I. Shamos, *Computational Geometry: An Introduction*. Springer-Verlag, New York, 1985.

PARALLEL ALGORITHMS FOR LABELING IMAGE COMPONENTS

Wen-Jing Hsu   Xiaola Lin

Department of Computer Science
Michigan State University
East Lansing, Michigan 48824

Summary

We study the problem of labeling connected components of an $N^{1/2} \times N^{1/2}$ image using parallel computers. The following new results are presented:

1). Based on the CRCW PRAM (Concurrent-Read Concurrent-Write Parallel Random Access Machine) model, we show that $\Theta(\frac{N}{logN})$ processors are necessary and sufficient to solve the image component problem in $O(logN)$ time. The best known algorithm uses N processors to achieve the same time bound.

2). We show that $\Theta(\frac{N}{log^2N})$ hypercube or shuffle processors are necessary and sufficient to solve the same problem in $O(log^2N)$ time. This new result improves the processor requirement of the best existing algorithm by a factor of $O(log^2N)$.

3). We present a new mesh computer algorithm aimed to improve the utilization of the processor resource while maintaining reasonable speedup over the sequential algorithm. The algorithm that is presented uses only $O(N^{1/2})$ processors and executes in $O(N^{1/2}logN)$ time. This time complexity is a factor of $O(logN)$ away from the optimal time complexity of $O(N^{1/2})$; however, the number of processors is reduced by a factor of $O(N^{1/2})$ from $O(N)$.

4). We show that these algorithms can be extended to compute components in k-dimension and to compute components based on generalized connectivities.

1. Introduction

Consider a binary image of $N=N^{1/2} \times N^{1/2}$ pixels, where a pixel takes the value of either 0 or 1. Two 1-valued pixels are said to be *adjacent* to each other if they share a vertical or horizontal boundary, and they are said to belong to the same *component* if there exists a sequence of adjacent pixels that connects one to the other. (In other words, we will consider 4-connectivity initially). The *Image Component Labeling Problem* is to mark each 1-valued pixel such that pixels on the same component consistently receive the same label. Because a connected component may correspond to an object of interest, this problem has received considerable attention in the image processing community (See, for example, [Cypher89], [Rosenfeld83], and [Tanimoto86]). Because the size of an image is typically in the order of 10^6- 10^8 pixels, it is justifiable to consider parallel algorithms. In a recent paper [Cypher89], Cypher, Sanz, and Snyder presented an algorithm for labeling the connected components of an $N^{1/2} \times N^{1/2}$ binary image using an N-processor hypercube or shuffle-exchange computer. Their algorithm was the very first to solve the problem in $O(log^2 N)$ time using the given computational models. We note that, in general, a straightforward simulation of the Cypher-Sanz-Snyder Algorithm does not result in an algorithm with optimal efficiency. We refer the reader to [Cypher87-89] for an account of other recent developments related to the problem.

Section 2 describes the PRAM algorithms which uses a basic scheme that will be followed in the algorithms for Hypercube (or Shuffle) computers. Section 3 describes the algorithms for the hypercube and shuffle-exchange computers, which constitute the main result of this paper. Section 4 presents the algorithm for mesh-connected computers. Section 5 discusses the extension of these results to the higher dimensions and different connectivity considerations. Section 6 concludes by summarizing the major results.

2.1. PRAM Algorithms

Assume that initially (*Stage 0*) an image of size N is evenly distributed over $N/logN$ processors, with each processor holding a sub-image (sub-window) of size $O(logN)$. These sub-images are then concurrently labeled using an optimal sequential algorithm, taking $O(logN)$ time. Subsequently (Stage 1), a set of $O(logN)$ sub-images are combined (*merged)* to form a larger image of size $O(log^2 N)$. Then (Stage 2) a set of $O(log^2 N)$ of the lager images are combined to form even larger images, and so on until the final image is of size N. In general, at Stage i, where $i>0$, the input is a set of $O((logN)^{2^{i-1}})$ sub-images, each of size $O((logN)^{2^{i-1}})$, and the output is a window of size $O((logN)^{2^i})$. It is clear that the key operation of this scheme lies in merging the partial results of the sub-images; there are at most $S= O(loglogN)$ stages of the window-merging process.

Two tasks can be identified when combining a set of sub-images:

    1). We must determine which components in different sub-windows belong to a same component, and

2). relabel the components using a common label.

The following observations are important in understanding our algorithm.

a). When there are multiple stages of window-merging involved, we may have to redefine the labels of sub-components during each stage. Therefore, if we relabel all pixels each time the labels of sub-components are changed, then there will be many redundant re-labeling. The key idea here is to defer the relabeling operations until the final values of the component labels are obtained.

b). (Therefore), we present a two-phase scheme where, in the first phase, called the *Chaining Phase*, the goal is to obtain the common labels via the boundary pixels and, in the second phase, (called the *Re-labeling Phase)*, the goal is to trace back the links provided by the first phase and finally relabel the pixels in a batch. Note that the set of boundary pixels will play two key roles in the two phases mentioned above: First in determining common labels for sub-components (in Phase 1) and then again in backward-propagating the global labels for renaming other pixels lying in the interior of a window (in Phase 2).

c). We will use the following data structure. (See also [Cypher89] and [Shiloach82]). At each stage, each connected component to be merged is represented as a star, which is a one-level tree with all the leaves pointing toward the root. Let C denote an (arbitrary) component in a given window W of Stage i, where $i \geq 0$. For the star representing C, the leaves consist of pixels of C that lie on the boundary of W; the root is simply a pixel in C that is selected to label (name) C. Let p denote a boundary pixel of C, the root of the star that contains p will be recorded in $Root(i,p)$. For example, the root of a component C may be chosen from one of the pixels in C that are located at the highest position in W. In fact, in the initial (sequential) labeling (Stage 0), the first pixel encountered in each component can be designated as the root. Then, in the merging stages, the root can be uniquely determined from the lexicographical order of the coordinates of the pixels.   □.

We refer to [Hsu89] for more detailed description of the two phases. It is rather straightforward to prove that the total space requirement for representing the components is linear. (See [Hsu89] for the proof). Using an inductive argument it is straightforward to show that the backward-propagation phase will correctly carry the global labels back to the smallest sub-windows. It remains to examine the time complexity of the algorithm.

We now show that $O(N/logN)$ processors are sufficient for carrying out the two phases in $O(logN)$ time. It can be easily shown ([Hsu89]) that the time complexity of Phase 2 is upper-bounded by that of Phase 1. Therefore, it suffices to analyze the time requirement of Phase 1.

For convenience, we use the following notation.

Notation:

$W(i)$: the size of the window considered at Stage i, where $0 \le i \le S$.

P: the total number of processors ($P = \dfrac{N}{logN}$ in the case of PRAM, and $P = \dfrac{N}{log^2N}$ in the case of hypercube/shuffle).

$P(i)$: the number of processors available for each window during Stage i, where $0 \le i \le S$. (For example, in the case of PRAM, at Stage 1, $P(1)=logN$ processors will determine the common labels for $logN$ windows from Stage 0 to form a window of size $W(1) = log^2N$).

$B(i)$: the total number of 1-valued pixels that lie on the boundary of a subwindow at Stage i, where $1 \le i \le S$.

$T(i)$: the time required at Stage i, where $0 \le i \le S$. Let $T = \displaystyle\sum_{0 \le i \le S} T(i)$ denote the total time required in executing the Phase 1 of the abovementioned algorithm.

It is straightforward to prove the following lemma (See [Hsu89] for detailed proofs).

Lemma 2.1

1). $W(i)=log^{2^i}N$, where $0 \le i <S$.

2). $S \le \lceil log(\dfrac{logN}{loglogN}) \rceil$.

3). $B(i) = O(log^{3(2^{i-2})}N)$, where $1 \le i \le S$.

4). $P(i) = log^{2^i-1}N$, where $1 \le i \le S$.

5). $T(i)= O(logB(i))$, where $2 \le i \le S$.

6). $T= O(logN)$.   □.

Table 1 shows a stage-by-stage analysis of the processor requirements versus the available processor resource during the first phase. (Note that a few entries in the second column (Stage 0) of the Table are not defined, hence are indicated by N/A (Not Applicable)). It is straightforward to verify that the number of processors are sufficient in all but the very first two stages.

Stages	0	1	2	$2<i<S$	$S=$ $\lceil log(\frac{logN}{loglogN})\rceil$
Actions	apply the sequential algorithm	Merge $logN$ subwindows of Stage 0	Merge log^2N subwindows of Stage 1	Merge $log^{2^{i-1}}N$ subwindows of Stage $i-1$	Merge $O(N^{1/2})$ subwindows of Stage $S-1$
W(i)	$logN$	log^2N	log^4N	$(logN)^{2^i}$	N
B(i)	N/A	$log^{1.5}N =$ $(logN)^{0.5}{\times}logN$	$log^3N =$ $logN{\times}log^2N$	$(logN)^{3{\times}2^{i-2}} =$ $(logN)^{2^{i-2}}{\times}(logN)^{2^{i-1}}$	$O(N^{3/4}) =$ $O(N^{1/4}){\times}O(N^{1/2})$
Q(i)	N/A	$O((logN)^{1.5})$	$O((logN)^3)$	$O((logN)^{(3{\times}2^{i-2})})$	$O(N^{(3/4)})$
P(i)	1	$logN$	log^3N	$(logN)^{2^i-1}$	$\frac{N}{logN}$
δ(i)	N/A	$\delta(1)=O((logN)^{0.5})$	$\delta(2)=O(1)$	$\delta(i)=O(1)$	$\delta(S)=O(1)$
T(i)	$O(logN)$	$O(\delta(1){\times}loglogN)$ $=O(logN)$	$O(\delta(2){\times}loglogN)$ $=O(loglogN)$	$O(2^iloglogN)$	$O(logN^{3/4})$ $=O(logN)$

Legend

W(i) : The size of a window considered at stage i, where $0{\leq}i{\leq}S$.

B(i) : The total number of pixels that lie on the boundaries of all the sub-windows of a window at stage i, where $0<i{\leq}S$.

N/A: (Not Applicable) An entry which is undefined.

Q(i) : The number of processors required to execute the Shiloach-Vishkin algorithm in $O(logB(i))$ time, where $0<i{\leq}S$.

P(i) : The actual number of processors available to each window at Stage i, where $0{\leq}i{\leq}S$.

δ(i): The slow-down factor caused by insufficient number of processors when applying the Shiloach-Vishkin's algorithm during Stage i, where $0<i{\leq}S$.

T(i) : The execution time required at Stage i, where $0{\leq}i{\leq}S$.

Table 1. Processing Times at Each Stage on the PRAM

Theorem 2.1 [Hsu89]

On a CRCW PRAM, $\Theta(N/logN)$ processors are necessary and sufficient to label connected components of a 2-dimensional binary image of size N in $O(logN)$ time.

[proof]

We have shown that $O(N/logN)$ processors are sufficient for labeling the components in $O(logN)$ time. A trivial lower bound for sequential algorithms is $\Omega(N)$. Therefore, p processors require $\Omega(N/p)$ time to solve the problem. In other words, to solve the problem in $O(logN)$ time, it is necessary to have at least $\Omega(N/logN)$ processors. $\quad\Box$.

3. Algorithms for Hypercubes and Perfect-Shuffles

Our new algorithms for the hypercube and shuffle machines use $O(N/log^2N)$ processors and solve the problem in $O(log^2N)$ time. The same divide-and-conquer approach (as described in Section 2) will be followed. Again, at the bottom of the recursion, an optimal sequential algorithm is applied to determine the components of a log^2N-pixel sub-window. When merging subwindows, however, Shiloach and Vishkin's algorithm [Shiloach82] will be *simulated* on the hypercube (or the shuffle). To simulate a concurrent READ/WRITE step of the CRCW PRAM, Nassimi and Sahni's fast sorting algorithm will be used. [Nassimi81]. The Nassimi-Sahni algorithm uses $O(M^{1+1/k})$ processors to sort M records in $O(klogM)$ time, where $1<k$ denotes a selected constant. With the simulation, an $O(logM)$-time PRAM algorithm can be executed in $O(k\times log^2M)$ time, where M denotes the size of the data involved. (This critical idea of the simulation of Shiloach-Vishkin's algorithm originated in [Cypher89]). The key issue here lies in showing that $O(\frac{N}{log^2N})$ processors are sufficient for merging all sub-windows in $O(log^2N)$ time. As is the case in the PRAM algorithm, the concurrent READ/WRITE operations of the Relabeling Phase constitute only a subset of the operations carried out in the Chaining Phase. Therefore, the time complexity of the algorithm is dominated by the Chaining Phase. In the following, we will only be concerned with the time complexity of the Chaining Phase.

We discuss the time complexity using both top-down (recursive) and bottom-up (iterative) analyses. At the top level of recursion, we have to merge $N^{1/2}$ windows, each of size $N^{1/4}\times N^{1/4}$. The amount of *boundary data* (i.e., the number of pixels on the boundaries of a window and the data representing the adjacency relation implied by these pixels) is bounded by $3\bullet N^{3/4}$. Therefore, to simulate the Shiloach-Vishkin's PRAM algorithm in $O(log^2(N^{3/4}))$ time, we need $O((N^{3/4(1+1/k)}$

Stages	0	1	2	$2<i<S$	$S = \lceil log(\frac{logN}{loglogN}) \rceil$
Actions	apply the sequential algorithm	Merge log^2N subwindows of Stage 0	Merge log^4N subwindows of Stage 1	Merge $log^{2^i}N$ subwindows of Stage $i-1$	Merge $O(N^{1/2})$ subwindows of Stage $S-1$
W(i)	log^2N	log^4N	log^8N	$(logN)^{2^{i+1}}$	N
B(i)	N/A	$log^3N =$ $logN \times log^2N$	$log^6N =$ $log^2N \times log^4N$	$(logN)^{3 \times 2^{i-1}} =$ $(logN)^{2^{i-1}} \times (logN)^{2^i}$	$O(N^{3/4}) =$ $O(N^{1/4}) \times O(N^{1/2})$
Q(i, k)	N/A	$O((logN)^{3\lambda})$ where $\lambda=1+1/k$	$O((logN)^{6\lambda})$ where $\lambda=1+1/k$	$O((logN)^{(3 \times 2^{i-1})\lambda})$ where $\lambda=1+1/k$	$O(N^{(3/4)\lambda})$ where $\lambda=1+1/k$
P(i)	1	log^2N	log^6N	$(logN)^{2^{i+1}-2}$	$\frac{N}{log^2N}$
δ(i)	N/A	$\delta(1)=O((logN)^{1+3/k})$	$\delta(2)=O((logN)^{6/k})$	$\delta(i)=O(1)$ if $k>6$	$\delta(S)=O(1)$
T(i)	$O(log^2N)$	$O(\delta(1) \times k \times (loglogN)^2)$ $= O(log^2N)$ if $k>3$	$O(\delta(2) \times k \times (loglogN)^2)$ $= O(log^2N)$ if $k>3$	$O(2^{2i} \times (loglogN)^2)$ $= O(log^2N)$ if $k>6$	$O(k \times (logN^{3/4})^2)$ $= O(log^2N)$ if $k>1$

Legend

W(i) : The size of a window considered at stage i, where $0 \leq i \leq S$.

B(i) : The total number of pixels that lie on the boundaries of all sub-windows of a window at stage i, where $0<i \leq S$.

Q(i, k) : The number of processors required to simulate the Shiloach-Vishkin algorithm in $O(k \times log^2B(i))$ time on the hypercube/shuffle computer by running the Nassimi-Sahni sorting algorithm, where $0<i \leq S$ and k denotes a design constant. (We shall let $k>6$).

P(i) : The actual number of processors available to each window of Stage i, where $0 \leq i \leq S$.

δ(i): The slow-down factor caused by insufficient number of processors during Stage i, where $0<i \leq S$.

T(i) : The execution time required at Stage i, where $0 \leq i \leq S$.

Table 2. Processing Times at Each Stage on the Hypercube/Shuffle

processors. However, for sufficiently large N and k, this is smaller than N/log^2N, the number of available processors. At the next level of recursion, we have $N^{1/2}$ windows, with each subdivided into $N^{1/4}$ subwindows of size $N^{1/4} = N^{1/8} \times N^{1/8}$. Thus the total size of the boundary data is bounded by $O(N^{1/2+1/4+1/8}) = O(N^{7/8})$, which implies that $O((N^{7/8})^{1+1/k})$ processors are required (if we are to finish combining these data in $O(log^2(N^{7/8}))$ time, which is our goal). But this processor requirement can be satisfied with the $O(N/log^2N)$ available processors. Based on essentially the same approach used in proving Lemma 2.1, it is straightforward to verify that $O(N/log^2N)$ processors are sufficient for all but a finite number of stages. (Recall that the *stages* are counted from the bottom level of the recursion). Table 2 shows a bottom-up, stage-by-stage analysis of the processor requirements versus the available processor resource, which is essentially the basis for the following theorem.

Theorem 3.1 [Hsu89]

On either a hypercube or a perfect shuffle, $\Theta(N/log^2N)$ processors are necessary and sufficient for labeling connected components of a 2-D binary image of size N in $O(log^2N)$ time.   □.

4. Mesh Algorithms

Currently, the best time complexity for mesh-connected computers is $O(N^{1/2})$, which requires $O(N)$ processors. Thus, the efficiency of processor utilization is $O(\frac{1}{N^{1/2}})$, which is relatively poor (even for a small N) [Stout85]. In this section, we show that $O(N^{1/2})$ processors are sufficient for solving the component labeling problem on a mesh within $O(N^{1/2}logN)$ time. Therefore, the new algorithm has a much improved efficiency of $O(\frac{1}{logN})$ while the time complexity loses only by a factor of $O(logN)$.

We will use the following sorting algorithm in the simulation of a concurrent READ (or concurrent WRITE) operation of the PRAM. It can be shown that the simulation of a step of the PRAM computer is bounded by the time complexity of the sorting algorithm. (We refer the reader to Nassimi and Sahni's original results [Nassimi82]).

Theorem 4.1 ([Akl86])

Given N data items evenly distributed on p mesh-connected processors, it takes $O((N/p) \bullet max(log(N/p), p^{1/2}))$ time to sort the data into non-descending (or non-ascending) order, with N/p data on each processor.   □.

Now, given $O(N^{1/2})$ processors, the image is divided into $O(N^{1/2})$ windows, each of size $O(N^{1/2})$ $= N^{1/4} \times N^{1/4}$. The windows are labeled by using an optimal sequential algorithm, taking $O(N^{1/2})$ time. We will use the Shiloach-Vishkin algorithm to merge all windows in one pass. Let M denote the total amount of the boundary data to be merged during this phase. Then $M = O(N^{3/4})$. Let T denote the total time required in simulating the Shiloach-Vishkin algorithm. Then $T = O(logM \bullet S(M,p))$, where $S(M,p)$ denotes the time required to sort a set of M records using p mesh-connected computer. Let $p = N^{1/2}$ and apply Theorem 4.1, we have $T = O(logN \bullet N^{1/2})$. Therefore, the total execution time, including the Chaining and Relabeling, can be bounded by $O(N^{1/2}logN)$.

Theorem 4.2

Using $O(N^{1/2})$ mesh-connected processors, the 2-dimensional binary image component labeling problem of size N can be solved in $O(N^{1/2}logN)$ time.

Remark

Unlike the recursive algorithms presented in the previous sections, the algorithm just presented applies Shiloach-Vishkin's algorithm for only one time. So far, algorithms based on the recursive scheme have resulted in an identical time complexity while requiring more processor resources. It is still unknown whether we can use $O(N^{1/2})$ mesh-connected processors to solve the problem in $O(N^{1/2})$ time, which would then be optimal.   □.

5. Extensions to K-Dimension and Generalized Connectivity

It is possible to extend the new results in several directions. For m-ary images, for example, if the components are defined to be composed of adjacent pixels of the same pixel value, then the algorithms for binary images can be modified to run m passes, each time erasing pixels that have been processed. The time complexity is bounded by $O(m \bullet T(N))$, where $T(N)$ denotes the complexity of the binary-image algorithm for a selected computational model.

For 3-dimensional images and k-dimensional images in general, the time complexities can be preserved as well. In the previous definition, two pixels are neighbors if and only if they are adjacent in horizontal and vertical positions. This definition of adjacency can be easily extended to k-dimension, where a pixel can be identified by a k-tuple.

Definition

A pixel at $(i_1,i_2,...,i_k)$ is said to be *adjacent* to a pixel at $(j_1,j_2,...,j_k)$ if and only if both of the following conditions are true:

1). $i_l \neq j_l$, for all $1 \leq l \leq k$, and

2). $\sum\limits_{1 \leq l \leq k} |i_l - j_l| = 1$.

Another extension is also based on the generalization of pixel connectivity.

Definition

Let $\beta \geq 0$ denote a constant. A pixel at $(i_1,i_2,....,i_k)$ is a *generalized* β-*neighbor* of the pixel at $(j_1,j_2,...,j_k)$ if and only if both of the following conditions are true:

1). $(i_1,i_2,....,i_k) \neq (j_1,j_2,...,j_k)$, and

2). $\sum\limits_{1 \leq l \leq k} |i_l - j_l| \leq \beta$.

In [Hsu89], it is shown that to determine components based on the generalized connectivity, the algorithms presented can be modified (in the part that tests neighborhood), and the time complexities will still be preserved. The following theorem summarizes the major results presented in this paper.

Theorem 5.4 [Hsu89]

Assume that k, m, and β are all bounded by a constant. Suppose that a k-dimensional m-ary image of size N is evenly distributed over $P(N)$ processors, with $O(\frac{N}{P(N)})$ pixels on each processor. The image components are to be determined according to the generalized β-neighborhood.

1). On a CRCW PRAM, $P(N)=\Theta(N/logN)$ processors are necessary and sufficient to label connected components in $O(logN)$ time.

2). On either a hypercube or a perfect shuffle, $P(N)=\Theta(N/log^2N)$ processors are necessary and sufficient to label connected components in $O(log^2N)$ time.

3). Using $P(N)=O(N^{1/2})$ mesh-connected processors, the image component problem can be solved in $O(N^{1/2}logN)$ time.

[proof] (Omitted).   □

417

6. Conclusion

We have presented three new parallel algorithms for the PRAMs, hypercube/perfect shuffle machines, and mesh-connected computers. Except the algorithm for the mesh-connected computers, the new algorithms are shown to require the same amount of execution time as the best existing algorithms while using reduced processor resources. We also show that these results can be extended to the k-dimensional space and generalized connectivities. It remains open whether the algorithm for the mesh computer can be further improved.

References

[Akl85] S. Akl, *Parallel Sorting*, Academic Press, Orlando, Fla., 1985.

[Batcher68] K. E. Batcher, Sorting Networks and Their Applications, Proceedings of the AFIPS 1968 Spring Joint Computer Conference, Atlantic City, New Jersey, pp. 307-314, AFIPS Press, Montvale, N. J., 1968.

[Cypher89] R. Cypher, J. L. C. Sanz, and L. Snyder, Hypercube and Shuffle-Exchange Algorithms for Image Component Labeling, J. of Algorithms, 10, pp. 140-150, (1989).

[Cypher87] R. Cypher, J. L. C. Sanz, and L. Snyder, EREW PRAM and Mesh Connected Computer Algorithms for Image Component Labeling, in "1987 Computer Society Workshop on Computer Architecture, Pattern Analysis and Machine Intelligence", pp. 122-130.

[Cypher87] R. Cypher, J. L. C. Sanz, and L. Snyder, Practical Algorithms for Image Component Labeling on SIMD Mesh Connected Computers, in Int. Conf. on Parallel Processing, 1987, pp. 722-779.

[Fishburn82] J. P. Fishburn and R. A. Finkel, Quotient Networks, IEEE TC (Transactions on Computers), Vol. C-31, No. 4, pp. 288-295 (1982).

[Hsu89] W. -J. Hsu and X. Lin, Parallel Algorithms for Labeling Image Components, TR. CPS-89-14, Dept. Comput. Sci., MSU, 1989.

[Hummel86] R. Hummel, Connected Component Labeling in Image Processing with MIMD Architectures, in "Intermediate-Level Image Processing", pp. 101-127, Academic Press, New York, 1986.

[Hung87] Y. Hung and A. Rosenfeld, Parallel Processing of Linear Quadtrees on a Mesh-Connected Computer, TR CAR-TR-278, Center for Automation Research, University of Maryland, March, 1987.

[Kumar86] V. K. P. Kumar and M. M. Eshaghian, Parallel Geometric Algorithms for Digitized Pictures on Mesh of Trees (preliminary version), in Proc., 1986 Int. Conf. on Parallel Processing, 697-699.

[Lim86] W. Lim, Fast Algorithms for Labeling Connected Components in 2-D Arrays, TR. 86.22, Thinking Machine Corp., Cambridge, MA, July 1986.

[Miller87] R. Miller and Q. Stout, Data Movement Techniques for the Pyramid Computer, SIAM J. Comput. 16, No. 1 (1987), 38-60.

[Nassimi80] D. Nassimi and S. Sahni, Finding Connected Components and Connected Ones on a Mesh-Connected Parallel Computer, SIAM J. Comput. 9, NO. 4 (1980), 744-757.

[Nassimi81] D. Nassimi and S. Sahni, Data Broadcasting in SIMD Computers, IEEE TC, C-30, No. 2 (1981), 101-107.

[Nassimi82] D. Nassimi and S. Sahni, Parallel Permutation and Sorting Algorithms and a New Generalized Connection Network, J. ACM, 29, No. 3 (1982), 642-667.

[Rosenfeld83] A. Rosenfeld, Parallel Image Processing Using Cellular Arrays, IEEE TC C-32 (1983), 14-20.

[Rosenfeld82] A Rosenfeld and A. Kak, "Digital Picture Processing", Vols. 1-2, Academic Press, New York, 1982.

[Shiloach82] Y. Shiloach and U. Vishkin, An $O(logN)$ Parallel Connectivity Algorithm, J. Algorithms 3 (1982), 57-67.

[Stone72] H. Stone, Parallel Processing with the Perfect Shuffle, IEEE TC, C-20, No. 2 (1972), 153-161.

[Stout85] Q. F. Stout, Properties of Divide-and Conquer Algorithms for Image Processing, in "1985 IEEE Computer Workshop on Computer Architecture for Pattern Analysis and Image Data Management", pp 203-209.

[Tanimoto86] S. Tanimoto, Architectural Issues for Intermediate-Level Vision, in "Intermediate-Level Image Processing", pp. 3-16. Academic Press, New York, 1986.

[Thompson77] C. D. Thompson and H. T. Kung, Sorting on a Mesh-Connected Parallel Computer, Comm. ACM, 20, 4 pp. 263-271, 1977.

A Hyperplane Incidence Problem

with Applications to Counting Distances[1]

Herbert Edelsbrunner[2]   Micha Sharir[3]

Abstract

This paper proves an $O(m^{2/3}n^{2/3} + m + n)$ upper bound on the number of incidences between m points and n hyperplanes in four dimensions, assuming all points lie on one side of each hyperplane and the points and hyperplanes satisfy certain natural general position conditions. This result has application to various three-dimensional combinatorial distance problems. For example, it implies the same upper bound for the number of bichromatic minimum distance pairs in a set of m blue and n red points in three-dimensional space. This improves the best previous bound for this problem.

1   Introduction

Combinatorial distance problems for finite point sets in Euclidean spaces are classical topics in discrete geometry. Most popular is probably the question how often the unit distance can occur in a set of m points in two or three dimensions; this question was originally studied by Erdös [8, 9]. The currently best upper bound in the plane is $O(m^{4/3})$ (see [13] and also [4]), and in three dimensions Clarkson et al. [4] proved $O(m^{3/2}\beta(m))$, where $\beta(m)$ is an extremely slowly growing function related to the inverse of Ackermann's function. No matching lower bounds are known.

This paper considers three distance problems in three dimensions and improves the best previous upper bound in each case. The results are as follows. The number of bichromatic minimum distance pairs in a set of m blue and n red points is $O(m^{2/3}n^{2/3} + m + n)$. The same bound holds for the number of bichromatic nearest neighbor pairs if no three points are collinear. For a (monochromatic) set of m points, the number of furthest neighbor pairs is $O(m^{4/3})$ if no three points are collinear. All three bounds are corollaries of the following more general result on the number of incidences between m points and n hyperplanes in four dimensions. If no three points are collinear, no three hyperplanes intersect in a common 2-flat, and each hyperplane bounds a closed half-space

---

[1]Research of the first author was supported by the National Science Foundation under grant CCR-8714565. Work of the second author was supported by Office of Naval Research Grants DCR-83-20085 and CCR-89-01484, and by grants from the U.S.-Israeli Binational Science Foundation, the NCRD – the Israeli National Council for Research and Development, and the Fund for Basic Research in Electronics, Computers and Communication administered by the Israeli Academy of Sciences.

[2]Department of Computer Science, University of Illinois at Urbana-Champaign, Urbana, Illinois 61801, USA.

[3]Courant Institute of Mathematical Sciences, New York University, New York, New York 10012, USA and School of Mathematical Sciences, Tel Aviv University, Tel Aviv 69978, Israel.

that contains all points, then the number of incidences is $O(m^{2/3}n^{2/3} + m + n)$. Besides the application of this result to three-dimensional distance problems we also consider an application to three-dimensional Delaunay triangulations.

This paper is organized as follows. Section 2 proves the upper bound on the hyperplane incidence problem, Section 3 explains how this problem relates to counting incidences between points and spheres in three dimensions, Section 4 discusses the combinatorial distance problems, and Section 5 concludes the paper.

2 The Hyperplane Incidence Problem

This section proves the main result of this paper, an upper bound for the following combinatorial geometry problem.

Problem Specification. Let P be a set of m points and H a set of n hyperplanes in four-dimensional Euclidean space satisfying the following conditions.

(H.i) No three points of P are collinear.

(H.ii) No three hyperplanes of H intersect in a common 2-flat.

(H.iii) Each hyperplane in H bounds a closed half-space that contains P.

What is the maximum number of incidences between the points and the hyperplanes, in terms of m and n, where the maximum is taken over all sets P of size m and H of size n satisfying the three conditions? We define $I(m,n)$ to be equal to this number.

We prove an upper bound on $I(m,n)$ by adapting the methods of Clarkson et. al [4]. This is done in two steps. First, bounds are obtained which are tight for the cases when m is much smaller than n ($m < \sqrt{n}$) and when n is much smaller than m ($n^2 < m$); using the terminology of [4] we call these bounds Canham thresholds. Second, these bounds together with the methods of [4] are used to establish an upper bound for the remaining case ($\sqrt{n} \le m \le n^2$). To prove the Canham thresholds (Lemma 2.1) we make use of the fact that the problem is self-dual, that is, if we apply a dual transform we arrive at the same problem, only with the roles of m and n interchanged. Indeed, choose the origin in the interior of the intersection of half-spaces that contains all points of P. For a point p different from the origin let

$$p^* = \{x \mid \langle x, p \rangle = 1\}$$

be the *polar hyperplane*, and for a hyperplane h avoiding the origin define the *polar point* h^* so that $h = (h^*)^*$. Using straightforward algebraic manipulations it can easily be verified that $p \in h$ iff $h^* \in p^*$. Furthermore, three points are collinear iff their polar hyperplanes intersect in a common 2-flat. In other words, if we map all points to their polar hyperplanes and all hyperplanes to their polar points we get a point/hyperplane incidence problem satisfying conditions (H.i) and (H.ii). In addition to the incidence preserving property the polar transform preserves sidedness relative to the origin, which implies that also condition (H.iii) is satisfied after polarization. Hence, $I(m,n) = I(n,m)$.

The Canham Thresholds. The intersection of half-spaces defined by the hyperplanes is a closed convex polyhedron C with at most n facets. The upper bound theorem (see [3] or [7]) implies that C has at most $O(n^2)$ vertices, edges, and ridges.[4] Let the *degree* of a face (vertex, edge, ridge or facet) be the number of hyperplanes in H that contain it. Clearly, the degree of any facet is 1, by assumption (H.ii) the degree of any ridge is 2, and the degree of any edge or vertex can be arbitrary. Still, the sum of degrees, over all faces, is at most $O(n^2)$ by the following argument. Take each hyperplane that contains a vertex or edge of C and push it inwards (towards the origin) by a tiny amount so that each such hyperplane now supports a facet. Call the new polytope C'. Next we slightly perturb the hyperplanes, on a much smaller scale than before, in order to transform C' into a simple polyhedron C'', again without decreasing the sum of face degrees. Now every edge has degree 3 and every vertex has degree 4 which implies the claim by applying the upper bound theorem to C''. After these introductory remarks we are ready to prove the Canham thresholds.

Lemma 2.1 $I(m,n) = O(m\sqrt{n} + n)$ and $I(m,n) = O(n\sqrt{m} + m)$.

Proof. Because of the self-dual property of the incidence problem we can restrict ourselves to proving $I(m,n) = O(n\sqrt{m} + m)$. Let P be the set of m points, H the set of n hyperplanes, and C the polyhedron defined above which contains P. If a point $p \in P$ lies in the interior of C it contributes 0 to the number of incidences, if it lies in the relative interior of a facet it contributes 1, and if it lies in the relative interior of a ridge it contributes 2. The total number of incidences involving such points is therefore $O(m)$. Each edge of C contains at most two points of P, by condition (H.i), and each vertex contains at most one point of P. By the claim established earlier, the total number of incidences involving points on edges and vertices of C is therefore $O(n^2)$ which gives a combined bound of $O(m+n^2)$. To get $I(m,n) = O(n\sqrt{m}+m)$ we simply partition H into about $\frac{n}{\sqrt{m}}$ subsets of size about $\sqrt{m}$ each. For each subset we get only $O(m)$ incidences with points of P and therefore we get at most $O(n\sqrt{m} + m)$ incidences altogether. □

Remarks. (1) Conditions (H.i) and (H.ii) can be relaxed to allow up to some constant number of points that are collinear and hyperplanes that meet in a common 2-flat without sacrificing the asymptotic bounds in Lemma 2.1. However, if we drop (H.i) or (H.ii) we can find examples with $\Omega(mn)$ incidences. The problem that results when we drop (H.iii) but not (H.i) and (H.ii) is more difficult. The best lower bound known to the authors is $\Omega(m^{4/3} \log\log m)$ for the case $m = n$ and is based on an example of Erdös [9] (see also [4]).

(2) Conditions (H.i) and (H.ii) imply that no three hyperplanes can all be incident to three common points. Using a standard extremal graph lemma (see [2] or [4]) this can be used to prove $O(mn^{2/3}+n)$ and $O(nm^{2/3}+m)$ as upper bounds on $I(m,n)$. These bounds are significantly weaker than the bounds of Lemma 2.1 which illustrates the importance of condition (H.iii) and the use of the upper bound theorem.

The Main Result. We need some notation. For a hyperplane $h \in H$ let h^+ be the closed

---

[4]In general, a *ridge* is a $(d-2)$-face of a d-polyhedron. Because C is a 4-polyhedron a ridge of C is a 2-face.

half-space bounded by h that contains the origin. Recall that we assume that the origin lies in the interior of C which implies that $C = \bigcap_{h \in H} h^+$. We now present a sequence of arguments that add up to a proof of the main result of this section.

First, choose a random sample $R \subseteq H$ of size r. Form the convex polyhedron $\mathcal{R} = \bigcap_{h \in R} h^+$ and note that $\mathcal{R}$ contains all points of P and also the origin. It will be convenient to assume that $\mathcal{R}$ is actually a polytope (that it, it is bounded) which can be achieved by intersecting it with a sufficiently large tetrahedron.

Second, we triangulate $\mathcal{R}$ as follows. Choose a directed line so that no hyperplane normal to this line contains two vertices of $\mathcal{R}$ and call the direction defined by this line *vertical*. It thus makes sense to talk about a point being *higher* than another point. A ridge (that is, 2-face) of $\mathcal{R}$ is triangulated by connecting its highest vertex to all other vertices. Each triangle is thus bounded by two edges incident to the highest vertex of the ridge and a third edge which is an original edge of the ridge. Similarly, a facet (that is, 3-face) of $\mathcal{R}$ is triangulated by connecting its highest vertex to all vertices, edges, and triangles in the (triangulated) boundary of the facet. Thus, each tetrahedron is incident to the highest vertex and is bounded by a triangle not incident to this vertex. Finally, the interior of $\mathcal{R}$ is triangulated by connecting the origin to all vertices, edges, triangles, and tetrahedra in the (triangulated) boundary of $\mathcal{R}$. The number of 4-simplices generated is equal to the number of tetrahedra in the boundary of $\mathcal{R}$. In turn, the number of tetrahedra is at most twice the number of triangles because any ridge is incident to only two facets. Finally, the number of triangles used to triangulate the ridges is bounded from above by the number of edge/ridge incidences. The number of ridges incident to a single edge is at most the degree of the edge, and we noted earlier that the sum of edge degrees is at most quadratic in the number of hyperplanes. Thus, if k is the number of 4-simplices in the triangulation then $k = O(r^2)$.

For the application of a probabilistic counting result due to Clarkson and Shor [5] in step 5 below it is important that each 4-simplex is uniquely determined by some constant number of hyperplanes. Indeed, a triangle lies in a plane (the intersection of two hyperplanes) and is defined by at most five lines (the intersection of this plane with at most five other hyperplanes). A tetrahedron connects a triangle to a vertex (the intersection of four hyperplanes one of which contains the triangle) and is thus defined by at most 10 hyperplanes. A 4-simplex just connects a tetrahedron with the origin and is therefore determined by at most 10 hyperplanes and a point that is fixed independent of $\mathcal{R}$.

Another necessary property for the application of the probabilistic counting result is that a 4-simplex defined by the origin and at most 10 hyperplanes in the way described above is in the triangulation *if and only if* if does not intersect any other hyperplane. Unfortunately, this is not strictly true because the hyperplanes are not in general position which allows for the possibility that different sets of at most 10 hyperplanes define the same 4-simplex. However, we can simulate an arbitrarily small perturbation of the hyperplanes to get them into general position and define the triangulation with respect to this perturbation. The perturbation is used merely for the purpose of assigning proper sets of hyperplanes to the 4-simplices of the triangulation – it is ignored as far as point/hyperplane incidences are concerned. Before we go on with the proof we remark that the origin is an arbitrary point in the interior of C. We may therefore assume that each point of P that does not lie on the boundary of $\mathcal{R}$ lies in the interior of a 4-simplex of the triangulation.

The third step of the proof bounds the number of incidences involving points in the interior of $\mathcal{R}$. For $1 \le i \le k$, let σ_i be the interior of the ith 4-simplex, set $m_i = |P \cap \sigma_i|$, and let n_i be the number of hyperplanes in H that have non-empty intersection with σ_i. Using the first bound in Lemma 2.1 we thus get $\sum_{i=1}^{k} O(m_i \sqrt{n_i} + n_i)$ as an upper bound for the number of incidences that happen in the interior of $\mathcal{R}$.

Fourth, we bound the number of incidences involving points on the boundary of $\mathcal{R}$. For a hyperplane $h \in R$ define $P_h = P \cap h$ and $m_h = |P_h|$. The convex hull of P_h is a convex polytope of dimension 3 or less. Because of its low dimensionality, it has at most $O(m_h)$ edges and 2-faces. Since no three points in P_h are collinear (condition (H.i)) and no three hyperplanes in H meet in a common 2-flat (condition (H.ii)), the total number of incidences involving points of P_h is $O(m_h + n)$. If we sum this bound over all r hyperplanes $h \in R$ we get $O(I(m, r) + nr)$ since a point p belongs to as many sets P_h as there are incident hyperplanes h in R. We have $I(m, r) = O(r\sqrt{m} + m)$ using the second bound in Lemma 2.1.

In the fifth and final step we make use of the probabilistic counting result of Clarkson and Shor [5]. It implies that the expected value of $\sum_{i=1}^{k} O(m_i \sqrt{n_i} + n_i)$ is $O(m\sqrt{\frac{n}{r}} + \frac{kn}{r})$. Note that $\frac{kn}{r} = O(nr)$ and that the above sum is an upper bound on the number of incidences that happen in the interior of $\mathcal{R}$. Thus, there exists a subset $R \subseteq H$ of size r so that the number of incidences in the interior of $\mathcal{R}$ is indeed at most $O(m\sqrt{\frac{n}{r}} + nr)$. Adding to this bound the incidences on the boundary of $\mathcal{R}$ we get

$$I(m, n) = O\left(m\sqrt{\frac{n}{r}} + nr + r\sqrt{m} + m\right)$$

as an upper bound on the total number of incidences. If we choose r about equal to $\frac{m^{2/3}}{n^{1/3}}$ this bound becomes $O(m^{2/3}n^{2/3} + \frac{m^{7/6}}{n^{1/3}} + m)$. This choice of r is meaningful if $\sqrt{n} \le m \le n^2$ in which case $\frac{m^{7/6}}{n^{1/3}} \le m$. For the remaining cases, $m < \sqrt{n}$ and $n^2 < m$, Lemma 2.1 shows that the number of incidences is at most $O(m + n)$. This concludes the proof of the main result of this section which we now state.

Theorem 2.2 The maximum number of incidences between m points and n hyperplanes in four dimensions satisfying conditions (H.i) through (H.iii) is $I(m, n) = O(m^{2/3}n^{2/3} + m + n)$.

Remarks. (1) Remark (1) after Lemma 2.1 also applies to the bound in Theorem 2.2.

(2) Note that the only way condition (H.ii) is exploited in the above proofs is that more than some constant number of points cannot be incident to more than some constant number of common hyperplanes. Another condition achieving the same goal is "(H.ii') no four points of P lie in a common 2-flat". In other words, Theorem 2.2 can also be shown if we replace (H.ii) by (H.ii').

(3) We state as an open problem to prove any superlinear lower bound for $I(m, n)$.

3 The Sphere Incidence Problem

Using Theorem 2.2 and a fairly standard geometric lifting transform (see for example [7]), we can derive a good upper bound for an incidence problem involving points and spheres in three dimensions. To simplify the notation we use the word *sphere* to either mean a sphere in the common Euclidean sense, or a plane. We define a *(generalized) ball* as a closed ball, a closed half-space, or the complement of an open ball in three dimensions. Thus, each sphere bounds two balls and for each ball there is a sphere that bounds it. This slightly non-standard use of terms will be restricted to the scope of this section.

Problem Specification. Let P be a set of m points and S a set of n spheres in three-dimensional Euclidean space satisfying the following conditions.

(S.i) No three spheres in S intersect in a common circle or line.

(S.ii) Each sphere $s \in S$ bounds a ball s^+ so that $P \subseteq s^+$.

What is the maximum number of incidences between the points and spheres, where the maximum is taken over all sets P of size m and S of size n satisfying (S.i) and (S.ii)?

We derive an upper bound on the number of incidences by mapping P to a set of m points and S to a set of n hyperplanes in four dimensions so that conditions (H.i) through (H.iii) are satisfied. More specifically, a point $p = (\pi_1, \pi_2, \pi_3)$ is mapped to the point $\bar{p} = (\pi_1, \pi_2, \pi_3, \pi_1^2 + \pi_2^2 + \pi_3^2)$. Note that $\bar{p}$ is the vertical projection of p, a point in the $x_1 x_2 x_3$-space, onto the paraboloid of revolution specified by the equation $x_4 = x_1^2 + x_2^2 + x_3^2$. Similarly, a sphere s in the $x_1 x_2 x_3$-space is mapped to a four-dimensional hyperplane $\bar{s}$ by vertically projecting each of its points onto the same paraboloid and defining $\bar{s}$ as the (unique) hyperplane that contains all these points. The crucial property of this transform is that a point p lies outside, on, or inside a sphere s iff $\bar{p}$ lies vertically above, on, or below $\bar{s}$. (If s is a plane then $\bar{s}$ is a vertical hyperplane (parallel to the x_4-axis) and sidedness is maintained as in the general case.) It thus follows that the sets $\bar{P} = \{\bar{p} \mid p \in P\}$ and $\bar{S} = \{\bar{s} \mid s \in S\}$ satisfy condition (H.iii). Condition (H.i) holds because no three points of the paraboloid are collinear, and (H.ii) follows from (S.i). What we said above includes as a special case that $p \in s$ iff $\bar{p} \in \bar{s}$ which implies the following upper bound for the number of point/sphere incidences.

Theorem 3.1 The number of incidences between m points and n spheres in three dimensions satisfying conditions (S.i) and (S.ii) is $O(m^{2/3} n^{2/3} + m + n)$.

Remarks. (1) In agreement with remark (1) after Lemma 2.1, condition (S.i) can be relaxed to allow up to some constant number of spheres intersecting in a common circle or line.

(2) As noted in remark (2) after Theorem 2.2, it is possible to replace condition (H.ii) by (H.ii') without sacrificing the $O(m^{2/3} n^{2/3} + m + n)$ upper bound. By the same reason it is possible to replace (S.i) by "(S.i') no four points of P lie on a common circle or line" without sacrificing the upper bound given in Theorem 3.1.

(3) No superlinear lower bound for the sphere incidence problem of this section is currently known to the authors; see also remark (3) after Theorem 2.2.

4   Combinatorial Distance Problems

In this section we apply Theorem 3.1 to get upper bounds on problems about repeated distances between points in three dimensional Euclidean space. Many such problems were originally posed by Paul Erdös and were considered before in the literature. We refer to the problem collection of Moser and Pach [12] as a general source of relevant information.

4.1   Bichromatic Minimum Distance

Given a set P of m blue points and a set Q of n red points in three-dimensional Euclidean space, what is the maximum number of pairs $(p, q) \in P \times Q$ that realize the minimum distance between points of different color? The best previous bound, derived in [4], is $O(m^{3/4}n^{3/4}\beta(m, n) + m + n)$, with $\beta(m, n) = 2^{\Theta(\alpha(m^3/n)^2)}$ and α the inverse of Ackermann's function. The reduction to Theorem 3.1 should be obvious: around each blue point draw a sphere with radius equal to the minimum bichromatic distance. Condition (S.i) is satisfied because all spheres are equally large, and condition (S.ii) holds because all red points lie on or outside all spheres.

Theorem 4.1 The number of bichromatic minimum distance pairs in a set of m blue and n red points in three-dimensional Euclidean space is $O(m^{2/3}n^{2/3} + m + n)$.

Remark. In the case of a monochromatic set of m points in three dimensions the maximum number of minimum distance pairs is $\Theta(m)$. This is because if 13 or more points have the same distance to a point p then at least two of them are closer to each other than to p. If follows that each point p belongs to at most 12 minimum distance pairs. This packing argument fails in the bichromatic case because points of the same color can be arbitrarily close to each other.

4.2   Bichromatic Nearest Neighbors

Given sets P and Q as before, call $(p, q) \in P \times Q$ a *(blue/red) nearest neighbor pair* and q a *nearest neighbor* of p if $q \in Q$ minimizes the Euclidean distance from p to Q. Let $N(p)$ be the number of nearest neighbors of p and consider $\sum_{p \in P} N(p)$. We will assume that either no three points of P are collinear or that no four points of Q are cocircular. The best previous upper bound for this sum is $O(m^{3/4}n^{3/4}\beta(m, n) + m + n)$, see [4]. By drawing a sphere around each blue point p, with radius equal to the distance between p and its nearest neighbors, we again reduce the distance counting problem to the problem of Section 3. If no three points of P are collinear we get condition (S.i) and if no four points of Q are cocircular we get condition (S.i') (see remark (2) after Theorem 3.1). In both cases we obtain the following upper bound.

Theorem 4.2 The number of blue/red nearest neighbor pairs in a set of m blue and n red points in three-dimensional Euclidean space is $O(m^{2/3}n^{2/3} + m + n)$ if no three blue points are collinear or no four red points are cocircular.

Remarks. (1) As in the case of repeated minimum distance pairs, the maximum number of nearest neighbor pairs is $\Theta(m)$ for a monochromatic set of m points in three dimensions. Again, the packing argument used to prove the upper bound fails in the bichromatic case.

(2) The case where all points of Q lie on a circle in three-dimensional space and all points of P lie on the axial line of this circle shows that the bound of Theorem 4.2 does not hold without restrictions on the locations of the points.

4.3 Furthest Neighbors

Let P be a (monochromatic) set of m points in three dimensions, and call (p,q) a *furthest neighbor pair* and q a *furthest neighbor* of p if $q \in P$ maximizes the Euclidean distance from p. For each $p \in P$ let $F(p)$ be the number of furthest neighbors and consider $\sum_{p \in P} F(p)$. Assuming no three points are collinear, the best previous bound on this sum can be found in [4] and is $O(m^{3/2}\beta(m))$, with $\beta(m) = 2^{\Theta(\alpha(m)^2)}$. If we draw around each point p the sphere with radius equal to the distance between p and its furthest neighbors we get an instance of the problem in Section 3. Indeed, condition (S.ii) is satisfied because no point lies outside any of these spheres, and we get condition (S.i) if no three points are collinear and (S.i') if no four points are cocircular. This implies the following result.

Theorem 4.3 The number of furthest neighbor pairs in a set of m points in three-dimensional Euclidean space is $O(m^{4/3})$ if no three points are collinear or no four points are cocircular.

Remarks. (1) An example similar to the one in remark (2) after Theorem 4.2 shows that the maximum number of furthest neighbor pairs is quadratic in m if no condition on the location of the points is imposed.

(2) It is worthwhile to note that the maximum number of maximum distance pairs is $\Theta(m)$, as shown independently in [10, 11, 14].

4.4 Delaunay Triangulations

There is a relation between three-dimensional Delaunay triangulations and the incidence problems considered in Sections 2 and 3. Let P be a set of m points in three dimensions. We call $P' \subseteq P$ a *proper Delaunay subset* if there is a unique sphere so that the points of P' lie on the sphere and all other points of P lie outside the sphere, and we call the sphere a *Delaunay sphere*. The *Delaunay triangulation* of P, denoted by $\mathcal{D}(P)$, is the cell complex whose bounded cells are the convex hulls of the proper Delaunay subsets of P and whose unbounded cell is the complement of the convex hull of P (see [6] or [7]). We note that $\mathcal{D}(P)$ is not necessarily a triangulation because there may be bounded cells that are not tetrahedra. In fact, the problem we study below is interesting only if many of the cells are not tetrahedral.

Let the *degree* of a cell of $\mathcal{D}(P)$ be its number of vertices; the number of edges and 2-faces of the cell is proportional to its degree. We consider the problem of bounding the sum of degrees of subsets of the collection of Delaunay cells. Let S be the set of all Delaunay spheres. By definition, all points of P lie on or outside any Delaunay sphere. In other words, P and S satisfy condition (S.ii) of Section 3. We now argue that S also satisfies (S.i), that is, no three Delaunay spheres intersect in a common circle. Suppose to the contrary that there are three Delaunay spheres, s_1, s_2, and s_3, that meet in a common circle. It follows that the centers of the three spheres lie on a common line. Assume that the center of s_2 lies between the other two centers. But then all points of s_2 that do not belong to the common circle lie either inside s_1 or inside s_3. Because no points of P can lie inside s_1 or s_3 it follows that all points of $P \cap s_2$ lie on a circle which contradicts the assumption that s_2 is a Delaunay sphere. Using Theorem 3.1 we thus get the following bound on the sum of degrees. Note that since the Delaunay triangulation of P has at most $O(m^2)$ cells, the term n in the bound of Theorem 3.1 is always subsumed by the leading term.

Theorem 4.4 The sum of degrees of n cells in the Delaunay triangulation of any m points in three-dimensional Euclidean space is $O(m^{2/3}n^{2/3} + m)$.

Remark. Because of the dual correspondence between Delaunay triangulations and Voronoi diagrams (see [7]), Theorem 4.4 implies that the total number of edges incident to n vertices of the Voronoi diagram of any m points in three dimensions is $O(m^{2/3}n^{2/3} + m)$.

5 Conclusions

This paper studies a number of combinatorial distance problems in three-dimensional Euclidean space and derives improved upper bounds for all problems considered. The bounds are direct applications of an $O(m^{2/3}n^{2/3} + m + n)$ bound on the number of incidences between m points and n hyperplanes in four dimensions which holds if no three points are collinear, no three hyperplanes intersect in a common 2-flat, and each hyperplane bounds a closed half-space that contains all points.

Coincidentally, the above upper bound is the same as for the number of incidences between m points and n lines in the plane, without restriction on the points and lines. However, unlike in the planar problem where the upper bound is known to be tight, there is no superlinear lower bound known for the hyperplane incidence problem. To close the gap between the current upper and lower bounds is the most important open problem suggested by the results of this paper.

We would like to point out that although the results of this paper are mainly combinatorial, the techniques have also algorithmic applications. For example, [1] gives a randomized algorithm inspired by our constructive proof that finds a bichromatic minimum distance pair for a set of m blue and red points in three dimensions in expected time $O(m^{4/3} \log^c m)$, for some constant c. This algorithm is used to construct a minimum spanning tree of m points in three dimensions in roughly the same amount of time.

References

[1] P. Agarwal, H. Edelsbrunner, O. Schwarzkopf and E. Welzl. Euclidean minimum spanning trees and bichromatic closest pairs. To appear in "Proc. 6th Ann. Sympos. Comput. Geom. 1990".

[2] B. Bollobás. *Extremal Graph Theory*. London Math. Soc. Monographs, No. 11, Academic Press, London, 1978.

[3] A. Brønsted. *An Introduction to Convex Polytopes*. Grad. Texts in Math., Springer-Verlag, New York, 1983.

[4] K. L. Clarkson, H. Edelsbrunner, L. J. Guibas, M. Sharir and E. Welzl. Combinatorial complexity bounds for arrangements of curves and spheres. *Discrete Comput. Geom.* **5** (1990).

[5] K. L. Clarkson and P. W. Shor. Applications of random sampling in computational geometry, II. *Discrete Comput. Geom.* **4** (1989), 387–421.

[6] B. Delaunay. Sur la sphère vide. *Izv. Akad. Nauk SSSR, Otdelenie Matematicheskii i Estestvennyka Nauk* **7** (1934), 793–800.

[7] H. Edelsbrunner. *Algorithms in Combinatorial Geometry*. Springer-Verlag, Heidelberg, Germany, 1987.

[8] P. Erdös. On sets of distances of n points. *Amer. Math. Monthly* **53** (1946), 248–250.

[9] P. Erdös. On sets of distances of n points in Euclidean space. *Magyar Tud. Akad. Mat. Kutaló Int. Kozl.* **5** (1960), 165–169.

[10] B. Grünbaum. A proof of Vázsonyi's conjecture. *Bull. Res. Council Israel Sect. A* **6** (1956), 77–78.

[11] A. Heppes. Beweis einer Vermutung von A. Vázsonyi. *Acta Math. Acad. Sci. Hungar.* **7** (1956), 463–466.

[12] W. O. J. Moser and J. Pach. Research problems in discrete geometry. Manuscript, Dept. Math., McGill Univ., Montreal, Quebec, 1986.

[13] J. Spencer, E. Szemerédi and W. T. Trotter, Jr. Unit distances in the Euclidean plane. In *Graph Theory and Combinatorics*, Academic Press, London, 1984, 293–303.

[14] S. Straszewicz. Sur un problème geometrique de P. Erdös. *Bull. Acad. Polon. Sci. Cl. III* **5** (1957), 39-40.

[15] E. Szemerédi and W. T. Trotter, Jr. Extremal problems in discrete geometry. *Combinatorica* **3** (1983), 381–392.

Splitting a Configuration in a Simplex

Kazumiti Numata[1]   Takeshi Tokuyama[2]

[1]Department of Computer Science and Information Mathematics, The Electro-Communication University, 1-5-1 Chofugaoka, Choufu-shi, Tokyo 182, Japan.

[2]IBM Research, Tokyo Research Laboratory, 5-11 Sanbancho, Chiyoda-ku, Tokyo 102, Japan.

1. Introduction

Partitioning a point set into subsets of (almost) equal size is a popular problem in computational geometry, since it is a key subroutine in divide-and-conquer algorithms as well as a useful technique in range search problems. In the two dimensional case, we can make a four-section partition with a pair of lines and a six-section partition with three concurrent lines [5]. However, it is known that a partition with hyperplanes cannot always generate a good partition of a point set in a space with more than four dimensions [4].

In this paper, we present another type of partition. Consider a configuration S consisting of n points located in a d-dimensional simplex T in E^d. For an interior point G of T, we can split T into $d+1$ simplices, each of which is spanned by G and a facet of T. We deal with the partition of S naturally associated with the above splitting. Avis and ElGindy [2] found a splitting point G (refered as a *splitter*) from S to partition the point set in such a way that none of the subsimplices contains more than $\frac{dn}{d+1}$ points. On the other hand, we find a splitter G called a π-*splitter* (or an *optimal splitter* if π is known from the context) from T to generate any given partition $\pi = (\pi_1, \pi_2, ..., \pi_{d+1})$ satisfying $\sum_{i=1}^{d+1} \pi_i = n$. For example, we can find a splitter such that each subsimplex contains at least $[\frac{n}{d+1}]$ points. The problem is a natural extension of the *median-finding problem* [1, 3], which can be regarded as its one-dimensional version. We design an $O(d^2 n \log n)$ time algorithm for a point set in a d-dimensional simplex, and an $O(n)$ time algorithm for one in a triangle. In the $O(d^2 n \log n)$ time algorithm, we find a π-splitter by changing the balance of the partition gradually. For the $O(n)$ time algorithm, we use a *prune-and-search* method [6] instead. The splitter finding has a deep relation to the scheduling problem originally [7]. From the viewpoint of computational geometry, we can design an efficient spatial triangulation algorithm by applying it.

2. Existence of an Optimal Splitter

Let S be a configuration consisting of n points in the interior of a simplex T in the Euclidean space E^d of d dimensions. We deal with a central splitting of the simplex T with respect to an interior point G, and an associated partition of S. Let F_i, $(i = 1, 2, .., d + 1)$ be the facets $((d - 1)$-dimensional faces) of T.

Definition 1. A sequence $\pi = (\pi_1, \pi_2, ..., \pi_{d+1})$ of non-negative integers is called an ordered partition of n if $\sum_{i=1}^{d+1} \pi_i = n$.

Definition 2. Given an ordered partition $\pi = (\pi_1, \pi_2, ..., \pi_{d+1})$ of n, a π-splitter of (S, T) is a point G in the simplex T that subdivides the point set S such that there are at most π_i points in the interior and at least π_i points in the closure of the simplex spanned by G and F_i. We often call a π-splitter an *optimal splitter* if π is given in the context.

A splitter G is called a *singular splitter* if there is at least a point on the boundary of one of the sub-simplices generated by the central splitting. Otherwise, G is called a *regular splitter*.

Proposition 1. *For any given point set in a simplex and an ordered partition π, there exists a π-splitter.*

We design an algorithm to find an optimal splitter in Section 4, which gives a constructive proof of the proposition.

For a $(d - 2)$-dimensional face f of the simplex T and a point x of S, $h(f,x)$ denotes the hyperplane through both f and x. $\mathcal{H}(S, T)$ is the set of all such hyperplanes. The configuration (S, T) is called *simple* if the following two conditions holds: (1) $h(f, x) = h(g, y)$ iff $g = f$ and $x = y$, (2) Any intersection of $d + 1$ distinct hyperplanes in $\mathcal{H}(S,T)$ is either vacant or a point of S.

We denote by $Arr(S, T)$ the *arrangement of hyperplanes* [4] associated with $\mathcal{H}(S, T)$

If (S, T) is simple, there is at least one *cell* (open region) in $Arr(S,T)$ any interior point of which is an optimal splitter for a given π. Thus, we obtain the following lemma.

<u>Lemma 1.</u> *If (S, T) is simple, there is a regular optimal splitter for any given π.*

Although the π-splitter is not unique in general, the partition of points is unique in the following sense:

<u>Proposition 2</u> *If (S, T) is simple, any two π-splitters G_1 and G_2 generate the same partition of S.*

Proof. If $d = 1$, the proposition is trivial. Since (S, T) is simple, we can assume G_2 is a regular splitter. We consider the splitting of the simplex by the splitter G_1. Then, the point G_2 is located in one of the sub-simplices, say T_1, spanned by G_1 and a facet F_i of T. In this case, the simplex T_2 spanned by F_i and G_2 is involved in T_1 because of convexity. Since both T_1 and T_2 contains π_i points of S, they contain the same subset S_1 of S. Let V be the opposite vertex to F_i in T and let $S_0 = S - S_1$. Pr denotes the projection onto F_i along the ray through the vertex V. Thus, $Pr(G_1)$ and $Pr(G_2)$ are both optimal splitters of $(Pr(S_0); F_i)$ for $(\pi_1, ..., \pi_{i-1}, \pi_{i+1}, ..., \pi_{d+1})$, and the proposition is proved by induction.

From now on, we assume that (S, T) is simple, since we can reduce a non-simple case to a simple case by using ε-perturbation [4]. We also remark that we can solve the problem even if some points of S are located on the boundary of T.

3. Finding an Optimal Splitter of a Configuration in a Triangle

In this section we first present an $O(n \log n)$ time algorithm to find an optimal splitter of a configuration S consisting of n points in a triangle $\triangle ABC$ for $\pi = (\pi_1, \pi_2, \pi_3)$. Although we will give a simple linear time algorithm in section 5, it is appropriate to give this algorithm here, since the method is applied to higher-dimensional cases.

For any splitter G, the points in $\triangle ABG$, $\triangle CAG$, and $\triangle BCG$ are colored *blue, yellow,* and *red* respectively. The numbers of points colored blue, yellow, and red are denoted by $N(blue)$, $N(yellow)$, and $N(red)$ respectively. If G is an optimal splitter, $N(blue) = \pi_1$, $N(yellow) = \pi_2$, and $N(red) = \pi_3$. The outline of the algorithm is as follows: We first find a splitter G on the edge BC such that $N(red) = 0$, $N(blue) = \pi_1 + [\frac{\pi_3 + 1}{2}]$, and $N(yellow) = \pi_2 + [\frac{\pi_3}{2}]$. Then, we raise the segment AG slowly to increase the number of red points one by one, keeping the difference between $N(blue) - \pi_1$ and $N(yellow) - \pi_2$ less than 2. Thus, we can find an optimal splitter after π_3 recursions. We provide the following sorted lists: L_A is a list of points of S sorted with respect to the counter-clockwise argument around the vertex A. L_B is a list of points of S sorted with respect to the counter-clockwise argument around the vertex B. L_C is a list of points of S sorted with respect to the clockwise argument around the vertex C. We maintain these lists dynamically by using *balanced trees* [1], so that we can find both the largest element and the smallest element with each color with respect to the argument in $O(\log n)$ time. We name this procedure PENDULUM_2, since the segment AG swings like a pendulum in the two-dimensional plane. We show below the procedure roughly written in Pidgin ALGOL [1]. We illustrate the behavior of the algorithm in Figure 1.

Procedure PENDULUM_2;
begin
{Initialization}
1:   Look up L_A to find its $(\pi_1 + [\, \frac{\pi_3 + 1}{2} \,])$-th element M;
2:   The intersection of the line $l = AM$ with the edge BC is G;
3:   Color the elements of S in $\triangle ABG$ blue, and those in $\triangle AGC$ yellow;
{Loop: increment red points}
4:   *while* $N(red) < \pi_3$ *do*
      begin
5:        Find the smallest blue element X (with respect to the argument) in L_B;
6:        Find the smallest yellow element Y in L_C;
7:        $X' = AG \cap BX$; {the intersection point}
8:        $Y' = AG \cap CY$;
9:        *if* $X'G \le Y'G$, *then* color X in red; $G = X'$;
10:       *else* color Y in red; $G = Y'$;
11:       *if* $|N(blue) - N(yellow) - \pi_1 + \pi_2| = 2$, *then*
            begin
12:              *call* BALANCE_2 (G,X,Y);
            end;
        end; {end of loop}
13: Output G
end

Procedure BALANCE_2 (G, X, Y);

{Without loss of generality, we assume that $N(blue) - \pi_1 = N(yellow) - \pi_2 + 2$. Thus, X is blue and Y is red. Y is on the boundary line CG between the red region and the yellow region. We will increase $N(yellow)$ by 1 and decrease $N(blue)$ by 1.}

begin

1: Find the largest blue element Z in L_A;

2: $X' = CY \cap BX$;

3: $Z' = CY \cap AZ$;

4: **if** $X'G \geq Z'G$, **then** $G = Z'$; color Z yellow;

5: **else** $G = X'$; color X red; color Y yellow;

end;

<u>Proposition 3.</u> *The procedure PENDULUM_2 finds a π-splitter in $O(n \log n)$ time and $O(n)$ space.*

The analysis is straightforward.

Figure 1. Befaviour of PENDULUM_2 for $\pi = (4,4,4)$.

$\odot$ = bule, $\oplus$ = yellow, $\bullet$ = red.

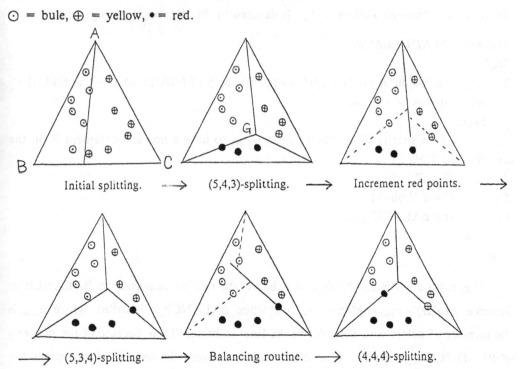

The procedure PENDULUM_2 outputs a singular splitter since at least one point of S is located on one of the edges AG, BG, and CG. In other words, G is on a line that passes through a vertex of the triangle $\triangle ABC$ and a point of S. However, we can find a regular optimal splitter by finding the corresponding open cell bounded by the point G in the arrangement $Arr(S, T)$. Since each cell of $Arr(S, T)$ has at most six boundary edges, this additional cost is $O(\log n)$.

4. Finding an Optimal Splitter of a Configuration in a Simplex

We consider a configuration S consisting of n points in a simplex T in the space E^d with d dimensions. For $\pi = (\pi_1, \pi_2, ..., \pi_{d+1})$, $\tilde{\pi} = (\tilde{\pi}_1, ..., \tilde{\pi}_d)$ is defined by $\tilde{\pi}_i = \pi_i + [\dfrac{\pi_{d+1} + d - i}{d}]$. Let $N(color)$ be the current number of points with a particular color. The region $T(color)$ of a color is spanned by the facet $F(color)$ and the splitter G. $V(color)$ is the opposite vertex to $F(color)$. The projection onto the facet $F(black) = F_{d+1}$ through $V(black) = V_{d+1}$ is denoted by Pr. We sketch our algorithm:

Procedure PENDULUM_d;
begin
1:     G = the $\tilde{\pi}$-splitter of $(Pr(S), F(black))$; {we apply PENDULUM_$d - 1$ to find G}
2:     *while* $N(black) < \pi_{d+1}$ *do*
        begin
3:          Find a splitter G' on the segment GV_{d+1} to have a non-black element Z on the boundary of $T(black)$;
4:          $G = G'$;
5:          Color Z black;
6:          *call* BALANCE_d;
        end;
end;

The subroutine BALANCE_d is a balancing routine analogous to BALANCE_2. Suppose $N(blue) - \pi_{blue} = N(red) - \pi_{red} + 2$ when BALANCE_d is called, where π_{color} is the number of points with the color in the final splitting. Then, because of the property of PENDULUM_d, a point Z on the boundary between the red and black regions has been just changed its color from red into black. The procedure BALANCE_d increases $N(red)$ by 1 and decreases $N(blue)$ by 1.

Procedure BALANCE_d;

begin

1: *if* non-black elements are balanced, *then* return;

    else begin

    {Suppose we will increase N(red) by 1 and decrease N(blue) by 1.}

2:     $C\_set = \{black, red\}$ is a set of colors;

3:     $f = F(red) \cap F(black)$;

4:     hp = the hyperplane spanned by f and G; {the hyperplane between the black region and red region}

5:     m = the line through $V(black)$ and $V(red)$;

6:     pl = the plane through m and G;

5:     $l = pl \cap hp$; { l is a line on which G lies.}

    repeat

7:     Find the nearest splitter G' to G on l such that a point colored in *color_new* that is not a member of C_set is on the boundary of a region with a color in C_set;

8:     Add *color_new* to C_set;

9:     $G = G'$;

10:     $f = f \cap F(color\_new)$;

11:     m = the affine space spanned by m and $V(color\_new)$;

12:     pl = the affine space through m and G ;

13:     hp = the affine space through f and G ; { $\dim(hp) + \dim(pl) = d + 1$.}

14:     $l = hp \cap pl$; { l is a line on which G lies.}

15:     *until color_new* is blue;

{ Now there is a bule boundary element.}

16:     Update the coloring of boundary elements such that $N(blue)$ is decreased by 1, $N(red)$ is increased by 1, and the numbers of other colored points are not changed;

17:     return;

    end ;

end ;

Theorem 1. *We can find an optimal splitter of (S, T) in $O(d^2 n \log n)$ time and $O(d^2 n)$ space.*

We omit the details of the analysis because of lack of space. The following is the key lemma:

Lemma 2. *A cycle of the while loop of PENDULUM_d is done in $O(d^2 \log n)$ time.*

Thus, $t(d, n) = t(d - 1, n) + O(\pi_{d+1} d^2 \log n)$ for the time complexity $t(d, n)$ of PENDULUM_d. Hence, the time complexity in Theorem 1 follows.

5. A Linear Time Algorithm for Planar Case

We present a simple linear time algorithm to find a π-splitter of a triangle adopting a prune-and-search technique, which was invented by Blum et al. [3] and developed by Megiddo [6]. Let $\pi = (\pi_1, \pi_2, \pi_3)$ be an ordered partition of n.

Theorem 2. *We can design an algorithm to find a π-splitter of $(S; \triangle ABC)$ in $O(n)$ time and $O(n)$ space.*

Proof. We can assume that $\pi_1 \geq \pi_2 \geq \pi_3$. We place $\pi_1 - \pi_2$ dummy points on the edge AC and $\pi_1 - \pi_3$ points on the edge BC, and denote the union of S and the set of dummy points by $\overline{S}$. Then, it is obvious that a (π_1, π_1, π_1)-splitter of $(\overline{S}; \triangle ABC)$ is a π-splitter of $(S; \triangle ABC)$. Since $3\pi_1 \leq 3n$, it suffices to show that we can find a (π_1, π_1, π_1)-splitter in linear time. Therefore, we assume $\pi_1 = \pi_2 = \pi_3 = \frac{n}{3}$ without loss of generality.

We illustrate the behavior of the algorithm in Figure 2. For a π-splitter G (Figure 2.1) of $(S; \triangle ABC)$, we color the points in $\triangle ABG$, $\triangle ACG$, and $\triangle BCG$ blue, yellow, and red. Because of Proposition 2, the coloring is independent of the choice of the splitter (this property holds though we have several points on the edges of the triangle.) First, we find a point P on BC such that $\triangle ABP$ contains $[\frac{n}{2}]$ points of S in its interior. Then, we find the points G_1 and G_2 on AP such that each of the triangles $\triangle ABG_1$ and $\triangle ACG_2$ contains $\pi_1 = \frac{n}{3}$ points. We immediately obtain the following claim:

Claim. *If G_1 is nearer to A than G_2 is, then points of S contained in $\triangle G_2 PC$ are colored red. Else, $\triangle G_1 BP$ consists of red points.*

Therefore, we find at least $\left[\frac{n}{6}\right]$ red points (Figure 2.1.) Similarly, we can find at least $\left[\frac{n}{6}\right]$ blue points (Figure 2.2) and $\left[\frac{n}{6}\right]$ yellow points (Figure 2.3.) If we use a linear time *k-th element finding algorithm* [1,3], we can find all these points in $c_0 n$ time for a constant c_0. Now, we prune away these points from S to obtain a point set S_1 with $3\left[\frac{n+3}{6}\right]$ points (Figure 2.4.) If $n \geq 6$, at least three points are pruned. By definition, G is a $(\left[\frac{n+3}{6}\right], \left[\frac{n+3}{6}\right], \left[\frac{n+3}{6}\right])$-splitter of $(S_1, \triangle ABC)$. Then, we apply the same technique to $(S_1, \triangle ABC)$ recursively until the number of the remained points becomes less than 6 (actually, it is 3). Thus, we obtain the partition of the points except 3 points with at most $m = \left[\log_2 n\right]$ recursions. Finally, we (applying PENDULUM_2) determine the color of the remained 3 points with $O(1)$ time. The sum of processing time for these recursions is less than $\sum_{i=0}^{m} c_0\{(\frac{1}{2})n + 6\} \leq c_0(2n + 6m + 6)$. Once we know the partition of the point set, we can find an optimal splitter in an additional $O(n)$ time. Thus, the total processing time is $O(n)$, and we complete the proof.

Figure 2. Prune and search method ($n = 12$).

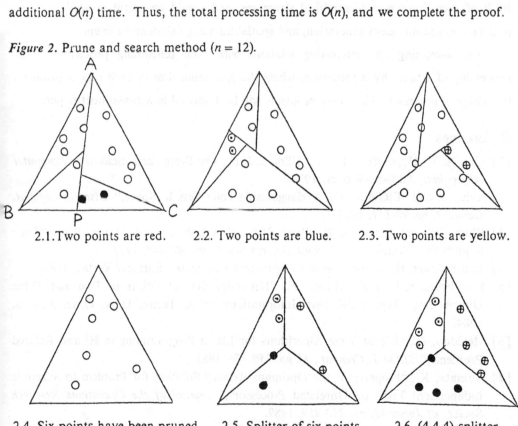

2.1. Two points are red. 2.2. Two points are blue. 2.3. Two points are yellow.

2.4. Six points have been pruned. 2.5. Splitter of six points. 2.6. (4,4,4)-splitter.

This prune-and-search method does not seem to be immediately applicable to a high-dimensional case, though we hope a linear time algorithm should exist there.

6. Applications

For a given configuration of $n = (d + 1)^k$ points in a simplex, we consider a splitting such that each subsimplex contains $(d + 1)^{k-1}$ points. If we apply this splitting recursively, we can construct a *spatial triangulation* of the simplex into n sub-simplices in $O(d^2 n (\log n)^2)$ time such that at most one point is included in each sub-simplex. For general n, we can similarly construct a spatial triangulation consisting of $n + d - 1$ sub-simplices in $O(d^2 n (\log n)^2)$ time by choosing suitable ordered partitions. The number of simplices is smaller than those of spatial Delaunay triangulation and Avis-ElGindy's triangulation [2]. Although the triangulation does not have S as the vertex set, its dual graph does. Both of them have nice hierarchical structures, and would be useful in application to pattern recognition, mesh generation, and spatial traveling salesman problem.

The π-splitting has interesting relations with such scheduling problems as the processing of n tasks by d processors where the processing time of each task depends on the assigned processor [7]. These problems will be discussed in a subsequent paper.

References

[1] Aho, A. V., Hopcroft, J. E., and Ullman, J. D., *The Design and Analysis of Computer Algorithms,* Addison-Wesley, 1974.

[2] Avis, D. and ElGindy, H., "Triangulating Point Sets in Space," *Discrete Comput. Geom. 2, pp. 99-111, 1987.*

[3] Blum, M. and Floyd, R.W., Pratt, V.R., Rivest, R.L., and Tarjan, R.E., "Time bounds for Selection," *J. Comput. System Sci. 7, pp. 448-461, 1972.*

[4] Edelsbrunner, H., *Algorithms in Combinatorial Geometry,* Springer Verlag, 1986.

[5] Edelsbrunner, H. and Huber, F., "Dissecting Sets of Points in Two and Three Dimensions," Rep. F138, Inst. Informationsverarb., Techn. Univ. Graz, Austria, 1984.

[6] Megiddo, N., "Linear Time Algorithms for Linear Programming in $\mathbb{R}^3$ and Related Problems," *SIAM J. Comput., 12 pp. 759-776, 1983.*

[7] Numata, K., "Property of the Optimum Relaxed Solution for Problem to Schedule Independent Tasks on Unrelated Processors," *Journal of the Operations Research Society of Japan 32, pp. 233-259, 1989.*

Weaving Patterns of Lines and Line Segments in Space

J. Pach *

Courant Institute, New York University

Mathematical Institute of the Hungarian Academy of Sciences

R. Pollack †

Courant Institute, New York University

E. Welzl‡

Freie Universität Berlin, Institut für Informatik

Abstract

A *weaving* W is a simple arrangement of lines (or line segments) in the plane together with a binary relation specifying which line is "above" the other. An m by n bipartite weaving consists of m horizontal and n vertical lines or line segments which mutually intersect. A system of lines (or line segments) in 3-space is called a *realization* of W, if its projection into the plane is W and the relative positions of the lines respect the "above" specifications. An equivalence class of weavings induced by the combinatorial equivalence of the underlying planar arrangement of lines is said to be a *weaving pattern*. A weaving pattern is *realizable* if at least one element of the equivalence class has a realization. A weaving (pattern) W is called *perfect*, if along each line of W, the lines intersecting it are alternately "above" and "below". We prove that (i) a perfect weaving pattern of n lines is realizable if and only if $n \leq 3$, (ii) a perfect m by n bipartite weaving pattern is realizable if and only if $\min(m, n) \leq 3$, (iii) if n is sufficiently large then almost all weaving patterns of n lines are nonrealizable.

1 Introduction

For scene analysis, hidden surface elimination or separability of objects in 3-space, it is important to consider the projection of the objects in some direction together with information about the relative position of objects having a common point in their projection. Not only are these objects and relations exceedingly complex but they remain complex even when we restrict our attention to the simplest and most basic case – when the objects are lines or line segments in 3-space. In contrast to the enormous literature

*Supported in part by Hungarian NFSR grant 1812, NSF grant CCR-8901484 and the Center for Discrete Mathematics and Theoretical Computer Science (DIMACS), a National Science Foundation Science and Technology Center, under NSF grant STC88-09648.

†Supported in part by NSA grant MDA904-89-H-2030, NSF grants DMS-85-01947 and CCR-8901484, and DIMACS.

‡Supported in part by the ESPRIT II Basic Research Actions Program of the EC under contract no. 3075 (project ALCOM) and DIMACS

about 2-dimensional arrangements of lines and line segments there have been very few results on arrangements of lines or line segments in space (see [13] [15] and [16]). Recently, however, due to the potential applications already mentioned, a vivid interest has developed in this subject (see [1], [2], [10] and [11]).

In this paper we consider whether there exists an arrangement of lines (or line segments) in 3-space whose orthogonal projection into some 2–dimensional plane has a given combinatorial type, and where the above–below relationship is also prescribed.

More precisely, let L and L' be two arrangements of lines (or line segments) in the plane. We say that L and L' are *combinatorially equivalent*, if the incidence structures of the vertices, edges and faces determined by L and by L' are isomorphic. A *weaving* is a pair $(L, \succ)$ where L is a 2–dimensional arrangement of lines (or line segments) and $\succ$ is an antisymmetric, nonreflexive binary relation on L. If $l_1 \succ l_2$ for some $l_1, l_2 \in L$, then we shall say that l_1 is *above* l_2 (or l_2 is *below* l_1). Two weavings $(L_1, \succ_1)$ and $(L_2, \succ_2)$ are *equivalent*, if there is a one-to-one correspondence between the elements of L_1 and L_2 under which L_1 and L_2 are combinatorially equivalent and $\succ_1$ and $\succ_2$ are isomorphic. The equivalence classes of weavings with respect to this equivalence relation are called *weaving patterns*.

Let Λ be a family of lines in 3-space, Π be an oriented 2−dimensional plane, and assume that no two lines of Λ cross each other, and none of them is perpendicular to Π. We say that $l_1 \in \Lambda$ is above $l_2 \in \Lambda$ if a positively oriented line normal to Π meets l_2 before it meets l_1. Obviously, the orthogonal projection of Λ to Π together with the above-below relations between the elements of Λ (with respect to Π) define a weaving $W = W(\Lambda, \Pi)$. All weavings that can be obtained in this way are called *realizable*. A *weaving pattern* is *realizable*, if some representative of it is realizable. In this paper we make the first steps towards a characterization of realizable weaving patterns. We will show that if n tends to infinity, then almost all weaving patterns of n lines are nonrealizable. It is a difficult problem to exhibit for any large n a nonrealizable weaving pattern with n lines which is minimal in the sense that deleting any of its lines it becomes realizable. In the next two sections we will consider two special classes of weavings. A weaving $W = (L, \succ)$ and the corresponding weaving pattern is called *perfect*, if (i) the lines belonging to L are in general position, i.e., no two are parallel and no three coincident; and (ii) according to the relation $\succ$, each line $l \in L$ is alternately "above" and "below" the other lines in the order they cross l. A system of $m + n$ line segments in the plane, $H = \{h_1, h_2, \ldots, h_m\}$ and $V = \{v_1, v_2, \ldots, v_n\}$, is called a *bipartite weaving*, if their arrangement is equivalent to an arrangement of m horizontal and n vertical line segments which mutually intersect. An m by n bipartite weaving $W = (H \cup V, \succ)$ (and the corresponding weaving pattern) is said to be *perfect*, if each segment $h_i \in H$ ($v_i \in V$) is alternately "above" and "below" the elements of V (H) with respect to the relation $\succ$. See figure 3.

Our principal results are that a perfect weaving pattern on n lines is realizable if and only if $n \leq 3$, and that a perfect m by n bipartite weaving pattern is realizable if and only if $\min(m, n) \leq 3$.

2 Perfect weaving patterns of lines

Theorem 1 *A perfect weaving pattern on n lines is realizable if and only if $n \leq 3$.*

Proof: Let $W = (L, \succ)$ be a perfect weaving of n lines. Assume, in order to obtain a contradiction, that $n > 3$ and $W = W(\Lambda, \Pi)$, i.e., it can be obtained by projecting the 3-dimensional arrangement of lines Λ into the plane Π, as described in section 1. Given any line $\lambda \in \Lambda$, we can move λ vertically up or down until λ meets another line $\lambda' \in \Lambda$. Call this new arrangement of lines Λ'. The lines λ and λ' now determine a plane Π'. By applying an affine transformation of space, we can assume that $\Pi = \Pi'$ is horizontal, and the weaving W does not change except for the incidence of the lines λ and λ'. We will refer to λ and λ' as the *axes* of the plane Π, and the intersection of the lines λ and λ' will be called the *origin*. Moreover, we will think of the positive z axis as directed orthogonally above Π. We can also assume without loss of generality that no non-axis line of Λ' is parallel to Π, so that we can orient all segments of these lines in the direction of increasing z coordinate. In this way L, the orthogonal projection of Λ into Π, becomes a directed graph, see figure 1 (some of the edges are really half- lines, and all segments belonging to the axes are undirected).   Finally, we shall color certain directed edges

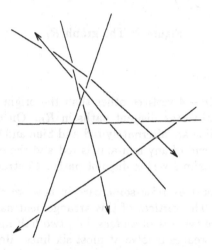

Figure 1: The directed arrangement L for a perfect weaving pattern on 6 lines

of L red, to obtain a directed red subgraph of L, which we denote by R_L. An edge is colored red if it either goes to an undercrossing or from an overcrossing (we mention both possibilities to include the edges which are half-lines, for these only one of these cases applies). All other directed edges are coloured blue. The graph R_L has the following properties (see figure 2).

1. Along any line of L, different from the axes, every second edge is blue.

2. Each vertex not on an axis is incident to exactly 2 red edges, it has indegree one and outdegree one in R_L.

3. Each vertex on an axis other than the origin is incident to exactly one red edge.

4. Along any directed path in R_L the z coordinate of a point moving on the corresponding segments of Λ is strictly increasing. Thus there are no directed cycles in R_L, and no directed path can begin and end at axis vertices.

5. Every maximal path begins and ends either at an axis vertex or with a half-line.

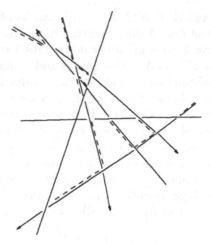

Figure 2: The graph R_L

We shall distinguish two cases.

n even: Since there are $2n - 4$ vertices other than the origin on the axes, there are at least $2n - 4$ maximal vertex disjoint paths in R_L. On the other hand, since the edges on a non-axis line are alternately red and blue and since n is even, one of the two half-lines at the end of any non-axis is red and the other is blue. Hence there are at most $n - 2$ maximal vertex disjoint paths. Contradiction.

n odd: In this case we must exercise some care in how we choose our axes. Look at the arrangement L. The vertices of this arrangement have at least three extreme points, and these are vertices of *wedges*, i.e., two half-lines incident to a common vertex. These three wedges involve at most six lines. Hence, there is either a line not involved with any wedge, or there are exactly five lines and one is involved with two wedges and each of the others is involved with one wedge. In either case we can choose the axes so that there is a wedge that is not adjacent to any of the axes. Again, it is clear that R_L contains at least $2n - 4$ vertex disjoint paths starting or ending at an axis. On the other hand, the two half-lines of any non-axis line have the same color. Hence, in order to avoid a contradiction all of these half-lines must be red. However, in this case a wedge formed by two half-lines not belonging to the axes constitutes a maximal directed path R_L. Hence, there can be at most $2n - 6$ vertex disjoint directed red paths which begin or end at an axis. Contradiction.

3 Perfect bipartite weaving patterns of segments

Theorem 2 *A perfect bipartite m by n weaving pattern is realizable if and only if*

$$\min(m, n) \leq 3.$$

Proof: It is not difficult to see that the lines

$$h_{2k} : x = 1 + t, y = 2k + t/4, z = 2k - 1/7 + (2k - 2/7)t$$

$$h_{2k+1} : x = 1 + t, y = 2k + 1 - t/4, z = 2k + 9/8 + (k + 1/4)t$$

$$v_{-1} : x = -1, z = -y$$

$$v_0 : x = 0, z = 0$$

$$v_1 : x = 1, z = y$$

realize a perfect bipartite $\aleph_0$ by 3 weaving of lines. Restricting this construction to some finite portion of space, we obtain a realization of the perfect m by 3 weaving for any m.

This family of lines is obtained by manipulating some of the lines

$$h_{y_0} : y = y_0, z = xy_0$$

and

$$v_{x_0} : x = x_0, z = x_0 y,$$

all of which lie on the hyperbolic paraboloid $z = xy$. Simply notice that, if we move the line h_{y_0} so that its projection in the x, y plane has a small positive (negative) slope while keeping it in contact with the lines v_{-1} and v_0, then the line will pass below (above) v_1. We then rotate it a bit in a vertical plane about its intersection with v_0 so that it passes below (above) v_{-1}. Finally, we translate it a little vertically up (down) so that it passes above (below) v_0.

Even though a perfect m by 3 weaving pattern is realizable for every m, it is not true that we can prescribe the underlying arrangement $H \cup V$. In fact, it is quite easy to see that, if $W = (H \cup V, \succ)$ is a 3 by 3 perfect weaving pattern with the property that the lines of H (and the lines of V) are pairwise parallel, then W is not realizable. It is equally easy to see that, if $W(\Lambda_H \cup \Lambda_V, \Pi)$ is a realization of a perfect 3 by 3 weaving, then the lines of Λ_H (and Λ_V) are pairwise skew, i.e., not coplanar.

In what follows, we will often refer ambiguously to the segments and to the lines containing them. Now we shall prove that a perfect bipartite 4 by 4 weaving pattern is not realizable. Suppose that such a pattern, $H \cup V = \{h_1, \ldots, h_4, v_1, \ldots, v_4\}$ had a realization $(\Lambda_H \cup \Lambda_V, \Pi)$, where $\Lambda_H = \{\lambda_1, \ldots, \lambda_4\}$, $\Lambda_V = \{\nu_1, \ldots, \nu_4\}$ and Π is horizontal. (See figure 3.)   By using a projective transformation, we may assume

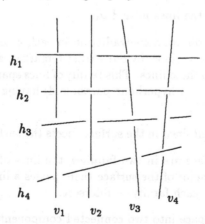

Figure 3: The perfect bipartite 4 by 4 weaving pattern

that h_1 and h_4 are parallel. Now, by performing a series of rotations for each line in its vertical plane, we can assume that the projection of $\Lambda_H \cup \Lambda_V$ into Π is the same as before, but now λ_i and ν_j intersect except for $(i,j) = (1,4),(4,4),(4,1),(1,1)$, where the above-below relations remain unchanged (see figure 4). Now we want to move ν_1 (as

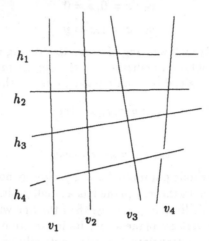

Figure 4: Perfect 4 by 4 weaving after rotations

we did to obtain the construction for the "if" part of the theorem) maintaining the fact that it intersects both λ_2 and λ_3 in such a way that the line v_1 meets the intersection of the lines v_2 and v_3. Either we succeed in this effort, or we are prevented by the fact that we create an incidence between the segments ν_1 and λ_1 or between the segments ν_1 and λ_4. In this way (after playing the same game with ν_4 as well) we can guarantee that at least 2 of the remaining four crossings become incidences. Thus, without loss of generality, we may assume that one of the following cases holds:

case 1: The lines v_1, v_2, v_3 are concurrent.

case 2: The line λ_1 meets the line ν_1 and λ_4 meets ν_4.

case 3: The line λ_1 meets the lines ν_1 and ν_4.

The method we employ to obtain a contradiction in each case is to choose three of the lines (which will necessarily be skew, by our earlier remarks), and consider the family of lines which are tranversal to these lines. This family of lines span a quadric surface (either a hyperboloid of one sheet or a hyperbolic paraboloid) having the important properties that

1. A line which is not contained in the surface meets the surface in at most two points.

2. The lines on the surface are in two families, the lines of each family are pairwise skew, through every point of the surface there passes a line of each family, and any two lines — one from each family — intersect.

3. The surface divides space into two connected components.

The detailed proof we be in the full version of this paper.

4 Most weaving patterns are not realizable

Theorem 3 *The number of realizable weaving patterns divided by total number of weaving patterns of n lines tends to 0 exponentially fast, as $n \to \infty$.*

Proof: Let $\Lambda = \{\lambda_i | 1 \leq i \leq n\}$ be a collection of lines in 3-space. Assume that λ_i is not perpendicular to the x-axis, and can be described by the equations

$$\lambda_i : y = a_i x + b_i, z = c_i x + d_i.$$

Whether or not the line λ_i is above or below the line λ_j ($i < j$) is determined by the sign of the polynomials

$$\phi_{i,j} = (c_i - c_j)(b_j - b_i) - (d_i - d_j)(a_j - a_i)$$

and

$$\psi_{i,j} = (a_j - a_i).$$

Thus, we have $2\binom{n}{2}$ polynomials of degree at most 2 in the the $4n$ variables $\{a_i, b_i, c_i, d_i\}$, $1 \leq i \leq n$. Warren's theorem ([14]) states that the number of sign patterns for m polynomials of degree at most d in k variables is at most $(\frac{4edm}{k})^k$. This is a slight improvement over the similar bound that can be obtained from the well-known theorems of Thom and Milnor ([12] and [8]). Applying this bound to our situation, we obtain that there are at most $n^{4n + O(\frac{n}{\log n})}$ different ways how the above-below relationship can be defined among n lines in a *realizable* weaving pattern. However the total number of possibilities is $2^{\binom{n}{2}}$, which yields that the fraction of realizable weaving patterns of n lines goes to 0 exponentially fast.

Note that, in fact, this fraction can be proved to be much smaller, taking into account that there are many feasible specifications of the above-below relation on Λ which are not consistent with a given combinatorial type of the underlying planar arrangement $L = \{a_i x + b_i | 1 \leq i \leq n\}$.

5 Remarks

We have answered the realizability question only for a few very special weaving patterns. It seems unlikely that there is a simple characterization of the realizable weaving patterns. It would be interesting to know, or estimate, the complexity of an algorithm to decide whether a given weaving pattern is realizable. We conjecture that this is an NP-hard probem in contrast to the polynomial problem of deciding whether a given weaving is realizable (which is reducible to linear programming).

6 Acknowledgements

We would like to acknowledge fruitful conversations with H. Edelsbrunner, J.R. Sack and V. Rödl.

References

[1] B. Chazelle, H. Edelsbrunner, L.J. Guibas and M. Sharir, Lines in space – combinatorics, algorithms and applications, Proc. 21st Ann. ACM Sympos. Theory Comput. (1989), 382-393.

[2] B. Chazelle, H. Edelsbrunner, L.J. Guibas, R. Pollack, R. Seidel, M. Sharir and J. Snoeyink, Counting and cutting cycles of lines and line segments in space (in preparation).

[3] H. Edelsbrunner, *Algorithms in Combinatorial Geometry.* Springer-Verlag, Heidelberg, Germany, 1987.

[4] H. Edelsbrunner, L.J. Guibas, J. Pach, R. Pollack, R. Seidel and M. Sharir, Arrangements of curves in the plane – Topology, Combinatorics and Algorithms, Proc. 15th Int. Colloq. on Automata, Languages and Programming (1988), 214-229.

[5] B. Grünbaum, Arrangements and Spreads, Reg. Conf. Series in Math. AMS, Providence, 1972.

[6] J.E. Goodman and R. Pollack, Semispaces of configurations, cell-complexes of arrangements, J. Comb. Theory, Ser. A 37, 1984, 257-293.

[7] H. Edelsbrunner, L.J. Guibas and M. Sharir, The complexity and construction of many faces in arrangements of lines and of segments, Discrete and Comp. Geom., 5, 1990, 161-196.

[8] J. Milnor, On the Betti numbers of real varieties, Proc. Amer. Math. Soc., 15 (1964), 275-280.

[9] M. McKenna and J. O'Rourke, Arrangements of lines in 3-space: A data structure with Applications, Proc. 4th Symp. on Computational Geometry (1988), 371-380.

[10] R. Penne, On line diagrams, (manuscript, 1989).

[11] R. Penne, Algorithms for line diagrams, (manuscript, 1989).

[12] R. Thom, Sur l'homologie des varietes algebriques reelles. Differential and Combinatorial Topology, Ed. S.S. Cairns (Princeton University Press, 1965).

[13] O.Ya. Viro, Topological problems concerning lines and points of three–dimensional space, Soviet Math. Dokl. 32 (1985), 528-531.

[14] H.E. Warren, Lower bounds for approximation by linear manifolds, Trans. Amer. Math. Soc. 133 (1968), 167-178.

[15] W. Whiteley, Weaving lines and tensegrity frameworks, (manuscript, 1988).

[16] W. Whiteley, Weaving and lifting plane line configurations, (manuscript, 1988).

EFFICIENT PARALLEL ALGORITHMS FOR
PATH PROBLEMS IN PLANAR DIRECTED GRAPHS

Andrzej Lingas
Department of Computer Science
Lund University, Box 118, 22100 Lund, Sweden

Abstract

We present a preprocessing of planar digraphs which yields processor-efficient, poly-log time algorithms for the single-source and all-pairs shortest-path problems, breadth-first search and reachability in planar digraphs.

1. Introduction

Almost all known parallel implementations of basic digraph problems, e.g. reachability, breadth-first search (BFS for short), shortest paths [KaR] use matrix multiplication. Therefore the processor-time costs of these algorithmic solutions are trivially at least the number of operations required to multiply two $n \times n$ Boolean matrices (the best known result is presently $O(n^{2.376})$, see [CW] and [AHU] p. 243). This situation is annoying in the case of sparse digraphs for which we can sequentially compute for instance the BFS tree in time proportional to the number of edges and vertices. The exception are the problems of computing strongly connected components of a planar digraph and constructing a depth-first search tree in a planar digraph. Kao and Shannon, and Kao and Klein, respectively, have recently obtained NC algorithms for these problems, using only linear number of processors [KS,KK].

The BFS problem can be seen as a special case of the single-source shortest-path problem (SSSPP for short) [AHU]. In [PR85], [PR89], Pan and Reif gave a reduction of SSSPP in planar undirected graphs to the problem of finding a family of separators. This reduction combined with a randomized parallel algorithm for a simple cycle $O(\sqrt{n})$ separator in planar two-connected graphs [GM] yields an efficient parallel method for SSSPP in planar undirected graphs. The method takes $O(\log^3 n)$ time in the probabilistic CRCW PRAM model with $O(n^{1.5}/\log n)$ processors (see [PR85], [PR89], [GM]). *It is not known whether this method could be efficiently generalized to include planar directed graphs* [Re].

In this paper, we develop a different parallel method for SSSPP in the planar case which works also in the more general directed case. The main idea is to augment the input graph G by some edges appropriately such that any directed path in G can be simulated by a path consisting of logarithmic number of edges, and then to apply a straight-forward algorithm for SSSPP in the augmented digraph. Given the augmented digraph, we can solve SSSPP for an arbitrary vertex v in G in time $O(\log^2 n)$ using an EREW PRAM with $O(n)$ processors. The construction of the augmented digraph non-trivially reduces to the problem of finding a family of simple cycle $O(\sqrt{n})$ separators in the corresponding *undirected* planar graph. Combining the above reduction with the randomized algorithm for finding a small simple cycle separator in two-connected planar graphs due to Gazit and Miller [GM], we obtain the following result:

1) We can preprocess a planar digraph with nonnegative integer edge lengths of polynomial size such that the single-source shortest-path problem for its arbitrary vertex can be solved in time $O(\log^2 n)$ using an EREW PRAM with $O(n)$ processors. Let $M(n)$ be the number of processors in the CRCW PRAM model required to compute the min +

product of two $n \times n$ matrices with nonnegative integer entries of polynomial size in time $O(\log^2 n)$. The preprocessing takes $O(\log^4 n)$ time using a probabilistic CRCW PRAM with $O(\min\{n^{1.5}, n^{\frac{1}{4}} M(\sqrt{n})\})$ processors.

The above result applied to planar digraphs with unit edge lengths immediately yields the following corollary:

2) For an arbitrary vertex v of a planar digraph with radius $O(n^\delta)$, a breadth-first search tree rooted at v can be constructed in time $O(\log^4 n)$ using a probabilistic CRCW PRAM with $O(\min\{n^{1.44+\delta}, n^{1.5}, n^{\frac{1}{4}} M(\sqrt{n})\})$ processors.

Using our method of preprocessing planar digraphs, we obtain the following, almost optimal solution to the all-pairs shortest-path problem for planar digraphs:

3) The all-pairs shortest-path problem for a planar digraph with integer lengths of polynomial size can be solved in time $O(\log^4 n)$ using a CRCW PRAM with $O(n^2/\log^2 n)$ processors.

Finally, by trivializing our solution to the shortest paths problems for planar digraphs, we obtain the following reachability results:

4) We can preprocess a planar digraph such that for any its vertex v, we can report all vertices reachable from v by directed paths in the digraph in time $O(\log^2 n)$ using an EREW PRAM with $O(n)$ processors. The preprocessing takes $O(\log^4 n)$ time using a probabilistic CRCW PRAM with $O(n^{1.44})$ processors. The transitive closure graph of a planar digraph can be constructed in time $O(\log^4 n)$ using a CRCW PRAM with $O(n^2/\log^2 n)$ processors.

Note that if we could replace the randomized parallel algorithm for a small simple cycle separator in planar digraphs due to Gazit and Miller with an equally efficient deterministic one, all our results could be rewritten in terms of deterministic PRAMs

The structure of this paper is as follows. In Section 2, we introduce the concept of a simple cycle separator tree for planar digraphs and derive a lemma on constructing such trees in parallel. In Section 3, we define the augmented digraph and estimate the cost of its construction in parallel. In Section 4, we apply the straight-forward SSSPP algorithm to the augmented digraph to obtain our main result on solving SSSPP in planar digraphs in parallel and the other results as corollaries in part.

2. Preliminaries

We shall use standard set and graph theoretic notation and definitions (see [AHU] for example). Specifically, we assume the following set and graph conventions.

1) For a finite set S, $|S|$ denotes the cardinality of S.

2) For a graph $G = (V, E)$, and a subset W of V, $G(W)$ denotes the induced subgraph of G on the set W.

3) A graph $G' = (V', E')$ is a *supergraph* of a graph $G = (V, E)$ if $V \subseteq V'$ and $E \subseteq E'$. For a positive real r, $G' = (V', E')$ is an *r-supergraph* of $G = (V, E)$ if G' is a supergraph of G and $|V'| \le r|V'|$, $|E'| \le r|E|$.

4) A subset S of V is a *separator* for $G = (V, E)$ if the remaining vertices of V can be partitioned into two sets V_1 and V_2 such that there are no edges between V_1 and V_2, and $|V_i| \le \frac{2}{3}|V|$ for $i = 1, 2$. We say that G has an $s(n)$ *separator* if there is a separator S for

G of size bounded by $s(|V|)$ and each of the induced subgraphs $G(V_1 \cup S)$ and $G(V_2 \cup S)$ has also an $s(n)$ separator (here, we slightly deviate from the standard definition that requires instead $G(V_1)$ and $G(V_2)$ to have $s(n)$ separator [LT,U]). Analogously, we say that G has a *simple cycle $s(n)$ separator* if there is a separator C for G of size bounded by $s(n)$ in the form of a simple cycle and each of the graphs $G(V_1 \cup S)$ and $G(V_2 \cup S)$ has also a simple cycle $s(n)$ separator.

A labelled tree T is an *$s(n)$ separator tree* for $G = (V, E)$ if it satisfies the following conditions. The root of T is labelled with V and a separator S for G of size $\leq s(|V|)$. If a node q in T is labelled with $V' \subset V$, $|V'| \geq 10$, and with a separator S' for $G(V')$ then q is the father of two sons whose first labels are respectively $V_1' \cup S'$ and $V_2' \cup S'$ (where V_1', V_2' are the two subsets of V' that are disconnected by the removal of S' from V'). The other labels are separators for $G_1(V_1' \cup S')$ and $G_2(V_2' \cup S')$ of size bounded by $s(|V_1' \cup S'|)$ and $s(|V_2' \cup S'|)$ provided that $|V_1' \cup S'| \geq 10$, $|V_2' \cup S'| \geq 10$ respectively. If each of the separators labelling T induces a simple cycle in G then T is called a *simple cycle $s(n)$ separator tree* for G.

It is well known that planar graphs have $O(\sqrt{n})$ separator [LT], and two-connected planar graphs have even simple cycle $O(\sqrt{n})$ separators if they can be embedded in the plane such that each face is of $O(1)$ size [GM]. In [GM], Gazit and Miller presented a randomized parallel algorithm for finding a simple cycle separator in such planar graphs. Their algorithm runs in time $O(\log^2 n)$ using a CRCW PRAM with $O(n^{1+\epsilon})$ processors, for a given arbitrary, positive ϵ. Hence, we obtain the following lemma.

Lemma 2.1: Let G' be a planar two-connected graph embedded in the plane such that each face is of size $O(1)$. Next, let ϵ be a positive rational. We can find a simple cycle $O(\sqrt{n})$ separator tree for G' in time $O(\log^3 n)$ using a probabilistic CRCW PRAM with $O(n^{1+\epsilon})$ processors.

A planar digraph not necessarily satisfies the requirements from the thesis of Lemma 2.1. For our purposes, it will be enough to construct a planar $O(1)$-supergraph of the digraph that does it.

Lemma 2.2: Let G be a planar connected graph on n vertices. We can find a planar two-connected $O(1)$-supergraph G' of G embedded in the plane, such that each face in the embedding of G' is of size $O(1)$, in time $O(\log n)$ using a CREW PRAM with $O(n)$ processors.

Sketch: Find a planar embedding of G by applying an algorithm due to Klein and Reif [KlR] which runs in time $O(\log n)$ on a CREW PRAM with $O(n)$ processors. Next, "triangulate" the graph embedded in the plane, adding dummy vertices on new edges in order to avoid creating a multigraph. Form the final adjacency lists by sorting the set of old and new edges, for instance, by using Cole's algorithm [C]. ∎

Combining the two above lemmas, we obtain the following key lemma.

Lemma 2.3: Let G be a planar connected graph on n vertices, and let ϵ be a positive rational. We can find a simple cycle $O(\sqrt{n})$ separator tree for a planar $O(1)$-supergraph of G in time $O(\log^3 n)$ using a probabilistic CRCW PRAM with $O(n^{1+\epsilon})$ processors.

3. Shortcuts

We shall specify a directed supergraph of the input planar digraph G such that any directed path in G can be simulated by a path in the supergraph which has a logarithmic number of edges. The supergraph can be efficiently constructed in parallel if a simple cycle $O(\sqrt{n})$ separator tree for a planar $O(1)$-supergraph of G (in particular G) is given. The basic idea is to add shortcuts in the form of single edges, appropriately defining their lengths, for all directed path connections in G between vertices of different cycle separators labelling T and not separated from each other by any other cycle separator labelling T. The supergraph is of size $O(n \log n)$. Throughout the remaining part of this paper, we shall assume that the planar digraphs G, G' etc. considered have nonnegative edge lengths specified.

Definition 3.1: Let $G = (V, E)$ be a planar digraph, and let T be a simple cycle $s(n)$ separator tree for a planar $O(1)$-supergraph $G' = (V', E')$ of G. Given an embedding of G in the plane, the graph G^T is recursively defined as follows.
Let q_0 be the root of T labelled with (V', S_0). Next, let V_l, V_r be the two subsets of V that are disconnected by the removal of the cycle S_0. Then, let $G_l = G(V_l \cup S_0 \cap V)$ and $G_r = G(V_r \cup S_0 \cap V)$. For $d \in \{l, r\}$, recursively compute the graph $G_d^{T_d}$ where T_d is the subtree of T rooted at the son of the root of T whose first label includes V_d. Let q_1 be the left son of q_0 in T. Find the unique path $q_0, q_1, ..., q_k$ down to the leaf q_l such that if S_j is the cycle separator labelling q_j, $j = 1, ..., k - 1$, then for $j = 1, ..., l - 1$, both S_0 and S_{j+1} simultaneously lie inside or outside the cycle S_j. Let q_1' be the right son of q_0 in T. Also, find the unique path $q_0, q_1' \ ... \ q_p'$ to the leaf q_p' where the cycle labels S_j' of q_j', $j = 1, ..., p - 1$, satisfy the analogous requirements. Now, construct the directed graph $L = ((S_0 \cup \bigcup_{j=1}^{l} S_j) \cap V, E_L)$ where (v, w) is in E_L if and only if $\{v, w\} \cap S_0$ is not empty and there is a directed path from v to w in G_l. Analogously, construct the graph $R = ((S_0 \cup \bigcup_{j=1}^{r} S_j') \cap V, E_R)$ where (v, w) is in E_R if and only if $\{v, w\} \cap S_0$ is not empty and there is a directed path from v to w in G_r. Compute the transitive closure $(L \cup R)^*$ of $(L \cup R)$. Set the edge length of each edge (v, w) in $(L \cup R)^*$ to the length of a shortest directed path from v to w in G. Set G^T to the union $G_l^{T_l} \cup G_r^{T_r} \cup (L \cup R)^*$. ∎

Theorem 3.1: Let G be a directed planar graph, and let T be a simple cycle $O(\sqrt{n})$ separator tree for a planar $O(1)$-supergraph of G.
1) For any pair v, w of vertices of G if there is a directed path from v to w in G which has length t then there is a directed path from v to w in G^T which has length not greater than t, and consists of $O(\log n)$ edges.
2) The all-pairs shortest-path problem (APSPP for short) for G is equivalent to that for G^T.

Proof: By Definition 3.1, G^T results from augmenting G by edges (v, w) of the transitive closures of some subgraphs of G whose lengths are equal to the lengths of shortest directed paths from v to w in G. This yields the second part of the thesis. The proof of the first part is as follows.
Let $G' = (V', E')$ be the planar $O(1)$-supergraph of G. Consider a directed path P from v to w in $G = (V, E)$. Let $l(v)$ be a leaf in T whose set label includes v, and let $l(w)$ be a leaf in T whose set label includes w. Next, let q_0 be the lowest common ancestor of $l(v)$ and $l(w)$ in T such that all the vertices of P are in the subset of V' that is the first set label of q_0. Consider the path $q_0, q_1, ..., q_m$ in T connecting q_0 with $q_m = l(v)$.

For $j = 0, ..., m$, let V_j, S_j be the labels of q_j. Although P has to pass through S_0 by the definition of the node q_0, it does not necessarily passes through all the cycles S_j, $j = m, ..., 0$. For $j = m-1, ..., 0$, we shall denote by P_j the maximal initial fragment of P wholly contained in $G'(V_{j+1})$ and ending at a vertex of S_j. Let k be the minimum index greater than 0 such that P touches or intersects S_k. By $S_k \subset V_{k+1}$ and the definition of k, P_k is an non-empty path. Also, none of the cycles S_{k-1} through S_1 can separate S_k from S_0. Therefore, the remaining part of P, i.e. $P - P_k$, can be simulated just by a single edge of G^T from the transitive closures of directed connections from $S_k \cap V$ to $S_0 \cap V$ in G, and directed connections between $S_k \cap V$ and $S_k \cap V$, and between $S_0 \cap V$ and $S_0 \cap V$ in G (see Definition 3.1). By Definition 3.1, the edge has length equal to the length of shortest directed path in G between its endpoints. Now, we can analogously argue for the minimum index l greater than k such that P touches or intersects S_l, showing that the path fragment $P_k - P_l$ can be simulated by a single edge of G^T, etc. It follows that there is a directed path Q_0 in G^T of the same starting and ending point as P_0 which has $O(m)$ edges, and length not greater than that of P_0. Note that $O(m) = O(\log n)$. Symmetrically, we can show that a final fragment of P, say P'_0, starting at a vertex of S_0 can be simulated by a path in G^T, say Q'_0, which has $O(\log n)$ edges, and length not greater than that of P'_0. The two paths Q_0 and Q'_0, if necessary, can be glued together by a single edge in G^T originating from the transitive closure of connections from $S_0 \cap V$ to $S_0 \cap V$ in G. ∎

We define the Γ functional by $\Gamma_s(n) = s(n) + \Gamma_s(\lceil \frac{2}{3} n \rceil + s(n))$, $\Gamma_s(k) = O(1)$ for $k \leq 10$.

Remark 3.1: Let T be a simple cycle $O(\sqrt{n})$ separator tree for a planar $O(1)$-supergraph of a planar digraph G. The graph G^T has $O(n \log n)$ edges.

Proof: Recall Definition 3.1. Note that the transitive closure of the union of the graphs L, R has $\Gamma_s(n)$ vertices and $O(\Gamma_s(n)^2)$ edges, where $s(n) = O(\sqrt{n})$. Observe that for $s(n) = O(\sqrt{n})$, $\Gamma_s(n) = O(\sqrt{n})$. It follows by induction that at the $j-th$ recursion level, G^T is augmented by $O(2^j \sqrt{n/2^j} \times \sqrt{n/2^j})$ edges. Consequently, G^T has $O(n \log n)$ edges. ∎

Before proving the following, key lemma, let us recall that $M(m)$ denotes the number of processors required in the model of CRCW PRAM to compute the min, + product of two $m \times m$ matrices with integer entries of polynomial in m size in time $O(\log^2 m)$.

Lemma 3.1: Let T be a simple cycle $O(\sqrt{n})$ separator tree for a planar $O(1)$-supergraph of a planar digraph G with nonnegative integer edge lengths of polynomial size. Given T, we can compute G^T in time $O(\log^4 n)$ using a CRCW PRAM with $O(n^{\frac{1}{4}} M(\sqrt{n}))$ processors.

Proof: Definition 3.1 specifies the way of computing G^T. Recursively compute $G_l^{T_l}$ and $G_r^{T_r}$. The two paths from the root of T to the leaves q_l, q_r, respectively labelled with the separators $S_0, S_1, ..., S_k$, and $S_0, S'_1, ..., S'_p$, can easily be found in time $O(\log n)$ since T is of logarithmic depth. Then, the edges of the graph L, and their edge lengths can be determined as follows (the construction of the graph R is analogous, and thus omitted). First, for each vertex v in $S_0 \cap V$, we find all vertices in $\bigcup_{j=1}^l S_j \cap V$ reachable from v in G_l, and the lengths of shortest directed paths from v to them. Let K be the set of $O(\sqrt{n})$ leaves in T_l labelled with sets containing the vertices in $S_0 \cap V$. Next, let W_l be a minimal subtree of T_l including all leaves from K, and all nodes of T_l labelled with the separators $S_1, ..., S_k$.

For each vertex v in V_l that occurs in a separator labelling a node m in W_l, we shall consider the set $R(v)$ of pairs (w, p) in V_l such that:

1) w occurs in separators labelling these nodes in W_l that are ancestors of m,

2) w is reachable from v in G_l,

3) p is the length of a shortest directed path from v to w in G_l.

Let h be the height of W_l. We shall construct the sets $R(v)$ by induction on $h - i$ where i is the level in W_l at which a separator containing v occurs.

First, for $i < j \le h$, we determine the set $U_j(v)$ of vertices in V_l that are direct successors of v in G_l, and occur in separators labelling the $j - th$ level of W_l. Note that the sets $R(v)$, $U_j(v)$, $i < j \le h$, are of size $O(\sqrt{n})$.

Now, suppose that the sets $U_j(v)$, $i < j \le h$, $R(w)$, $w \in U_j(v)$, are given for all vertices v in V_l that occur in separators labelling the $i - th$ level of W_l. To determine the sets $R(v)$, for the above vertices v, first, we compute the union $\{(w, length(v, w)) \mid \exists i < j \le h \, \exists w \in U_j(v)\} \cup \{(u, p + length(v, w)) \mid \exists i < j \le h \, \exists w \in U_j(v) \, (u, p) \in R(w)\}$. Then, we lexicographically sort the union, and delete all pairs that do not minimize the second coordinate among the pairs with the same first coordinate. To compute the union, we determine the sets $S(v, j) = \{(u, p + length(v, w)) \mid \exists w \in U_j(v) \, (u, p) \in R(w)\}$, where $i < j \le h$, v is in V_l and occurs in a separator labelling the $i - th$ level of W_l, as follows. For each node m of W_l on the $j - th$ level, we distinguish the set $B(m)$ of vertices v that occur in separators labelling the nodes on the $i - th$ level of W_l for which m is an ancestor node. Let $s(j) = \left(\frac{2}{3}\right)^{h-j}$. Note that the set $\bigcup_{v \in B(m)} U_j(v)$ has, as a subset of the separator labelling the node m, size $O(\sqrt{s(j)})$. Further, the set $\{(u, p) \mid \exists v \in B(m) \, \exists w \in U_j(v) \, (u, p) \in R(w) \, \& \, p$ is the minimum length of a path from v to u in $G_l \}$ is of size $O(\sqrt{n})$ as the first coordinates of pairs are vertices from separators labelling the path from m to the root in W_l. Therefore, computing this set easily reduces to a min + product of two matrices of sizes $| B(m) | \times O(\sqrt{s(j)})$ and $O(\sqrt{s(j)}) \times O(\sqrt{n})$ respectively.

Let $mi = \min\{| B(m) |, \sqrt{s(j)}, \sqrt{n}\}$. Since $s(j) \le n$, we have $mi \le \sqrt{n}$. The matrix multiplication can be done in time $O(\log^2 n)$ using a CRCW PRAM with $O((| B(m) | /mi) \cdot (\sqrt{s(j)}/mi)^2 (\sqrt{n}/mi) M(mi))$ processors. Since $M(m) = \Omega(m^2)$, we may estimate the latter term from above by $O(| B(m) | \sqrt{n}/s(j)) M(\sqrt{s(j)}))$.

Further, we shall use the following observation:

(*) Let $m(k) = \min\{O(\sqrt{n}), 2^k\}$. The sum of cardinalities of the separators labelling the $i - th$ level of W_l can be estimated by $O(m(i)\sqrt{n/m(i)})$, and consequently by $O(\sqrt{m(i)}\sqrt{n})$.

As the total size of the sets $B(m)$ over all nodes m on the $j - th$ level of W_l is $O(\sqrt{m(i)}\sqrt{n})$, we can compute all the sets $\bigcup_{w \in U_j(v)} R(w)$ for the vertices in V_l associated with the $i - th$ level of W_l in time $O(\log^2 n)$ using a CRCW PRAM with $O(n\sqrt{m(i)}M(\sqrt{s(j)})/s(j)))$ processors. By $M(m) = O(m^{1.5})$, the monotonicity of $M(\cdot)$ and straight-forward calculations, the maximum of the upper bound on the number of required processors is achieved for $h - i = 0.5 \log n + O(1)$, and $h - j = O(1)$ and turns out to be $O(n^{\frac{1}{4}} M(\sqrt{n}))$. As W_l has $O(\log n)$ levels, it takes $O(\log^3 n)$ time, and $O(n^{\frac{1}{4}} M(\sqrt{n}))$ processors totally.

Note that computing $R(v)$ for v in $V \cap S_0$ solves the problem of determining the lengths of shortest paths originating from v and terminating in $S_1 \cap V_1, ..., S_k \cap V$ only for these S_j which label nodes in W_l on the path from the leaf associated with v to the root of W_l. In general, by the proof of Theorem 3.1, a directed path from v to a vertex u in $S_j \cap V$ in G_l can be partitioned into two directed paths P_1 and P_2 as follows. P_1 leads from v to a vertex w in a separator labelling a node of Q that is an ancestor of the node labelled by S_j in W_l. Note that w occurs in the first coordinate of some pair in $R(v)$.

Let G'_l be the directed graph resulting from reversing the directions of edges in G_l. Define the set $R'(u)$ for G'_l analogously as the set $R(u)$ for G_l. Note that w occurs also in the first coordinate of some pair in $R'(u)$. Let $U = \bigcup_{i=1}^{k} S_j \cap V$. It follows that to find out whether u is reachable from v in G_l, and to determine the length of a shortest directed path from v to u in G_l it is enough to compute $\min\{r \mid \exists w \in U \exists p, q \in N \ r = p + q \ \& \ (w, p) \in R(v) \ \& \ (w, q) \in R'(u)\}$. We assume that the minimum of an empty set is $+\infty$. It is clear that we can compute the sets $R'(u)$, $u \in S_1 \cap V, ..., S_k \cap V$, applying to G'_l a procedure similar to that used to compute the sets $R(v)$ for vertices v in $S_0 \cap V_l$ associated with the leaves of W_l. It takes $O(\log^3 n)$ time on a CRCW PRAM with only $O(M(\sqrt{n}))$ processors, since S_1 through S_k label a single path from the root to a leaf in W_l.

Now, it remains to expand the sets $R(v)$ by pairs (u, r') where $u \in U$, $r' \neq +\infty$, $r' = \min\{r \mid \exists w \in U \exists p, q \in N \ r = p + q \ \& \ (w, p) \in R(v) \ \& \ (w, q) \in R'(u)\}$, for the $O(\sqrt{n})$ vertices v in S_0 occurring in the sets labelling the leaves of W_l.

Observe that the union $\bigcup_{u \in U} R'(u)$ is of cardinality $O(\sqrt{n})$. Recall also that the sets $R(v)$ are of size $O(\sqrt{n})$. Therefore, we can compute the extensions of $R(v)$ for the vertices v in $S_0 \cap V$, by running $O(\sqrt{n})$ times sorting algorithm on integer sequences of length $O(\sqrt{n})$. It can be done in time $O(\log n)$ on an EREW PRAM with $O(\sqrt{n} \cdot \sqrt{n})$ processors [C]. Summarizing, we can find the lengths of shortest directed paths starting from $S_0 \cap V$ and ending in $S_1 \cap, ..., S_k \cap V$ in G_l in time $O(\log^3 n)$ using a CREW PRAM with $O(n^{\frac{1}{4}} M(\sqrt{n}))$ processors. The dual problems of finding the lengths of shortest directed paths starting from $S_1 \cap V, ..., S_k \cap V$ and ending in $S_0 \cap V$ in G_l can be solved within the same resource bounds by using the same method in G'_l instead of G_l.

Finally, the lengths of shortest directed paths starting from $S_0 \cap V$ and ending in G_l can be solved as follows. First, we determine the sets $R'(v)$ for vertices in $S_0 \cap V$ associated with the leaves of W_l in time $O(\log^3 n)$ using a CREW PRAM with $O(n^{\frac{1}{4}} M(\sqrt{n}))$ processors. For each node m of W_l, we compute the set $C(m)$ of pairs $((v, u), r)$ such that $r \neq +\infty$, $\exists w \in V \ \exists p, q \in N$, $r = p + q \ \& \ (w, p) \in R(v) \ \& \ (w, q) \in R'(u)$, and m is an ancestor of the leaves in W_l whose labels include v and u, respectively. Recall that the labels of the leaves in W_l are of size $O(1)$. Let t be the number of leaves in W_l that are descendents of the node m. It follows that $C(m)$ can be determined by computing the min, + product of two $O(t) \times O(\sqrt{n})$ matrices. It can be done in time $O(\log^2 n)$ using a CRCW PRAM with $O((\sqrt{n}/t)^2 M(t))$ processors. Since W_l has $O(\sqrt{n})$ leaves, and $O(\log n)$ levels, all the sets $C(m)$ can be computed in time $O(\log^2 n)$ using a CRCW PRAM with $O(M(\sqrt{n}) \log n)$ processors.

We conclude that the whole graph L, and analogously the graph R, can be determined in time $O(\log^3 n)$, using a CRCW PRAM with $O(n^{\frac{1}{4}} M(\sqrt{n}))$ processors. Since each of these graphs has $O(\sqrt{n})$ vertices, solving APSPP for their union takes $O(\log^3 n)$ time using a CRCW PRAM with $O(M(\sqrt{n}))$ processors (recall the definition of $M(\cdot)$ and see [KaR]). Putting everything together, G^T can be recursively computed in time $O(\log^4 n)$, using a CRCW PRAM with $O(n^{\frac{1}{4}}M(\sqrt{n}))$ processors. ∎

Now, combining the above lemma and observation with Lemma 2.3, we obtain the following corollary.

Theorem 3.2: Let G be a planar digraph with nonnegative integer edge lengths of polynomial size which is connected in the undirected graph sense. We can compute a simple cycle $O(\sqrt{n})$ separator tree T for a planar $O(1)$-supergraph of G and the digraph G^T in time $O(\log^4 n)$ using a CRCW PRAM with $O(n^{\frac{1}{4}}M(\sqrt{n}))$ processors.

Proof: By Lemma 2.3, we can compute the tree T in time $O(\log^3 n)$ using a CRCW PRAM with $O(n^{1+\epsilon})$ processors, for a given, arbitrary positive ϵ. On the other hand, by Lemma 3.1, the graph G^T in the planar case can be computed in time $O(\log^4 n)$ using a CRCW PRAM with $O(n^{\frac{1}{4}}M(\sqrt{n}))$ processors. By choosing $\epsilon < 0.24$, we obtain the thesis. ∎

4. SSSPP

For a digraph $G = (V, E)$, by the *relative radius* of G, we shall mean the maximum of the minimum number of edges on a directed path from v to w taken over all pairs v, $w \in V$ such that there is a directed path from v to w in V.

By Theorem 3.1, G^T has a small, logarithmic relative radius. This makes it possible to apply the following straight-forward algorithm for SSSPP in G^T.

Algorithm 1: SSSPP

Input: a directed graph G given by tables $P_i(k)$, $W_i(k)$, $i = 1, ..., n$, $k = 1, ..., j_i$, such that j_i is the number of direct predecessors of the $i - th$ vertex in G, for $k = 1, ..., j_i$, $P_i(k)$ is the $k - th$ direct predecessor of the $i - th$ vertex in G, and $W_i(k)$ is the length of the edge directed from $P_i(k)$ to the $i - th$ vertex; an upper bound l on the relative radius of G.

Output: for $i = 1, ..., n$, the length of a shortest path from the first vertex to the $i - th$ vertex in G.

$m(1) \leftarrow 0$;
for $i = 2, ..., n$ **do**
$m(i) \leftarrow +\infty$;
for $m = 1, ..., l$ **do**
begin
    for $i = 2, ..., n$ **do in parallel**
    $m(i) \leftarrow \min_{k=1}^{j_i}(m(P_i(k)) + W_i(k))$
end

Remark 4.1: Let G be a directed graph on n vertices and m edges, with the relative radius bounded by l from above. SSSPP for a given vertex of G can be solved in time $O(l \log n)$ using an EREW PRAM with $O((n + m)/\log n)$ processors.

Proof: Create the input tables $P_i(k)$, $W_i(k)$ to run Algorithm 1, by sorting the list of edges of G in time $O(\log n)$ using an EREW PRAM with $O(n)$ processors [C]. It is clear that in each iteration of the body of the main loop, $m(i)'s$ can be recomputed in $O(\log n)$ time using a CREW PRAM with $\sum_{i=2}^{n} j_i$ processors. The use of concurrent read can be eliminated by duplicating the values to read in additional $O(\log n)$ time

[KaR]. The number of processors can be decreased by a $O(\log n)$ factor using the known method of grouping arguments of symmetric functions into blocks with size logarithmically bounded. For each such a block, we sequentially compute the minimum, and then we finish computing $m(i)$ in a standard tree-like fashion. The grouping into blocks of logarithmic size can be done using a standard parallel algorithm for prefix sums running in $O(\log n)$ time on an EREW PRAM with $O(n/\log n)$ processors [KaR]. ∎

First of all, we can apply Algorithm 1 to parts of G^T already constructed in order to build recursively G^T. In this way, we obtain another upper bound on the time-processor cost of constructing G^T, completing that derived in Lemma 3.1.

Lemma 4.1: Let $G = (V, E)$ be a planar digraph with nonnegative integer edge lengths of polynomial size. Given a simple cycle $O(\sqrt{n})$ separator tree T for a planar $O(1)$-supergraph of G, we can compute G^T in time $O(\log^3 n)$ using a CREW PRAM with $O(n^{1.5}\log n)$ processors.

Proof: We follow Definition 3.1 to compute G^T. The two paths in T from the root to the leaves q_l, q_r can easily be found in time $O(\log n)$ since T is of logarithmic depth. Recursively compute $G_l^{T_l}$ and $G_r^{T_r}$. Then, for each vertex s in $(S_0 \cup \bigcup_{j=1}^l S_j) \cap V$ solve the SSSP problem in G_l, by running in parallel $\Gamma_s(n)$ times Algorithm 1. Analogously, for each vertex s' in $(S_0 \cup \bigcup_{j=1}^r S_j') \cap V$ solve the SSSP problem in G_r. In this way, we can produce the graphs L and R. Since both G_l and G_r have logarithmic relative radius, the multi-running of Algorithm 1 takes $O(\log^2 n)$ time and totally $O(size(G^T)\Gamma_s(n))$ processors by Remark 4.1. Note that each of the graphs L, R has at most $\Gamma_s(n)$ vertices. Therefore, by definition of $M(\cdot)$ and the known relatioship between min + product and APSPP (see [KaR]), solving APSPP for the union of L and R takes $O(\log^2 n)$ time and $O(M(\Gamma_s(n)))$ processors. It is known that $M(n) = O(n^{1.5})$ (see [PR89] p. 502). Now it suffices to recall that G^T has size $O(n\log n)$ by Remark 3.1, and $\Gamma_s(n) = O(\sqrt{n})$ to complete the proof of the lemma. ∎

Applying Algorithm 1 to G^T, where T is a simple cycle $O(\sqrt{n})$ separator tree for a planar $O(1)$-supergraph of G, we can efficiently solve SSSPP for a distinguished vertex of G. Hence, we obtain the first main result of our paper.

Theorem 4.1: Let G be a planar digraph with nonnegative integer edge lengths of polynomial size. We can preprocess G in time $O(\log^4 n)$ using a CRCW PRAM with $O(\min\{n^{1.5}, n^{\frac{1}{4}}M(\sqrt{n})\})$ processors such that given a vertex v of G, we can solve SSSPP for v in time $O(\log^2 n)$ using an EREW PRAM with $O(n)$ processors.

Proof: Assume first that the undirected graph corresponding to G is connected. Then, we can proceed as follows. By Theorem 3.2 and Lemma 4.1, given a simple cycle $O(\sqrt{n})$ separator tree T for a planar $O(1)$-supergraph of G, we can construct the graph G^T in time $O(\log^4 n)$ using a CRCW PRAM with $O(\min\{n^{\frac{1}{4}}M(\sqrt{n}), n^{1.5}\})$ processors. We get rid of the $\log n$ factor at $n^{1.5}$ by slowing down the method of Lemma 4.1 by an $\Omega(\log n)$ factor. Recall that G^T has $O(n\log n)$ edges. Now, by Theorem 3.1, it is sufficient to apply Algorithm 1 to G^T with v as the first vertex in order to solve SSSPP in G. By Remark 4.1, it takes time $O(\log^2 n)$ on an EREW PRAM with $O(n)$ processors.

If the undirected graph corresponding to G is not connected, we can find its connected components in time $O(\log n)$ using a CRCW PRAM with $O(n)$ processors [SV]. Now, the original problem trivially reduces to SSSPP for v restricted to the digraph induced by this component. ∎

Our result applied to planar digraphs with unit edge lengths immediately yields the corresponding result on bredth-first search. We can even substantially decrease the number of processors used in the preprocessing for bredth-first search in case the input digraph has a radius of length $O(n^\delta)$ where $\delta < 0.04$. Simply, then any shortest path in the digraph is of length $O(n^\delta)$, and in the proof of Lemma 3.1 we need compute min + products only for matrices with nonnegative integer entries bounded by $O(n^\delta)$. By using Romani's reduction of the problem of computing the min + matrix product to that of the arithmetic one [Ro], this can be done using $O(n^{2.367+\delta})$ bit operations (see [M] p. 156 and [CW]). In particular, it can be done in time $O(\log^2 n)$ on a CREW PRAM with $O(n^{2.367+\delta})$ processors. Putting everything together, and observing that $2.367/2 + 0.25 < 1.44$, we obtain our second main result.

Theorem 4.2: For an arbitrary vertex v of planar digraph with radius $O(n^\delta)$, a breadth-first search tree rooted at v can be constructed in time $O(\log^4 n)$ using a probabilistic CRCW PRAM with $O(\min\{n^{1.44+\delta}, n^{1.5}, n^{\frac{1}{4}}M(\sqrt{n})\}$ processors.

By solving SSSPP for all vertices in a digraph, we can solve APSPP for the digraph. In this way, by slightly modifying the proof of Theorem 4.1, we obtain our third main result.

Theorem 4.3: Let G be a planar digraph with nonnegative integer edge lengths of polynomial size. We can solve the APSPP problem for G in time $O(\log^4 n)$ using $O(n^2/\log^2 n)$ processors.

Proof: The proof is similar to that of Theorem 4.1. The difference is that we apply n times Algorithm 1 in G^T starting from the n vertices of G in parallel. By Remark 4.1, this increases the processor count to $O(n \cdot n)$ and takes $O(\log^2 n)$ time. By slowing Algorithm 1 by an $\Omega(\log^2 n)$ factor, we can reduce the number of processors to $O(n^2/\log^2 n)$. If the undirected graph corresponding to G is not connected than we find its connected components as in the proof of Theorem 4.1 and apply Algorithm 1 to the digraphs induced by them. ∎

Note that our bounds for APSPP planar digraphs match those from [PR85, PR89] for undirected planar graphs.

If we neglect the distance aspect in the proof of Theorem 4.1 and Theorem 4.3, we can derive parallel theorems on the problem of reachability in planar graphs. Note that then we can replace the min + matrix product by the Boolean one in the proof of Lemma 3.1. Since the latter product efficiently reduces to the arithmetic one over the ring Z_{n+1} (see [AHU] p. 243, [KaR]), we can decrease the number of processors used in the preprocessing to $O(n^{1.44})$. Hence, our fourth main theorem is as follows.

Theorem 4.4 : We can preprocess a planar digraph such that for any its vertex v, we can report all vertices reachable from v by directed paths in the digraph in time $O(\log^2 n)$ using an EREW PRAM with $O(n)$ processors. The preprocessing takes $O(\log^4 n)$ time using a probabilistic CRCW PRAM with $O(n^{1.44})$ processors. The transitive closure graph of a planar digraph can be constructed in time $O(\log^4 n)$ using a CRCW PRAM with $O(n^2/\log^2 n)$ processors.

Acknowledgements: I would like to express my appreciation to John Gilbert for telling me about the matrix multiplication and transitive closure bottleneck in parallel algorithms for digraphs, and to Alok Aggarwal, Ming Kao for and John Reif for their valuable comments.

References

[AHU] A.V. Aho, J.E. Hopcroft and J.D. Ullman, *The Design and Analysis of Computer Algorithms* (Addison-Wesley, Reading, Massachusetts, 1974).

[C] R. Cole, *Parallel Merge Sort*, Proc. 27th Ann. Symp. on Foundations of Computer Science.

[CW] D. Coppersmith, S. Winograd, *Matrix Multiplication via Arithmetic Progressions*, Proc. 28th Annual ACM Symp. on Theory of Computing, 1987, pp. 1-6.

[G] H. Gazit, *personnal communication*, 1989.

[GM] H. Gazit and G.L. Miller, *A Parallel Algorithm for Finding a Separator in Planar Graphs*, Proc. 28th Symp. on Foundations of Computer Science, 1987.

[KaR] R.M. Karp and V. Ramachandran, *A Survey of Parallel Algorithms for Shared-Memory Machines* , Technical Report UCB-CSD 88-408, Computer Science Division, EECS, University of California at Berkeley, 1988. To appear in the Handbook of Theoretical Computer Science, North-Holland.

[KK] M.Y. Kao, P.N. Klein, *Towards overcoming the transitive-closure bottleneck: Efficient parallel algorithms for planar digraphs*, to appear in Proc. 22nd Annual ACM Symp. on Theory of Computing, 1990.

[KlR] P.N. Klein, J.H. Reif, *An Efficient Parallel Algorithm for Planarity*, Proc. 27th Symp. on Foundations on Computer Science, 1986.

[KS] M.Y. Kao, G. Shannon, *Local Reorientation, Global Order, and Planar Topology*, Proc. 21st Annual ACM Symp. on Theory of Computing, Seattle, 1989.

[LT] R.J. Lipton and R.E. Tarjan, *A Separator Theorem for Planar Graphs*, SIAM J. Appl. Math. 36 (1979), pp. 177-189.

[M] K. Mehlhorn, *Data Structures and Algorithms 2: Graph Algorithms and NP-Completeness* Springer Verlag, 1984, pp. 155-158.

[Re] J. Reif, *personnal communication*, 1989.

[Ro] F. Romani, *Shortest-path problem is not harder than matrix multiplication*, Information Processing Letters, 11(1980), pp. 134-136.

[PR85] V.P. Pan and J. Reif, *Extension of the Parallel Nested Dissection Algorithm to the Path Algebra Problems*, Technical Report 85-9, Comput. Sci. Dept., SUNYA, Albany, NY, 1985.

[PR89] V.P. Pan and J. Reif, *Fast and Efficient Solution of Path Algebra Problems*, JCSS 38 (1989), pp. 494-510.

[SV] Y. Shiloach and U. Vishkin, *An $O(\log n)$ Parallel Connectivity Algorithm*, J. Algorithms 3, 1, 57-67.

Parallel Algorithms for Finding Steiner Forests in Planar Graphs
(A preliminaly version)

Hitoshi Suzuki, Chiseko Yamanaka and Takao Nishizeki

Department of Information Engineering, Faculty of Engineering
Tohoku University, Sendai 980, Japan

Abstract. Given an unweighted planar graph G together with nets of terminals, our problem is to find a Steiner forest, i.e., vertex-disjoint trees, each of which interconnects all the terminals of a net. This paper presents four parallel algorithms for the Steiner forest problem and a related one. The first algorithm solves the problem for the case all the terminals are located on the outer boundary of G in $O(\log^2 n)$ time using $O(n^3/\log n)$ processors on a CREW PRAM, where n is the number of vertices in G. The second algorithm solves the problem for the case all terminals of each net lie on one of a fixed number of face boundaries in poly-log time using a polynomial number of processors. The third solves the problem for the case all terminals lie on two face boundaries. The fourth finds a maximum number of internally disjoint paths between two specified vertices in planar graphs. Both the third and fourth run either in $O(\log^2 n)$ time using $O(n^6/\log n)$ processors or in $(\log^3 n)$ time using $O(n^3/\log n)$ processors.

1. Introduction

Several routing problems such as VLSI river routing and single-layer routing can be formulated as a problem of finding a Steiner forest in a planar graph. A set of terminals (vertices) that are all to be interconnected is called a *net*. Given an undirected graph G together with nets of terminals, our problem is to find a *Steiner forest*, i.e., a set of vertex-disjoint trees, each of which interconnects all the terminals of a net. The well-known Steiner tree problem is to find a minimum tree interconnecting all specified terminals in a given weighted graph. However our Steiner forest is not required to minimize the total weight (or number) of edges. Thus our problem is not a generalization of the Steiner tree problem but of the disjoint path problem. Since the disjoint path problem is NP-complete even for planar graphs [Lyn] or plane grids [KL,Ric], so is our problem if there is no restriction on the location of terminals.

Robertson and Seymour [RS] showed that the Steiner forest problem can be solved sequentially in polynomial time if G is planar and all the terminals are located on two face boundaries. Suzuki, Akama and Nishizeki [SAN] gave an efficient sequential algorithm which runs in $O(n\log n)$ time, where n is the number of vertices in G. On the other hand Schrijver showed that a Steiner forest of a planar graph can be found in $O(n^{h+2}\log^2 n)$ time if all the terminals lie on a fixed number h of face boundaries [Sch1,Sch2].

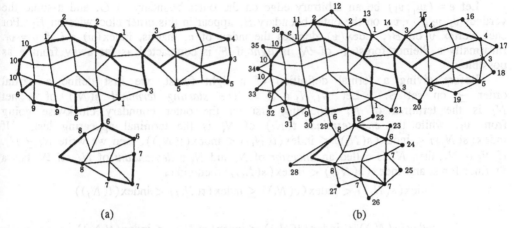

Fig. 1 (a) A planar graph G with a net set on the outer boundary, and (b) a modified
graph obtained from G by applying step (1) in FOREST1.

In this paper, we present four parallel algorithms for the Steiner forest problem and a
related one on planar graphs. The first algorithm FOREST1 finds a Steiner forest of a
planar graph G for the case that all terminals are located only on the outer boundary of G.
Fig. 1(a) depicts a planar graph G and a Steiner forest, where all the terminals of ten nets
lie only on the outer boundary, and a Steiner forest of ten disjoint trees is drawn in thick
lines. FOREST1 runs in $O(\log^2 n)$ time and uses $O(n^3 / \log n)$ processors on a CREW
PRAM model. The other three algorithms use FOREST1 as a subroutine. The second
algorithm FORESTh finds a Steiner forest of a planar graph for the case that all terminals
of each net lie on one of a fixed number h of face boundaries. The algorithm runs in
poly-log time using a polynomial number of processors. The third algorithm FOREST2
finds a Steiner forest of a planar graph for the case that all terminals lie on two face
boundaries. The fourth algorithm finds a maximum number of internally disjoint paths
between two specified vertices in a planar graph. Both the third and fourth algorithms
run either in $O(\log^2 n)$ time using $O(n^6 / \log n)$ processors or in $O(\log^3 n)$ time using
$O(n^3 / \log n)$ processors.

2. Tight Steiner forests and lemmas

In this section we define some terms and present several lemmas on which our
algorithms are based.

Let $G = (V, E)$ be an undirected planar graph with vertex set V and edge set E. We
sometimes write $V = V(G)$ and $E = E(G)$. Let n be the number of vertices in G, that
is, $n = |V|$. We assume that G is connected and embedded in the plane. We denote
by B_1 the outer boundary of G. For two graphs G and G', $G + G'$ means a graph
$(V(G) \cup V(G'), E(G) \cup E(G'))$. A set of vertices on G are designated as *terminals*. A
net is a set of terminals that are all to be interconnected. A *net set* $S = \{N_1, N_2, ..., N_k\}$ is
a partition of the set of terminals. Then a *network* $\mathcal{N} = (G, S)$ is a pair of a planar graph
G and a net set S. A *Steiner forest of network* $\mathcal{N}$ is a forest $F = T_1 + T_2 + \cdots + T_k$ in G
such that $N_i \subseteq V(T_i)$ for each tree T_i in F. Hereafter we assume that there exists a Steiner
forest in a given network $\mathcal{N}$. In this and next sections, we assume that all the terminals lie
on the outer boundary B_1. Furthermore in this section, we assume that the outer boundary
of G is a simple cycle and every vertex on the outer boundary is designated as a terminal.
We will show later in Section 3 that this assumption does not lose any generality.

Let $e = (v_{b_1}, v_1)$ be an arbitrary edge on the outer boundary of G, and assume the vertices $v_1, v_2, \cdots, v_{b_1}$ on the outer boundary B_1 appear in this order clockwise on B_1. For each vertex v on B_1, index(v) denotes the index of v, that is, index(v) = i if $v = v_i$. Informally a Steiner forest F of $\mathcal{N}$ is *tight* if F is compacted as far away from e as possible.

Before giving a formal definition of a tight forest, we first define a partial order $\preceq$ on set S. Let $N_i, N_j \in S$. The *starting terminal* $s(N_i)$ of a net N_i is the terminal of N_i appearing first on the outer boundary clockwise going from v_1, while the *end terminal* $t(N_i)$ of N_i is the terminal appearing last. If index($s(N_j)$) $\leq$ index($s(N_i)$) $\leq$ index($t(N_i)$) $\leq$ index($t(N_j)$), then we write $N_i \preceq N_j$. If $N_i \prec N_j$, then N_j is called an *ancestor* of N_i and N_i a *descendant* of N_j. If $\mathcal{N}$ has a Steiner forest and index($s(N_j)$) $<$ index($s(N_i)$), then either

$$\text{index}(s(N_j)) < \text{index}(s(N_i)) \leq \text{index}(t(N_i)) < \text{index}(t(N_j))$$

or

$$\text{index}(s(N_j)) \leq \text{index}(t(N_j)) < \text{index}(s(N_i)) \leq \text{index}(t(N_i)).$$

The minimum (youngest) ancestor of N_i, if exists, is called the *parent* of N_i and N_i is called a *child* of the parent.

We are now ready to define a tight forest. A Steiner forest F of $\mathcal{N}$ is *tight for edge e* or simply *tight* if, for every net $N_i \in S$ and every vertex $v \in V(T_i) - V(B_1)$, there is a face whose boundary contains both v and a vertex on trees of N_i's children. One can easily prove the following lemma.

LEMMA 1. If a network $\mathcal{N} = (G, S)$ has a Steiner forest and e is any edge on the outer boundary of G, then $\mathcal{N}$ has a Steiner forest tight for e. ∎

Suzuki, Akama and Nishizeki [SAN] gave a linear-time sequential algorithm for finding a Steiner forest of network $\mathcal{N} = (G, S)$. The algorithm finds a Steiner forest by repeating the following steps (1) and (2):

(1) for a net N_i which has no child, find the walk on the outer boundary of G going clockwise from $s(N_i)$ to $t(N_i)$, and let T_i be a spanning tree of the walk;

(2) remove the walk from G, and let G be the resulting graph.

The algorithm above can be rather easily implemented to run sequentially in linear time [SAN]. However, it seems that the algorithm cannot be easily transformed into a poly-log time algorithm on a PRAM. In the next section we will give a poly-log time algorithm which finds all trees T_i in parallel using a shortest path algorithm.

For two vertices u and u' in a planar graph G, a *vf-path* between u and u' is a sequence of vertices $w_0 w_1 \cdots w_m$ such that $w_0 = u$ and $w_m = u'$, and w_i and w_{i+1} lie on the same inner face boundary for every i, $0 \leq i < m$. The *length* of a vf-path $w_0 w_1 \cdots w_m$ is m. The *distance* $d(u, u')$ between two vertices u and u' is the minimum length of a vf-path between u and u'.

For every terminal v in a net N_j, we denote by anc(v, i) the ith ancestor of the net N_j. Thus, if N_j has exactly l ancestors, then

$$\text{anc}(v, i) = \begin{cases} N_j, & i = 0; \\ \text{the parent of anc}(v, i-1), & 1 \leq i \leq l; \\ N_\infty, & l < i, \end{cases}$$

where $N_\infty \notin S$ is a virtual net such that $N \preceq N_\infty$ for every net $N \in S$.

For every vertex $v \in V(G)$, $F(v)$ denotes the youngest net (under $\preceq$) among all the nets anc($u, d(u, v)$), $u \in V(B_1)$.

The following two lemmas guarantee that $F(v)$ is well-defined and immediately yields a Steiner forest of $\mathcal{N}$.

LEMMA 2. If $\mathcal{N} = (G, S)$ has a Steiner forest, then for every vertex $v \in V$ the set $C(v) = \{\mathrm{anc}(u, d(u, v)) | u \in V(B_1)\}$ is totally ordered under the relation $\preceq$, that is, either $\mathrm{anc}(u, d(u, v)) \preceq \mathrm{anc}(u', d(u', v))$ or $\mathrm{anc}(u', d(u', v)) \preceq \mathrm{anc}(u, d(u, v))$ holds for every two vertices u and u' on the outer boundary of G.

PROOF. Suppose, for a contradiction, that $C(v)$ contains two distinct nets N_i and N_j such that neither $N_i \preceq N_j$ nor $N_j \preceq N_i$. Let $N_i = \mathrm{anc}(u, d(u, v))$ and $N_j = \mathrm{anc}(u', d(u', v))$. Let R be a vf-path between u and v of length $d(u, v)$, and R' be a vf-path between u' and v of length $d(u', v)$. Let $R^* = R + R'$ be a vf-path connecting u and u'. Then $|V(R^*)| = d(u, v) + d(u', v) + 1$.

On the other hand, each net $N_l \in \{\mathrm{anc}(u, i) | 0 \le i \le d(u, v)\} \cup \{\mathrm{anc}(u', i) | 0 \le i \le d(u', v)\}$ is separated by the vf-path R^*, that is, any tree connecting N_l must occupy a vertex on R^*. Since $\{\mathrm{anc}(u, i) | 0 \le i \le d(u, v)\} \cap \{\mathrm{anc}(u', i) | 0 \le i \le d(u', v)\} = \phi$, the vertices on R^* must be occupied by $d(u, v) + d(u', v) + 2$ different trees, a contradiction. ∎

LEMMA 3. For every net N_i, all the terminals in N_i belong to a single component of the subgraph of G induced by $V_i = \{v \in V | F(v) = N_i\}$.

PROOF. Let F_t be a Steiner forest of $\mathcal{N}$, tight for edge e. Let T_i be the tree of F_t connecting N_i, and let v be any vertex on T_i. It suffices to prove $F(v) = N_i$ under the assumption that $F(w) = N_j$ for every descendant N_j of N_i and every vertex w on the tree T_j in F_t.

We first prove that $F(v) \preceq N_i$. If $v \in V(B_1)$, then $F(v) \preceq \mathrm{anc}(v, d(v, v)) = \mathrm{anc}(v, 0) = N_i$. On the other hand, if $v \notin V(B_1)$, then there is a face whose boundary contains both v and a vertex w on the tree of a child N_j of N_i. Since there is a vertex u on the outer boundary such that $\mathrm{anc}(u, d(u, w)) = N_j$ and since $d(u, v) \le d(u, w) + 1$, $F(v) \preceq \mathrm{anc}(u, d(u, v)) \preceq N_i$.

We then prove that $F(v) = N_i$. Suppose to the contrary that $F(v) \prec N_i$. Let $F(v) = N_j$, let u' be a vertex on the outer boundary such that $\mathrm{anc}(u', d(u', v)) = N_j$, and let Q be a vf-path between u' and v of length $d(u', v)$. Since u' is a terminal of a descendant of N_j, $Q - \{v\}$ intersects the tree T_j of N_j at a vertex w. However by the assumption above, we have $F(w) = N_j$. Therefore $F(v) \succ N_j$, a contradiction. ∎

3. Algorithm and complexity

The input to our problem is an embedding list of the given planar graph G together with a net set S. We find a Steiner forest of a network $\mathcal{N} = (G, S)$ as follows:

procedure FOREST1($\mathcal{N}$);
begin
(1)   modify the given graph G so that the outer boundary is a simple cycle and every vertices on the boundary is designated as a terminal, and compute index(v) for every vertex v on the outer boundary;
(2)   compute $\mathrm{anc}(v, i)$ for every vertex v on B_1 and every integer i, $0 \le i \le n$;
(3)   compute $d(u, v)$ for every two vertices u and v;
(4)   compute $F(v)$ for every vertex v on G;
(5)   construct the subgraph $G_s = (V, \{(u, v) \in E | F(u) = F(v)\})$, find a spanning forest F in G_s, remove from F all the trees containing no terminals, and output the resulting graph as a Steiner forest of $\mathcal{N}$.
end.

We now show the detail of each step in FOREST1.

(1) We first rank the edges of the outer boundary in clockwise order with starting from $e = (v_{b_1}, v_1)$. Assign to each terminal v the minimum rank of edges which join v and a clockwise next vertex on the outer boundary. Assume that there are m terminals in $\mathcal{N}$ and let $r_1, r_2, ..., r_m$ be their ranks listed in increasing order. For each two terminals v and v' with ranks r_i and r_{i+1}, add a new vertex to G, join the vertex with v and v', and add to net set S a new net consisting of only the vertex, where $r_{m+1} = r_1$ (see Fig. 1(b)). Compute index(v) for every vertex v on the new outer boundary. This step (1) can be done total in $O(\log n)$ time using $O(n)$ processors.

(2) For each net N_i, compute $s(N_i)$ and $t(N_i)$, and compute the parent $p(N_i)$ of N_i as follows. Initially set $p(N_i)$ to be the net N_j which contains the terminal v with index$(v) = $ index$(t(N_i)) + 1$ if index$(t(N_i)) < b_1$; and set $p(N_i) = N_\infty$ if index$(t(N_i)) = b_1$, where b_1 is the number of vertices on the outer boundary B_1 (or the number of terminals in S). Then repeat the following procedure $\log n$ times:

> **for** each net N_i **in parallel do**
> **if** $p(N_i) \not\vdash N_i$ **then** $p(N_i) := p(p(N_i))$.

Using $p(N_i)$ computed above, one can easily compute anc(v, i) for every vertex v on the outer boundary and every integer i, $0 \leq i < n$. This step (2) can be done total in $O(\log n)$ time using $O(n^2)$ processors.

(3) For every face of G, find all the vertices on the boundary of the face. Then one can compute $d(u, v)$ for all pairs of vertices u and v in G by using an algorithm for finding shortest paths between all pairs of vertices. The algorithm runs in $O(\log^2 n)$ time using $O(n^3 / \log n)$ processors [GR].

(4) Using the results of (2) and (3) we can compute $F(v)$ for every vertex $v \in V$ in $O(\log n)$ time using $O(n^2)$ processors.

(5) One can construct G_s in $O(\log n)$ time using $O(n)$ processors. Furthermore one can find a spanning forest of G_s in $O(\log^2 n)$ time using $O(n^2 / \log^2 n)$ processors [CLC].

Thus we have the following theorem.

THEOREM 1. A Steiner forest in a plnanar graph can be found in $O(\log^2 n)$ time using $O(n^3 / \log n)$ processors on a CREW PRAM if all the terminals lie on the outer face boundary. ∎

4. Forest connecting terminals on a fixed number of face boundaries

In this section, we present an algorithm FORESTh for finding a Steiner forest in a network in which all the terminals of every net are located on one of h face boundaries $B_1, B_2, ..., B_h$. We can find a Steiner forest in such a network $\mathcal{N} = (G, S)$ by using the following algorithm FORESTh.

procedure FORESTh;
begin
    let $S_i \subseteq S$ be the net set on B_i, $1 \leq i \leq h$;
    for each permutation $B_{i_1}, B_{i_2}, ..., B_{i_h}$ of the h face boundaries and each combination of edges $e_{i_1} \in E(B_{i_1}), e_{i_2} \in E(B_{i_2}), ..., e_{i_{h-1}} \in E(B_{i_{h-1}})$ **in parallel do**
        begin
            for each j, $1 \leq j \leq h - 1$ **do**
                begin

find a Steiner forest F_{i_j}, tight for edge e_{i_j}, in network $\mathcal{N}_{i_j} = (G, S_{i_j})$
by using FOREST1;
remove the forest F_{i_j} from the graph G, that is, $G := G - F_{i_j}$
    end;
find a Steiner forest F_{i_h} tight for an arbitrary edge $e_{i_h} \in E(B_{i_h})$ in network
$\mathcal{N}_{i_h} = (G, S_{i_h})$ by using FOREST1;
check whether $F = F_1 + F_2 + \cdots + F_h$ is a Steiner forest of $\mathcal{N}$ or not
    end
end;

One can easily prove that a Steiner forest of $\mathcal{N}$ can be found in at least one of the parallel executions in FORESTh if $\mathcal{N}$ has a Steiner forest. Thus we have the following theorem.

THEOREM 2. If all terminals of each net lie on one of h face boundaries, then a Steiner forest of $\mathcal{N}$ can be found in $O(h \log^2 n)$ time using $O(n^{h+2} / \log n)$ processors, and hence in poly-log time using a polynomial number of processors if h is constant. ∎

5. Forests connecting terminals on two face boundaries

In this section we present algorithms for finding a Steiner forest of a network in which all the terminals lie on two specified face boundaries B_1 and B_2. We first present an algorithm for the special case in which every net has exactly two terminals, one on B_1 and the other on B_2, and then present an algorithm FOREST2 for a general case.

Consider a network $\mathcal{N} = (G, S)$ with $S = \{\{s_1, t_1\}, \{s_2, t_2\}, ..., \{s_k, t_k\}\}$. One may assume that terminals $s_1, s_2, ..., s_k$ appear clockwise on B_1 in this order, and $t_1, t_2, ..., t_k$ appear clockwise on B_2 in this order. See Fig. 2(a). One can find a Steiner forest of $\mathcal{N}$ by appling FOREST1 to new networks $\mathcal{N}_i^* = (G^*, S_i)$ constructed from $\mathcal{N}$ as follows.

Let R be a shortest vf-path going from s_1 to t_1, and let the vertices $v_1 = s_1, v_2, v_3, ..., v_l = t_1$ appear on R in this order. We may assume that R has the minimum length l among the shortest vf-paths between s_j and t_j for all j, $1 \le j \le k$.

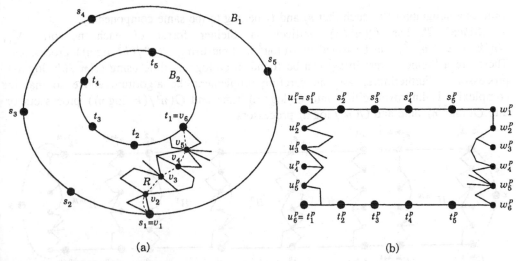

(a)   (b)

Fig. 2 (a) A graph G with a set of nets each of which has exactly two terminals, one on B_1 and the other on B_2, and (b) graph H^p obtained from G.

Otherwise, renumber the indices of s_j and t_j. Then we have $l \leq n/k$ because, otherwise, there would exist no Steiner forest in $\mathcal{N}$. For each vertex v_q on R, replace v_q by two new vertices u_q and w_q, replace each edge (v_q, v') which is incident with v_q and is to the left of R by (u_q, v'), and replace each of the other edges (v_q, v') by (w_q, v'). Fig. 2(b) depicts the modified graph obtained from the graph in Fig. 2(a). Denote by H^p, $-3n/k \leq p \leq 3n/k$, multiple copies of the modified graph. We write each vertex in H^p with attaching superscript p, like w_q^p. Identify w_q^p in H^p and u_q^{p+1} in H^{p+1} for every p, $-3n/k \leq p \leq 3n/k - 1$, and q, $1 \leq q \leq l$. Add to the resulting graph two new vertices x and y, join x with $s_1^{-3n/k}$ and $t_1^{-3n/k}$, and join y with $s_k^{3n/k}$ and $t_k^{3n/k}$. Let G^* be the resulting graph. See Fig. 3. For an integer i let

$$
\begin{aligned}
S_i^* = &\{\{s_j^p, t_j^{p+i}\} | 1 \leq j \leq k \text{ and } |p| \leq 2n/k\} \\
&\cup \{\{s_j^p\} | 1 \leq j \leq k \text{ and } |p| > 2n/k\} \\
&\cup \{\{t_j^p\} | 1 \leq j \leq k \text{ and } |p - i| > 2n/k\} \\
&\cup \{\{x\}\} \cup \{\{y\}\}.
\end{aligned}
$$

Then network $\mathcal{N}_i^* = (G^*, S_i^*)$ has all the terminals on the outer face boundary. One can prove the following two lemmas.

LEMMA 4. If $\mathcal{N}$ has a Steiner forest, then at least one of $\mathcal{N}_i^*$, $-n/k^2 \leq i \leq n/k^2$, has a Steiner forest. ∎

LEMMA 5. Assume that $F(v)$ for every vertex v in G^* is computed for edge $e = (x, s_1^{-3n/k})$ by applying FOREST1 to $\mathcal{N}_i^*$. For every q, $1 \leq q \leq l$, if $F(u_q^0) = \{s_j^p, t_j^{p+i}\}$ then $F(w_q^0) = \{s_j^{p+1}, t_j^{p+i+1}\}$. ∎

Therefore, for each j, $1 \leq j \leq k$, the vertex set

$$\{v \in V(G) | v^0 \in V(H^0), F(v^0) = \{s_j^p, t_j^{p+i}\} \text{ for some } p, -n/k \leq p \leq n/k\}$$

induces a subgraph of G such that s_j and t_j belong to the same component.

Since G^* has $O(n^2/k)$ vertices, a Steiner forest of each network $\mathcal{N}_i^*$, $-n/k^2 \leq i \leq n/k^2$, can be found in $O(\log^2 n)$ time using $O(n^6/(k^3 \log n))$ processors. Therefore a Steiner forest in $\mathcal{N}$ can be found in $O(\log^2 n)$ time using $O(n^7/(k^3 \log n))$ processors. Furthermore, we can carefully implement the algorithm above so that the complexity is improved either into $O(\log^2 n)$ time and $O(n^6/(k^3 \log n))$ processors or into $O(\log^3 n)$ time and $O(n^3/\log n)$ processors.

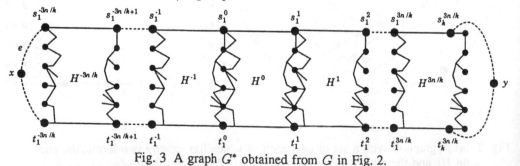

Fig. 3 A graph G^* obtained from G in Fig. 2.

Using the algorithm above, we next construct an algorithm FOREST2 for finding a Steiner forest in a network $\mathcal{N} = (G, S)$ for the case all the terminals are located on two face boundaries B_1 and B_2 and at least one net intersects with both boundaries. Note that a net in S may three or more terminals.

procedure FOREST2;
begin
(0)   Let (S_1, S_2, S_{12}) be a partition of the net set S such that

$$N \in S_1 \Rightarrow N \subseteq V(B_1),$$

$$N \in S_2 \Rightarrow N \subseteq V(B_2), \text{ and}$$

$$N \in S_{12} \Rightarrow N \cap V(B_1) \neq \phi \text{ and } N \cap V(B_2) \neq \phi.$$

(1)   Let N be a net in S_{12}, and let $e_1 \in E(B_1)$ and $e_2 \in E(B_2)$ be edges which are incident with terminals in N. Find a Steiner forest F_1 tight for e_1 in network $\mathcal{N}_1 = (G, S_1)$ and also find a Steiner forest F_2 tight for e_2 in network $\mathcal{N}_2 = (G, S_2)$. Remove F_1 and F_2 from G.

(2)   Let $S_1' = \{N \cap V(B_1) | N \in S_{12}\}$ and $S_2' = \{N \cap V(B_2) | N \in S_{12}\}$. Then G has a set of disjoint walks on B_1 which is a Steiner forest F_1' of $\mathcal{N}_1' = (G, S_1')$, and also has a set of disjoint walks on B_2 which is a Steiner forest F_2' of $\mathcal{N}_2' = (G, S_2')$. Contract all the edges in F_1' and F_2'. Then, in the resulting network $\mathcal{N}'$, every net has exactly two terminals, one on the outer boundary and the other on an inner face boundary.

(3)   Find $|S_{12}|$ disjoint paths in network $\mathcal{N}'$ by applying the algorithm presented in this section.
end;

See [SAN] for the correctness of the algorithm above. The following lemma holds.

THEOREM 4. If every terminal lie on two face boundaries, then a Steiner forest can be found either in $O(\log^2 n)$ time using $O(n^6/(k^3 \log n))$ processors or in $O(\log^3 n)$ time using $O(n^3/\log n)$. ∎

6. Internally disjoint paths

In this section, we briefly explain an algorithm for finding a maximam number of internally disjoint paths between two specified vertices s and t in a planar graph G. First we compute the maximam number of paths. The following theorem holds.

THEOREM 5 [Men]. Let s and t be two non-adjacent vertices of an undirected graph G. Then G has k internally disjoint paths if and only if there is no s-t separating set X with $|X| < k$. ∎

We may assume without loss of generality that s lies on the outer face boundary of G. Furthermore we may assume that s and t are not adjacent. If s and t are adjacent, then a maximum number of internally disjoint paths in G can be constructed from a maximum number of internally disjoint paths in graph $G - (s, t)$ by adding the path of a single edge (s, t). The minimum number of vertices in an s-t separating set can be computed by computing all pair shortest vf-paths in a certain graph similar to H^p in the previous

section. Therefore the maximum number k of internally disjoint paths can be computed in $O(\log^2 n)$ time using $O(n^3/\log n)$ processors.

We next show how to find k internally disjoint paths in G. The problem can be reduced to a Steiner forest problem in a new graph G' constructed from G as follows. Let $u_1, u_2, ..., u_{b_1}$ be the vertices adjacent with s in counterclockwise order, and let $w_1, w_2, ..., w_{b_2}$ be the vertices adjacent with t in counterclockwise order. Assume that graph $G - \{s, t\}$ is embedded in a plane doughnut $\{[x, y] \in \mathbb{R}^2 | 2 \le (x^2 + y^2)^{1/2} \le 3\}$, and that $u_i = [3\cos(2\pi i/b_1), 3\sin(2\pi i/b_1)] \in \mathbb{R}^2$ and $w_i = [2\cos(2\pi i/b_2), 2\sin(2\pi i/b_2)] \in \mathbb{R}^2$ for every u_i and v_i. Let

$$u'_{i,j} = [(3 + j/(2k))\cos(\pi i/b_1), (3 + j/(2k))\sin(\pi i/b_1)], \text{ and}$$
$$w'_{i,j} = [(2 - j/(2k))\cos(\pi i/b_2), (2 - j/(2k))\sin(\pi i/b_2)]$$

for every $i, j \in \mathbb{Z}$. Note that $u'_{2i,0} = u_i$ and $w'_{2i,0} = w_i$. Let

$$
\begin{aligned}
V_1 &= \{u'_{2i,j} | 1 \le i \le b_1, 0 \le j \le 2k\} \cup \{u'_{2i-1,j} | 1 \le i \le b_1, 1 \le j \le 2k\}, \\
V_2 &= \{w'_{2i,j} | 1 \le i \le b_2, 0 \le j \le 2k\} \cup \{w'_{2i-1,j} | 1 \le i \le b_2, 1 \le j \le 2k\}, \\
E_1 &= \{(u'_{i,j}, u'_{i+1,j}) | 1 \le i \le 2b_1, 1 \le j \le 2k\} \\
&\quad \cup \{(u'_{2i,j}, u'_{2i,j-1}) | 1 \le i \le b_1, 1 \le j \le 2k\} \\
&\quad \cup \{(u'_{2i-1,j}, u'_{2i-1,j-1}) | 1 \le i \le b_1, 2 \le j \le 2k\}, \text{ and} \\
E_2 &= \{(w'_{i,j}, w'_{i+1,j}) | 1 \le i \le 2b_2, 1 \le j \le 2k\} \\
&\quad \cup \{(w'_{2i,j}, w'_{2i,j-1}) | 1 \le i \le b_2, 1 \le j \le 2k\} \\
&\quad \cup \{(w'_{2i-1,j}, w'_{2i-1,j-1}) | 1 \le i \le b_2, 2 \le j \le 2k\},
\end{aligned}
$$

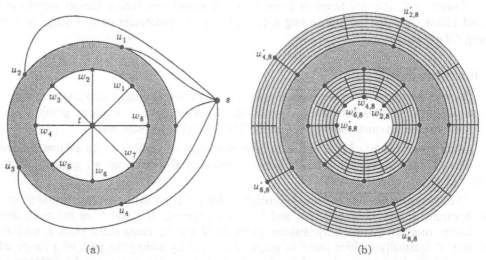

(a) (b)

Fig. 4 (a) A graph G with two specified vertices s and t and (b) a graph G' obtained from G with net set $\{\{u'_{2,8}, w'_{2,8}\}, \{u'_{4,8}, w'_{4,8}\}, \{u'_{6,8}, w'_{6,8}\}, \{u'_{8,8}, w'_{8,8}\}\}$, where $k = 4$.

where $u'_{2b_1+1,j} = u'_{1,j}$ and $w'_{2b_2+1,j} = w'_{1,j}$. Then $G' = (V(G - \{s,t\}) \cup V_1 \cup V_2, E(G - \{s,t\}) \cup E_1 \cup E_2)$. Fig. 4 depicts a given graph G and graph G'. Let $S' = \{\{u'_{2i,2k}, w'_{2i,2k}\} | 1 \leq i \leq k\}$, and let $\mathcal{N}' = (G', S')$. Then one can prove the following lemma.

LEMMA 6. A planar graph G has k internally disjoint paths between two specified vertices s and t if and only if network $\mathcal{N}'$ has a Steiner forest, that is, k disjoint paths. ∎

One can easily obtain k internally disjoint paths in G from a Steiner forest of $\mathcal{N}'$. Since graph G' has $O(nk)$ vertices, algorithm FOREST2 finds a Steiner forest either in $O(\log^2 n)$ time using $O(n^6 k^3 / \log n)$ processors or in $O(\log^3 n)$ time using $O(n^3 k^3 / \log n)$ processors. Furthermore, we can carefully implement the algorithm so that the number of processors are reduced into $O(n^6 / (k^3 \log n))$ and $O(n^3 / \log n))$, respectively.

Khuller and Schieber [KS] gave an algorithm which, given a positive integer k, finds k internally disjoint paths between two specified vertices in a general undirected graph. Their algorithm runs in $O(\log n)$ time using $O(m\alpha(m, n) / \log n)$ processors if k is constant, where m is the number of edges in a graph and $\alpha(m, n)$ is the functional inverse of Ackermann's function. On the other hand, our algorithm finds a *maximam number* of internally disjoint paths in poly-log time, although the algorithm is valid only for planar graphs.

Acknowledgment

We would like to thank Professor N. Saito for fruitful discussions.

References

[CLC] F. Y. Chin, J. Lam and I. Chen, Efficient parallel algorithms for some graph problems, Communications of the ACM 25, 9 (1982).

[GR] A. Gibbons and W. Rytter, Efficient Parallel Algorithms, Cambridge University Press, Cambridge (1988).

[KS] S. Khuller and B. Schieber, Efficient parallel algorithms for testing connectivity and finding disjoint s-t paths in graphs, Proc. 30th IEEE Symp. on Foundations of Computer Science, pp. 288-293 (1989).

[KL] M. R. Kramer and J. van Leeuwen, Wire-routing is NP-complete, Report No. RUU-CS-82-4, Department of Computer Science, University of Utrecht, Utrecht, the Netherlands (1982).

[Lyn] J. F. Lynch, The equivalence of theorem proving and the interconnection problem, ACM SIGDA Newsletter 5:3, pp. 31-65 (1975).

[Men] K. Menger, Zur allgemeinen Kurventheorie, Fund. Math., 10, pp. 95-115 (1927).

[RS] N. Robertson and P. D. Seymour, Graph minors. VI. Disjoint paths across a disc, Journal of Combinatorial Theory, Series B, 41, pp. 115-138 (1986).

[Sch1] A. Schrijver, Disjoint homotopic trees in a planar graph, manuscript(1988).

[Sch2] A. Schrijver, personal communication (1988).

[SAN] H. Suzuki, T. Akama and T. Nishizeki, Finding Steiner forests in planar graphs, Proc. 1st ACM-SIAM Symp. on Discrete Algorithms, pp. 444-453(1990).

Optimally Managing the History of an Evolving Forest

Vassilis J. Tsotras *
Department of Electrical Engineering
Center for Telecommunications Research
Columbia University
NY, NY 10027

B. Gopinath
Dept. of Electrical and Computer Engineering
and Department of Computer Science
Rutgers University
New Brunswick, NJ 08903

George W. Hart
Department of Electrical Engineering
Center for Telecommunications Research
Columbia University
NY, NY 10027

ABSTRACT

There are many applications where data to be archived have the form of an evolving forest. In this paper we address the problem of efficiently reconstructing the history of a forest evolving in time. A forest is a set of trees and at each time instant a constant number of changes occur in the forest. We propose an optimal, dynamic algorithm that reconstructs the state of any subtree in the forest at a given instant t in the past, in $O(|s(t)| + \log\log T)$ time where $|s(t)|$ is the cardinality of the requested subtree and T is the total length of the evolution. The space used is linear to the total number of changes that occurred during the forest's evolution. If a total of $O((\log T)^{1/c})$ parallel processors is available, where c is a constant ($c \geq 1$), the $\log\log T$ factor is eliminated from the previous time bound. Finally, we optimally address the following query: given a path p in the forest, find all the lifetimes of this path during the forest's evolution.

----

* *This research was supported in part by National Science Foundation Grant CDR 881111.*

1. Introduction

For many databases that incorporate versioning, and for object oriented databases that implicitly or explicitly maintain the history of persistent objects, it becomes necessary to manage the history of data structures such as trees. The problem of efficiently reconstructing the history of a time-evolving forest is presented in this paper. A forest is a set of trees, and each tree is a collection of edges. For the purposes of this paper, time is assumed to be discrete and described by a succession of nonnegative integers: $1, ..., T$. At each time instant a constant number of changes may occur on the evolving forest. A change is the result of one of the following two operations: a) addition of a new edge in the forest, or b) the deletion of an already existing edge. The deletion of an edge also results in the deletion of the whole subtree (if any) under the deleted edge. The time instants corresponding to the addition of a new edge in the forest and its later deletion (if any), create a lifetime interval for this edge. For any time instant t, if there is a lifetime interval of a given edge that contains t, then this edge is called existing at t. If a new edge e is added under an edge f, then edge e (respectively f) is called a son (father) of edge f (e). Observe that the trees in our model of evolution have a very general form. Trees with constant number of son edges per edge (binary, etc.) are then special cases.

At any time instant, there is a name uniqueness condition satisfied in the forest. Under this condition, each edge has a unique name among the edges that share the same father edge. Moreover, rebirths of an edge with name e are allowed. As a result of the name uniqueness condition, the lifetime intervals of the same edge do not overlap. If we name each path in the forest by the sequence of the names of its edges, then at each time instant the set of pathnames in the forest contains no duplicates. The existing edges in the forest at time t, constitute the state $s(t)$ of the forest at that instant. Our goal in this paper is to provide a data structure and an algorithm for efficiently reconstructing the state of the forest at any given time in the past. Throughout this paper we use Yao's cell-probe model for storage [14]. The length of a word (cell) of memory is assumed to be $\log T$ bits, where T is the length of the evolution. If the state $s(t)$ was stored for every t in an array of T entries, then reconstruction of any past state would be possible in $O(1 + |s(t)|)$ time, just by accessing the appropriate entry of the array and writing its contains on the output. But such an approach would require space proportional to $O(T^2)$, as births of new edges may continuously enlarge the system's state.

Intuitively, the minimal information needed to reconstruct any forest's state would consist of the changes that occurred in the forest evolution, starting with an empty (or some initial) state at time zero. This straightforward approach, even if it uses minimal space, increases the reconstruction time to $O(\min(t, T\text{-}t))$, as the reconstruction can be forward or backward. In this paper we show that managing the history of an evolving forest in space linear to the number of changes is possible in a less expensive way. We present an *on-line* algorithm that reconstructs the state $s(t)$ in time $O(|s(t)| + \log\log T)$ while using only linear space. We actually prove more, i.e., the reconstruction time of our algorithm is optimal among the algorithms that use the same space requirements. Moreover it is observed that if a total of $O((\log T)^{1/c})$ parallel processors is available, where c is a positive constant ($c \geq 1$), the $\log\log T$ factor can be disregarded from the time bound.

One of the advantages of our approach is its on-line nature. The proposed algorithm has two phases, the Update and the Query phases. In the update phase the algorithm produces and updates its data structures by "observing" the forest evolution. The input to the algorithm in this phase consists of the changes occurring in the evolving forest. By the on-line behavior of our algorithm, we mean that for any change occurring at some path p in the forest evolution, the algorithm needs at most $O(l)$ time to update its data-structures, where l is the length of path p. Since any algorithm that is fed only with the changes occurring in the forest needs at least $O(l)$ time just for reading its input, our algorithm achieves the lower bound for the processing in the update phase. In the query phase, the input to the algorithm consists of a pathname p and a time instant t. The output is the subtree under path p as it was existing at time t. If the null pathname is provided, the whole forest at t is reproduced.

Previous work in history searching has been focused in the problem of reconstructing the state of an evolving system, where generally a system is a set of elements. The usual approach is by applying results from the field of computational geometry, as time can be transformed to the x-axis of the plane (see [2], [3], [6] and [8]). The disadvantage of these methods is the use of balanced tree schemes as data structures, forcing a non-on-line update phase and (or) a logarithmic query time. The case of more complicated structures representing the evolving system, such as a forest, has not yet been examined to the best of our knowledge.

Our approach is based on an optimal algorithm for the evolving system problem that appears in [11]. Section 2 provides the preliminaries of this algorithm, which for convenience we call the "one-level" algorithm. In section 3 we provide a direct extension of the one-level algorithm; however this algorithm does not achieve optimal query time. The proposed optimal algorithm appears in section 4, while section 5 provides an optimal approach to the query of finding the lifetimes of a given path. Finally section 6 contains conclusions and problems for further research.

2. Preliminaries

We proceed by discussing the one-level algorithm. The evolving system can be understood as an "evolving set". Starting with an empty set, at every time instant one of the following operations (or none) is possible: the addition of a new element in the system, or the deletion of an already existing element. The time instants corresponding to the addition of a new element and its later deletion (if any), create a lifetime interval for this element. At any time, the names of the existing elements are distinct. Rebirths of deleted elements are allowed. The state $s(t)$ of the system at time t is just the set of the elements that are existing at time t. Let T be the total length of the evolution, then our goal is to reproduce the state of the system at any given $t \leq T$. The one-level algorithm optimally answers the above query in $O(|s(t)| + \log\log T)$ time, using space linear to the number of changes occurred in the system's evolution. The algorithm has two phases, an update phase where it "observes" the evolution of the system and updates its data structures, and the query phase where the state of the system is reproduced for the given query t.

The data-structures used by the algorithm are the following: a doubly-linked list L, a queue Q, and array A that keeps the times of element additions to the system's state. There is also a *dynamic hashing function* [5], called *dhf-a*, used to store the different names of the ever added elements. The list L and queue Q are needed for creating the basic data-structure of the one-level algorithm, which for simplicity is called the *one-level forest*. As the name implies, this structure has the form of a forest. When a new element is added in the system's state, the algorithm creates an *entity-record*, where it keeps the information about the lifetime of this element. The name of the newly added element is placed at the end of array A and list L. Based on the element's lifetime, each entity record is placed in the one-level forest. An entity record contains the following fields: the name of the corresponding element, its birth time (the time of its addition in the system's state), its deletion time, a pointer to its father entity in the one-level forest, a pointer P_s to its own list of son entities in the same forest (called the SN list) and a pointer P_L to its position in the list L.

If an element, say with name w, ever gets deleted, the algorithm locates the element's entity record in $O(1)$ through the *dhf-a* and updates the deletion entry. Based on the element's position in list L (which is found through pointer P_L), the entity record is placed in the one-level forest. If there is another element f before w in the list, then it is placed as a son of element f, and w's father entry is updated accordingly. Similarly the name w is added on the son list of f. If the deleted element is in the beginning of the list, it is placed in the current end of queue Q. It then gets deleted from list L (see Fig. 1). It is shown in [11] that the one-level forest is updated in $O(1)$ time per change. The algorithm is able to reconstruct any past state in time $O(|s(t)| + \log\log T)$. The query phase has two steps. Firstly, given the query t, the birth time closest to it is located inside array A; this task is performed in $O(\log\log T)$ by using an on-line version of Willard's algorithm [13]. To implement this algorithm, we use a set of $\log T$ *dhf's* [5], therefore the update phase is expected on-line. The expectation is over a randomization and does not assume any distribution of the input data. Then by travelling along the one-level forest, the state $s(t)$ is reconstructed in $2\, s(t)$ steps, thus establishing the above bound.

3. An Extended Algorithm

The described one-level algorithm will be used as a module for managing the history of more general evolving systems. From the previous discussion it is obvious that the history of an evolving system can be managed and reconstructed efficiently if the memory addresses of the following data-structures are known: the *dhf-a* which accesses the system's elements by name, the array A which keeps the elements' births, list L, and queue Q. We call a record that keeps pointers to the above memory locations the "history-record" of the system. The history of multiple, independently evolving systems can be accessed individually, if every system has a history-record of each own.

Returning to the evolving forest problem, we should note that a deletion operation applied on an edge e results in the deletion of the whole subtree under e, if any, but such possible multiple deletions count as only one change in our model of evolution. Firstly we will describe an algorithm, called the "extended" one, which is a direct extension of the one-level algorithm and achieves an $O(|s_a(t)| \log\log T)$ query time. Here $s_a(t)$ is the set of all edges in the given path and underneath it, that were existing at the query time t. The space complexity is linear to the total number of changes.

The algorithm has two phases, the update and the query one. During the update phase it produces and updates its data structures, which we call the *access forest*, in order to distinguish from the actual evolving forest. The access forest contains information about all the history of the actual forest's evolution. The input of the algorithm in the update phase is nonempty whenever a change takes place in the actual forest and consists of a pathname followed by the kind of operation applied to it.

By applying the one-level algorithm as a module for the extended algorithm, we can visualize each edge to be a system by itself and its son edges playing the role of the system's elements. Therefore each edge has its own history-record for managing and reconstructing the history of its son edges. Similarly, there is a history-record, called $h(root)$, which accesses the history of the forest's roots. Record $h(root)$ contains a pointer to a dynamic hashing function, $dhf(root)$, that accesses the history-record of each root by name. Another pointer leads to array $A(root)$ where the births (and rebirths) of the forest's roots are kept. Similarly there are pointers for list $L(root)$ and queue $Q(root)$, which play the same role as list L and queue Q of the one-level case, storing now only names of the roots. Figure 2 provides an example of the access forest.

When a root with name a is firstly born at time t_1, a $find(a)$ operation is applied to $dhf(root)$ providing a negative answer. Operation $insert(a)$ will return a free memory location where a pointer to the history-record of the root with name a is stored. Since rebirths of root a are allowed, the history-record contains one more pointer that points to an array called $life(a)$, from where the current and future birthtimes of root a are accessible in increasing birthtime order. Array $life(a)$ enables the extended algorithm to check if a given edge a was existing at a given t. As there is always exactly one copy of a given edge existing at any time, the lifetime intervals of an edge's copies do not overlap. If t falls inside such an interval, then a copy of that edge was present at time t.

Returning to our example, after the history-record $h(a)$ is created, a record called the entity-record for the current root a, is also made. An entity-record contains the name of root a, its birth time, its deletion time, a pointer to its father in the forest structure that the one-level algorithm will produce for the evolution of the roots (therefore this father is another root, taken from list $L(root)$), pointers P_s and P_L, and two more pointers $lastQ$ and $lastL$, whose use will be explained later. After the creation of the entity-record for the root with name a, the next free locations of array $A(root)$, array $life(a)$ and list $L(root)$ store a pointer towards this entity-record. The birth time entry of this record is filled with the current time t_1 and pointer P_L points to the position of a inside list $L(root)$. The same procedure is followed when a son edge b is created (Fig. 2).

Let us assume that at time t_3 root a is deleted. Then the input of the algorithm consists of: t_3, a, deletion. By applying $find(a)$ on the $dhf(root)$, the algorithm locates the history record of root a, and then the entity record of the last copy of a is visited, through array $life(a)$. There the deletion time entry is updated with t_3. By using the P_L pointer of the entity record, the location of root a inside list $L(root)$ is found and the father entry is appropriately filled. The deletion of root a in the actual forest also results in the deletion of the whole subtree under a. To update the entities of all the edges under the deleted edge about their deletion might be very costly, destroying the on-line behavior of our algorithm. Therefore we choose to update only the edge given as deleted in the algorithm's input. Such a choice does not deteriorate the effectiveness of our algorithm in providing the correct answer. A basic property of the actual evolving tree provides the justification for our strategy: whenever an edge is existing, all the edges in the path leading to this edge, are also existing. The negation of the above leads to a simple property of the algorithm's access structure: if an edge is found non-existing at a given t, no other edge in its subtree was existing at time t.

Future rebirths of an edge will only provide new entity records. Therefore, if at some later instant t_4 root a is reborn, the algorithm will locate again the history record of the roots with name a, and by visiting the last location in $life(a)$ can check that this occurrence of a is a new one. Then a new entity record will be created for the new copy of root a, and pointers to it will be stored in the next free locations of arrays $life(a)$ and $A(root)$.

The entries *lastQ* and *lastL* of the entity record will store pointers to the currently last positions of $L(a)$ and $Q(a)$. Such pointers allow the algorithm to use the same list $L(a)$ and queue $Q(a)$ for all the above son edges, as they define which part of $L(a)$ and $Q(a)$ corresponds to every copy of edge a. The one-level algorithm applied on a new copy of edge a, will visualize $L(a)$ and $Q(a)$ as starting from positions *lastL* and *lastQ* respectively.

The update phase continues similarly for every change in the actual forest. The basic data structure of our algorithm, named before as the access forest has actually the form of a tree. The root contains the history record *h(root)*, while the son edges are provided by all the elements stored in the *dhf(root)*. Branching continues similarly through the *dhf* of each root and so on. Roughly speaking, the access forest is the projection of all the different shapes that the actual forest obtained during its evolution. An obvious difference between them is that the actual forest can grow or shrink, while the access one can only grow (as it stores the whole history and not only the currently existing forest).

In the query phase, the given path p is firstly checked for consisting edges existing at the given time t. Again the algorithm applies the name of each edge in the path, on the appropriate *dhf*, starting from the path's root. If such an edge was ever created, the *dhf* will locate the *life()* array of that edge and in $O(\log\log T)$ steps the algorithm certifies if a copy of that edge was existing at the given time. The same procedure continues until a non-existing edge of the path or the final edge of it, say e, is reached. In the second case, the algorithm should also find if there were any edges in the subtree under edge e, existing at t. Such a search can be accomplished by recursive applications of the one-level algorithm. By locating time t in the array $A(e)$, in $O(\log\log T)$, the one-level algorithm can then provide the son edges of e, that were existing at t, by performing a number of steps proportional to the number of them. For each of the existing son edges, a subsequent application of the one-level algorithm will provide their existing son edges and so on.

The above strategy will exhaustively search the subtree under path p, finding all the existing edges at the given t. Since for each existing edge the one-level module will require time $O(|sons(t)| + \log\log T)$, with *sons(t)* being the set of its existing sons, the extended algorithm requires a total of $O(|s_a(t)| \log\log T)$ for reconstructing the subtree under path p. The time complexity stated before is thus established. The space complexity is also established by the previous discussion.

4. The Optimal Forest Algorithm

We will now present the "forest" algorithm which uses space linear to the total number of changes, has an update phase that is on-line in an amortized sense and can reconstruct the subtree under some given path in time $O(|s_a(t)| + \log\log T)$. The data structures used in the forest algorithm are basically the same as for the extended one. For simplicity of exposition, we firstly present a way to reconstruct the state of the whole forest at some time instant t. We assume also that there are only additions during the evolution. The case of deletions is examined later. Consider the example presented in Figure 3, where only the array $A(e)$ of each born edge e is included. When a new son edge is born under edge e, a new entry is created in array $A(e)$. Array $A(e)$ is actually an array of pointers to entity records. In this figure, for simplicity, we only provide the birth times and the names of these records. Furthermore, when the $(Ci + 1)th$ new entry (where C : constant and $i = 1,2, ...$) is created for array $A(e)$, a new entry is also created for array $A(f)$, edge f being the father of edge e.

The new entry in $A(f)$ is actually an artificial one as it does not correspond to any change in the evolution of the actual forest. This new entry contains the deletion of edge e and its simultaneous birth. Both the artificial deletion and birth are stamped with the current time. The algorithm also creates a pointer from the artificial entry of $A(f)$ to the son entry of $A(e)$ that caused it. We call such an update process "backward" updating, as it propagates a change that occurred on the environment of a son, to its father. The backward updating technique is adapted from the works by Chazelle[4] and Cole[5]. The constant C is called the "accumulating" factor as it serves as a threshold to the number of entries accumulated in an edge's array, before a backward update is allowed. It might as well be the case that every edge's array $A()$ on a path leading to a new edge is already at the threshold point, resulting to a "batch" of artificial edges, one per edge in the path. The time needed for such a backward updating is proportional to the path given at the input, thus keeping the on-line behavior of the forest algorithm. The space used by the algorithm is still linear to the number of changes Y, as we can prove [10] that the total number of artificial entries is $O(Y)$. Note that as the value of C increases, the length of the backward updating decreases.

We proceed our presentation with the modifications needed in the data structures of the extended algorithm in order to support the forest algorithm. We firstly note that since an $A(e)$ array is keeping pointers to the entity records of the son edges of the edge e, when such a son is deleted no new entry is created in array $A(e)$. The result of a deletion operation is simply the updating of the deleted edge's entity record. Therefore deletions in the actual evolution do not affect array $A(e)$ and the description of the forest algorithm. We will now prove that the query time of the forest algorithm in order to reconstruct the state of the whole forest at some given time t, is $O(|s(t)| + \log\log T)$. In order to deduct the above query time, it is sufficient firstly to prove that the artificial deletions and births of an edge do not affect the one-level algorithm which is used again as a module, and secondly, locating the given time t in the history is now done only once (using Willard's algorithm). For the one-level algorithm that is applied on array $A(f)$, list $L(f)$ and queue $Q(f)$, of the father edge of some edge e, the artificial deletion and birth of edge e at the same time, is "understood" as the deletion of an element with name e and the rebirth of that element at that time. But the one-level algorithm is capable of handling rebirths, as long as there is only one copy of every ball existing at any particular time, a condition that holds here. At the query phase the one-level algorithm will thus provide only one copy of each existing edge.

Applying a predecessor search for the query time t on array $A(root)$, the forest algorithm locates the last entry of the array that was created before or at time t. From $A(root)$, list $L(root)$ and queue $Q(root)$ the one-level algorithm will provide the entities of the roots existing at the given t. Each such entity r is accompanied by a pointer f_A to its array $A(r)$. Following this pointer, the algorithm visits an entry which by construction was created at most C units of time away from the given t. Since C is constant, time t is located in array $A(r)$ in constant time. Having located t in $A(r)$, the one-level algorithm is able to provide the sons of r that were alive at time t. The same procedure continues on the existing sons and so on. As a result the query time for reconstruction of the whole forest is $O(|s(t)| + \log\log T)$, where $s(t)$ is the set of all the existing edges at time t.

If the subtree under a path p is only needed, the answer is provided in $O(|s_s(t)| + \log\log T)$. Again $s_s(t)$ is the set of the edges in the given path and underneath it existing at t. As a result our algorithm is able to access the history of any subtree without having to reconstruct the history of the whole forest. We are thus able to prove the following theorem (for complete proof see [10]):

T h e o r e m : There is an on-line algorithm, that manages the history of an evolving forest in linear space and reconstructs the subtree under a given path, at a given instant t, in optimal time $O(|s_a(t)| + \log\log T)$.

A final note concerns the case where a set of $O((\log T)^{1/c})$ parallel processors is available, where c is a constant $(c \geq 1)$. It is shown in [18] that with such a processor set, the query time of Willard's algorithm is reduced to $O(1)$. As a result any predecessor query used in the forest algorithm can be answered in $O(1)$ time. Thus the algorithm's query time is reduced to $O(|s(t)|)$.

5. Finding the Lifetimes of a Given Path

Another query that the above algorithm can address efficiently, is the following: given a path p, find all of the time intervals (lifetimes) of this path during the forest's evolution. From the description of the previous algorithms we note that in order for the update phase to be on-line, when a deletion of an edge occurs in the actual forest, the algorithm updates this deletion only on the entity record of the deleted edge. The edges underneath are also deleted in the actual forest but are not updated in the algorithm's access structure. The current version of the forest algorithm can efficiently provide only the birth times of the paths lifetimes. If the path has ever existed in the forest's evolution, a *dhf* for each of its edges has already been created. Following these *dhfs*, the array *life()* of the path's last edge is visited, where pointers to the birth times of the path's lifetimes are stored in increasing order. In order to locate the deletion times, the algorithm has to compute the intersection of the lifetimes of all the edges in p. Such a computation can be very costly. We will prove that some extra constant update per change can solve the above query efficiently, i.e., in time $O(l + s)$, where l is the length of the path and s is the cardinality of the answer set. This query time is optimal as any algorithm solving the same problem needs at least the time to read the path and to provide the answer. The following simple strategy provides the solution: when an edge is reborn, update the previous copy of it if not yet updated about its deletion.

To facilitate this update strategy, every actual entity record of an edge has a pointer to the actual entity record of its father entity in the actual forest (note the difference of the actual father entity and the father that the one-level algorithm produces). The actual entity record of the previous copy of that edge is found through the previous entry of array *life()*. The following discussion is applied merely on the actual entity records pointed by the entries of arrays *life()*, thus is not affected by the artificial births and deletions of the forest algorithm presented above (which affect the *A()* arrays). Consider the example of Figure 4, (arrays *life()* are omitted): when at time 5, root a is deleted, the deletion times of the first entity records of edges b, c, and d are not informed. When later on at time 7, edge b is reborn, the entity record of its previous copy is visited and the deletion time of its father (time 5) is also stored in that entity. Similarly at time 10, the rebirth of edge c, updates the deletion time of the previous copy of c. The extra updating is still constant per change in the state of the actual forest.

Note that at any given instant only the last copy of a particular path remains unupdated. Therefore when the lifetimes of path p are requested, the algorithm locates the array *life()* of the last edge of path p, in the access structure. The entries of this array point to the actual entities of all the copies of this edge in increasing order of birth times. The lifetimes of path p are exactly the lifetimes of these entities. If the deletion time in the last entity's record is still empty, the algorithm has to check backwards the entity record of its actual father and so on, until a deletion time is found or if the root of the path is encountered (meaning that a copy of the path is still existing at the end of the evolution). These checks are at most l, the length of the path p, thus establishing the query time mentioned above.

6. Conclusions

In this paper we presented an optimal on-line algorithm that reconstructs any past state of an evolving forest in time $O(|s(t)| + \log\log T)$, where $|s(t)|$ is the cardinality of the answer set and T is the total evolution length. The space used is linear to the number of changes that occurred during the forest's evolution. We also optimally addressed the problem of finding all the lifetimes of a given path in the forest. The allowed operations on the evolving forest were the addition or the deletion of an edge. We are currently investigating efficient algorithms that manage the history of forests evolving under more complicated operations and the case of evolving graphs.

References

[1] M. Ajtai, "A Lower Bound for Finding Predecessors in Yao's Cell Probe Model", *Combinatorica*, to appear.
[2] B. Chazelle, "How to Search in History", *Inform. and Control*, 1985, Vol. 64, pp 77-99.
[3] B. Chazelle, "Filtering Search: A new Approach to Query Answering", 24th *IEEE FOCS*, 1983, pp 122-132.
[4] R. Cole, "Searching and Storing Similar Lists", *J. Algorithms*, Vol 7, 1986, pp 202-220.
[5] M. Dietzfelbinger, A. Karlin, K. Mehlhorn, F. Meyer, H. Rohnhert, R. Tarjan, "Dynamic Perfect Hashing: Upper and Lower Bounds", 29th *IEEE FOCS*, 1988.
[6] D.P. Dobkin, J.L. Munro, "Efficient Uses of the Past", 21st *IEEE FOCS*, 1980.
[7] M.L. Fredman, J. Komlos, E. Szemeredi, "Storing a Sparse Table with O(1) Worst Case Access Time", *JACM*, July 1984, Vol.31, pp 538-544.
[8] M. Overmars, "The Design of Dynamic Data Structures", *Springer Lecture Notes in Comp. Sc.* 156, Springer Verlag, Berlin, 1983.
[9] M. Overmars, "Range Searching on a Grid", *J. Algorithms*, Vol 9, 1988, pp 254-275.
[10] V.J. Tsotras, B. Gopinath, G. Hart, "Optimally Managing the History of an Evolving Forest", *TR 177 90-07 CTR-Columbia University, TM Rutgers University*, 1990.
[11] V.J. Tsotras, B. Gopinath, "Managing the History of Evolving Sets", submitted for publication, also *TR 176 90-06 CTR-Columbia University, TM Rutgers University*, 1990.
[12] V.J. Tsotras, B. Gopinath, G. Hart, "A New Bound on Parallel Searching", *IEEE 4th Ann. Symp. Parallel Processing*, April 4-6, 1990, Vol 2, pp 613-622.
[13] D.E. Willard, "Log-logarithmic Worst case Queries are Possible in Space O(n)", *Information Processing Letters*, 1983, Vol 17, pp 81-84.
[14] A.C. Yao, "Should Tables Be Sorted ?", *JACM*, July 1981, Vol.28, pp 615-628.

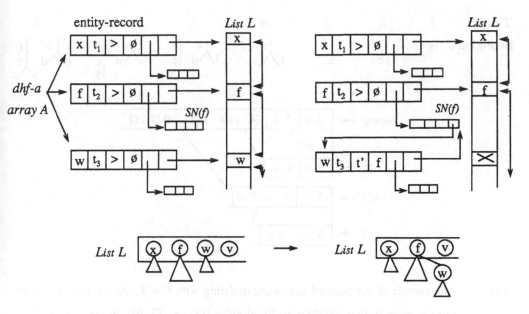

Fig. 1 : The data structures of the one-level algorithm, before and after a deletion

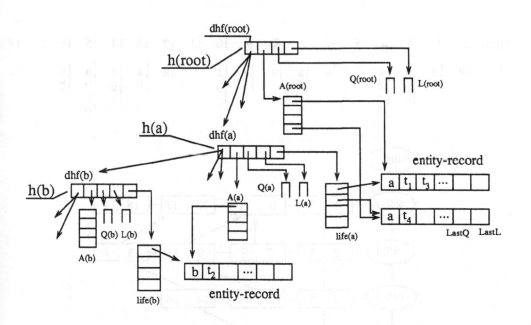

Fig. 2 : An example of the Access Forest

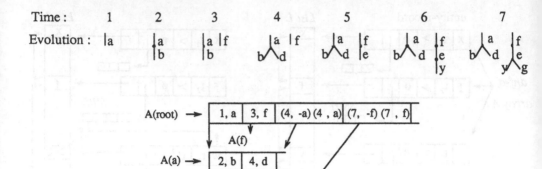

Fig. 3 : An example of the batched backward updating with $C = 2$. At $t = 7$, edge e gives birth to edge g, thus creating artificial entry $[(7, -e), (7, e)]$, in array $A(f)$. This entry overflows $A(f)$, resulting in the artificial entry $[(7, -f), (7, f)]$ in $A(root)$.

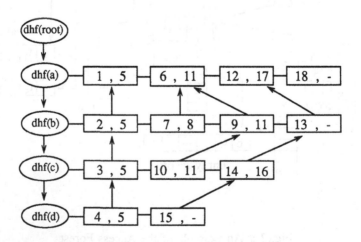

Fig. 4 : An example of the previous-entity updating strategy of the forest algorithm

Author Index

Vol. 408: M. Leeser, G. Brown (Eds.),Hardware Specification, Verification and Synthesis: Mathematical Aspects. Proceedings, 1989. VI, 402 pages. 1990.

Vol. 409: A. Buchmann, O. Günther, T. R. Smith, Y.-F. Wang (Eds.), Design and Implementation of Large Spatial Databases. Proceedings, 1989. IX, 364 pages. 1990.

Vol. 410: F. Pichler, R. Moreno-Diaz (Eds.), Computer Aided Systems Theory – EUROCAST '89. Proceedings, 1989. VII, 427 pages. 1990.

Vol. 411: M. Nagl (Ed.), Graph-Theoretic Concepts in Computer Science. Proceedings, 1989. VII, 374 pages. 1990.

Vol. 412: L. B. Almeida, C. J. Wellekens (Eds.), Neural Networks. Proceedings, 1990. IX, 276 pages. 1990.

Vol. 413: R. Lenz, Group Theoretical Methods in Image Processing. VIII, 139 pages. 1990.

Vol. 414: A.Kreczmar, A. Salwicki, M. Warpechowski, LOGLAN '88 – Report on the Programming Language. X, 133 pages. 1990.

Vol. 415: C. Choffrut, T. Lengauer (Eds.), STACS 90. Proceedings, 1990. VI, 312 pages. 1990.

Vol. 416: F. Bancilhon, C. Thanos, D. Tsichritzis (Eds.), Advances in Database Technology – EDBT '90. Proceedings, 1990. IX, 452 pages. 1990.

Vol. 417: P. Martin-Löf, G. Mints (Eds.), COLOG-88. International Conference on Computer Logic. Proceedings, 1988. VI, 338 pages. 1990.

Vol. 418: K. H. Bläsius, U. Hedtstück, C.-R. Rollinger (Eds.), Sorts and Types in Artificial Intelligence. Proceedings, 1989. VIII, 307 pages. 1990. (Subseries LNAI).

Vol. 419: K. Weichselberger, S. Pöhlmann, A Methodology for Uncertainty in Knowledge-Based Systems. VIII, 136 pages. 1990 (Subseries LNAI).

Vol. 420: Z. Michalewicz (Ed.), Statistical and Scientific Database Management, V SSDBM. Proceedings, 1990. V, 256 pages. 1990.

Vol. 421: T. Onodera, S. Kawai, A Formal Model of Visualization in Computer Graphics Systems. X, 100 pages. 1990.

Vol. 422: B. Nebel, Reasoning and Revision in Hybrid Representation Systems. XII, 270 pages. 1990 (Subseries LNAI).

Vol. 423: L. E. Deimel (Ed.), Software Engineering Education. Proceedings, 1990. VI, 164 pages. 1990.

Vol. 424: G. Rozenberg (Ed.), Advances in Petri Nets 1989. VI, 524 pages. 1990.

Vol. 425: C. H. Bergman, R. D. Maddux, D. L. Pigozzi (Eds.), Algebraic Logic and Universal Algebra in Computer Science. Proceedings, 1988. XI, 292 pages. 1990.

Vol. 426: N. Houbak, SIL – a Simulation Language. VII, 192 pages. 1990.

Vol. 427: O. Faugeras (Ed.), Computer Vision – ECCV 90. Proceedings, 1990. XII, 619 pages. 1990.

Vol. 428: D. Bjørner, C. A. R. Hoare, H. Langmaack (Eds.), VDM '90. VDM and Z – Formal Methods in Software Development. Proceedings, 1990. XVII, 580 pages. 1990.

Vol. 429: A. Miola (Ed.), Design and Implementation of Symbolic Computation Systems. Proceedings, 1990. XII, 284 pages. 1990.

Vol. 430: J. W. de Bakker, W.-P. de Roever, G. Rozenberg (Eds.), Stepwise Refinement of Distributed Systems. Models, Formalisms, Correctness. Proceedings, 1989. X, 808 pages. 1990.

Vol. 431: A. Arnold (Ed.), CAAP '90. Proceedings, 1990. VI, 285 pages. 1990.

Vol. 432: N. Jones (Ed.), ESOP '90. Proceedings, 1990. IX, 436 pages. 1990.

Vol. 433: W. Schröder-Preikschat, W. Zimmer (Eds.), Progress in Distributed Operating Systems and Distributed Systems Management. Proceedings, 1989. V, 206 pages. 1990.

Vol. 435: G. Brassard (Ed.), Advances in Cryptology – CRYPTO '89. Proceedings, 1989. XIII, 634 pages. 1990.

Vol. 436: B. Steinholtz, A. Sølvberg, L. Bergman (Eds.), Advanced Information Systems Engineering. Proceedings, 1990. X, 392 pages. 1990.

Vol. 437: D. Kumar (Ed.), Current Trends in SNePS – Semantic Network Processing System. Proceedings, 1989. VII, 162 pages. 1990. (Subseries LNAI).

Vol. 438: D. H. Norrie, H.-W. Six (Eds.), Computer Assisted Learning – ICCAL '90. Proceedings, 1990. VII, 467 pages. 1990.

Vol. 439: P. Gorny, M. Tauber (Eds.), Visualization in Human-Computer Interaction. Proceedings, 1988. VI, 274 pages. 1990.

Vol. 440: E.Börger, H. Kleine Büning, M. M. Richter (Eds.), CSL '89. Proceedings, 1989. VI, 437 pages. 1990.

Vol. 441: T. Ito, R. H. Halstead, Jr. (Eds.), Parallel Lisp: Languages and Systems. Proceedings, 1989. XII, 364 pages. 1990.

Vol. 442: M. Main, A. Melton, M. Mislove, D. Schmidt (Eds.), Mathematical Foundations of Programming Semantics. Proceedings, 1989. VI, 439 pages. 1990.

Vol. 443: M. S. Paterson (Ed.), Automata, Languages and Programming. Proceedings, 1990. IX, 781 pages. 1990.

Vol. 444: S. Ramani, R. Chandrasekar, K. S. R. Anjaneyulu (Eds.), Knowledge Based Computer Systems. Proceedings, 1989. X, 546 pages. 1990. (Subseries LNAI).

Vol. 445: A. J. M. van Gasteren, On the Shape of Mathematical Arguments. VIII, 181 pages. 1990.

Vol. 446: L. Plümer, Termination Proofs for Logic Programs. VIII, 142 pages. 1990. (Subseries LNAI).

Vol. 447: J. R. Gilbert, R. Karlsson (Eds.), SWAT 90. 2nd Scandinavian Workshop on Algorithm Theory. Proceedings, 1990. VI, 417 pages. 1990.

Vol. 449: M. E. Stickel (Ed.), 10th International Conference on Automated Deduction. Proceedings, 1990. XVI, 688 pages. 1990. (Subseries LNAI).

Vol. 450: T. Asano, T. Ibaraki, H. Imai, T. Nishizeki (Eds.), Algorithms. Proceedings, 1990. VIII, 479 pages. 1990.